Handbook of Acoustic Emission

Handbook of Acoustic Emission

Edited by **Sonny Lin**

Clanrye International

New Jersey

Published by Clanrye International,
55 Van Reypen Street,
Jersey City, NJ 07306, USA
www.clanryeinternational.com

Handbook of Acoustic Emission
Edited by Sonny Lin

International Standard Book Number: 978-1-63240-253-0 (Hardback)

Printed in China.

Contents

Preface

In my initial years as a student, I used to run to the library at every possible instance to grab a book and learn something new. Books were my primary source of knowledge and I would not have come such a long way without all that I learnt from them. Thus, when I was approached to edit this book; I became understandably nostalgic. It was an absolute honor to be considered worthy of guiding the current generation as well as those to come. I put all my knowledge and hard work into making this book most beneficial for its readers.

The aim of this book is to provide state-of-the-art information regarding the concept of Acoustic Emission (AE). AE is a significant and non-harmful testing procedure for substances, constructions and machines. It is defined as the transient elastic energy that is spontaneously released when substances undergo deformation, fracture or both. It presents significant applications of AE procedure like the evaluation of machinery and civil structures condition, fatigue and fracture substances research, detection of substance defects and deformities, monitoring of stress and ageing in substances, chemical reactions and phase transitions research, and earthquake prediction.

I wish to thank my publisher for supporting me at every step. I would also like to thank all the authors who have contributed their researches in this book. I hope this book will be a valuable contribution to the progress of the field.

Editor

1

Acoustic Emission for Civil Structures

Paul Ziehl and Adrian Pollock
University of South Carolina and Mistras Group, Inc., USA

1. Introduction

"Acoustic emission" (AE) is the name given to the transient stress waves that are generated by crack growth and many other kinds of material degradation and deterioration. The phenomenon has been known intuitively since the beginnings of knowledge, and studied scientifically for at least a century. The cracking of ice and the snapping of twigs are commonplace examples. In recent decades, acoustic emission has been used as a nondestructive testing method. A substantial body of technique has developed to allow its application to the monitoring of bridges, pressure vessels, storage tanks, etc. These devices have an amazing sensitivity to high-frequency motion. At the frequencies most commonly used for AE testing, 100-300kHz, the AE sensor can give a detectable signal for surface movements of 10^{-13} m or less, a thousand times smaller than the size of an atom.

As a monitoring device for structural integrity, the acoustic emission sensor is effective over distances from a few inches to tens of feet. It can be compared to accelerometers that are often used to assess the condition of bridges, for example, through the techniques of modal analysis. Accelerometers are also piezoelectric devices, but they operate at much lower frequencies (typically tens or hundreds of Hz instead of hundreds of kHz). Both the AE sensor and the accelerometer are used to sense movements. However, the motion sensed by the accelerometer has a wavelength on the order of tens to hundreds of feet. It is thus measuring the movement of the structure as a whole and is not sensitive to small point disturbances. The AE sensor inspects a local part of the structure, and is very sensitive to point disturbances. Specifically, it can be used to sense damage processes at the moment they occur.

Acoustic emission has been used as a formalized structural evaluation method since the early 1980's. The formalized acoustic emission evaluation methods that are in place today are the result of a significant number of structural failures of fiber reinforced polymeric (FRP) vessels that took place in the preceding decade (Fowler and Gray, 1979; Fowler et al., 1989). Many of the failures could be attributed to manufacturing defects and/or inappropriate design procedures with a particular emphasis on the discontinuity regions of the vessels. Acoustic emission data was gathered on actual vessels and trends in the data were analyzed, resulting in a standardized loading and evaluation procedure (CARP, 1982). The concepts developed were later incorporated into the ASME Boiler and Pressure Vessel Code and are used today (ASME 2010a and 2010b). AE remains as the preferred method of evaluation for one-of-a-kind vessels prior to implementation and also for evaluation of in-service vessels.

For FRP vessels AE is used as a primary means of evaluation. It is generally not combined with strain gages or other sensing devices. This may be partially due to the nature of damage in FRP vessels and the resulting brittle failure modes. In these structures the localized damage can lead to catastrophic failure without visibly detected warning. One of the key advantages of acoustic emission in this industry is its ability to rate the significance of the damage. Once a damaged region has been located with AE, more localized follow-up methods are often used to further assess and map the damage. In contrast to evaluation of buildings and bridges, detailed calculations of the entire structure are generally not a part of the standardized loading and evaluation procedures.

Similar AE procedures have been developed for other polymeric devices such as manlifts (Ternowchek and Mitchell, 1992; Pollock and Ternowcheck, 1992); metallic railroad cars and other vessels (AAR, 1999; AAR, 2002; Fowler et al., 1989). AE has also been proposed for reinforced concrete structures (Ohtsu et al., 2002; JSNDI, 2000) and has been employed in the field (Golaski et al., 2000). Acoustic emission has been incorporated into the design process itself for FRP vessels as described in ASME Section X (ASME, 2010a; Ziehl and Fowler, 2003; ASTM, 2006) and has been related to fatigue behavior in FRP pipes (Ramirez et al., 2004). An overview of the AE method and its applications is given in Pollock (Pollock, 2008).

The use of AE as the primary method of evaluation in many industries differs from the evaluation of civil engineering structures. For civil evaluations calculations are generally combined with information gathered from strain gages and other sensing devices under applied or ambient loading conditions. An example of the calculation based approach is conventional load rating of existing bridges. With this approach the geometry and structural aspects of the bridge must be known (such as depth of girders, connections between girders and deck, reinforcing steel details, strength of reinforcing steel, etc.). With this information and assuming boundary and support conditions along with lateral load distribution characteristics, a beam-line analysis may then be conducted and load rating factors developed (AASTHO, 1994). This approach has inherent limitations, including its dependence on the assumptions made regarding materials and boundary conditions. The assumptions can be minimized through diagnostic load testing (Schulz, 1993; Goble et al., 1990 and 1992). This approach combines strain response under known loading conditions with numerical models of the bridge.

The differences between the evaluation approaches for civil structures and other structural systems discussed earlier are significant. For many non-civil structural systems, acoustic emission is used as a primary means of evaluation and the results are categorized ('minor damage', 'intermediate damage', 'severe damage') in the absence of detailed calculation procedures. For civil structures detailed calculations and numerical simulations are an integral part of the evaluation process and therefore in-depth knowledge of the structure is required. For civil structures AE has been sparingly used and is rarely, if ever, used as the primary means of evaluation. However, AE has recently seen an increase in the evaluation of civil structures and the sensitivity and non-invasive nature of the method are clear advantages for many civil applications (Ziehl, 2008).

This chapter describes recent work on the application of acoustic emission to civil structures. The applications are grouped according to the primary material type of interest including steel, reinforced concrete, and fiber reinforced polymers. The civil structures monitored or evaluated include both buildings and bridges. In some cases acoustic emission is used as a means of passive monitoring only and in other cases it is used to complement data collected and evaluated as part of a load testing procedure. One of the more recent and promising developments of AE for reinforced concrete structures is the in-situ monitoring of active corrosion and the results of preliminary research in this area are discussed.

2. Steel structures

A major threat to mechanical integrity of steel civil structures is cracking, in particular fatigue cracking. In-service steel bridges are reaching their design fatigue lives each year. There is a growing need to evaluate fatigue damage and predict remaining fatigue life. AE techniques have been extensively used in nondestructive testing and structural health monitoring (Gong et al., 1992; Martin et al., 1995; Chen and Choi, 2004). In the nature of fatigue cracks, energy arising from plastic deformation and fracture events is transmitted as stress waves that can be detected by remote sensors. The high sensitivity of AE techniques (Ghorbanpoor and Vannoy, 1988; Kohn et al., 1992; Bassim et al., 1994) offers demonstrated reliability for the detection of active cracks. For AE monitoring applications the cracking location does not need to be precisely known, sensors together with appropriate algorithms are capable of locating and quantifying active cracks. Correlation between AE and corresponding crack growth behavior is the basis for interpretation of acquired AE signals for the evaluation of fatigue damage and prediction of remaining fatigue life.

Two driving forces, maximum stress intensity K_{max} and stress intensity range ΔK, govern fatigue crack growth behavior (Sadananda and Vasudevan, 1997). For a specific material and set of test conditions, ΔK is equal to (1-R) K_{max} where R is the load ratio. The driving forces have their thresholds, K_{maxTH} and ΔK_{TH}. Fatigue cracks will not develop if K_{max} or ΔK in the actual structure is below the threshold. The fatigue lifetime is conventionally divided into three stages. In Stage I which lasts for most of the lifetime, the crack is initiating. In Stage II, the crack propagation rate depends strongly on ΔK and also to some extent on $K_{max.}$ Thus under constant-amplitude cyclic loads, the crack propagation rate increases as the crack advances. If K_{max} reaches the fracture toughness K_{IC}, the crack will come into the stage of unstable propagation (Stage III). Failure occurs after a relatively small number of cycles, and may be catastrophic or not, depending on the structural geometry.

Thus, in the area of interest to us, the first requirement is that K_{max} and ΔK are higher than their thresholds K_{maxTH} and ΔK_{TH}. Next, we are especially interested when K_{max} exceeds K_{IC} corresponding to the critical transition from Stage II (stable propagation) to Stage III (unstable propagation). The AE behavior takes a distinctive upturn at this transition, an example of which is shown in Figure 1 along with a compact tensile test specimen that was utilized to generate the fatigue crack (Yu et al., 2011).

In acoustic emission monitoring of ductile metals such as the structural steel for bridge construction, it would be nice if all acoustic emission events were simply related to rapid extension of the crack. However, this is not always the case and a good body of work exists that attempts to address the source of acoustic emission events. Mechanisms include crack extension, 'fretting' or 'friction' of the crack surfaces, yielding ahead of the plastic zone, the fracturing of brittle inclusions, separation of ligaments by internal necking, and others. Because acoustic emission is an in-situ method of evaluation, a one-to-one correlation between received data and actual internal mechanisms is not available and

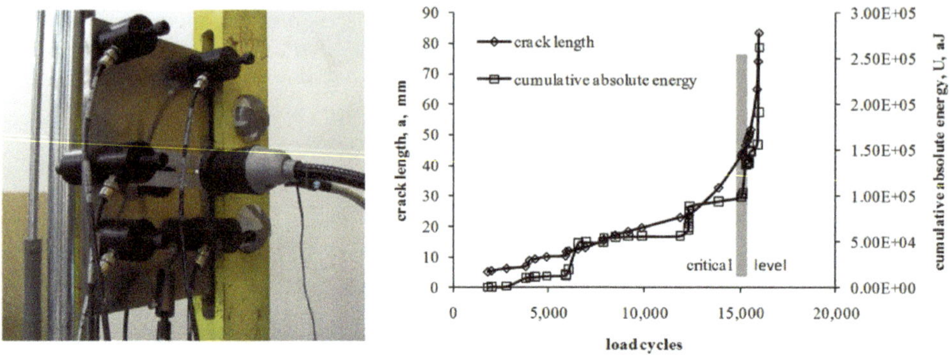

Fig. 1. CT Specimen with Five AE Sensors and Resulting Data

therefore the debate related to source mechanisms is an open topic. Guidance and further discussion may be found in Ohira and Pao, 1986; Han et al., 2011; Huang, 1998, and Ono, 2005.

Sison et al., 1996 includes a summary of the dozen earliest efforts to apply AE to steel bridges. By the mid 1990's sufficient knowledge had been gained to attempt transfer of the technology to the body of state highway inspectors. A project was undertaken, sponsored by the Federal Highway Administration, to define a focused technical approach and provide written guidelines for its application (Pollock and Carlyle, 1995). As always in AE inspection, there were strategic choices to be made about the monitoring approach. These choices included short-term versus long-term monitoring, controlled loading versus normal service loading, and wide area inspection versus local area monitoring. At the outset of this project, careful consideration was given to what kind of AE test could most likely find a useful place in the day-to-day operations of the state highway departments. It was recognized that evaluation of fatigue cracks in welded details in steel bridges is a substantial part of the integrity-related work of the state highway departments. The concept emerged that sometimes a state highway engineer, considering what to do about a known flaw, might want to get more information about it - and in this situation, it would be helpful to know whether it was acoustically active or not. Thus, a potentially useful kind of AE test would be to assess specific welded details, probably containing known cracks, as quickly and economically as possible and to return information promptly to the bridge owner's engineering staff.

In pursuit of this concept, a dozen flaws on four different bridges were inspected with AE and appropriate technology for an efficient inspection was developed. It was found that on the busy bridges selected for this study, monitoring for an hour would give a sufficiently representative sample of the flaw's AE activity. It is well known that heavy vehicles, much more than passenger car traffic, are responsible for fatigue damage in bridges. So the main criterion for choosing the monitoring period is that it should include a representative amount of heavy vehicle traffic. Also, of course, one must avoid adverse weather conditions such as rain, which produces unacceptable background noise.

The study included fatigue cracks in welded details, fatigue cracks in rolled sections, and several other conditions. An example is illustrated in Figure 2. This is a small crack in a floor beam flange adjacent to a rivet in the Brooklyn Bridge in New York City. The sensor is a 300 kHz resonant type (PAC µ30). A small sensor such as this is convenient for local area monitoring, and the relatively high frequency is good for reducing background noise. Background noise in bridge monitoring is produced by the passing traffic, not directly but indirectly. The traffic loading produces rubbing of structural members, generating acoustic emission at places that could be remote from the traffic but close enough to the inspection area to be detected. The study showed the effectiveness of guard sensor techniques for avoiding problems from this kind of noise.

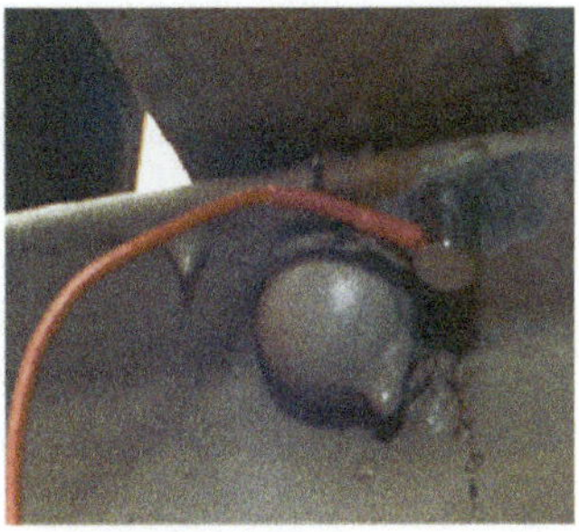

Fig. 2. AE Sensor on Brooklyn Bridge Floor Beam

The guard sensor concept is to have a "data sensor" close to the flaw being monitored, then to surround that spot with several "guard sensors". If the flaw emits, the "data" channel will be hit first. If noise comes into the inspection area from outside, one of the "guard" channels will be hit first. On this basis the AE signal can be either accepted or rejected. Figure 3 shows a set of four guard sensors surrounding a data sensor on a welded detail. In the test shown in Figure 2 guard sensors were also installed but they were backside of the beam, unseen in the photograph.

Fig. 3. Data and Guard Sensors on I-10 Mississippi River Bridge (Baton Rouge, LA)

The effect of using guard sensors is illustrated in Figure 4. Here a flaw was monitored with two data sensors 8 inches apart in a linear location array, surrounded by several guard sensors. It was possible to record all hits on all sensors, then to examine the data post test, either using (left) or not using (right) the guards. Figure 4 shows how when guards are used, there is a very clear indication of the flaw, a spike in the location plot standing out clearly from the residual background noise (some noise still defeats the guards). However when the guards are not used, many more external noise events appear to locate between the sensors and there are even some peaks that are comparable in size with the peak from the flaw.

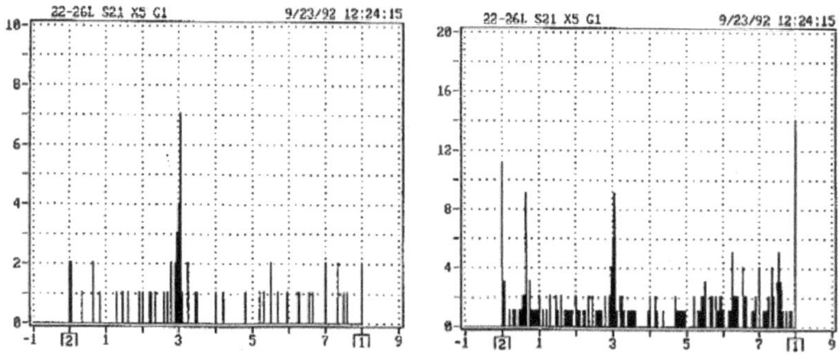

Fig. 4. Linear Location Plot with (left) and without (right) Guard Sensors, Bryte Bend Bridge, Portland, OR

Although these events originate outside the area of interest, their received waveforms are such that their Δt's place them apparently between the data sensors. It was concluded from this study that even though it required additional channels, the use of guard sensors was the most straightforward technique for discounting this kind of noise to permit valid data evaluation.

By the end of this project, a dozen flaws of several different kinds had been monitored on several different bridges, using essentially the same monitoring conditions and equipment setup. A table could be drawn up showing the activity recorded from these flaws, starting with the most active and working down to the least active in terms of located events per minute, during normal traffic loading. This table is shown below (Table 1).

These results were satisfying in that they showed AE activity ranging through more than three orders of magnitude, as the flaws went from code-rejectable inclusions and large cracks, to "nothing". The inclusions at the top of the list were characterized by ultrasonic testing; in general, inclusions can serve as starters for fatigue cracks. The activity of these inclusions is in very strong contrast to the minimal activity of previously discovered discontinuities in an electroslag weld, detailed at the bottom of the table. A "league table" of this kind puts AE activity into a meaningful context and can certainly help the bridge engineer to decide what to do about these flaws. The table also shows some results with repairs and retrofits. A retrofit, such as might be applied to reinforce a cracked area after drilling an arrester hole, is intended to hold the area tight so that it does not move and a new crack does not start from the repair. If the retrofit is not tight, it will not do its job and there may be further crack growth. A loose retrofit gives additional AE, as can be seen by comparing the 5th and 8th lines of Table 1. With further work along these lines, AE became recognized as a method for checking the effectiveness of repairs and retrofits.

A simple report form was developed so that the bridge engineer could get a summary of the test results on one page. Figure 5 shows the front side of this one-page form which carried standardized information; the back side would carry free-form test-specific information

Bridge**	Flaw	Time (min)	Total Events	Events/min
WW	Code reject able inclusion in web weld	60	1600	26.7
WW	17" crack in floor beam, intersection of top flange with web, more active end	60	800	13.3
WW	Code rejectable inclusion in bottom tensile flange	60	200	3.3
BB	0.03" crack discovered at web-to-stiffener weld	30	57	1.9
MS	13" crack growing at floor beam / truss panel joint - retrofit removed	92	90	1
BB	Crack at web-to-stiffener weld	30	25	0.8
WW	17" crack in floor beam, intersection of top flange with web, less active end	60	20	0.3
MS	13" crack arrested at floor beam / truss panel joint - retrofit operational	61	18	0.3
BB	No cracks, stiffener/web/top flange/floor beam area	30	1	0.03
WR	4" ultrasonic indication in electroslag weld	120	0	<0.01
WR	13" ultrasonic indication in electroslag weld	120	0	<0.01

** WW = Woodrow Wilson (DC), BB = Bryte Bend (Sacramento), MS = Mississippi R. (I10 Baton Rouge), WR = Willamette R. (Portland OR)

Table 1. Flaws on Steel Bridges and their Respective Levels of AE Activity

such as photographs and any AE data graphs of particular interest. The simplicity of this form would help to keep the cost of the test down, as well as expediting communication of the results. A table such as Table 1 would be used on site to explain the test to bridge owner personnel. This approach became a useful starting point for test services to inspect limited areas of interest.

A feature of this approach was that to keep it simple, one did not get into the difficult question of whether the emissions were coming from friction (crack face interference) or actual crack growth. In fact, the great majority of detected AE is likely coming from friction. In arguing that this relates to the severity of the flaw, the position can be taken that any local movement is bad because it implies changing strains and stresses, which are likely to be driving crack growth. This position: "a quiet piece of structure is a good piece of structure, any AE is bad" may be simplistic, but it has much to commend it in terms of practicality. On the next level of technological sophistication, more advanced AE analysis attempts to tell the difference between frictional AE and actual crack growth events. This kind of advanced analysis will lead to more precise diagnostic and prognostic capabilities in the future.

Fatigue cracking is a threat to railway bridges as well as to highway bridges. Significant work on the integration of AE into the qualification processes for railway bridges has been reported by (Gong et al., 1992). In Gong's program, the severity of cracks (and the level of urgency assigned to their further inspection and eventual repair) is based on an engineering assessment of the range of stress intensity factor, ΔK, to which they are exposed in service. ΔK is the important factor to consider because it is this that governs the crack growth rate. Correlations were established between AE and ΔK, thus allowing AE to be used in the evaluation of the severity of the crack and its subsequent disposition.

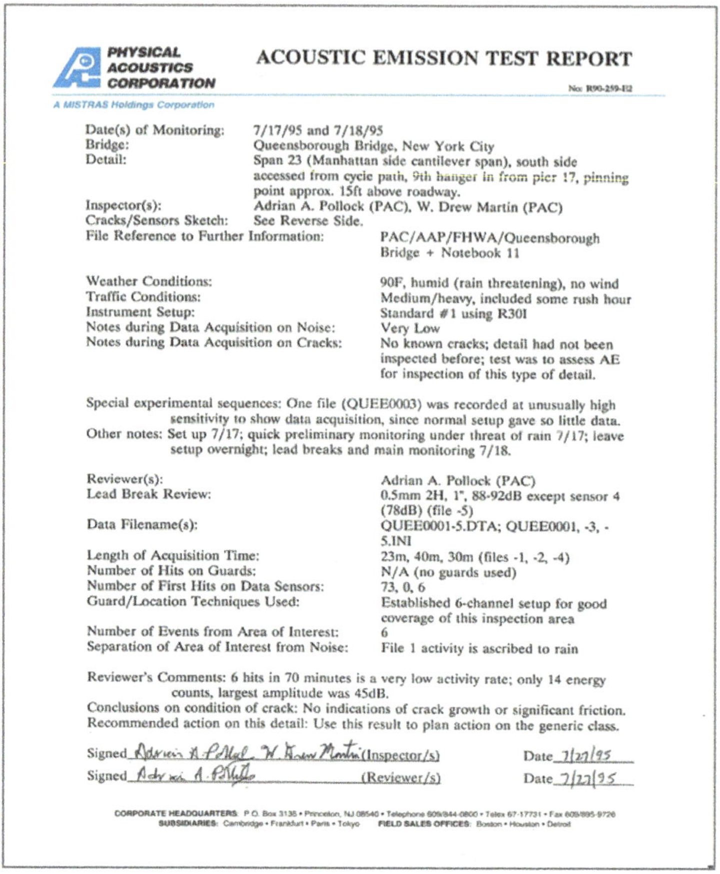

PHYSICAL ACOUSTICS CORPORATION
A MISTRAS Holdings Corporation

ACOUSTIC EMISSION TEST REPORT

No: R90-259-I12

Date(s) of Monitoring: 7/17/95 and 7/18/95
Bridge: Queensborough Bridge, New York City
Detail: Span 23 (Manhattan side cantilever span), south side accessed from cycle path, 9th hanger in from pier 17, pinning point approx. 15ft above roadway.
Inspector(s): Adrian A. Pollock (PAC), W. Drew Martin (PAC)
Cracks/Sensors Sketch: See Reverse Side.
File Reference to Further Information: PAC/AAP/FHWA/Queensborough Bridge + Notebook 11

Weather Conditions: 90F, humid (rain threatening), no wind
Traffic Conditions: Medium/heavy, included some rush hour
Instrument Setup: Standard #1 using R301
Notes during Data Acquisition on Noise: Very Low
Notes during Data Acquisition on Cracks: No known cracks; detail had not been inspected before; test was to assess AE for inspection of this type of detail.

Special experimental sequences: One file (QUEE0003) was recorded at unusually high sensitivity to show data acquisition, since normal setup gave so little data.
Other notes: Set up 7/17; quick preliminary monitoring under threat of rain 7/17; leave setup overnight; lead breaks and main monitoring 7/18.

Reviewer(s): Adrian A. Pollock (PAC)
Lead Break Review: 0.5mm 2H, 1", 88-92dB except sensor 4 (78dB) (file -5)
Data Filename(s): QUEE0001-5.DTA; QUEE0001, -3, -5.INI
Length of Acquisition Time: 23m, 40m, 30m (files -1, -2, -4)
Number of Hits on Guards: N/A (no guards used)
Number of First Hits on Data Sensors: 73, 0, 6
Guard/Location Techniques Used: Established 6-channel setup for good coverage of this inspection area
Number of Events from Area of Interest: 6
Separation of Area of Interest from Noise: File 1 activity is ascribed to rain

Reviewer's Comments: 6 hits in 70 minutes is a very low activity rate; only 14 energy counts, largest amplitude was 45dB.
Conclusions on condition of crack: No indications of crack growth or significant friction.
Recommended action on this detail: Use this result to plan action on the generic class.

Signed ____________ (Inspector/s) Date 7/27/95
Signed ____________ (Reviewer/s) Date 7/27/95

CORPORATE HEADQUARTERS: P.O. Box 3135 • Princeton, NJ 08540 • Telephone 609/844-0800 • Telex 67-17731 • Fax 609/895-9726
SUBSIDIARIES: Cambridge • Frankfurt • Paris • Tokyo FIELD SALES OFFICES: Boston • Houston • Detroit

Fig. 5. A Simple Form for Reporting AE Tests on Welded Details

Before leaving the topic of fatigue cracks, it should be mentioned that in all such work on bridges, a crucial consideration is whether or not the crack is in a "critical member". Depending on the geometry of the load-bearing structure in the neighborhood of the crack, the growth of a crack can end in either of two ways. Some fatigue cracks follow a path that has them accelerating towards a catastrophic failure of major structure. Other cracks follow a path such that the load is redistributed to other members that manage to carry it, while the crack finds its way to a low-stress region and practically stops propagating. An understanding of the bridge structure and load paths is needed to tell the crucial difference between these cases.

A second major threat to mechanical integrity of steel civil structures is corrosion. A leading example of this is the slow deterioration of the cables of suspension bridges and cable stay bridges. Corrosion of the inner strands of the cables leads to a slow reduction of its strength. This is a real concern in old bridges where procedures for maintaining a good chemical environment for the cable strands (oil injection, etc.) have been neglected and water ingress has occurred during many decades of service. Diagnosis and prognosis of structural health are very important for long term planning of the maintenance, and ultimately the replacement of these slowly deteriorating structures.

A role that AE can play in addressing this problem is the detection of individual wires breaking in the corroded parts of the cable. A main cable may contain upwards of ten thousand bundled wires so a few breaks do not amount to much, but over a period of time it is the cumulative breaking of many wires that would lead to the failure of the cable. Systems for monitoring wire breaks with AE have been in place for a number of years on some well known bridges. The main challenges are installation (Figure 6), maintenance and data interpretation (discrimination of wire breaks from background noise). Special algorithms have been developed for recognizing wire breaks, using an economically viable number of sensors, even in the presence of major background noise sources such as trains passing over the bridge.

Fig. 6. Installing Sensors for Long Term Monitoring of a Suspension Bridge Cable

To understand better the mechanisms of cable deterioration, a specially fabricated cable was installed in a test cell (Figure 7) at Columbia University (New York City), in a joint project with MISTRAS Group and Parsons, the well known bridge engineering company. This cable had 50 implanted sensors for measuring environmental conditions (temperature, humidity, pH) as well as corrosion vulnerability using several different kinds of corrosion sensor. The sensors were deployed in normal regions and in regions where the normal protective layers were disrupted. The cable could be pulled in tension, heated and cooled, sprayed with simulated acid rain and so forth. The purpose of this study was to understand the intra-cable environment and the dependence of corrosion on external challenges and maintenance variables. Information was collected for input to predictive models that could tell the degradation of a cable's strength over long periods of time as a function of environment and cable condition. The design of the monitoring system emphasized data fusion. While it included AE detection of wire breaks, this was not the only purpose of the system, not even its main purpose. More to the point in today's system design is the recording of the many assorted variables that are pertinent to structural integrity, whether they be challenges, responses or measures of condition. This kind of fusion of data from many sources has become a theme in the emerging technology of structural health monitoring.

The practical utilization of AE on civil structures has been substantially assisted by the development of wireless systems during the first decade of the 21st century, and this development will yield even greater benefits as wireless techniques continue to improve.

Fig. 7. Suspension Bridge Cable Test Cell with Multiple Sensor Types

Traditionally, the running of long cables between the sensor/preamplifier and the main AE instrument has always been a major part of the effort involved in any AE field test. And in any consideration of permanent installations, the cost of professional conduiting of the cables would typically be several times greater than the cost of the AE electronics.

Replacement of these cables with wireless systems is a longstanding dream that has become a growing reality in recent years, thanks to the burgeoning of digital communications infrastructure in our society in general. Wireless hardware for AE first took the form of a node for sending parametric information (i.e. slowly varying quantities, not needing significant bandwidth). Next came a node with onboard signal measurement capabilities that could transmit the measured signal features to a receiving station near the central computer. By 2011, a four-channel node (Figure 8) was introduced that could also transmit the full AE waveforms; source location also became possible, with the four channels associated with a single clock for good measurement of the arrival time differences.

Fig. 8. Low-Power, Four-Channel Wireless Node with Feature Extraction, Full Waveform Transmission, and Inputs for Strain Guages and Six Parametrics

An associated costs-cutting development is the introduction of self powered (energy harvesting) AE systems (Karami et al., 2012). With wireless, you don't have to run so many cables; with self-power, you can leave the system there permanently. At first sight this may seem like an indulgence, but when all the costs of lane closure are taken into account it becomes clear that it is cheaper to leave the equipment in place than to close the lane to go and remove it. For these reasons self powered, wireless systems are expected to bring about a major improvement in the practical applicability of AE technology to both steel and concrete structures.

3. Reinforced concrete

Genuine source mechanisms for acoustic emission in concrete structures are numerous and include cracking of the concrete (crack extension), rubbing of crack surfaces during crack closure, debonding of the reinforcing steel from the surrounding concrete, and localized cracking in the vicinity of the reinforcement due to doweling action. Sources of non-genuine AE in concrete bridges and structures are not as severe as in steel bridges due to the attenuating nature of the concrete itself. Nonetheless, non-genuine sources do exist and need to be taken into account. These sources include movement of supports including bearings and the customary environmental noise sources such as wind-borne debris and rain.

The safe load carrying capacity of reinforced concrete structures can come into question for a number of reasons including a change in use or occupancy, questions regarding details of construction such as missing or misplaced steel reinforcement, and in some cases the use of newer materials and systems that may not be addressed in existing codes and standards. A variety of load test methods exist for both buildings and bridges. For buildings, the cyclic load test (CLT), as described in Appendix A of ACI 437R-03 (ACI 437, 2003), is a recently introduced in-situ evaluation method. This method has the potential to reduce the time of the load test in comparison to the 24-hour load (24-h LT) test method described in chapter 20 of ACI 318-08 (ACI 318, 2008) while simultaneously providing improved insight to the response of the structure. The typical instrumentation utilized for load testing of buildings and bridges consists of displacement and rotation gages and in some cases these may be supplemented with strain gages. These devices lack the sensitivity of acoustic emission. Because many of the damage mechanisms that are present in reinforced concrete manifest themselves as cracking of the concrete prior to structural collapse, AE would seem to be an ideal method for the evaluation of reinforced concrete structures during load testing. Additionally, the loading pattern that is specified for the CLT method (Figure 9) is serendipitously reminiscent of loading patterns that have been used for many decades for the evaluation of fiber reinforced polymeric pressure vessels and tanks (ASME, 2010a and 2010b).

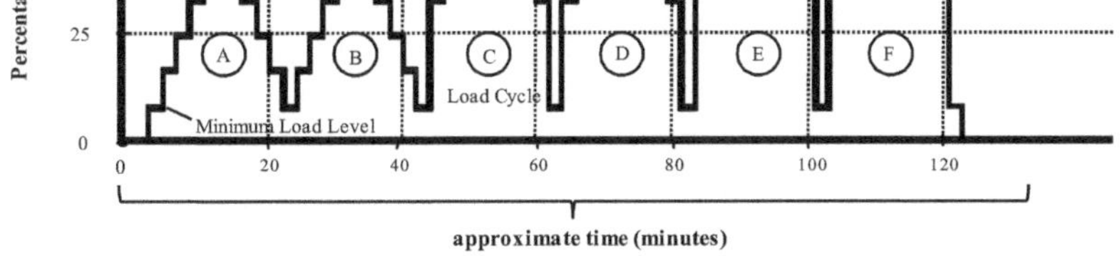

Fig. 9. Loading and Unloading Protocol associated with the Cyclic Load Test (after ACI 437R-03, Appendix A)

For prestressed and post-tensioned concrete flexural elements, cracking of the component under service level loads can be considered as a failure of the serviceability criteria. In such cases acoustic emission can be useful to minimize damage to the element during the load test itself. Another area in which acoustic emission can provide significant advantages over rotation, displacement, and strain gages is in the assessment of shear-dominated failure mechanisms. In such cases the current load testing approaches may not be useful until very significant damage has been done to the web and/or the bond of the reinforcing bars or strands near the ends of the girders (Figure 10).

Fig. 10. Shear-Tensile Failure in Prestressed Girder Specimen (after Xu, 2008)

For acoustic emission evaluation, several investigators have reported correlation of AE parameters to damage levels of reinforced concrete structures. One of the most widely implemented damage assessment methods is the correlation of calm ratio and load ratio (also referred to as Felicity ratio) (Yuyama et al., 1999; Ohtsu et al., 2002; JSNDI, 2000). Another approach makes use of severity versus historic index, known as 'Intensity Analysis', as a measure of deterioration for concrete bridges (Golaski et al., 2002). This method is directly related to assessment of fiber reinforced polymeric vessels. "Relaxation ratio" has likewise been used to quantify the residual strength of reinforced concrete beams (Colombo et al., 2005).

While the AE method is clearly useful for detection of cracking in reinforced concrete, the AE method of structural evaluation for reinforced concrete alone is not likely to be accepted at this time. This is due to a lack of widespread implementation combined with the inescapable fact that data interpretation methods are conducted on a case by case basis without the benefit of a governing code or standard for basic settings such as test and evaluation thresholds and noise rejection methodologies. Therefore, much of the effort to date has been placed on a combined inspection approach, wherein the acoustic emission data is used in combination with data gathered through more conventional instrumentation such as displacement gages. (Galati et al., 2008; Ziehl et al., 2008).

While the application of controlled loading such as that shown in Figure 9 is possible and even customary for building applications, this is not commonly the case for highway structures. For highway structures it is much more common and practical to use loading trucks with known wheel weights. In many applications a determination of the load carrying capacity is of interest whereas in others the interest is simply in the detection of active cracking, or lack thereof, under particular loading scenarios. One such application is the determination of locations of active cracking in a prestressed girder system in the shear region of prestressed girders. In this particular case a combination of factors resulted in severe cracking at the support region. AE sensors were arranged in an array of six in the web region of each girder and loading trucks were positioned to maximize the shear in the girders. A sensor array and a related plot of acoustic emission activity related to active crack growth are shown in Figure 11.

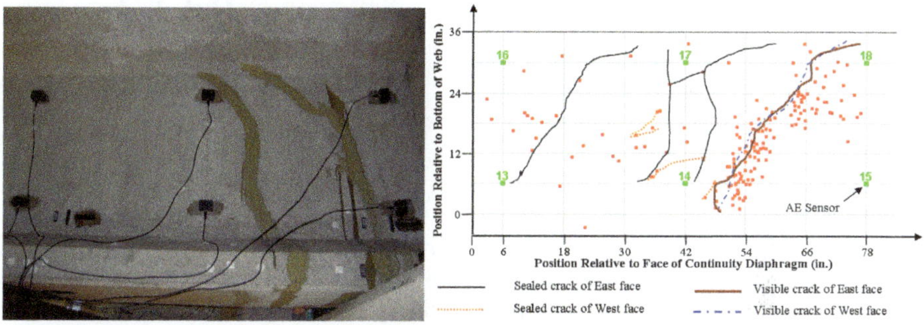

Fig. 11. AE for Detection of Active Cracking in Prestressed Girder (after Xu, 2008)

Due to the extreme sensitivity of AE, the monitoring of the passage of superloads presents another common and informative application (Grimson et al., 2008). In most cases the permitting process involves the development of computer models of the bridges to assure that damage is not done during the passage of the superload. Due to the many simplifying assumptions made in the development of computer models, such as simple supports, lateral load distributions factors, degree of composite action between the girders and the deck, and the longitudinal distribution of the superload itself, the actual response of the bridge may differ from the computer models. One of the largest superloads in the state of Louisiana crossed the Bonnet Carre' bridge (Figure 12). This superload passage resulted in acoustic emission activity that was significantly increased in comparison to the AE activity during the passage of normal traffic (Figure 13).

Fig. 12. Superload Crossing of Bonnet Carre' Bridge (after Grimson et al., 2008)

In addition to the use of acoustic emission for the evaluation of load carrying capacity and serviceability, AE has more recently been used to directly assess the presence of corrosion in both reinforced and prestressed concrete structures. It is intellectually clear that the crack growth activity generated by the expansive products associated with corrosion will produce acoustic emission. The use of AE is particularly attractive for the detection and monitoring of corrosion rates because the existing electro-chemical methods are invasive by nature and are generally limited to the evaluation of corrosion at a particular point within the structure. The fact that AE sensors can be applied as a global network combined with the inherent ease of installation is appealing.

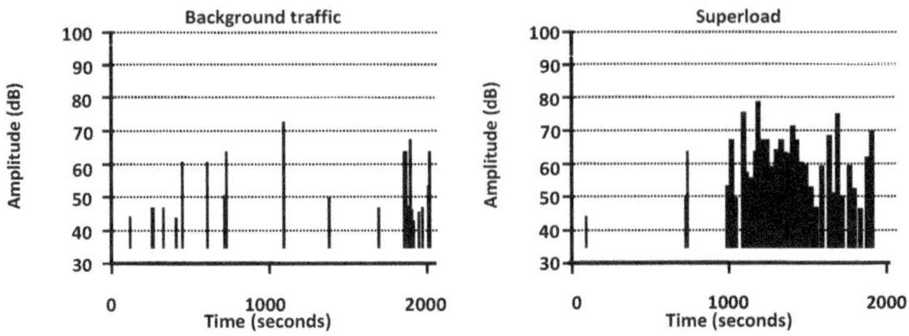

Fig. 13. AE Data related to Superload Crossing (after Grimson et al., 2008)

While much of the focus for acoustic emission has been placed on passively reinforced concrete, prestressed concrete construction represents a large portion of bridge construction and surpasses traditional reinforced concrete (NBI, 2011). The use of prestressed concrete in parking garages and buildings is also prevalent. Prestressing is generally selected due to its low initial cost, minimum downtime for on-site construction, and long life expectancy. In spite of the good overall performance record of prestressed concrete elements, it has been reported that approximately 30,000 prestressed bridges have some sort of deficiency (NBI, 2011). Furthermore, many bridges are rapidly approaching their design lives.

Corrosion of steel reinforcement is a primary contributor to deterioration and is of particular concern in marine environments and where deicing salts are used. The annual cost of bridge corrosion is $13.3 billion and life-cycle analysis estimates indirect costs at more than 10 times the direct cost (Hart et al., 2004; Koch et al., 2006). For prestressed construction the cracking is in many cases not allowed by the governing design codes under service level loading. However, cracks occur nonetheless due to temperature effects during the initial stages of curing, during transportation and erection, and due to overloading. Cracks in prestressed concrete can increase the rate of water penetration. Corrosion is particularly common at midspan of highway bridges where a collision has occurred and the concrete has been patched. It is also common at the support region of the girders where the beneficial effect of prestressing is not fully developed (Klieger and Lamond, 1994) and the steel strands are sometimes exposed. In prestressed concrete piling, corrosion is pervasive in the tidal zone where wet-dry cycling takes place.

Electrochemical methods have been developed to assess the degree of corrosion. While these techniques are useful for mapping the general areas of corrosion, they generally have the drawbacks of being intrusive and time consuming. Furthermore, in many cases they require the use of on-site experts for operation of the equipment (Baboian, 2005; ICRI 1996). In contrast to this acoustic emission monitoring makes use of the fact that the expansive nature of the corrosion process initiates micro-cracking of the surrounding concrete and this micro-cracking is readily detectable with AE sensors. A prime benefit of AE monitoring is that the sensors can be simply affixed to the surface of the concrete member without a need to access the embedded reinforcing steel. This can be accomplished in either a localized or more global array, depending upon the scale of the structural component and the suspected extent of corrosion. The sensors themselves can be used for assessment over a particular period of time and then moved to another structure for monitoring, which is particularly important for the assessment of a large inventory of structures as is typical for bridge owners. This stands in contrast to other methods of corrosion monitoring wherein sacrificial anodes are used and the probe containing the anodes must be in close proximity to the steel reinforcement, thereby requiring drilling of the element to be monitored.

From a structural point of view, changes in failure and serviceability mechanisms such as cracking, debonding, and strand rupture due to corrosion have been investigated with acoustic emission (Yoon et al., 2000; Austin et al., 2004). For the direct monitoring of corrosion activity in the absence of load the AE method has proven to be more sensitive than electro-chemical methods and therefore holds promise for the quantification of corrosion rate, and this information can then be incorporated into projections for the remaining serviceability of components or systems.

Due to the promise of the AE method for detection of corrosion in steel strands, recent detailed studies have been conducted under accelerated corrosion with the express purpose of verifying the potential for detecting the onset of corrosion and the rate of corrosion. To this end the primary sources of acoustic emission activity during the onset and progression of corrosion were located and the results in terms of acoustic emission activity compared to electrochemical methods. Both the AE and electro-chemical results were verified with visual inspection.

The investigations to this point have focused primarily on relatively small specimens, 4.5 × 4.5 × 20 in. (114 × 114 × 508 mm), Figure 14. Two inches of cover was provided for the specimens as this is generally representative of field construction. Each specimen was cast with a single 270 ksi (1.9 GPa), ½ inch (13 mm) diameter seven-wire low relaxation prestressing strand. All specimens were cast using concrete with a design compressive strength of 6 ksi (41 MPa) at 28-days with a maximum coarse aggregate size of ½ inch (13 mm).

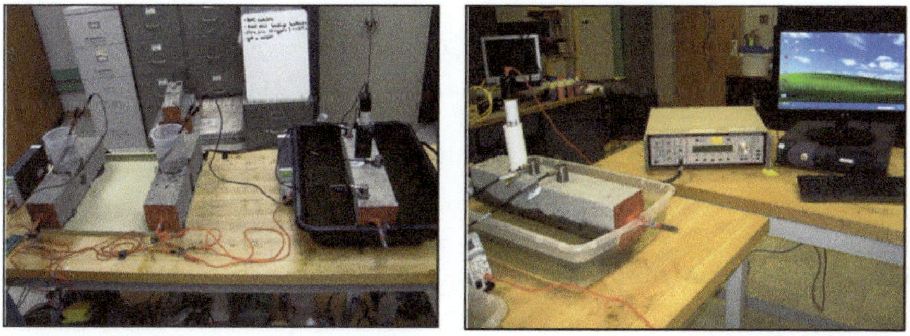

Fig. 14. Test Setup for Accelerated Corrosion of Prestressed Concrete Specimens (after Mangual et al., 2011)

The presence or absence of cracking is a critical parameter in acoustic emission monitoring in general, and this is particularly the case of corrosion monitoring due to the relatively low energy sources involved at the corrosion initiation stage. Cracking has two significant effects on the received data; a) the cracks form an additional reflective surface that can complicate the AE data interpretation, and b) the cracks provide a means of energy release for the expansive product itself. The second item tends to reduce the energy of the recorded signals in cracked concrete structures. All specimens discussed were intentionally cracked to account for this condition which may well be present in actual field structures. The crack was kept to a reasonable size of 0.016 in. (0.4 mm), which is above the threshold value from which rapid chloride can permeate and depassivate the reinforcing steel (Tutti, 1982).

Specimens were placed in a container filled with 3% NaCl to a level 0.25 in. (7 mm) below the level of the strand. An electrochemical cell was formed with a copper plate brought into contact with the steel strand. A constant potential was applied with a current range dependent on the resistivity due to the concrete and pore solution.

Potential measurements were taken in the vicinity of the initial crack. The applied voltage lowered the potential of the steel below -350 mV and depassivation took place after approximately five hours of testing. Values more negative than -350 mV are considered indicative of a 90% probability of corrosion in the area interrogated (ASTM, C876).

While broadly applied in field applications, half-cell potential measurements are not intended to quantify corrosion or corrosion rate. Therefore while this method provides quantitative readings the interpretation of the data is limited and is not particularly helpful for predicting the remaining service life of a structure. In contrast to the 'corrosion/no corrosion' information obtained from half-cell potential readings, acoustic emission can provide equal or better sensitivity combined with the ability to monitor the rate of the corrosion process. A plot of representative data is shown in Figure 15.

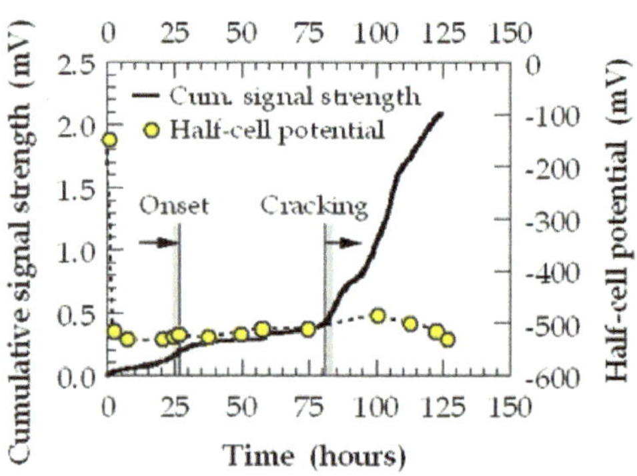

Fig. 15. Comparison of AE and Half-Cell Potential Data vs. Time (after Mangual et al., 2011)

4. Fiber reinforced polymers

Fiber reinforced polymers offer promise for civil engineering structures due to their inherent lack of susceptibility to corrosion and high strength. Acoustic emission is well-established for FRP materials and, as discussed above, has its roots in the FRP vessel industry (Fowler and Gray, 1979). Damage mechanisms in FRP structures include fiber breakage, matrix cracking, delamination, and fiber-matrix debonding. Of these mechanisms fiber breakage is relatively easily discriminated from the others due to the high energy of this source mechanism and this is particularly the case for breakage of carbon fiber bundles.

In some FRP structures, the material degradation mechanisms are of secondary importance to the connections between disparate systems. Such is the case for FRP honeycomb (FRPH) structures such as those used for panel type bridge construction and AE has been successfully implemented to evaluate fatigue damage for such a system prior to implementation (Cole et al., 2006). A similar investigative approach but with focus on degradation and/or manufacturing defects (such as internal delamination between plies) has recently been undertaken as part of a quality assurance program for two hybrid FRP/reinforced concrete bridges constructed near Corpus Christi, Texas. In both cases AE sensors were affixed to a pre-determined number of bridge beams for evaluation prior to implementation. Specialized loading procedures for the girders were developed in general conformance with those implemented for FRP vessels when evaluating against pre-determined acoustic emission evaluation criteria (Ulloa et al., 2004; Ramirez et al., 2009; Chen et al., 2009). In one case AE evaluation indicated potential intra-ply delamination in a girder and the location of the indication was later followed up with ultrasonic inspection. A photograph of FRP girders in place at one of the two bridge sites prior to placement of the concrete deck is shown in Figure 16 (Ramirez et al., 2009).

Fig. 16. Hybrid FRP/Reinforced Concrete Bridge Girders Evaluated with Acoustic Emission (after Ramirez et al., 2009)

Because FRP is a relatively new material for civil construction, it is sometimes prudent to utilize the sensitivity of acoustic emission for field evaluation after the structure has been opened to traffic. For one of the FRP/reinforced concrete bridges mentioned above, load testing was performed with AE each six months over a two year time span (Ziehl et al., 2009). In such cases it is very important to carefully weigh the axles of the loading trucks prior to evaluating the resulting AE data. This is because even a slight overload in relation to a previous loading can result in copious amounts of AE data in fiber reinforced polymeric systems due the Kaiser effect. However, much of the resulting data in such cases is of little consequence. Another factor to carefully consider for field applications in general is the potential effect of wind-borne debris (such as sand) on the AE data and how best to discriminate between such debris and actual AE data. For such cases the use of broadband sensors may be useful for clustering of noise versus genuine data based on frequency content.

Another aspect that is rarely considered, but should be in certain applications, is the effect of temperature on the AE evaluation criteria (Chen et al., 2007). This is not an issue in most environments, but may be important in very hot or very cold climates. In general the effect of an increase in temperature is to decrease the acoustic emission activity, and this may result in the inappropriate passing of evaluation criteria if those criteria were developed for ambient conditions. This effect is due to the viscous nature of the matrices commonly used for FRP construction.

5. Conclusions

Acoustic emission is a useful method of evaluation for many different materials used for civil construction including steel, reinforced concrete, and fiber reinforced polymers. Each of these materials offers certain advantages and challenges from the standpoint of acoustic emission monitoring.

Steel construction is typically achieved with highly ductile materials and the source mechanism itself is not well understood at this time. This challenge is combined with the low attenuation characteristics of the material which leads to a good deal of emission due to fretting of the crack surface. A primary challenge for the assessment of crack growth in steel structures therefore is how best to discriminate between fretting and other emission, such as that associated with crack extension.

Reinforced, prestressed and post-tensioned concrete have a different body of opportunities and challenges. Concrete is one of the least studied materials from the standpoint of acoustic emission. This material is characterized by high attenuation coupled with large amounts of emission due to its brittle nature. In terms of evaluation criteria reinforced concrete behaves very differently than prestressed or post-tensioned concrete due to the active nature of the reinforcement in the latter cases, which leads to significant friction in the cracks during unloading. One of the newer and more promising developments for reinforced and prestressed concrete is the use of AE for detection and monitoring of active corrosion.

Fiber reinforced polymers are perhaps the most widely studied of the three materials from the standpoint of acoustic emission. This is due to the large body of work that was conducted during the 1980's on FRP tanks and vessels. Because the materials used in tanks and vessels are generally reinforced with glass fibers many of the evaluation criteria and loading protocols bear a close relation to those for glass fiber reinforced girders and bridge decks.

An increase in temperature may result in non-conservative evaluations for structures fabricated with fiber-reinforced polymers. Further study is warranted for steel, reinforced concrete, and FRP structures with respect to the effect of temperature on acoustic emission evaluation criteria.

6. Acknowledgements

Portions of the work described were performed under the support of the U.S. Department of Commerce, National Institute of Standards and Technology, Technology Innovation Program, Cooperative Agreement Number 70NANB9H9007.

We wish to offer thanks to Drs. G. Ramirez, R. Barnes, and J. Xu for contribution of figures and technical suggestions. We also gratefully acknowledge the contributions of J. Mangual, M. ElBatanouny, W. Velez, and Drs. F. Matta and J. Yu at the University of South Carolina.

7. References

American Association of State Highway Transportation Officials (AASHTO), "Manual for the Condition Evaluation of Bridges," Washington, D.C., (1994).

Association of American Railroads, "Procedure for Acoustic Emission Evaluation of Tank Cars and IM101 Tanks", *Issue 8, Washington, DC, Association of American Railroads,* (1999).

Association of American Railroads, "Procedure for Structural Integrity Inspection of Tank Cars Using Acoustic Emission", *Issue 1, Revision 2, Washington, DC, Association of American Railroads,* (2002).

ASME Boiler and Pressure Vessel Committee, "Section X: Fiber Reinforced Plastic Pressure Vessels," *Subcommittee on Fiber Reinforced Plastic Pressure Vessels, 2004 Edition with subsequent addenda,* ASME, New York, (2010a).

ASME Boiler and Pressure Vessel Committee, "Section V: Nondestructive Examination," *Subcommittee on Fiber Reinforced Plastic Pressure Vessels, 2004 Edition with subsequent addenda,* ASME, New York, (2010b).

ACI Committee 437, "Strength Evaluation of Existing Concrete Buildings (ACI 437R-03)", American Concrete Institute, Farmington Hills, Michigan, 2003, 28 pp.

ACI Committee 318, "Building Code Requirements for Structural Concrete (ACI 318-05) and Commentary (318R-08)", American Concrete Institute, Farmington Hills, Mich., 2008, 430 pp.

ASTM C876, "Standard test method for half-cell potentials of uncoated reinforcing steel in concrete", *American Standard for Testing and Materials,* 2009, pp. 1-6.

ASTM E2478a, "Standard Practice for Determining Damage-Based Design Stress for Fiberglass Reinforced Plastic (FRP) Materials Using Acoustic Emission", *American Society of Testing and Materials.* West Conshohocken, PA (2006).

Austin, S., Lyons, R., and Ing, M., "Electrochemical Behavior of Steel-Reinforced Concrete during accelerated corrosion testing," *Corrosion,* V. 60, No. 2, 2004, pp. 203-212.

Baboian, R.; "Corrosion Tests and Standards: Applications and Interpretations", *ASTM International,* 2005, pp. 188-193.

Bassim, M.N., Lawrence, S.S. and Liu, C.D., "Detection of the onset of fatigue crack growth in rail steels using acoustic emission", Engineering Fracture Mechanics, 47(2), 207-214 (1994).

Chen, Y., Ziehl, P., Ramirez, G., and Fowler, T., *"Effect of Temperature on the Acoustic Emission Evaluation of FRP Vessels (Tensile Tests)", ASME Journal of Pressure Vessel Technology, Vol. 129, Issue 3, 2007, pp. 516-524.*

Cole, T., Lopez, M., and Ziehl, P., "Fatigue Behavior and Nondestructive Evaluation of FRP Honeycomb Bridge Specimen", *ASCE Journal of Bridge Engineering,* Volume 11, Issue 4, 2006, pp. 420-429.

The Committee on Acoustic Emission from Reinforced Plastics (CARP), (1982), "Recommended Practice for Acoustic Emission Evaluation of Fiber Reinforced Plastic (FRP) Tanks and Pressure Vessels," *a Division of the Technical Council of The American Society for Nondestructive Testing, Inc.*

Chen, H. L. and Choi, J.H., "Acoustic Emission Study of Fatigue Cracks in Materials Used for AVLB", Journal of Nondestructive Evaluation, 23(4), 133-151(2004).

Chen, Y., Ziehl, P., and Harrison, K., (2009), "Experimental Characterization and Optimization of Hybrid FRP/RC Bridge Superstructure System", *ASCE Journal of Bridge Engineering*, Vol. 14, No. 1, pp. 1-10.

Colombo, S., Forde, M., Main, I., Shigeishi, M., "Predicting the Ultimate Bending Capacity of Concrete Beams from the 'Relaxation Ratio' Analysis of AE Signals". *Construction and Building Materials*, 19, 2005, pp. 746-754.

Fowler, T. J., and Gray, E., (1979), "Development of an Acoustic Emission Test for FRP Equipment," *American Society of Civil Engineers Convention and Exposition*, Preprint 3538, Boston, April 1979, pp. 1–22.

Fowler, T.J., Blessing, J., Conlisk, P., and Swanson, T., "The MONPAC Procedure," *Journal of Acoustic Emission*, 8(3), 1-8, (1989).

Fowler, T. J. and Gray, E., "Development of an Acoustic Emission Test for FRP Equipment," *American Society of Civil Engineers Convention and Exposition*, Preprint 3538, Boston, 1-22, (1979).

Fowler, T. J., Blessing, J. A., and Conlisk, P. J., "New Directions in Testing", Third International Symposium on Acoustic Emission from Composite Materials, July 17-21, Paris, France, 16-27, (1989).

Ghorbanpoor, A. and Vannoy, D.W., "Laboratory acoustic emission investigation of full size ASTM A-588 bridge beams", Journal of Acoustic Emission, Special Supplement, S307-S310(1985).

Goble, G., Commander, B., Schulz, J., "Simple Load Capacity Tests for Bridges to Determine Safe Posting Levels," *Final Report FHWA-PA-90-009-86-09*, University of Colorado, Boulder, Colorado, (1990).

Goble, G., Schulz, J., and Commander, B., "Load Prediction and Structural Response," *Final Report, FHWA DTFH61-88-C-00053*, University of Colorado, Boulder, Colorado, (1992).

Golaski, L., Gebski, P., and Ono, K., "Diagnostics of Reinforced Concrete Bridges by Acoustic Emission", *Journal of Acoustic Emission*, 20, 2002, pp. 83-98.

Galati, N., Nanni, A., Tumialan, G., and Ziehl, P., (2008), "In-Situ Evaluation of Two RC Slab Systems - Part I: Load Determination and Loading Procedure ", *ASCE Journal of Performance of Constructed Facilities*, Volume 22, No. 4, pp. 207-216.

Grimson, J., Commander, B., and Ziehl, P., (2008), "Superload Evaluation of the Bonnet Carre' Spillway Bridge", *ASCE Journal of Performance of Constructed Facilities*, Volume 22, No. 4, pp. 253-263.

Gong, Z., Nyborg, E.O. and Oommen, G., "Acoustic emission monitoring of steel railroad bridges", Materials Evaluation, 50(7), 883-887(1992).

Golaski, L., Gebski, P., and Ono, K., "Diagnostics of Reinforced Concrete Bridges by Acoustic Emission," *Journal of Acoustic Emission*, 20, 83-98, (2002).

Han, Z., Luo, H., Cao, J., and Wang, H. Acoustic Emission during Fatigue Crack Propagation in a Micro-Alloyed Steel and Welds, Mat. Sci.& Eng: A 528 (25-26), 2011, pp. 7751-7756.

Hartt, W., Powers, G., Leroux, V, and Lysogorski, D., "A Critical Literature Review of High-Performance Corrosion Reinforcements in Concrete Bridge Applications", FHWA-HRT-04-093, July 2004, pp.1-19.

Huang, M., Using acoustic emission in fatigue and fracture materials research, *JOM* 50 (1998) (11), pp. 1–14.

ICRI, "Explanatory Note on Half-cell Potential Testing", *ICRI Concrete repair bulletin*, V.9, No. 4, 1996, pp. 7.

JSNDI, "Recommended Practice for In Situ Monitoring of Concrete Structures by Acoustic Emission," *Japanese Society for Nondestructive Inspection*, NDIS 2421, 6 pp., (2000).

Karami, M., Farmer, J., and Inman, D., "Nonlinear Dynamics of the bi-Stable Piezoelectric Wind Energy Harvester", SPIE Conference - Smart Structures and Materials and Nondestructive Evaluation and Health Monitoring, San Diego, California, 11-15 March, 2012.

Koch, G. H.; Brongers, P. H.; Thompson, N. G.; Virmani, Y. P. and Payer, J. H.; "Corrosion Costs and Prevention Startegies in the United States", Report No. FHWA-RD-01-156; Federal Highway Administration; March, 2002.

Klieger, P., and Lamond, J., "Significance of Tests and Properties of Concrete and Concrete-Making Materials", *ASTM International*, 1994, pp. 206.

Kohn, D.H., Ducheyne, P. and Awerbuch, J., "Acoustic emission during fatigue of TI-6AL-4V incipient fatigue crack detection limits and generalized data analysis methodology", Journal of Materials Science, 27, 3133-3142(1992).

Mangual, J., ElBatanouny, M., Ziehl, P., and Matta, F., (2011) "Determining Corrosion Rate of Prestressing Strands Using Acoustic Emission Technique", *SPIE Smart Structures/NDE*, March 6-11, San Diego, CA.

Martin, C.A., Van Way, C.B., Lockyer, A.J., and Kudva, J.N., "Acoustic emission testing on an F/A-18 E/F titanium bulkhead", Proc. SPIE 2444, 204-211(1995).

National Bridge Inventory (NBI), "Condition of Bridges by State and Highway System", http://www.fhwa.dot.gov/bridge/nbi.htm, 2011.

Ohira, T., and Pao, Y., Microcrack initiation and acoustic emission during fracture toughness tests of A533B steel, metallurgical and Materials Transactions A, Volume 17, Number 5, 1986, 843-852

Ono, K. Current understanding of mechanisms of acoustic emission, The Journal of Strain Analysis for Engineering Design, 2005 40: 1

Ohtsu, M., Uchida, M., Okamoto, T., and Yuyama, S., "Damage Assessment of Reinforced Concrete Beams Qualified by Acoustic Emission," *ACI Structural Journal*, 99(4), 411-417, (2002).

Ohtsu, M., Uchida, M., Okamoto, T., and Yuyama, S., "Damage Assessment of Reinforced Concrete Beams Qualified by Acoustic Emission", *ACI Structural Journal*, Vol. 99, No. 4, Aug. 2002, pp. 411-417.

Pollock, A., and Carlyle, J., "Acoustic Emission for Bridge Inspection - Application Guidelines" Technical Report No. TR-103-12-6/95, Physical Acoustics Corporation/MISTRAS Group Inc., Princeton Junction, NJ, Prepared for U.S. Department of Transportation, Federal Highway Administration, Washington DC.

Pollock, A., and Ternowcheck, S., "Personnel Qualifications and Certification for Acoustic Emission Testing of Aerial Devices", *ASTM Special Technical Publication, No. 1139*, Philadelphia, Pennsylvania, 57-67, (1992).

Pollock, A. "Loading and Stress in Acoustic Emission Testing", American Society of Nondestructive Testing, Back to Basics Archive, http://www.asnt.org/publications/materialseval/basics/mar04basics/mar04basics.htm, (2008).

Ramirez, G., Ziehl, P., and Fowler, T., (2004), "Nondestructive Evaluation of FRP Design Criteria with Primary Consideration to Fatigue Loading," *ASME Journal of Pressure Vessel Technology*, 126(2), 216-228.

Ramirez, G., Ziehl, P., and Fowler, T., (2009), "Effect of Temperature on Acoustic Emission Evaluation of FRP Vessels Subjected to Fluid Pressure", *ASME Journal of Pressure Vessel Technology*, Vol. 131, No. 9, pp. 1-1-6.

Schulz, J., "In Search of Better Load Ratings," Civil Engineering, ASCE, 63(9), 62-65, (1993).

Sison, M., Duke, J., Clemeña, G., and Lozev, M., "Acoustic Emission: A Tool for the Bridge Engineer", Materials Evaluation pp 888-900, August 1996.

Ternowchek, S., and Mitchell, J., "Advances in AE Technology for Testing of Aerial Devices," *ASTM Special Technical Publication, No. 1139,* Philadelphia, Pennsylvania, 68-82, (1992).

Tuutti, K.," Corrosion of steel in concrete", *CBI Swedish Cement and Concrete Research Institute*, S-100 44, Stockholm, 1982, pp. 159.

Ulloa, F., Medlock, R., Ziehl, P., and Fowler, T., "A Hybrid FRP Bridge for Texas", *Concrete International*, Volume 26, Issue 5, 2004, pp. 38-43.

Xu, J., "Nondestructive evaluation of prestressed concrete structures by means of acoustic emission monitoring." Dissertation, Dept. of Civil Engineering, Univ. of Auburn, Auburn, Alabama, (2008).

Yu, J., Ziehl, P., Zarate, B., and Caicedo, J., "Prediction of Fatigue Crack Growth in Steel Bridge Components using Acoustic Emission", *Journal of Constructional Steel Research*, Vol. 67, No. 11, 2011, pp. 1254-1260.

Yuyama, S.; Okamoto, T.; Shingeishi, M.; Ohtsu, M.; and Kishi, T., "A Proposed Standard for Evaluating Structural Integrity of Reinforced Concrete Beams By Acoustic Emission", *Acoustic Emission: Standards and Technology Update*, ASTM STP 1353, 1999, pp. 25-40.

Yoon, D., Weiss, W., and Shah, S., "Assessing Damage in Corroded Reinforced Concrete Using Acoustic Emission," *Journal of Engineering Mechanics*, V. 126, No. 3, 2000, pp. 273-283.

Ziehl, P., Engelhardt, M., Fowler, T., Ulloa, F., Medlock, R., and Schell, E., (2009), "Design and Live Load Evaluation of a Hybrid FRP/RC Bridge Superstructure System", *ASCE Journal of Bridge Engineering*, Vol. 14, No. 5, pp. 309-318.

Ziehl, P. and Fowler, T., "Fiber Reinforced Polymer Vessel Design with a Damage Approach," *Journal of Composite Structures*, 61(4), 395-411, (2003).

Ziehl, P.., "Applications of Acoustic Emission Evaluation for Civil Infrastructure", Proc. SPIE 6934,1-8(2008).

Ziehl, P., Galati, N., Tumialan, G., and Nanni, A., (2008), "In-Situ Evaluation of Two RC Slab Systems - Part II: Evaluation Criteria", *ASCE Journal of Performance of Constructed Facilities*, Volume 22, No. 4, pp. 217-227.

2

Exploring Plastic Deformation of Metallic Materials by the Acoustic Emission Technique

Kristián Máthis and František Chmelík
Department of Physics of Materials, Faculty of Mathematics and Physics, Charles University, Prague, Czech Republic

1. Introduction

In determination of direct correlation between material properties and parameters of testing environment, the non-destructive *in-situ* methods play an important role. Non-destructive experimental techniques facilitate correlation of material parameters to those of the testing environment in real time. Among the tools emerging in the last decades, the acoustic emission (AE) technique belongs to the most powerful and reliable ones. The AE method allows to determine, which external parameters are critical for behaviour of the investigated material. Thus the results of AE measurements could give an integral hint for further *post mortem* investigations (e.g. scanning or transmission electron microscopy), in which the material properties are evaluated just after the exposure of the specimen to the external environment.

The AE is generally defined as elastic energy spontaneously released during local, dynamic and irreversible changes of the (micro)structure of the materials. Various nature phenomena as earthquake, avalanche, landslide or crack propagation in ice cover are accompanied by AE. Therefore it is often called in non-professional terminology as "the sound of the danger". AE is also observed during plastic deformation and phase transformations in materials. Another kind of AE appears during leak testing, when not the material itself, but the fluid or gas leakage generates detectable sound waves

This chapter is focused on AE occurring during plastic deformation of metallic materials. In this case, it responds to dislocation motion and twinning and therefore yields information on the dynamic processes involved in plastic deformation of metals and alloys. The first documented observation of AE in metals is the so-called tin cry, the audible AE produced by mechanical twinning of pure tin. After the pioneering works of Czochralski (Czochralski,1916) and Joffe (Joffe,1928) from 1920s, the systematic investigation of the AE phenomenon started just after the World War II. In 1948, Mason and his co-workers observed intensive AE signals in a wide frequency range during deformation of thin tin plates. The most important milestone in the history of the AE was the PhD work of Josef Kaiser in 1950 dedicated to the systematic study of AE during forming of various metallic and non-metallic specimens (Kaiser,1950). He described typical AE response of several metallic materials and discovered the irreversibility phenomenon, which is called the *Kaiser-effect*. In 1957, Clement A. Tatro formulated the two main tasks of the AE research: (1)

discovering the physical background of the AE; (2) developing AE systems for industrial use (Tatro,1957). The latter task was fulfilled first by Harold L. Dunegan in 1963, who tested safety of pressure vessels by the AE technique (Dunegan,1963). Thanks to the fast progress in semiconductor industry, the number of the industrial applications in last three decades increased dramatically.

2. Principles of acoustic emission monitoring

The AE permits real-time monitoring of structural changes in the investigated material. Nevertheless, the interpretation of the results is often difficult. To this very day it does not exist an exact theory for characterization of AE, since the description of AE sources and the wave propagation sets a lot of unsolved mathematical problems. All the same works of Lord (Lord,1981) and Vinogradov (Vinogradov,2001) foreshow the possible way of solving at least some model situations. Fortunately, the combination of the outputs of AE experiments and the cognitions about the materials' structure resulted in several empirical and semi-empirical models that give a guideline for the everyday AE practice.

2.1 Features of the acoustic emission signals

The detection of AE is based on its physical nature, namely the elastic energy release due to some irreversible change of the (micro)structure. Near the AE source, the released energy forms a stress pulse, which propagates through the material bulk as transient elastic waves. Depending on the geometrical configuration and the dimensions of the specimen, the stress pulse is transformed into a specific wave mode. The wave component perpendicular to the surface may be detected by piezoelectric transducers, which are coupled to the specimen surface either mechanically or through a waveguide. Electrical signal from the transducer is first preamplified and then led to the input of the measuring system, where it is again amplified and analysed. Obviously, the waveform of the AE signal is significantly affected by the source mechanism, propagation through the specimen volume and detection by the transducer.

Principally, two basic types of AE signals can be recognized. In the case of the *continuous signal,* the amplitude does not fall under a certain threshold level during relatively long period of time. The continuous signal is connected with micromechanisms of plastic deformation (e.g. dislocation glide (Heiple&Carpenter,1987a, Heiple&Carpenter,1987b)) and relaxation of large internal stresses (recrystallization, phase transitions etc.). A number of spurious signals have also continuous character, as friction or machine noise. On the contrary, the *discontinuous signals* (also called as *discrete* or *burst type* AE) are distinctly separated in time. Their generation is avalanche-like and the released energy is by several orders larger than that of continuous signal. Characteristic sources of discontinuous signals are nucleation of twins, crack initiation and propagation or corrosion processes. Stephen and Pollock (Stephens&Pollock,1971) assumed a pulse character of elastic waves in discontinuous signal and a step-like movement of the elementary parts of the material. They described the spectral distribution of the AE energy as follows:

$$A(f) = \frac{y_0^2 E}{c_L} \exp\left(-\frac{(2\pi f\, T)^2}{2} \right), \tag{1}$$

where y_0 is the displacement of the elementary volume of the material due to the stress impulse, E is the Young 's modulus, c_L is the longitudinal sound speed, f is the frequency and T is the duration of the pulse. This equation results in a normal distribution and describes well the experimental results.

The continuous signals were long time considered as a result of overlapping discontinuous signals. The study of Hatano (Hatano,1975) first showed, that the continuous signal reflects a sum of random AE pulses.

2.2 Basic terms and parameters of acoustic emission signals

The *AE event* is a single dynamic process releasing elastic energy. The local process producing AE event is called the *AE source*. The electric signal detected at the output of the transducer is termed the *AE signal.*

The AE measurements are usually carried out in the presence of continuous background noise. Therefore a *threshold level* is set in order to filter out the signal noise.

In case of *discontinuous signal* every AE event is evaluated separately. The onset of the AE event is characterized by time t_s, when the AE signal first exceeds the threshold level (Fig. 1). The end of the event could be defined by using a *dead-time* t_d (*hit definition time - HDT* in Physical Acoustic Corporation devices). The time of the AE event termination t_e is set in, when the signal did not exceed the threshold level during the last t_d period. The *duration* of the event is defined as $t_e - t_s$.

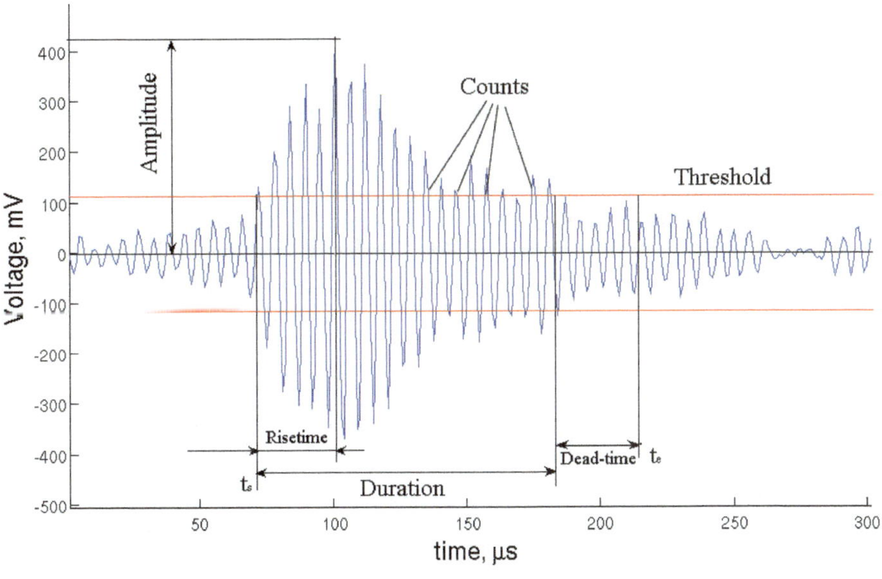

Fig. 1. Definition of discontinuous signal parameters.

The parameters of the discontinuous AE signal are the following (Fig. 1):

- *Amplitude* – maximum amplitude of the signal during its duration

- *Rise time* – time elapsed between the onset of the event and achieving the maximum amplitude.
- *Count* - the number of times the AE signal exceeds the threshold level
- *Duration* – the definition see above
- *Event energy* – is the area under the acoustic emission waveform. Usually, the evaluation systems calculates the electrical energy, defined as follows:

$$E = \frac{1}{R}\int_{t_s}^{t_e} V^2(t)dt\,, \tag{2}$$

where R is the electrical resistance of the measuring set up, V is the measured voltage and t is the time.

Cumulative representation of these parameters could be also defined, e.g. cumulative counts or cumulative energy.

The *continuous signal* is mostly evaluated in terms of the root *mean square value (RMS)* of the signal voltage and the *count rate* – the number of threshold crossings per a time unit.

Since the reliability of the AE measurements is very sensitive to the proper setting of the set up parameters (threshold, dead-time), the up-to-date AE systems enable continuous sampling and storing of AE signal, so-called data streaming. In this case the controlling computer acquires the raw data for later evaluation by the operator. The advantage of this approach is that it avoids any data loss due to incorrect system settings. Furthermore, the evaluation parameters could be fit more precisely. On the other hand, the data files are large (~1 Gb/min), which requires long computing time during the post-processing.

3. Microscopic origin of acoustic emission during plastic deformation of metals

The AE method is suitable for monitoring of collective dislocation motion, as well as for the investigation of deformation twinning. Both processes are associated with the dislocation dynamics, and they are significant sources of AE (Heiple&Carpenter,1987a). Therefore it is essential to understand the physical background of the AE generation by dislocations.

3.1 Dislocation model of acoustic emission

The dislocation model of AE has to include several key paragraphs. First of all it is necessary to identify the processes that generate AE. The aspects of the AE detection should be also discussed.

Hirth and Lothe (Hirth&Lothe,1982) have shown that a dynamic stress field has to exist in the material, when AE is generated. There are three basic dislocation mechanisms (Heiple&Carpenter,1987a) generating AE:

- *Relaxation of stress field caused by passage of dislocations* – the transient part of stress field of a moving dislocation causes AE.
- *Annihilation of dislocations* – disappearance of stress fields of annihilated dislocations causes AE

- *Bremsstrahlung* – acoustic radiation of accelerated (decelerated) dislocations

In practice, during the AE experiments the surface displacement is detected . Therefore it is essential to estimate the magnitude of the surface displacement produced by dislocation passage. The simplest configuration, i.e. surface displacement Δu from a longitudinal wave caused by growing dislocation loop in an isotropic sample, was calculated by Scruby et al. (Scruby,1981). They assumed a growing dislocation loop lying in depth D below the surface under angle 45° to the surface normal. If it reaches a maximal radius r at a constant growing velocity v, the maximal displacement can be written as:

$$\Delta u = \frac{brc_T^2}{Dc_L^3}, \tag{3}$$

Where b is the Burgers vector, c_L and c_T are the longitudinal and transversal wave speeds, respectively. Heiple and Carpenter (Heiple&Carpenter,1987a) calculated the maximal displacement in an aluminum single crystal with 4 mm in diameter. The substitution of typical values (c_L = 6400 ms^{-1}, c_T = 3200 ms^{-1}, D = 4 cm, b = 0.29 nm, v = 200 ms^{-1}) yielded a displacement of Δu = 10^{-13} m. Despite of the fact that the sensitivity of sensors are of the order of 10^{-15} m (Miller&Hill,2005), such a displacement is hardly detectable. In a real crystal, where the dislocation movement is hindered by various obstacles (grain boundaries, stress field of other dislocations etc.), the maximal surface displacement decreases by several orders (Heiple&Carpenter,1987a). Thus it could be concluded, that the signal generated by a single dislocation loop is not detectable.

The production of AE by annihilation of dislocations was investigated in several studies by Natsik et al. They investigated mechanisms of passing of dislocations at the surface (Natsik&Burkanov,1972), annihilation of moving dislocations (Natsik&Chishko,1972) and the activity of Frank-Read sources (Natsik&Chishko,1975, Natsik&Chishko,1978). Instead of assuming maximal surface displacement, they compared the amount of released energy as a consequence of annihilation of dislocations and the change of the grain shape caused by moving dislocation, respectively. The released energy caused by annihilation of a dislocation pair was set as (Natsik&Chishko,1972):

$$E_{an} = (\rho b^2 u^2 / 8\pi)\gamma \ln d / b, \tag{4}$$

where ρ is the density of the material, u is the velocity of the dislocations in the moment of the annihilation (in the original paper $u = 0.1 - 1\, c_T$ is recommended), γ = 1 for screw dislocations and $1 + (c_T / c_L)^4$ for edge dislocations, and d is the grain size. In case of Al crystal (ρ = 2700 kg m^{-3}, u = 1600 m s^{-1}, d = 80 μm) we get E_{an} = 2.9 × 10^{-10} J m^{-1}. According to Tetelman (Tetelman,1972) the released energy during deformation of a cube-shaped grain (d is the size of the cube's edge) caused by dislocation motion could be written as:

$$E_g = b\tau d, \tag{5}$$

where τ is the applied stress. For Al single crystal (τ = 4 MPa) E_g = 9.3 × 10^{-8} J m^{-1}. It means that during the dislocation annihilation the released energy is approximately three hundred times smaller than that for dislocation slip.

The *bremsstrahlung* was studied first by Eshelby (Eshelby,1962) and Kosevich (Kosevich,1964), who investigated oscillating dislocations (Eshelby,1962) and accelerating dislocation groups (Kosevich,1964). Kiesewetter and Schiller (Kiesewetter&Schiller,1976) described the nucleation, moving and pinning of dislocations as a periodic process, whose maximal amplitude is given by the distance source - pinning site. The radiated energy per second of a screw dislocation having an oscillation frequency of $\omega_0 / 2\pi$ is:

$$P_B = \frac{Gb^2 l^2 A_d^2 \omega_0^4}{20\pi c_T^3}, \tag{5}$$

where G is the shear modulus, l is the length of the dislocation line, $A_d = 2\pi\, v/\omega_0$ is the moving amplitude, v is the velocity of the dislocation. Again for Al, where $G = 2.64 \times 10^4$ MPa, $l = 80$ μm, $A_d = 40$ μm, $\omega_0 = 2\pi v/A_d = 3.1 \times 10^7$ s^{-1}, the value of $P_B = 10^{-8}$ W. Consequently, the bremstrahlung energy radiated during one period is:

$$E_B = \frac{2\pi P_B}{l\omega_0} = 5\times10^{-11}\,\mathrm{Jm^{-1}}. \tag{6}$$

This value is even smaller than that for dislocation annihilation.

According to the described results it can be concluded that the AE generated by a single dislocation is not detectable. Thus, the occurrence of detectable AE during deformation of metallic materials indicates a cooperative character of dislocation motion. In virtue of equations (3)-(5) it can be predicted that at least 100 moving dislocations or 10000 concurrent annihilations of dislocation pairs are necessary for generating detectable AE. Naturally, that is true only ideally, since part of the released energy is transformed to heat due to different interactions (e.g. with phonons) and another part is lost without being detected. Heiple and Carpenter (Heiple&Carpenter,1987b) and Gillis and Hamstad (Gillis&Hamstad,1974) estimated the part of the released energy, which will transfer to detectable AE to 0.7% and 1%, respectively,. However, as the number of dislocations in the pile-ups is very large, the detection of AE at proper conditions is possible. Sedgwick (Sedgwick,1968) assumed that the fast operating dislocation sources and the sudden release of dislocation pile-ups are in the background of detectable AE.

James and Carpenter (James&Carpenter,1971) investigated several parameters of plastic deformation of LiF crystals, as increment of dislocation density, increase of density of the mobile dislocations and changing rate of the density of the mobile dislocations. They demonstrated that the AE event rate is proportional to the change in the density of the mobile dislocations, i.e. it is connected with the conversion of immobile dislocations to the mobile ones. They proposed an avalanche-like dislocation breakaway model, in which the dislocations are pinned at obstacles. The increasing stress and/or the thermal activation cause breakaway of several dislocations. This process actuates a stress wave that causes further dislocation de-pinning. Finally, an avalanche-like process arises, accompanied with an intensive stress pulse.

Agarwal et al. (Agarwal,1970) examined the second AE source mechanism proposed by Sedgwick (Sedgwick,1968), i.e. the fast multiplication of dislocations. They assumed that the AE is generated, when the deformation stress reaches the value necessary for activation of the first dislocation source. Subsequently, the dislocations nucleate fastly and move towards the next obstacle. According to their estimation in case of 100 μm^2 large dislocation free area nucleation of at least 0.01 cm long dislocation line is necessary for generation of detectable AE.

The internal friction model worked out by Granato and Lücke (Granato&Lücke,1956) could reveal, which further mechanisms are linked with AE. The amplitude independent part of the damping curve is proportional to the fourth power of the average dislocation loop length between the pinning points and directly proportional to the dislocation density. It can be concluded that the dislocation breakaway process, which raises the average dislocation length increases the internal friction more than the dislocation generation. Thus a concurrent measurement of internal friction and AE could clarify , whether the breakaway or the dislocation multiplication process is the dominant AE source (Heiple&Adams,1976, Higgins&Carpenter,1979).

In practice, the above described processes will likely contribute to AE concurrently which makes difficult to distinguish between them.

3.2 Acoustic emission response of twinning activity

Twinning, which is a deformation mechanism observed mostly in structures with lower symmetry (e.g. hexagonal closed packed, hcp, structure) belongs to the first identified sources of AE. Even though, there has been no model describing this process in detail. Assuming the mechanisms discussed in the §3.1 we could attempt a rough approximation. It could be suggested that the source mechanisms of AE during twinning are relaxation of stress fields in the crystal and the bremsstrahlung, respectively. A pole mechanism for twin formation in hcp structures was proposed by Thomson and Millard (Thompson & Millard, 1952), who suggested the following dislocation reaction to form a pole source of twinning:

$$[0001] \rightarrow \alpha\left[10\bar{1}0\right] + \left[10\bar{1}0\right], \tag{7}$$

where α is a small fraction, ranging from 1/12 to ¼. This mechanism is in essence a modification of the Frank-Read source; only the rotating edge dislocation is replaced by a twinning one. The Thomson-Millard source reels off twinning dislocations, which are accelerated by local stresses to supersonic speed. This process causes similar surface displacement as a dislocation loop described by Eq. (3). Even a single twin dislocation (r = 40μm - the average length of twins are comparable with the grain size, v = 2000 ms^{-1}) produce a detectable surface displacement of $\Delta u = 10^{-14}$ m. Furthermore, the twin nucleation is a result of collective motion of several hundred dislocations. Thus, it is not surprising that the twin nucleation is a significant AE source. On the other hand, the twin growth does not generate detectable AE. Since the growth velocity of an elliptical twin is in order of 10^{-3} ms^{-1} (Papirov,1984), the surface displacement caused by twin growth is only $\Delta u = 5 \times 10^{-22}$ m. Carpenter and Chen (Carpenter&Chen,1988) proved experimentally the above mentioned theory. During deformation of zinc single crystal they found a strong correlation between the number of AE events and the number of twins. Furthermore, they clearly pointed out that the contribution of the twin growth is negligible compared to the twin nucleation.

3.3 Intensity of acoustic emission signal

As discussed above, there are numerous processes during plastic deformation that can generate AE. Their efficiency is summarized in Table 1.

Mechanism of plastic deformation	Strength of AE signal
Frank-Read source	strong
Twin nucleation	strong
Yield phenomenon	strong
Cutting of coherent precipitates by dislocations	strong
Orowan bowing	weak
Twin growth and thickening	negligible
Grain boundary sliding without cracking	negligible

Table 1. Intensity of AE signal due to various processes of plastic deformation

Generally, the AE signal is influenced by various experimental parameters: e.g. crystal structure, grain size, preferred crystal orientation, solute alloying, phase composition and structure, mechanical work history; strain rate, testing temperature, testing environment and irradiation. In the following chapters, the influence of some of these parameters on the AE response will be discussed in detail for various materials with hexagonal closed packed structure

4. Acoustic emission measuring system used in our experiments

In the experiments described in the following chapters the DAKEL-XEDO-3 AE facility by the DAKEL-ZD Rpety, Czech Republic was used. The AE signal was acquired at sampling rate of 4 MHz and evaluated in real time using a two threshold level detection. This procedure yields a simple amplitude discrimination and two AE count rates (Count rate at level 1 and Count rate at level 2). As the first threshold level was set directly above the peak values of noise, the AE count at this level was assumed to be the sum of the response of all deformation mechanisms detectable by AE. The burst AE count, measured at the second threshold level, was used for gathering signals coming from strong effects (e.g. twinning).

5. Acoustic emission during plastic deformation of magnesium alloys and composites

Magnesium-based alloys, as one of the lightest structural materials, have been the focus of researchers' and engineers' interest for many years. Their unique properties - such as high specific strength (the ratio of the yield stress to density), superior damping capacity and high thermal conductivity - predetermine these materials for use in many structural applications, such as in the automotive and aerospace industries. The main restriction to a wider use of magnesium alloys is their limited ductility and poor formability at ambient temperature, poor creep and corrosion resistance, and a degradation of their mechanical properties at higher temperatures (above 200°C) (Avedesian&Baker,1999). Part of the above listed disadvantages can be eliminated by using Mg-based metal matrix composites (MMCs), e.g. creep resistance or high temperature stability (Sklenicka&Langdon,2004, Trojanova,2010). The microstructure and mechanical properties of MMCs are influenced significantly by the interfaces between the matrix and the reinforcement. When a MMC is submitted to temperature changes (cooling down from the temperature of fabrication, cyclic temperature changes during operation of structural parts) thermal stresses arises at the interfaces owing to a considerable mismatch of the thermal expansion coefficient of the matrix and that of the reinforcement. Thus the right choice of processing and operating temperature of Mg-based MMCs is a key question.

The poor ductility of magnesium is due to its hexagonal closed packed (hcp) structure. In hcp metals there are three main slip systems (Fig. 2): basal, prismatic and pyramidal of first and second order.

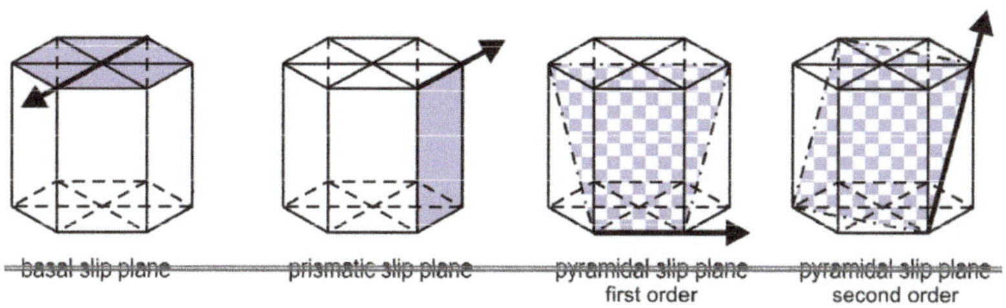

Fig. 2. Main slip systems in the hexagonal-closed packed structure

The mechanisms of plastic deformation significantly depend on the ratio of crystallographic axes *c/a*. If this ratio is larger than the ideal value of $\sqrt{8/3}$ (Zn), the basal plane has the largest atomic density and the basal slip will dominantly operate . If the *c/a* ratio is smaller than $\sqrt{8/3}$ (e.g. Ti), the primary slip occurs in the prismatic and pyramidal planes. Magnesium belongs to the third group, since its *c/a* ratio is nearly ideal. Therefore in certain cases the basal slip could be replaced by pyramidal or prismatic slip. The activation of non-basal systems depends on several parameters, e.g. on alloying content or testing temperature.

As it was shown first by von Mises (von Mises,1928), the activation of at least five independent slip systems is necessary for homogeneous plastic deformation of polycrystals. The slip systems in hcp metals are summarized in the Table 2.

	Burgers vector	**Slip direction**	**Slip plane**	**Independent slip system**
Basal	a	$\langle 11\bar{2}0 \rangle$	(0001)	2
Prismatic 1st order	a	$\langle 11\bar{2}0 \rangle$	$\{10\bar{1}0\}$	2
Pyramidal 1st order	a	$\langle 11\bar{2}0 \rangle$	$\{10\bar{1}1\}$	4
Pyramidal 2nd order	$c + a$	$\langle 11\bar{2}3 \rangle$	$\{11\bar{2}2\}$	5
Prismatic 1st order	c	$\langle 0001 \rangle$	$\{10\bar{1}0\}$	2
Prismatic 2nd order	c	$\langle 0001 \rangle$	$\{11\bar{2}0\}$	2

Table 2. Slip systems in the hexagonal closed-packed materials

As it is evident from the Table 2, the basal and prismatic systems provide only a total of four independent slip systems; none of them can produce a strain parallel to the *c*–axis. It is clear that the activation of second order pyramidal slip system is necessary for plastic deformation. Nevertheless, the size of the Burgers vector of the $c + a$ slip is large in comparison to the interplanar crystal spacing, therefore the dislocation width is small and its gliding is difficult (Hutchinson,1977). The $c + a$ dislocations were extensively studied in single crystals (Ando&Tonda,2000, Blish&Vreeland,1969, Obara,1973, Paton&Backofen,1970, Stohr&Poirier,1972). Generally, it could be concluded that the critical resolved shear stress (CRSS) for the second order pyramidal slip system decreases with the increasing temperature. Nevertheless, in case of magnesium single crystal, the results are ambiguous (Ando&Tonda,2000, Chapuis&Driver,2011, Obara,1973, Stohr&Poirier,1972). The reason could originate from the inaccuracy of the crystal orientation alignment. If the orientation difference between the *c*-axis of the crystal and the direction of the applied stress is larger than 5 angular minutes, the basal slip and twinning disturb the measurement (Stohr&Poirier,1972). However, the latest electron backscatter diffraction (EBSD) measurements (Chapuis&Driver,2011) show a monotonic decrease of CRSS for the $c + a$ slip with temperature.

There are only limited data about pyramidal slip in magnesium polycrystals. Agnew and his co-workers investigated Mg-Li alloys by means of transmission electron microscopy (TEM) and X-ray diffraction (Agnew,2001). They found that the Li addition decreases the CRSS of the pyramidal slip and the $c + a$ dislocation could be activated already at room temperature. Detailed study of temperature dependence of the $c + a$ slip activity in cast magnesium was presented in our previous work (Mathis,2004). We showed by means of X-ray diffraction peak profile analysis, that during the tensile test the $c + a$ slip becomes dominant deformation mechanism above 200°C, whereas the basal slip loses its importance.

Besides the slip in the second order pyramidal slip system, which is obviously a very difficult slip direction, twinning is the only active deformation mode that can provide straining along the *c*–axis at room temperature (Agnew&Duygulu,2005). Generally, those hcp metals that exhibit substantial twinning (e.g. Ti and Zr) possess higher ductility (Yoo,1981). The $\{10\bar{1}2\}\langle10\bar{1}\bar{1}\rangle$ twinning mode has been found as the most significant one in magnesium alloys (Barnett,2007). This twinning mode can accommodate extensions along the *c*-axis of the hexagonal lattice but not contractions along the same direction. In tension those grains undergo $\{10\bar{1}2\}$ twinning whose *c*-axis is nearly parallel to the loading direction (Fig. 3). On the contrary, in compression the *c*-axis must be perpendicular to the loading direction for easy activation of $\{10\bar{1}2\}\langle10\bar{1}\bar{1}\rangle$ twinning mode (Gharghouri,1999).

In case, when compression acts along the *c*-axis, complex double-twinned structures have been observed, where the twins in the $\{10\bar{1}1\}$ or $\{10\bar{1}3\}$ planes play the key role (Yoshinaga,1973).

Twinning of magnesium alloys is significantly dependent on the microstructural parameters, as e.g. grain size, initial texture and alloying content. The burst character of the AE response due to the twin nucleation enables the *in-situ* monitoring of twinning during the deformation tests. The advantage of the AE method stands upon monitoring the entire sample volume in real time, whereas in e.g. electron microscopy or electron backscattered diffraction , only a small volume of the specimen is investigated in *post-mortem* state.

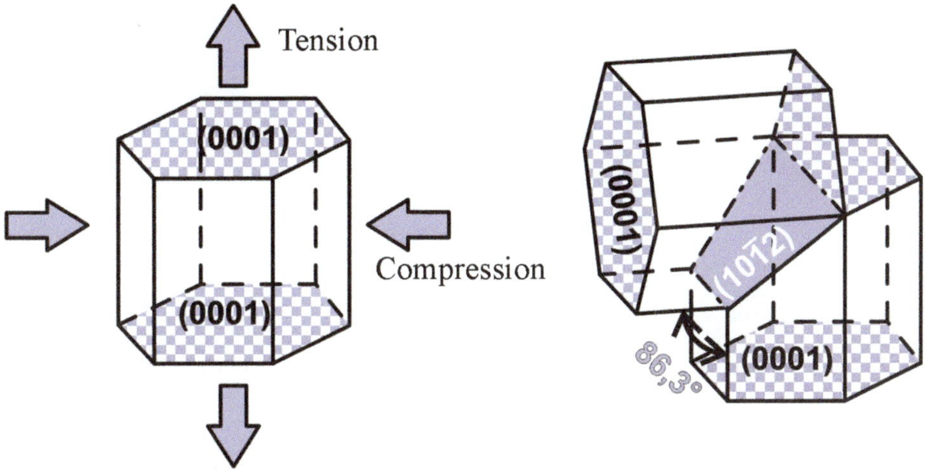

Fig. 3. Scheme of the twinning mechanism with respect to the loading direction.

5.1 Grain size dependence of acoustic emission in polycrystalline magnesium

The effect of grain size on the twinning activity or rather the AE response was studied in texture-free polycrystalline magnesium with 0.04 wt%, 0.15 wt% and 0.35 wt% Zr. (further referred to as Mg04, Mg15 and Mg35, respectively). As a consequence of the different Zr content, the specimens had different grain sizes: 550 μm (Mg04), 360 μm (Mg15) and 170 μm (Mg35). Tensile testing was carried out using specimens with a rectangular cross-section of 5 x 5 mm^2 and a gauge length of 25 mm. Compression testing was performed using prism samples 25 mm long and 15 x 15 mm^2 in cross-section. Tensile specimens were tested to fracture, while in the compression tests the specimens were deformed either to fracture or the tests was stopped after achieving of 20% strain in order to avoid the destruction of the AE sensor. All tests were performed at room temperature and at an initial strain rate of 10^{-3} s^{-1}. The specimens were tested in as-cast condition. AE was monitored directly using a computer-controlled DAKEL-XEDO-3 device. The facility incorporated a high temperature S9215 transducer (Physical Acoustic Corporation) with a flat response between 50 and 650 kHz and a preamplifier giving a gain of 30 dB. Total gain was about 94 dB. As described earlier, the AE analyzer detects the AE signals at two settable threshold levels, which corresponded to voltages of 730 mV for the total AE count 1 and 1450 mV for the burst AE count 2 (total range of the A/D converter is 2400 mV). The first level was used for the detection of the total AE signal (count rate 1). By setting the second level, we recorded only strong signals caused largely by twins (count rate 2).

The initial microstructure of all investigated specimens exhibited a nearly equiaxed grain structure. The tensile and compression true stress-true strain curves as a function of grain size are shown in Fig. 4. As expected, both yield stress and strength increase with decreasing grain size.

Fig. 5 shows the grain size dependence of the count rate 1 in compression. The largest AE response was observed for the Mg04 sample because of the larger grain size. As it was discussed before, the most important deformation mechanisms in magnesium at room

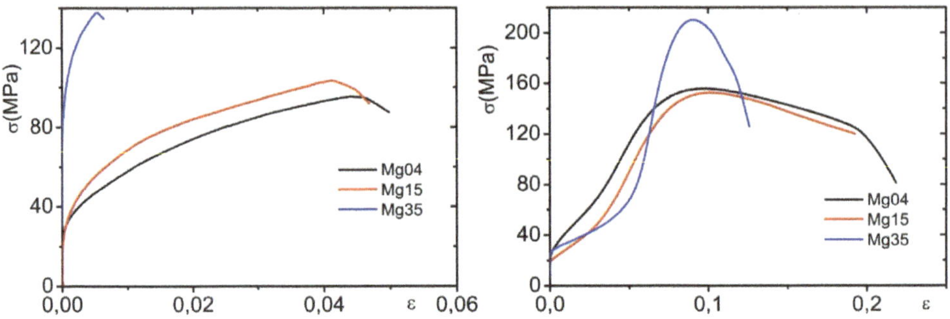

Fig. 4. Grain size dependence of true stress-true strain curves in a) tension; b) compression

temperature are slip of basal dislocations and twinning in system $\{10\bar{1}2\}\langle 10\bar{1}1\rangle$. These mechanisms were identified by Friesel and Carpenter (Friesel&Carpenter,1984) as the main sources of AE during plastic deformation of magnesium polycrystals. The size of grains, their mutual orientation and the structure of grain boundaries are very important for mechanical properties. The plastic deformation is influenced by dislocation pile-ups at the grain boundaries. The grain size dependence of the AE activity is not monotonic, as it is indicated by the numerous experimental data (Heiple&Carpenter,1987a). The largest AE activity was observed in single crystals and the large grained ($d \geq 10^3$ μm) materials, respectively. These observations could be explained in terms of large mean free path of long dislocation lines. In case of smaller grain sizes the AE is influenced by several mechanisms. The larger number of grain boundaries decreases the mean free path of dislocations. Thus the detectable AE activity decreases. On the other hand the number of dislocation sources in the grain boundaries increases (Kishi&Kuribayashi,1977). The velocity of moving dislocation could also increase due to the higher acting stress on dislocations. Nevertheless, the latter two mechanisms are negligible in the grain size range of several hundreds μm (James&Carpenter,1971, Kiesewetter&Schiller,1976) and the AE activity similarly increases with increasing grain size as it can be seen in Fig. 5.

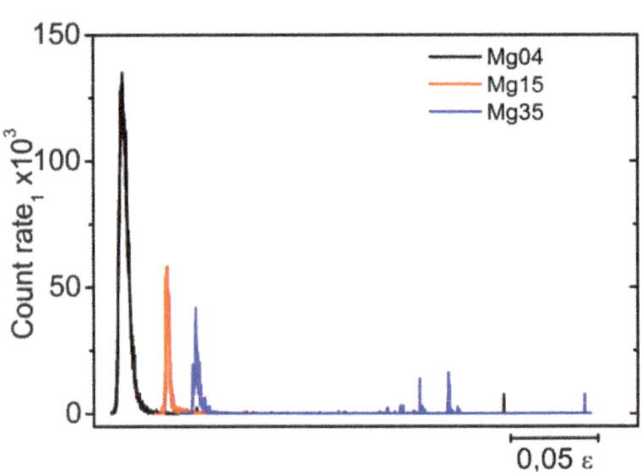

Fig. 5. Grain size dependence of Count rate 1 in compression. Due to better visibility, the curves are shifted along the x-axis.

Kim and Kishi (Kim&Kishi,1979) investigated pure aluminum with different grain sizes in range from 26 to 156 μm. They described relation between the RMS voltage and the grain size as:

$$U_{RMS}^2 = B\ \exp\left(Cd^{-1}\right), \tag{8}$$

where B and C are constants. Our data set it not sufficient to get good statistics, nevertheless the peak values of the count rate 1 follow an exponential curve (c.f. Fig. 5). The grain size dependence of twinning could be monitored at the second threshold level. The peak areas of the count rate 1 and the count rate 2 for Mg04 and Mg35 are depicted in Fig. 6. It can be seen that for Mg04 approximately 60% of the events crossing the first threshold cross the second threshold as well. On the other hand this fraction for Mg35 specimen is significantly smaller (~20%). It indicated a higher twinning activity and larger twin volume in coarse grained specimen. This result is in accordance with the EBSD observation of Beyerlein et al. (Beyerlein,2010), who found that the number of twins per grain and the twinned fraction increase with grain area.

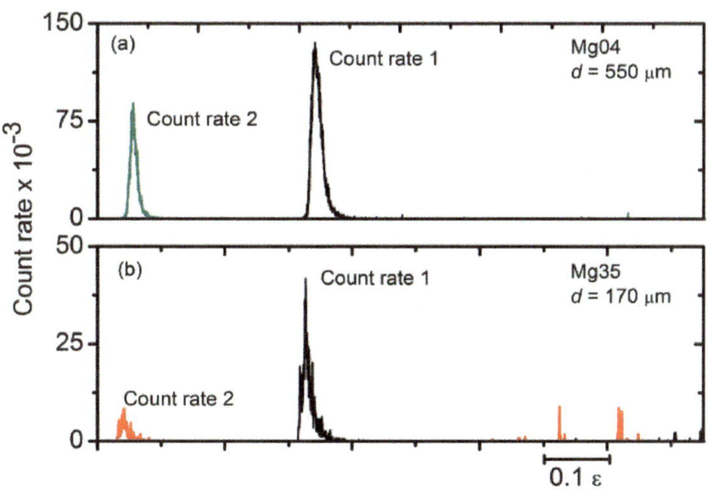

Fig. 6. Comparison of the count rate 1 and the count rate 2 for a) Mg04; b) Mg35 specimens. Due to better visibility, the curves are shifted along the x-axis.

5.2 Influence of loading mode and initial texture on the acoustic emission response in various magnesium alloys

As it is evident from Fig. 4, the shape of deformation curves depends on the loading mode. In compression S-shaped curves are observed, i.e. after a low stress plateau a significant hardening takes place. On the other hand, the tension curves are convex and the yield stress is markedly higher than that for the compressed samples (e.g. for Mg04: 58 MPa in tension and 32 MPa in compression). A similar behaviour, called tension-compression asymmetry, has been observed by several authors (Barnett,2007, Jain&Agnew,2007) and is connected with mechanical twinning (Barnett,2007). The comparison of the AE response for tension and compression depicted in Fig. 7 for Mg35 specimen revealed a different evolution of twinning.

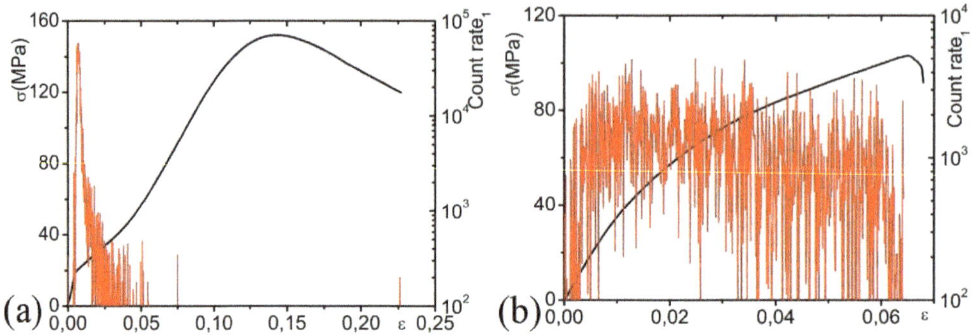

Fig. 7. Dependence of acoustic emission signal on loading mode (a) - compression, (b) - tension).

The signal recorded *in compression* (Fig. 7a) reached its maximum at approx. 1% strain in the low stress plateau range. Similar twinning evolution has been observed in compressed cast magnesium by Klimanek and Pötzsch (Klimanek&Pötzsch,2002). Using light microscopy they found a maximum in the apparent mean volume fraction of the twins in the strain range of 0.05 - 0.1ε. The rapid decrease of the AE signal after reaching the peak indicates the growth of the already present twins rather than nucleation of new twins. Actually, in compression, fast growth of $\{10\bar{1}2\}$ twins was observed (Knezevic,2010). The AE method is sensitive only to twin nucleation, not to twin growth (Friesel&Carpenter,1984), consequently the AE signal decreased. In further course of deformation the non-basal slip systems are activated (Agnew&Duygulu,2005, Clausen,2008). This mechanism increases the dislocation density considerably and thereby it reduces the mean free path of moving dislocations. Since the latter parameter is directly proportional to the AE intensity, the AE signal decreased under the detectable limit. During *tensile straining* burst AE signals were observed. After reaching the maximum, the signal level decreases gradually (Fig. 7b). This behaviour indicates that twin nucleation took place during the entire test. In tension, the nucleation of twins requires higher stresses and their growth is limited (Koike,2005). Our recent work (Máthis,2011), reporting on concurrent measurements of AE and neutron diffraction in deformed Mg - 1 wt.% Zr. supports these assumptions. We found a strong increase in the value of the integrated intensity of twin planes in the vicinity of the yield point. The neutron diffraction method is also capable of monitoring the twin growth. Using this feature we found that despite of different evolution of twinning, the total volume twinned during the tensile and compression deformation does not differ within the experimental error.

The effect of initial texture on the mechanical properties and the AE response could be demonstrated on a rolled sheet from magnesium alloy AZ31 (3 wt.% Al, 1 wt.% Zn, balance Mg). The samples used in this study had a gauge length of 25 mm and 5 x 1.6 mm^2 in cross section. The samples were in H24 temper condition (strain hardened and stress relieved state). The samples exhibited a basal texture with a significant concentration of basal planes oriented into normal direction. Tensile tests were carried out parallel (RD), perpendicular (TD) and under 45° to the rolling direction at a constant strain rate of 4×10^{-4} s^{-1} and at room temperature. Fig. 8a shows the development of engineering stress and the corresponding

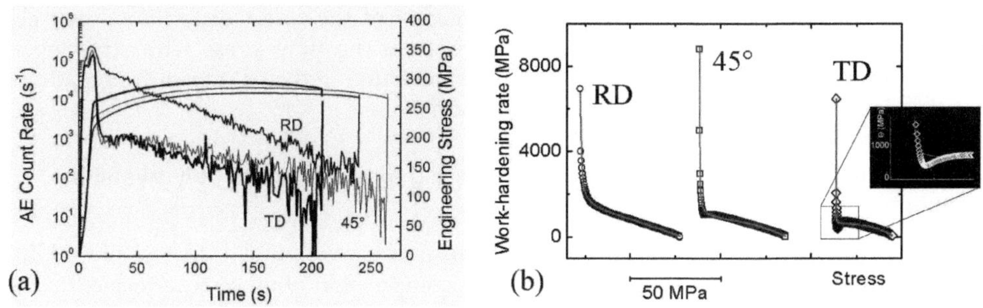

Fig. 8. (a) Development of engineering stress and AE count rate 1 with time, as a function of sample orientation; (b) Work hardening curves corresponding to the deformation curves in the adjacent figure.

AE count rate 1 with time. It is seen that the strength increases towards the TD direction. The course of AE curves exhibits a characteristic maximum related to the macroscopic yield point, which is followed by a decrease. The count rate differs with respect to the direction of the measurement. The decrease of the count rate is faster for TD and 45° specimens. The difference in the mechanical properties among the specimens could be explained in terms of different activation of deformation mechanisms during the testing. In RD, the slip of basal ***a*** dislocations is the dominant mechanism. Since this mechanism does not fulfil the von Mises criterion for plastic deformation, $\{10\bar{1}2\}$ twinning has also to take place (Bohlen,2004). As it has been shown by Yoo (Yoo,1981), the twin boundary is impenetrable for basal dislocations. Thus twinning decreases spacing between the obstacles of non-dislocation type and the AE signal decreases. In later stages of deformation non-basal slip systems (prismatic and/or pyramidal) are activated (Agnew,2003). The orientation relations in TD and 45° samples do not favour the basal slip. Thus, besides twinning, slip in prismatic and pyramidal planes is initiated at the onset of plastic deformation (Agnew,2003, Bohlen,2004). The evidence of non-basal slip is seen from the work hardening curves (Fig. 8b), where a characteristic lift appears in the vicinity of the yield point. Only the pyramidal ***c*** **+** ***a*** dislocations could cross the twin boundaries (Yoo,1981). If the glide ***c*** **+** ***a*** dislocations interact with ***a*** dislocations in the basal plane, immobile ***c*** dislocations may be produced in the basal plane according to the following reaction (Lukac,1985):

$$\tfrac{1}{3}\langle 2\bar{1}\bar{1}3\rangle + \tfrac{1}{3}\langle \bar{2}110\rangle \rightarrow \langle 0001\rangle , \qquad (9)$$

Another reaction may yield a sessile ***c*** **+** ***a*** dislocation.

$$\tfrac{1}{3}\langle 2\bar{1}\bar{1}3\rangle + \tfrac{1}{3}\langle \bar{1}2\bar{1}0\rangle \rightarrow \tfrac{1}{3}\langle 11\bar{2}3\rangle , \qquad (10)$$

A combination of two glide ***c*** **+** ***a*** dislocations gives rise to a sessile dislocation of ***a*** type that lies along the intersection of the second-order pyramidal planes according to the following reaction:

$$\tfrac{1}{3}\langle 2\bar{1}\bar{1}3\rangle + \tfrac{1}{3}\langle \bar{1}2\bar{1}\bar{3}\rangle \rightarrow \tfrac{1}{3}\langle 11\bar{2}0\rangle , \qquad (11)$$

Production of sessile dislocations increases the density of the forest dislocations that are obstacles for moving dislocations. Thus, the increase in the flow stress with straining is caused by increasing amount of obstacles. Furthermore, reduced mean free path of dislocation causes a rapid drop of the AE signal activity.

5.3 Dependence of acoustic emission on the testing temperature in polycrystalline magnesium

The influence of testing temperature on the AE response was studied on the alloy AZ31 prepared by continuous casting (Vidrich,2008). The compression tests were performed in the temperature range from 20°C to 300°C at an initial strain rate of 10^{-3} s^{-1}. The specimen size and the AE transducer were the same, as in experiments described in the section 5.1. The AE transducer was fixed directly on the sample. The AE count rate curves as a function of temperature for samples with a grain size of 150 μm showed a distinct maximum at 200°C (c.f. Fig. 9). The strong temperature dependence of AE was likely caused by the influence of thermally-activated processes. We believe that it was the activity of non-basal slip systems. It has been shown recently by Chapuis and Driver (Chapuis&Driver,2011), that the CRSS for $\{10\bar{1}2\}$ twinning and pyramidal slip decreases with increasing testing temperature. The maximum count rate, observed at 200°C, may have been a result of the synergistic effect of both mechanisms. Twinning activity probably reached its maximum at this point. Pyramidal slip systems are also active at this temperature, as we showed by X-ray methods (Mathis,2004). Decreasing AE activity above this temperature is influenced by both recrystallization and the activity of non-basal slip systems (Chapuis&Driver,2011, Mathis,2004). Since at 300°C the CRSS for pyramidal slip system is lower than that for twinning (Yoshinaga&Horiuchi,1964), its activation is more favorable than twinning. The decrease of number of twins at 300°C in pure magnesium and AZ31 magnesium alloy was observed also by Sitdikov (Sitdikov,2003) and Myshlyaev (Myshlyaev,2002). Nevertheless, the twinning does not disappear completely - the AE activity still involves strong bursts. Myshlyaev (Myshlyaev,2002) assumed that at high temperature dislocations in twin boundaries quickly annihilate. Furthermore the twins could interact with non-basal dislocations and decompose. This mechanism could alter the mutual orientation of the twin and the matrix. Dislocations piled-up at new matrix-twin boundaries

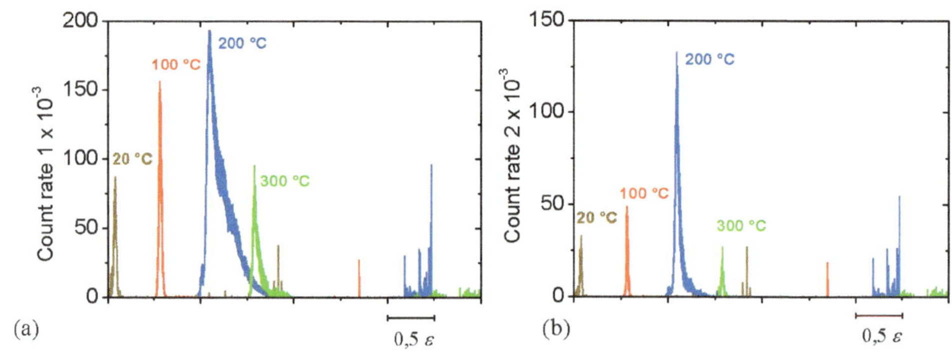

Fig. 9. Temperature dependence of the count rates 1 and 2 for the AZ31 alloy (compression test)

cause a stress field that could lead to activation of additional non-basal dislocations. Screw components of $c + a$ (and also a) dislocations may move to the parallel slip planes by double cross slip and annihilate, which decreases dislocation density and results in softening. Due to the lower twinning activity and the non-collective character of the softening, a strong decrease in the AE count rate was observed.

5.4 Acoustic emission from magnesium-based composites

Magnesium-based MMCs have been developed and manufactured over the last two decades for potential use as lightweight high-performance materials especially in the transportation industry. The standard operating conditions for most MMCs involve thermal and/or mechanical loading. This may induce microstructural changes and matrix plastic deformation characterized by dislocation generation and motion. In addition, for the case of MMCs thermal loading induces internal stresses owing to the often substantial mismatch between the thermal expansion coefficients of the matrix and the reinforcement. Under higher load and/or long-term exposure, structural damage (for example, interface de-cohesion, fiber fracture) may also occur. The main advantage of monitoring AE during either processing or straining of MMCs is in revealing the structural damages earlier than with other non-destructive methods (Miller&Hill,2005). However, there are only limited results demonstrating the use of AE in investigation of magnesium-based MMCs.

The matrix material used in this investigation was Mg of commercial purity. The matrix was reinforced with 20 vol. % of Saffil® δ-Al_2O_3 short fibers through the squeeze-casting procedure in which the molten Mg is infiltrated under pressure into short fiber pre-forms. The fabricated MMCs showed a planar isotropic fiber distribution (fiber diameters of ~3-5 μm and fiber lengths up to ~150 μm). There was a slight chemical reaction between the matrix and the reinforcement which led to chemical bonding at the interfaces. In addition, the matrix in the vicinity of the interfaces became enriched in aluminum. Specimens were machined for thermal cycling in the form of rods with lengths of 50 mm and diameters of 5 mm: the reinforcement planes in these samples were parallel to the longitudinal axes. Thermal cycling was conducted *in-situ* by placing the specimens within a dilatometer equipped with a radiant furnace permitting temperatures from ambient up to 400°C. The residual strain was measured after each cycle and the AE signal was transmitted through a quartz rod in contact with the specimen.

Figure 10a shows the variation with time of the AE count rate 1, $\dot{N}_{C1}$, and the specimen deformation, Δl, with the temperature, T, during a single temperature cycle having upper temperatures, T_{top}, of 300°C and 400°C, respectively. It is apparent that there is moderate AE during the heating within a temperature range from ~140°C to ~220°C and significant AE during the cooling of the sample at temperatures from ~180°C to room temperature. Following the cycles, a residual contraction was measured in the samples. The behaviour of the composite during thermal cycling was characterised in detail by using a stepped incremental temperature technique. The results are documented in Fig. 10b where the residual strain, $\Delta l/l_o$, and the AE count per cycle, N_{C1} are plotted against the upper cycle temperature, T_{top}, where l_o is the original length of the sample and one cycle was performed for each upper temperature corresponding to each separate experimental point recorded for the residual strain. This plot demonstrates that there is no residual strain up to a certain upper temperature, followed by a slight tendency for a compressive contraction. The residual elongation prevails in a certain temperature interval and for higher upper temperatures there is a residual contraction that increases in magnitude with increasing values of T_{top}. The AE response also depends on the upper temperature, increasing significantly at a critical value of T_{top} and exhibiting intense AE bursts which appear with a further increase in T_{top}.

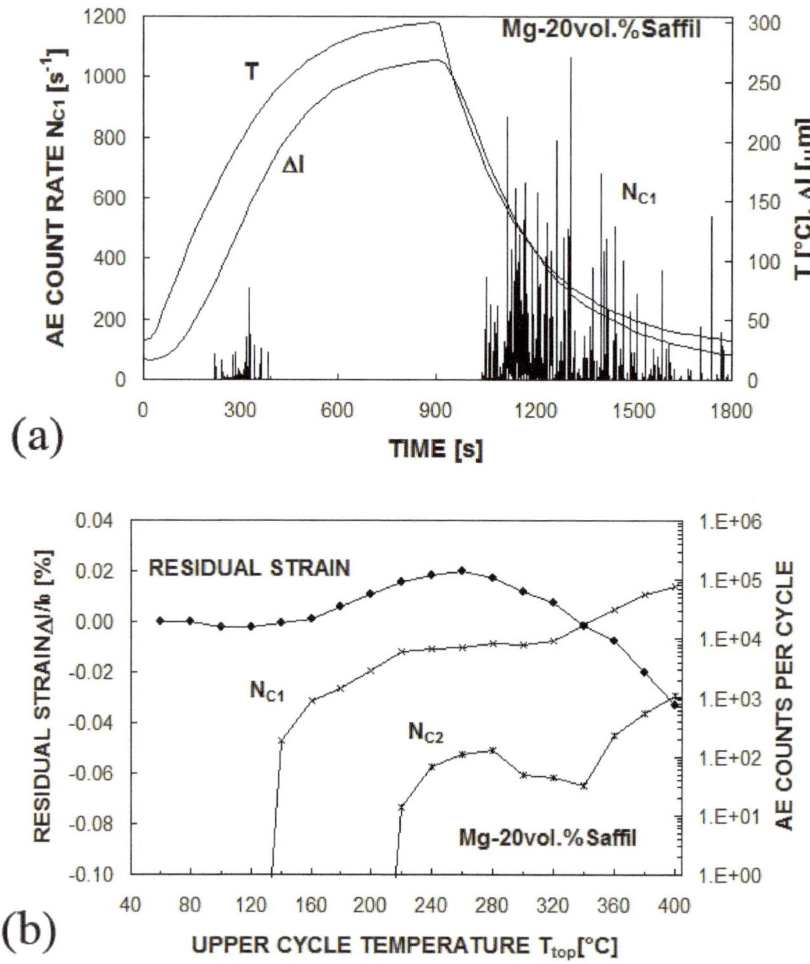

Fig. 10. (a) Value of the AE count rate 1, temperature and sample deformation (Δl) during temperature cycles to upper temperature of 300°C. (b) Residual strain, $\Delta l / l_0$, and AE counts for two different detection levels, N_{C1} and N_{C2}, versus the upper cycling temperature, T_{top}: the residual strain was estimated with reference to a temperature of 30°C and the AE counts were evaluated for the entire cooling period of each cycle.

It has been shown that, under conditions similar to those used in these experiments, more than 1000 thermal cycles are needed in order to produce any measurable damage in the samples (Kiehn,1994). Thus, the AE counts recorded in these experiments must be attributed to the occurrence of structural changes in the matrix and to any associated plastic deformation. Since the MMCs were fabricated by squeeze-casting at an elevated temperature, the composites contain thermal residual stresses at room temperature due to the mismatch in the thermal expansion coefficients between the matrix and the reinforcement (Arsenault&Taya,1987, Xia&Langdon,1994) and the magnitude of these stresses is of the order of the minimum stress required for creep in the matrix. In practice, the matrix in these MMCs experiences tensile stresses whereas the fiber reinforcements experience compressive stresses. Therefore, when the MMC is heated, the internal tensile stress acting on the matrix reduces to zero and, on further heating, there is a build up of compressive stresses; whereas on cooling, the internal stresses behave in the opposite sense. It is anticipated these stresses will be concentrated near the matrix-fiber interfaces and at the ends of the reinforcing fibers. These thermal stresses may also exceed the matrix yield stress within discrete temperature ranges and relaxation will then occur through the generation of new dislocations and plastic deformation within the matrix. This plastic deformation may appear as dislocation glide, as twinning or possibly as grain boundary sliding at the higher temperatures depending upon the precise temperature and the crystallographic structure of the matrix. In general, it is reasonable to anticipate that the compressive deformation that appears on heating will give some form of diffusion-controlled high temperature creep whereas the tensile deformation appearing on cooling will lead to dislocation glide and twinning. Thus, and in support of the experimental observations, a larger AE is expected during cooling at the lower temperatures.

It was suggested earlier that it may be possible to correlate the AE response and the internal thermal stresses produced during thermal cycling (Chmelik,1997). With respect to the dependence of the residual strain on the upper cycle temperature as plotted in Fig. 10a, a quantitative analysis of internal thermal stresses has been developed for a short-fiber reinforced aluminum composite exhibiting chemical bonding at the interfaces (Urreta,1996) and more recently this approach has been further developed and applied to experimental data (Carreno-Morelli,2000). The analysis predicts that the thermal stresses, σ_{TS}, produced by a temperature change at the interfaces of ΔT are given by

$$\sigma_{TS} = \frac{E_f E_M}{E_f f + E_M(1-f)} f \Delta\alpha \Delta T\,, \tag{12}$$

where E_f and E_M are the values of Young's modulus for the reinforcement and the matrix, respectively, f is the volume fraction of fibres and $\Delta\alpha$ is the difference between thermal expansion coefficients of the matrix and the reinforcement between the matrix and the reinforcement associated with the coefficients of thermal expansion.

For the composite used in these experiments, E_f = 300 GPa and E_M = 45 GPa at room temperature, f = 0.20, $\Delta\alpha = 20 \times 10^{-6}$ K^{-1} and the decrease in E_M with increasing temperature is ~50 MPa K^{-1}. Thus, Eq. 12 predicts that a temperature change by 1°C produces an increment in the thermal stress of ~0.6 MPa and at temperatures above ~250°C this increment decreases to ~0.4 MPa K^{-1}.

It is shown in Figure 10b that slight but measurable compressive deformation occurs in Mg-Saffil during cycling at upper temperatures above ~100°C and tensile deformation appears during cycling at upper temperatures above ~140°C. It follows from Eq. 12 that a temperature change of 70°C produces a thermal stress of ~40 MPa and, by comparison with the reported compressive acoustic yield (micro-yield) stress of ~20 MPa for the squeeze-cast unreinforced Mg at 100°C (Szaraz&Trojanova,2011), this implies the presence of an initial internal tensile stress of ~20 MPa. A similar calculation may be performed for cooling of the sample from ~140°C since, on heating, a temperature change of 110°C produces a thermal stress of ~60 MPa so that the matrix experiences a compressive stress of ~40 MPa at 140°C. This stress exceeds the matrix yield stress at this temperature and is relaxed by matrix plastic deformation (cf. observed AE) to values below ~15 - 20 MPa. Cooling of the composite to room temperature produces an estimated thermal stress of ~70 MPa. Consequently, a tensile stress approaching ~40 MPa should appear at temperatures near to room temperature and this will correspond to the macroscopic yield point. This effect will lead to AE, as observed experimentally. It is noteworthy that the disappearance of AE during heating at 220°C indicates a change in the deformation mechanism towards high temperature creep. Further details about the AE response of Mg-based MMCs during tensile test could be found in works of Trojanová et al. (Trojanova,2010, Trojanova,2011).

6. Continuous signal sampling - new approach in the acoustic emission measurement

Thanks to the fast developments in computing technology, a qualitatively new approach to the AE measurements can be proposed. Continuous sampling and storage of AE signal (called *data streaming* in Physical Acoustic Corp. terminology) offer a possibility of post-processing of the raw signal. The advantage of this approach is that the results are not influenced by setting up experimental parameters, e.g. threshold, dead-time etc. During the post-processing it is possible to run various successive analyses, evaluate individual time moments with respect to the properties of the whole data set and perform time consuming analyses that would not be possible in real time, i.e. during the experiment. To our knowledge, there are two AE systems on the market offering this feature: (1) New generation DAKEL-CONTI-4 AE system allows continuous sampling and storage of AE signal from 1 up to 4 channels with 2 MHz sampling frequency. The sampled AE signal is continuously stored on a dedicated hard disk connected to the measuring unit via a high speed interface, i.e. no data processing by the computer used to control the measuring unit takes place and a high data transfer speed is ensured. The software enables both the evaluation of individual AE events in a standard way and the analysis of the complete data set based on statistical method (Kral,2007). (2) The on-board PCI-2, 2-channel data acquisition and digital signal processing AE system is manufactured by Physical Acoustic Corporation. The system proceeds with real time AE waveform and signal processing features. The AE Data Streaming function allows continuous recording of AE waveforms to the hard disk at up to 10 MSamples/s rate. The system includes 2 parametric inputs for recording external experimental parameters (e.g. load, extension).

The method offers e.g. a new experimental insight regarding the intermittency character of plastic deformation (crackling plasticity). It was observed that the collective dislocation dynamics self-organizes into a scale-free pattern of dislocation avalanches characterized by intermittency, power law distributions of avalanche sizes (Groma,2003), aftershock triggering (Weiss&Miguel,2004) as well as fractal patterns (Weiss&Miguel,2004). This crackling noise (Bailey,2000) suggests reconsidering dislocation-driven plasticity within a close-to-criticality non-equilibrium framework. These observations are supported by discrete dislocation dynamics (Groma,2003), continuum (Zaiser,2006) and phase-field (Koslowski,2004) numerical models, which indicate that the observed picture might be of generic nature in plastic deformation of crystalline materials. Direct observation of dislocation avalanches in real time is rather difficult, usually micron-size samples are necessary (Ispanovity,2010). We have shown in our previous work (Richeton,2007) on Cd and Zn-0.08 wt.%Al single crystals that the above described effects could be monitored through continuous AE signal sampling. All stages of plastic deformation in both hcp-structured specimens were characterized by a strong intermittent AE activity (Fig. 11a). The AE signals recorded during stage I (easy basal glide) where twinning is absent, indicate the occurrence of dislocation glide avalanches, as it was observed previously in ice (Richeton,2005). The relative proportion of twinning events vs. slip avalanches increased from stage I to stage III, as expected. Whatever the material or the stage of deformation, a power law distribution of probability density of AE energy was observed, i.e. $P(E) \sim E^{-\tau_E}$, with the exponent $\tau_E = 1.6 \pm 0.1$ (Fig. 11b). This demonstrates that the generic character of crackling plasticity is not restricted to single-slip systems (such as ice), but is also observed in the presence of forest hardening or twinning. Twinning and slip avalanches can be discriminated from the AE waveforms and both populations are characterized by the same power law distribution (Richeton,2006). Plastic instabilities are clustered in time, as manifested by the presence of aftershocks. Moreover, it was shown that twinning events can trigger slip avalanches, or the reverse, suggesting that both types of instabilities participate to the same global critical-like dynamics.

The intrinsic structure of AE events during jerky flow in an aluminium alloy was also examined by this method with success (Weiss,2007).

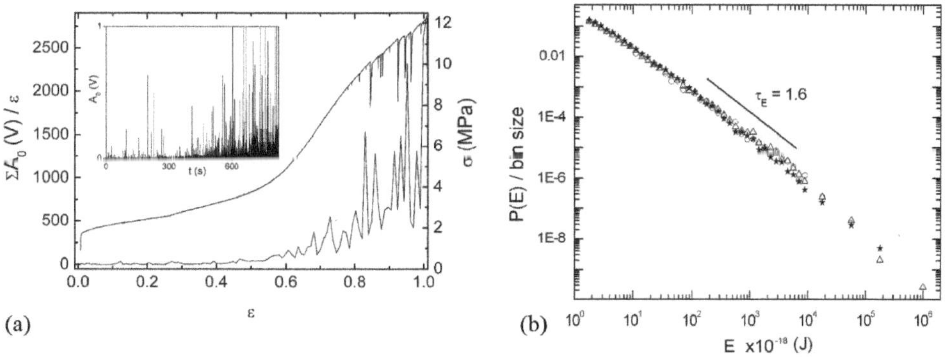

Fig. 11. (a) The stress-strain curve and the evolution of the AE activity during deformation of the sample Zn-0.08 wt.%Al. The curve represents the amplitudes of the AE events (A_0) cumulated over a time window of 7 s and normalized by the corresponding strain. The inset shows the recorded AE signal. (b) Distribution of AE energies recorded during the plastic deformation of hcp metallic single crystals. Triangles: Cd, strong forest hardening + twinning. Circles: Cd, stage I, easy basal glide. Stars: Zn-0.08wt.%Al. In courtesy of Dr. Patrik Dobroň.

7. Conclusions

In this chapter the potential of the AE technique as *in-situ* tool for monitoring of plastic deformation processes of hexagonal closed packed materials was presented. It was shown that both deformation twinning and dislocation glide are the major sources of AE. The capability of the AE measurements for investigating of magnesium-based metal matrix composites and of the intermittency character of plastic deformation was also discussed.

8. Acknowledgment

This work is part of the research program of MSM 0021620834 financed by the Ministry of Education, Youth and Sports of the Czech Republic. The authors are grateful for the financial support of the Czech Science Foundation under the contract P108/11/1267. Assistance of our students Zuzana Zdražilová and Jan Čapek from Charles University during experiments is also acknowledged.

9. References

Agarwal, A. B. L., Frederic, J. R. and Felbeck, D. K. (1970). Detection of plastic microstrain in aluminum by acoustic emission. *Met. Trans.*, Vol.1, No.4, pp. (1069-1071), 0026-086X

Agnew, S. R. and Duygulu, O. (2005). Plastic anisotropy and the role of non-basal slip in magnesium alloy AZ31B. *International Journal of Plasticity*, Vol.21, No.6, pp. (1161-1193), 0749-6419

Agnew, S. R., Tome, C. N., Brown, D. W., Holden, T. M. and Vogel, S. C. (2003). Study of slip mechanisms in a magnesium alloy by neutron diffraction and modeling. *Sci. Mat.*, Vol.48, No.8, (Apr), pp. (1003-1008), 1359-6462

Agnew, S. R., Yoo, M. H. and Tome, C. N. (2001). Application of texture simulation to understanding mechanical behavior of Mg and solid solution alloys containing Li or Y. *Acta Mater.*, Vol.49, No.20, (Dec), pp. (4277-4289), 1359-6454

Ando, S. and Tonda, H. (2000). Non-basal slip in magnesium-lithium alloy single crystals. *Materials Transactions Jim*, Vol.41, No.9, (Sep), pp. (1188-1191), 0916-1821

Arsenault, R. J. and Taya, M. (1987). Thermal-resiudual stress in metal matrix composite. *Acta Metall.*, Vol.35, No.3, (Mar), pp. (651-659), 0001-6160

Avedesian, M. M. and Baker, H. *Magnesium and Magnesium Alloys (ASM Specialty Handbook)* ASM International, 0-87170-657-1,

Bailey, N. P., Sethna, J. P. and Myers, C. R. (2000). Dislocation mobility in two-dimensional Lennard-Jones material In: *Multiscale Phenomena in Materials-Experiments and Modeling*, I. M. Robertson, D. H. Lassila, B. Devincre and R. Phillips, pp. (249-254), Materials Research Society Symposium Proceedings, 0272-9172,

Barnett, M. R. (2007). Twinning and the ductility of magnesium alloys Part I: "Tension" twins. *Mater. Sci. Eng. A*, Vol.464, No.1-2, (Aug), pp. (1-7), 0921-5093

Beyerlein, I. J., Capolungo, L., Marshall, P. E., McCabe, R. J. and Tome, C. N. (2010). Statistical analyses of deformation twinning in magnesium. *Phil. Mag.*, Vol.90, No.16, pp. (2161-2190), 1478-6435

Blish, R. C. and Vreeland, T. (1969). Dislocation velocity (1-2-12) 12-13 slip systems of zinc. *Journal of Applied Physics*, Vol.40, No.2, pp. (884-890), 0021-8979

Bohlen, J., Chmelik, F., Dobron, P., Kaiser, F., Letzig, D., Lukac, P. and Kainer, K. U. (2004). Orientation effects on acoustic emission during tensile deformation of hot rolled magnesium alloy AZ31. *J. Alloy. Compd.*, Vol.378, No.1-2, (Sep), pp. (207-213), 0925-8388

Carpenter, S. H. and Chen, C. M. (1988). The Acoustic Emission Generated during the Plastic Deformation of High Purity Zinc. *J. Acoust. Em.*, Vol.7, No.4, pp. (161-166), 0730-0050

Carreno-Morelli, E., Urreta, S. E. and Schaller, R. (2000). Mechanical spectroscopy of thermal stress relaxation at metal-ceramic interfaces in aluminum-based composites. *Acta Mater.*, Vol.48, No.18-19, (Dec), pp. (4725-4733), 1359-6454

Chapuis, A. and Driver, J. H. (2011). Temperature dependency of slip and twinning in plane strain compressed magensium single crystals. *Acta Mater.*, Vol.59, pp. (1986-1994),

Chmelik, F., Kiehn, J., Lukac, P., Kainer, K. U. and Mordike, B. L. (1997). Acoustic emission and dilatometry for non-destructive characterisation of microstructural changes in Mg based metal matrix composites submitted to thermal cycling. *Scripta Mat.*, Vol.38, No.1, (Dec), pp. (81-87), 1359-6462

Clausen, B., Tome, C. N., Brown, D. W. and Agnew, S. R. (2008). Reorientation and stress relaxation due to twinning: Modeling and experimental characterization for Mg. *Acta Mater.*, Vol.56, No.11, (Jun), pp. (2456-2468), 1359-6454

Czochralski, J. (1916). Metallographie des Zinns und die Theorie der Formänderung bildsamer Metalle. *Metall und Erz*, Vol.22, pp. (381-393), 1011-4602

Dunegan, H. L. (1963). *Acoustic Emission: A Promising Technique,* UCID-4643, Lawrence Radiation Laboratory, Livermore, CA.

Eshelby, J. D. (1962). Interaction of Kinks and Elastic Waves. *Proceedings of the Royal Society of London Series a-Mathematical and Physical Sciences,* Vol.266, No.1325, pp. (222-246),

Friesel, M. and Carpenter, S. H. (1984). Determination of the Source of Acoustic Emission Generated during the Deformation of Magnesium. *J. Acoust. Em.*, Vol.6, pp. (11-18),

Gharghouri, M. A., Weatherly, G. C., Embury, J. D. and Root, J. (1999). Study of the mechanical properties of Mg-7.7at.% Al by in-situ neutron diffraction. *Philosophical Magazine a-Physics of Condensed Matter Structure Defects and Mechanical Properties,* Vol.79, No.7, (Jul), pp. (1671-1695), 0141-8610

Gillis, P. P. and Hamstad, M. A. (1974). Some fundamental aspects of theory of acoustic-emission. *Mater. Sci. Eng. A*, Vol.14, No.2, pp. (103-108), 0025-5416

Granato, A. and Lücke, K. (1956). Theory of mechanical damping due to dislocations. *Journal of Applied Physics*, Vol.27, No.6, pp. (583-593), 0021-8979

Groma, I., Csikor, F. F. and Zaiser, M. (2003). Spatial correlations and higher-order gradient terms in a continuum description of dislocation dynamics. *Acta Mater.*, Vol.51, No.5, (Mar), pp. (1271-1281), 1359-6454

Hatano, H. (1975). Quantitative measurements of acoustic emission related to its microscopic mechanisms *Journal of Acoustic Society of America,* Vol.57, No.3, pp. (639-645), 0001-4966

Heiple, C. R. and Adams, R. O. (1976) Acoustic Emission from Beryllium *Proceedings of Third Acoustic Emission Symposium,* Tokyo,

Heiple, C. R. and Carpenter, S. H. (1987a). Acoustic Emission Produced by Deformation of Metals and Alloys - A Review: Part I. *J. Acoust. Em.*, Vol.6, No.2, pp. (177-204),

Heiple, C. R. and Carpenter, S. H. (1987b). Acoustic Emission Produced by Deformation of Metals and Alloys - A Review: Part II. *J. Acoust. Em.*, Vol.6, No.4, pp. (215-237),

Higgins, F. P. and Carpenter, S. H. (1979). Damping in iron during and after deformation. *Mater. Sci. Eng. A*, Vol.37, No.2, pp. (173-178), 0025-5416

Hirth, J. P. and Lothe, J. *Theory of dislocations* John Wiley & Sons, 0-471-09125-1, New York

Hutchinson, J. W. (1977). Creep and plasticity of hexagonal polycrystals as related to single-crystal slip. *Metall. Trans. A*, Vol.8, No.9, pp. (1465-1469), 0360-2133

Ispanovity, P. D., Groma, I., Gyorgyi, G., Csikor, F. F. and Weygand, D. (2010). Submicron Plasticity: Yield Stress, Dislocation Avalanches, and Velocity Distribution. *Phys. Rev. Lett.*, Vol.105, No.8, (Aug), pp. 0031-9007

Jain, A. and Agnew, S. R. (2007). Modeling the temperature dependent effect of twinning on the behavior of magnesium alloy AZ31B sheet. *Mater. Sci. Eng. A*, Vol.462, No.1-2, (Jul), pp. (29-36), 0921-5093

James, D. R. and Carpenter, S. H. (1971). Relationship between acoustic emission and dislocation kinetics in crystalline solids. *Journal of Applied Physics*, Vol.42, No.12, pp. (4685-4697), 0021-8979

Joffe, A. F. (June 1928). *The Physics of Crystals* McGraw-Hill, New York

Kaiser, J. (1950). *Untersuchung über das Auftreten von Geräuschen beim Zugversuch*, PhD thesis, Technische Universität München München.

Kiehn, J., Kohler, C. and Kainer, K. U. (1994). Deformation of short-fiber-reinforced Mg alloys caused by thermally-induced stresses In: *Plasticity of Metals and Alloys - Ispma 6*, P. Lukac, pp. (37-41), Key Engineering Materials, Trans Tech Publications, 0252-1059, Clausthal Zellerfeld

Kiesewetter, N. and Schiller, P. (1976). Acoustic-Emission from moving dislocations in aluminum. *Phys. Stat. Sol. a*, Vol.38, No.2, pp. (569-576), 0031-8965

Kim, H. C. and Kishi, T. (1979). Grain-size and flow-stress dependence of acoustic emission energy-release in polycrystalline aluminum. *Phys. Stat. Sol. a*, Vol.55, No.1, pp. (189-195), 0031-8965

Kishi, T. and Kuribayashi, K. (1977). Acoustic emission in the process of plastic deformation and its interpretation. *Kinzoku*, Vol.47, No.7, pp. (67-72),

Klimanek, P. and Pötzsch, A. (2002). Microstructure evolution under compressive plastic deformation of magnesium at different temperatures and strain rates. *Mater. Sci. Eng. A-Struct. Mater. Prop. Microstruct. Process.*, Vol.324, No.1-2, (Feb), pp. (145-150), 0921-5093

Knezevic, M., Levinson, A., Harris, R., Mishra, R. K., Doherty, R. D. and Kalidindi, S. R. (2010). Deformation twinning in AZ31: Influence on strain hardening and texture evolution. *Acta Mater.*, Vol.58, No.19, (Nov), pp. (6230-6242), 1359-6454

Koike, J. (2005). Enhanced deformation mechanisms by anisotropic plasticity in polycrystalline Mg alloys at room temperature. *Metall. Mater. Trans. A*, Vol.36A, No.7, (Jul), pp. (1689-1696), 1073-5623

Kosevich, A. M. (1964). Dinamicheskaya teoriya dislokacii. *Uspekhi Fizicheskikh Nauk*, Vol.84, No.4, pp. (579-609), 0042-1294

Koslowski, M., LeSar, R. and Thomson, R. (2004). Avalanches and scaling in plastic deformation. *Phys. Rev. Lett.*, Vol.93, No.12, (Sep), pp. 0031-9007

Kral, R., Dobron, P., Chmelik, F., Koula, V., Rydlo, M. and Janecek, M. (2007). A qualitatively new approach to acoustic emission measurements and its application to pure aluminium and Mg-Al alloys. *Kov. Mater.-Met. Mater.*, Vol.45, No.3, pp. (159-163), 0023-432X

Lord, A. E. (1981). Acoustic Emission - An Update In: *Physical Acoustic, Principles and Methods*, W. P. Mason and R. N. Thurson, pp. (295-360), Academic Press, 012-477-915-8, New York

Lukac, P. (1985). Hardening and softening during plastic-deformation of hexagonal metals. *Czechoslovak Journal of Physics*, Vol.35, No.3, pp. (275-285), 0011-4626

Máthis, K., Beran, P., Čapek, J. and Lukáš, P. (2011). In-situ neutron diffraction and acoustic emission investigation of twinning activity in magnesium *Journal of Physics-Conference series*, pp. (in press), 1742-659
Mathis, K., Nyilas, K., Axt, A., Dragomir-Cernatescu, I., Ungar, T. and Lukac, P. (2004). The evolution of non-basal dislocations as a function of deformation temperature in pure magnesium determined by X-ray diffraction. *Acta Mater.*, Vol.52, No.10, (Jun 7), pp. (2889-2894), 1359-6454
Miller, R. K. and Hill, E. v. K. *Acoustic Emission Testing* American Society for Nondestructive Testing, 1-57117-106-1, Columbus
Myshlyaev, M. M., McQueen, H. J., Mwembela, A. and Konopleva, E. (2002). Twinning, dynamic recovery and recrystallization in hot worked Mg-Al-Zn alloy. *Mater. Sci. Eng. A*, Vol.337, No.1-2, (Nov), pp. (121-133), 0921-5093
Natsik, V. D. and Burkanov, A. N. (1972). Radiation of Rayleigh Waves by an Edge Dislocation Emerged on Crystal Surface. *Fizika Tverdogo Tela,* Vol.14, No.5, pp. (1289-1296), 0367-3294
Natsik, V. D. and Chishko, K. A. (1972). Sound Radiation during Annihilation of Dislocations. *Fizika Tverdogo Tela*, Vol.14, No.11, pp. (3126-3132), 0367-3294
Natsik, V. D. and Chishko, K. A. (1975). Dynamics and Sound Radiation of Dislocation Frank-Read Source. *Fizika Tverdogo Tela*, Vol.17, No.1, pp. (342-345), 0367-3294
Natsik, V. D. and Chishko, K. A. (1978). Sound Radiation of Dislocations Moving near a Surface of Crystal. *Fizika Tverdogo Tela*, Vol.20, No.2, pp. (457-465), 0367-3294
Obara, T., Yoshinaga, H. and Morozumi, S. (1973). 11-22 (11-23) slip system in magnesium. *Acta Metall.*, Vol.21, No.7, pp. (845-853), 0001-6160
Papirov, I. I., Karpov, E. S., Palatnik, M. I. and Mileshkin, M. B. (1984). Acoustic-Emission during Plastic and Superplastic Deformation of a Zn-0.4-percent Al-Alloy. *Metal Science and Heat Treatment*, Vol.26, No.11-1, pp. (887-891), 0026-0673
Paton, N. E. and Backofen, W. A. (1970). Plastic deformation of titanium at elevated temperatures. *Metall. Trans.*, Vol.1, No.10, pp. (2839-2846), 0026-086X
Richeton, T., Dobron, P., Chmelik, F., Weiss, J. and Louchet, F. (2006). On the critical character of plasticity in metallic single crystals. *Mater. Sci. Eng. A*, Vol.424, No.1-2, (May), pp. (190-195), 0921-5093
Richeton, T., Weiss, J. and Louchet, F. (2005). Dislocation avalanches: Role of temperature, grain size and strain hardening. *Acta Mater.*, Vol.53, No.16, (Sep), pp. (4463-4471), 1359-6454
Richeton, T., Weiss, J., Louchet, F., Dobron, P. and Chmelik, F. (2007). Critical character of plasticity from AE experiments in hcp and fcc metals. *Kov. Mater.-Met. Mater.*, Vol.45, No.3, pp. (149-152), 0023-432X
Scruby, C., Wadley, H. and Sinclair, J. E. (1981). The oirigin of acoustic-emission during deformation of aluminum and an aluminum-magnesium alloy. *Philos. Mag. A*, Vol.44, No.2, pp. (249-274), 0141-8610
Sedgwick, R. T. (1968). Acoustic emission from single crystals of LiF and KC1. *Journal of Applied Physics*, Vol.39, No.3, pp. (1728-1740), 0021-8979
Sitdikov, O., Kaibyshev, R. and Sakai, T. (2003). Dynamic recrystallization based on twinning in coarse-grained Mg In: *Magnesium Alloys 2003, Pts 1 and 2,* Y. Kojima, T. Aizawa, K. Higashi and S. Kamado, pp. (521-526), Materials Science Forum, Trans Tech Publications Ltd, 0255-5476, Zurich-Uetikon
Sklenicka, V. and Langdon, T. G. (2004). Creep properties of a fiber-reinforced magnesium alloy. *Journal of Materials Science*, Vol.39, No.5, (Mar), pp. (1647-1652), 0022-2461

Stephens, R. W. B. and Pollock, A. A. (1971). Waveforms and frequency spectra of acoustic emissions. *Journal of Acoustic Society of America*, Vol.3, No.2, pp. (904-910), 0001-4966

Stohr, J. F. and Poirier, J. P. (1972). Electron-microscope study of pyramidal slip [11-22] [11-23] in magnesium. *Phil. Mag.*, Vol.25, No.6, pp. (1313-1330), 0031-8086

Szaraz, Z. and Trojanova, Z. (2011). Enhanced Plasticity of WE54/SiC Composite Prepared by Powder Metallurgy In: *Materials Structure & Micromechanics of Fracture*, P. Sandera, pp. (419-422), Key Engineering Materials, 1013-9826,

Tatro, C. A. (1957). Sonic technique in the detection of crystal slip in metal. *Engineering Research*, Vol.1, pp. (23-28),

Tetelman, A. S. (1972) Acoustic emission and fracture mechanics *Proceedings of US-Japan Joint Symposium on Acoustic Emission*, Tokyo,

Thompson, N. and Millard, D. J. (1952). Twin formation in cadmium. *Phil. Mag.*, Vol.43, No.339, pp. (422-440), 0031-8086

Trojanova, Z., Szaraz, Z., Chmelik, F. and Lukac, P. (2010). Acoustic emission from deformed Mg-Y-Nd alloy and this alloy reinforced with SiC particles. *J. Alloy. Compd.*, Vol.504, No.2, (Aug), pp. (L28-L30), 0925-8388

Trojanova, Z., Szaraz, Z., Chmelik, F. and Lukac, P. (2011). Acoustic emission from deformed magnesium alloy based composites. *Mater. Sci. Eng. A*, Vol.528, No.6, (Mar), pp. (2479-2483), 0921-5093

Urreta, S. E., Schaller, R., CarrenoMorelli, E. and Gabella, L. (1996). The internal damping of Al-Al2O3(f) composites during thermal cycling: The effect of fibre content and matrix strength. Vol.6, No.C8, (Dec), pp. (747-750), 1155-4339

Vidrich, G. (2008). *Grain refinement and dispersion-strengthening with finest ceramic particles*, PhD thesis, Technical University Clausthal, Clausthal-Zellerfeld.

Vinogradov, A. V., Patlan, V. and Hashimoto, S. (2001). Spectral analysis of acoustic emission during cyclic deformation of copper single crystals. *Phil. Mag. A*, Vol.81, No.6, (Jun), pp. (1427-1446), 0141-8610

von Mises, R. (1928). Mechanics of the ductile form changes of crystals. *Zeitschrift Fur Angewandte Mathematik Und Mechanik*, Vol.8, pp. (161-185), 0044-2267

Weiss, J. and Miguel, M. C. (2004). Dislocation avalanche correlations. *Mater. Sci. Eng. A*, Vol.387, pp. (292-296), 0921-5093

Weiss, J., Richeton, T., Louchet, F., Chmelik, F., Dobron, P., Entemeyer, D., Lebyodkin, M., Lebedkina, T., Fressengeas, C. and McDonald, R. J. (2007). Evidence for universal intermittent crystal plasticity from acoustic emission and high-resolution extensometry experiments. *Physical Review B*, Vol.76, No.22, (Dec), pp. 1098-0121

Xia, K. N. and Langdon, T. G. (1994). The thoughening and strengthening of ceramic materials through discontinuous reinforcement. *J Mat. Sci.*, Vol.29, No.20, (Oct), pp. (5219-5231), 0022-2461

Yoo, M. H. (1981). Slip, twinning and fracture in hexagonal closed packed metals. *Metall. Trans. A*, Vol.12, No.3, pp. (409-418), 0360-2133

Yoshinaga, H. and Horiuchi, R. (1964). On nonbasal slip in magnesium crystals. *Trans. JIM*, Vol.5, No.1, pp. (14-21), 0021-4434

Yoshinaga, H., Obara, T. and Morozumi, S. (1973). Twinning deformation in magnesium compressed along c-axis. *Mater. Sci. Eng.*, Vol.12, No.5-6, pp. (255-264), 0025-5416

Zaiser, M. (2006). Scale invariance in plastic flow of crystalline solids. *Adv. Phys.*, Vol.55, No.1-2, (Jan-Apr), pp. (185-245), 0001-8732

3

Advanced AE Technology for High-Power Microwave Radar Tubes

Narayan R. Joshi, Stephen D. Russell,
Ayax D. Ramirez and David W. Brock
HCC-NW, Houston, Texas and SPAWARSYSCEN PACIFIC, San Diego, California, USA

1. Introduction

Acoustic emission (AE) and micro-seismic activity are naturally occurring phenomena. Almost all materials emit sound or acoustic emission when they are sufficiently stressed. Wood and rock produce AE signals in audible frequency ranges when stressed. It was also believed that AE signal generation could exist in the ultrasonic frequency range during deformation of materials, but it was not until 1950 when Joseph Kaiser reported the first comprehensive investigations on acoustic emission phenomenon in conventional engineering materials using electronic equipment and tensile testing machines. Kaiser also observed that AE activity was irreversible. Acoustic emissions are not generated during the reloading of a material until the stress level has exceeded its previous highest value. This AE irreversible phenomenon is now known as the Kaiser Effect. He also proposed a distinction between burst and continuous AE activity. The use of piezoelectric sensors and electronic amplifiers to observe high-frequency AE activity gradually led to the definition of acoustic emission. According to the ASNT Handbook (1987), acoustic emission refers to the generation of transient elastic stress (strain) waves due to the rapid release of energy from a localized source within a material undergoing some kind of deformation. The kind of stress applied to materials under testing could be tensile, compressive or shear. The transient elastic stress waves of AE have frequencies ranging from 20 kHz (kilohertz) to 1 MHz (megahertz). Green (1980) has listed many mechanisms that produce acoustic emission activity in materials. Among them, the principal mechanisms are mechanical deformation, fracture, crack propagation, dislocation motion and multiplication, twin formation, phase transformation, corrosion, friction and internal magnetic processes. Mechanical loading is not the only way to generate AE activity (phonon signals). Thermal shock loading and electrical sparking are also known to cause AE activity. Generation of AE activity during chemical reactions has also been observed. It was realized quite earlier that AE activity appears in two types, burst signals and continuous signals.

2. Acoustic emission as a nondestructive evaluation technique

Applications of acoustic emission technology are growing rapidly with the improvement of AE instrumentation through several generations of refinement. It has reached a stage in which it is now used as a reliable nondestructive testing technique (NDT) in industry along with conventional techniques of ultrasonics, magnetic particles testing, liquid penetrants testing and eddy current testing. In certain industrial applications in which performance monitoring of systems occurs around the clock, acoustic emission is the only alternative available. Currently, acoustic emission testing offers high level of performance and user convenience. Although most of AE research and development took place in metallurgy and materials science in the early years, its usefulness was soon observed in many fields beyond metallurgical applications.

AE signals cover a wide range of energy levels and frequencies, from several hertz up to tens of megahertz, with two basic types of signals generally observed—burst and continuous. Some stimulus is necessary to trigger acoustic emissions. Stress, a major type of stimuli, may be mechanically applied, thermally generated, or caused by a changing magnetic field. AE sensors (transducers) convert stress wave parameters of the source event into electrical voltages corresponding to the detected event. The most popular sensor for acoustic emission experiments is a piezoelectric ceramic element, which produces a detectable response to a quick displacement change of 10^{-12} inch. The amplification of the detected signal is usually conducted in two steps to attain gains of up to 120 dB (decibels). The signals are also passed through appropriate filters for elimination of a large part of the background noise. Depending on the application, the subsequent signal may be analyzed either in the time domain or in the frequency domain. For many applications requiring frequency analysis in the range of 0.1 to 2 MHz, a broadband transducer is used. However, such transducers have a lower signal-to-noise ratio when compared with the resonant narrow-band transducers.

The propagation medium which causes attenuation, reflections and mode conversions, greatly affects the frequency content, which is also affected by the transducer and electronic system used in the experiment. Often, it is impossible to measure quantitatively the magnitude of the source of the AE signal. However, AE activity from an undamaged structure differs from the one undergoing damage over time. The relative changes in AE activity can be used to estimate changes in structural integrity of the system. From this point of view, AE technique has established itself as one of the advanced NDT (nondestructive testing) techniques in different engineering fields.

3. Applications of acoustic emission to electromagnetics

Recently, acoustic emission was used to detect electrical arcing from a sputtering target. The conventional arc detection technique involves the detection of voltage or current spikes occurring in the power supply line when the arc is initiated. This technique is not immune to unwanted electrical noise signals and may generate a false count. Therefore, the implementation of the nonelectric arc detection and counting technique may provide some advantage by improving noise immunity. Leybovich and Ferdinand (2001) used an arc detection and counting technique that monitors the acoustic emission signals generated by the arcing event. They observed that the AE response of the genuine arc generates a relatively long ringing radio frequency (RF) pulse with a significant number of cycles exceeding the threshold level. The prolonged cycling makes it possible to differentiate true arcing from false indications. The fundamental principle of this method is based on the phenomenon of generating an acoustic pulse when the arc generated shock wave strikes a solid. Sakoda and Nieda (2001) measured the characteristics of AE signals generated by a single pulse corona discharge and used them as a diagnostic tool to estimate insulation deterioration in oil immersed pole transformers. Boczar (1999 and 2001) used frequency analysis of AE signals generated by partial discharges to distinguish them from AE signals generated by corona discharges. Vahaviolos and Monsoor (1976) observed acoustic emission during laser spot welding. When the laser beam is absorbed by the surface of the material during welding, thermal stresses are generated leading to a stress wave packet due to the elastic behavior of the material. Tsukamoto and Iwasa (1986) used acoustic emission as a diagnostic and monitoring technique for arcing in a superconducting magnet. Commercial nondestructive testing companies are using acoustic emission technology to test power transformers for partial discharge, hot spots, arcing, loose connections and core clamping problems.

4. Problems with high-power microwave radar tubes

Microwave tubes used in high-power radar and communications systems are expensive. They have an operating life of a few thousand hours. When an expensive tube fails, it is generally impossible to determine the sequence of events that contributed to its failure. Three types of transmitting devices have been dominant in radar for many years. They are the magnetron, the klystron and the traveling wave tube. All are based on vacuum-tube technology. They are high vacuum devices (10^{-7} to 10^{-8} Torr). Their electrode voltages can run in excess of 10 kV (kilovolts). High heat dissipation (100 W (watts) to 10 kW (kilowatts)) is involved. Currently, microprocessor-based systems with as many as 11 sensors are designed to monitor tube performance, provide tube protection and record a comprehensive tube failure history. Watson and Troy (1987) discussed major limitations with these systems that results from the small amount of time available during the inter-pulse period of the tube for data buffering and fault analysis. The current monitoring systems work well if the microwave tube is operated at 200 pulses per second or less. However, most radar tubes are operated at up to 1000 pulses per second with pulse duration of a microsecond. Increasing data acquisition speed will, in some case, make the situation worse, since it will increase the amount of data that must be transferred and analyzed during the limited time interval available.

5. Acoustic emission as in-situ performance monitoring technique

Existing performance monitoring systems do not satisfy the modern requirements of NDT of high-power microwave tubes. Sensors that are self-contained, that could be battery powered or integrated with wireless telemetry are preferable. They should be nonintrusive and electrically isolated. They are expected to be inexpensive and compact. Two NDT techniques considered for microwave tubes included ultrasonics and acoustic emission. Ultrasonics is an active technique where high-frequency sound waves or pulses are injected into a test object and then echoes of the incident signal are monitored. Since the microwave tubes are high-power electronic devices, an active technique like ultrasonics may interfere with tube operation. On the other hand, AE is a passive technique where elastic stress waves of high frequencies generated by a device in operation can be monitored safely and externally without interfering with the normal operation of the radar tubes. The concept behind the in-situ performance monitoring of a high-power radar tube is as follows: Every time the tube generates a high–power radio-frequency pulse, it also produces a shock pulse in the tube body. An acoustic emission sensor coupled to the tube body then detects the pulse. As the tube ages, it starts to occasionally malfunctioning during pulse generation usually due to arcing, generating abnormal AE signals. Its progressive aging is characterized by monitoring of the increasingly abnormal AE signals. This technique enables decision-making to determine whether to determine whether to keep the tube in operation only a few more months or for several more years. In short, recording of AE activity is like listening to the tube, which in turn, provides an elegant, simple and inexpensive NDT tool for in-situ performance monitoring of high power microwave radar tubes.

The fundamental principle of this method is based on the phenomenon of the generation of an acoustic pulse when the arc-generated shock wave strikes a solid. In a recent experiment, characteristics of AE signals generated due to a single pulse corona discharge were measured and used as a diagnostic tool to estimate insulation deterioration in oil-immersed pole transformers. Frequency analysis of AE signals generated by partial discharges was used to distinguish them from AE signals generated by corona discharges.

Theobald et al. (2001) pointed out the difference between the AE event energy (AE_{energy}) and the AE energy descriptor (E_{AE}). The AE event energy (AE_{energy}) is given by

$$AE_{energy} \propto \int_0^T V(t)$$

Generally, AE stress waves disperse throughout the medium until interaction with an interface or boundary leads to the production of a reverberating field. Although this energy will be mostly absorbed by the medium, some of the energy can be detected by the use of a high-frequency piezoelectric ceramic transducer. Assuming all mechanisms of energy loss in the structure and measurement system are constant, then the measured electrical signal energy from the transducer could be said to be proportional to the AE event energy (AE_{energy}). Here T is the time length of transient signal produced and V(t) is the transient voltage. The measured electrical signal energy is often referred to as the AE energy descriptor (E_{AE}) and is written as

$$E_{AE} \propto \left(\frac{1}{R}\right)\int_0^T V(t)^2 dt\ ,$$

where E_{AE} refers to the AE energy measured in the transducer and R is the impedance of the complete measuring circuit. The power of the acoustic emission signal of the detected event is thus proportional to the power of the source event. The advantage of energy measurement over ring down counting is that energy measurements can be directly related to important physical parameters without having to model the acoustic emission signal. Energy measurements also improve the acoustic emission measurement when emission signal amplitudes are low.

6. Acoustic emission equipment

A new AE technique was developed, for performance monitoring purposes, to detect and measure the AE activity of radar tubes in operation. This section describes the equipment used for such a technique and some calibration procedures. Subsequent sections describe the methodology employed.

6.1 Acoustic emission transducers

Four transducers are used in this study. They are A, B, and C bought from a well-known commercial company located in the USA and D from a second company. A truncated conical aluminum piece is prepared as a match between the curved surface of the anode of the magnetron and the flat front surface of the sensors. The A sensor has peaks at 100, 300 and 500 kHz. The B sensor shows peak responses at frequencies 130, 300 and 390 kHz according to data from the manufacturer. The C sensor is highly damped with tungsten powder backing. The fourth sensor, D, was a broadband sensor. The characteristics of the four transducers or sensors were independently determined in this work and they corresponded to those supplied by their manufacturer.

These AE transducers are physically coupled to the device under test using a viscous or gel-type coupling compound.

Transducer	Description
A	General purpose transducer for monitoring acoustic emission (AE) activity in the range of 100 to 700 kHz
B	General purpose transducer for monitoring acoustic emission (AE) activity in the range of 200kHz to 1MHz
C	Wideband displacement sensor with flat response over the frequency range of 20 kHz to 1 MHz
D	Broadband sensor with flat response over the frequency range of 1 kHz to1.5 MHz

Table 1. Transducers chosen for research.

6.2 Acoustic Emission pre-amplifier

The 2/4/6 preamplifier used in this work has three gain settings of 20, 40, and 60 dB with a high-input impedance (greater than 20 MΩ (megohms). It can produce a large output signal up to 20 V (volts) peak to peak in 50 Ω (ohms). The useful frequency range of the preamplifier used in this work is 100 kHz to 1 MHz.

6.3 Acoustic Emission post-amplifier

The Model AE1A was used as the post-amplifier. The total gain achievable in 3-dB steps is 41 dB. The input impedance at the terminal "AE INPUT" is 50 ohms and at the terminal "AE OUT" is 50 ohms. The useful frequency range as tested in this work is 100 kHz to 1 MHz. The pre-amplifier receives about 21 volts DC from the post- amplifier on the same cable that carries the AE signal from the preamplifier to the postamplifier.

6.4 The oscilloscope

The oscilloscope used to monitor AE activity was a LeCroy Model 9354AM. Each of the oscilloscope's four input channels is equipped with a 500 MS/s (million samples per second), 8 bit A/D converter. Acquisition memories of up to 2 million points simplify transient capture by providing long waveform records that capture waveforms even when trigger timing or signal speed is uncertain. The important feature of the oscilloscope is its

500-MHz bandwidth. The data from the screen can be dumped and stored on a 3.5-inch floppy disk. This completes the description of the equipment setup.

6.5 Calibration of transducers

The literature provides a standardization method for calibrating AE sensors by using a capacitive transducer to capture the initial arrival of a displacement signal produced by the fracture of a glass capillary. This procedure allowed the development of quantitative comparison of AE experiment with theories of the wave propagation. AE activity may be of discrete or continuous nature. Attempts were made to generate standard pulses, which for discrete ones, correspond to the delta function, whereas in case of continuous emission, to white noise. The standard signal should be characterized by parameters determined in time and frequency domains as close to the AE signal source as possible. In the current work, it was decided to use the mechanical pencil lead breaking procedure to simulate an AE source of discrete nature. This procedure was performed on a circular aluminum plate a half-inch thick and six inches in diameter. The lead was broken on the plate at a distance of one half-inch from the AE transducer coupled with silicone grease to the same surface of the plate. AE signals were monitored on the oscilloscope. All four transducer showed satisfactory results.

7. Application of AE technology to high power microwave tube magnetron

The high-power microwave tube used in this experiment was a magnetron (Model 2J55). The magnetron oscillator is small, lightweight, operates at reasonable voltages, and has good efficiency, rugged construction and long life. The magnetron was the first practical high-power pulsed radar source used at microwave frequencies, and it is still rather widely used today. The magnetron converts energy extracted from a pulsed electric current to RF energy. The magnetron consists essentially of a copper ring into which resonant cavities have been machined. These cavities connect with the center of the ring via slots. A cathode is mounted at the center of the rings and an axial magnetic field is applied to the whole system by permanent magnets. The application of a high voltage to the ring anode with respect to cathode results in the emission of electrons from the cathode. Thus, electric field E and magnetic field H are crossed in the interaction region. Hence, magnetron is called a crossed-field device. In traveling towards the anode, paths of electrons are curved by the longitudinal magnetic field, and their trajectory takes them past the slots cut in the outer ring. If the conditions are chosen correctly, the electrons will give up some kinetic energy as they pass the slots. The energy given up will increase the amplitude of the oscillations in the cavity.

When the magnetron is oscillating, the cavities are coupled together by oscillatory E and H fields and energy is therefore extracted from all cavities by a loop within a single cavity. In an 8-cavity magnetron, these cavities form four spokes centered on the cathode. The magnetron is a kind of diode and Figure 1(a) shows its simplified schematic. A cutaway view of one possible magnetron configuration is shown in Figure 1(b) after T. C. Edwards (1984). The principal parts of magnetron are the cathode, the interaction region, the cavities, and the output coupling. The voltage/current relationship for a microwave magnetron is nonlinear and hence the magnetron is often represented by an equivalent circuit consisting of a biased diode with the magnetron's stray capacitance represented by a capacitor connected in parallel. An appreciable current is not drawn by a conventional magnetron until a critical voltage, approximately 90% of the operating voltage is reached. This voltage is called Hartree voltage. Low levels of RF output power may be generated by the tube even below the Hartree voltage. Therefore, care should be taken to ensure that voltage is promptly removed after the desired pulse and not reapplied, even at low levels.

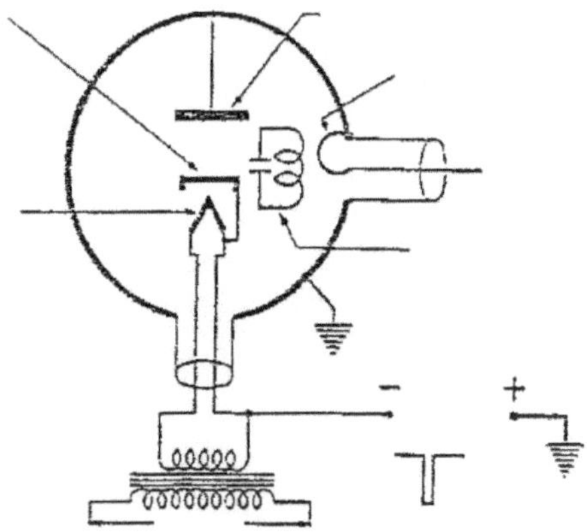

Fig. 1.(a) Simple schematic of Magnetron Tube.

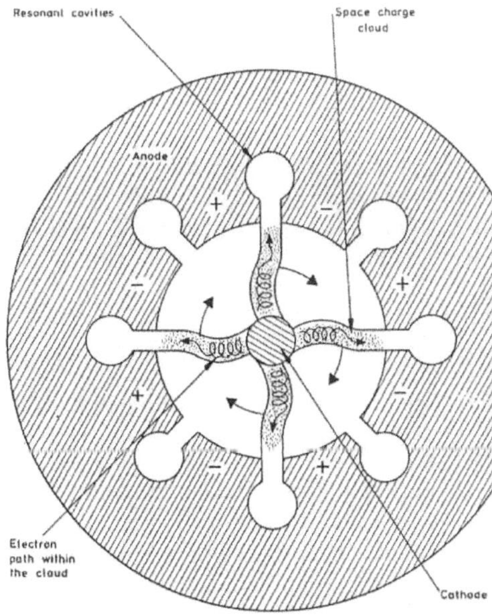

Fig. 1.(b) Electron paths in Magnetron

7.1 Magnetron operating characteristics

When a voltage pulse of the proper magnitude (-15 kilovolt DC) and a magnetic field of the correct strength (1900 to 2100 gausses) are applied, the magnetron will oscillate at a frequency determined by the magnetron cavity which acts like a tuned circuit. The RF (radio frequency) generated by the magnetron used in this work was 9.4 GHz (gigahertz). The size of the cavities may be around 2 mm. The cavity is coupled inductively to a pick-up loop which is connected to the inner conductor of a waveguide. Thus, the output of the magnetron was connected to a waveguide terminated in a load cooled by fins and air blower. The peak input power to the magnetron was 210 kW and the peak output power was 50 kW. The average current drawn by the magnetron is about 1.2 mA (milliampere) but the current during 1-μs (microsecond) pulse is about 20 A (amperes). The duty cycle (width of pulse/period between pulse = 1 μs/0.01666 s) for 60 pulses per second is 0.006 percent.

One problem with magnetron oscillators is that the interaction space in a magnetron will support more than one possible mode of oscillation. In well-designed magnetrons, the various modes are well separated in both voltage and frequency and the magnetron will operate in the desired mode in a stable manner. However, certain parameters, notably the characteristics of the modulator, such as the rate of rise of voltage, can strongly influence the mode selection in magnetrons. Operation in other than the normal mode (called moding) is usually undesirable, because oscillations are separated from the normal operating frequency and the power does not couple out of the tube in the proper fashion. Double moding can be detected by the appearance of a smaller pulse, which varies in amplitude underneath the main pulse. Double moding could happen due to a change in loading, in magnetic field strength, or in applied voltage.

Another property of magnetrons, which is probably related to moding, is the tendency of such tubes to produce an occasional RF output pulse of reduced amplitude or to fail to produce any RF output pulse. In a well-designed tube, such missing pulses would typically occur less often than 1 in 10^4. As with any high-power tube, the possibility of internal tube arcing always exists. Particularly in case of magnetrons, internal arcing is one factor that must be given attention in the design of the modulator or pulser to drive the tube. The outside of the magnetron is a grounded shell to which the plate of the magnetron is connected (Figure 1(a)). The cathode has a negative potential with respect to the anode. The cathode of the tube is heated indirectly and connected to one side of the filament. The filament is fed by the transformer, which is highly insulated to withstand the peak modulator pulse voltage applied to the cathode. The high-voltage pulse is applied to the magnetron through a high-voltage concentric bushing.

The output frequency varies about 0.05 MHz per degree change in magnetron temperature. Thus, a gradual frequency shift may be observed during initial warm-up period. An air blower cools the magnetron during normal operation and a temperature sensor connected to a digital readout device monitors the temperature. As mentioned before, the resistance of the magnetron is nonlinear. At low plate voltage values, changes of large plate voltage result in minor changes in plate current. In the normal operating mode, however, a 10% increase in plate voltage may produce a 100% increase in plate current, depending on the characteristics of the tube under use at a given value of the magnetic field strength. A decrease in magnetic flux density with constant plate voltage causes a large increase in plate current. A good operating procedure requires decrease in plate voltage to go with decrease in flux density and no operation of magnetron without a magnetic field.

7.2 Waveguides and VSWR

A waveguide can be used to minimize losses and produce high-power transmission at microwave frequencies. It consists of a single metallic conductor in the shape of rectangular box or cylinder, through which the electromagnetic waves are propagated. Such guided waves have field configurations somewhat different from those of two wire open lines. The waveguide transmission is characterized by two important features. First, below a minimum frequency a given waveguide will not transmit the wave. This is called the cut-off frequency, f_c, and is directly related to the size of the waveguide used. Second, a component of E or H always exists along the direction of propagation. The guided waves may be propagated along the waveguide with different field patterns called 'modes.' Modes are mainly of two types: those that have an E component along the direction of propagation are called TM or E waves, and those with an H component along the direction of propagation, which are called TE or H waves. Rectangular waveguides have losses less than those of circular waveguides and are less prone to mode changing. Hence, they are commonly used. The magnetron in the current work was also connected to a rectangular waveguide. Waveguides are usually made of brass, copper, or aluminum in various standard sizes corresponding to microwave frequencies used. Waveguide sections are usually coupled together by flanged assemblies that are bolted together and supported periodically on metal stands. The parameters associated with the rectangular or circular waveguide are impedance Z and cutoff wavelength λ_c, and they are closely related to dimensions of the waveguide, while attenuation losses depend on these factors as well as on the inner surface finish and metal of the waveguide walls. Just to give an example, some figures for the TE_{10} wave mode in rectangular waveguide are given next,

$$Z_{TE}\,(\text{Ohms}) = 377\lambda_g / \lambda \text{ and } \lambda_c = 2a,$$

where λ_g is the guide wavelength, λ is the free space wavelength, λ_c is the cutoff wavelength, and a is the width of the waveguide. See Connor (1972).

The magnetron can be attached to a waveguide to inject microwaves into it where they will propagate as a sinusoidal wave. If the load absorbs all of the wave, then there is no reflection. This occurs when the resistive load and the source load have equal impedance. This condition is called a matched load. But if there is a mismatch, then not all of the incident power is absorbed by a load and some of it is reflected back. The reflected wave then combines with the transmitted wave, resulting in standing wave. The ratio of the maximum and minimum amplitudes is called the voltage standing wave ratio or VSWR. It is important that the VSWR of the load is known, because if a VSWR is too high, an appreciable amount of energy will be reflected back to magnetron, which could cause severe damage. The VSWR can be written as

$$\text{VSWR} = V_{max} / V_{min} = (Vi + Vr) / (Vi - Vr),$$

where Vi and Vr are the voltage due to the incident and reflective wave, respectively. The VSWR can be measured with a network analyzer or with the slotted line method using a probe and VSWR meter. See Love (1995).

7.3 Experimental setup

Figure 2 shows the schematic of the experimental apparatus. The photo of the actual magnetron unit is presented in the report of Joshi et al. (2000). The AE signal is fed to one channel of the LeCroy digital oscilloscope. The cathode current pulse is fed to the second channel. The cathode voltage through a high-voltage probe (Tektronix, Model 015-049) is connected to the third channel. The high-voltage pulses have negative polarity since they are fed to the cathode while the anode is grounded. In Figure 2, an Interstate P25 Pulse generator is shown under the oscilloscope. Its pulsed output goes to the external trigger on the oscilloscope and to the pulser trigger on the left-hand cabinet of the modulator for the magnetron. The oscilloscope displays four (processed) traces A, B, C, and D and is capable of mathematical analysis and zooming of each trace. The scope had the capacity to test a waveform against a predefined mask, which made it possible to catch on the screen abnormal current pulses and run mathematical operations such as FFT (Fast Fourier Transform) of AE signals, which could help show abnormal functioning of the magnetron.

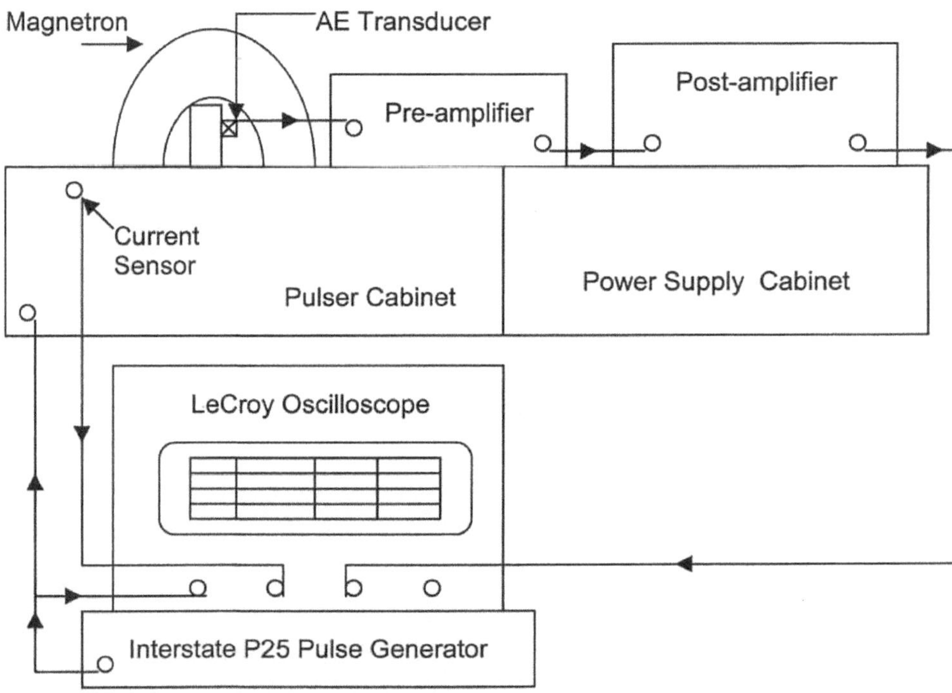

Fig. 2. Schematic of Experimental Equipment.

The magnetron was purposely stressed by changing parameters from their normal values prescribed for stable operation. This allowed accelerated aging and generated abnormal cathode current pulses leading to abnormal RF pulses. The magnitude of the FFT of the AE activity picked up during abnormal functioning of the tube was computed and all traces from the screen were saved to the disk. These traces were compared with those obtained during the stable or normal functioning of the tube. This testing procedure showed whether the acoustic emissions generated by the pulsating magnetron could be used as a noninvasive method for monitoring the normal and abnormal functioning of the tube.

7.4 Experimental procedure

The proper operation of the magnetron depends on the amplitude of the voltage pulse applied, the strength of the magnetic field, and the proper loading of the magnetron by the

RF system. The normal operating parameters of the magnetron tube (2J56) used in this work are described below. These parameters were intentionally changed to stress the magnetron tube leading to its abnormal functioning. The effect on AE during normal and abnormal functioning of the magnetron was observed and recorded.

The normal filament voltage of the oscillator tube is 6.3 volts AC obtained through a step down transformer whose primary operates at 115 volts AC. By decreasing the voltage of the primary from 115 to 60 V, the cathode emission can be decreased to produce abnormal current pulse. This is the first deviation from the normal parameter.

The normal operating high-voltage pulse rate is 1000 pulses per second, which can be increased up to 1500 pulses per second or decreased to 100 pulses per second. This is the second deviation from the normal parameter. The pulse width is maintained constant at 1 μs. In this work, the pulse rate was changed within the range of 100 to 1000 pulses per second. With 1000 pulses per second, the average power output was around 50 W. With 100 pulses per second, it reduces to 5 W.

The ambient room temperature is about 20 °C. With 100 pulses per second, the temperature of the magnetron rises to 35 °C. Air from a fan cools the magnetron. A second fan keeps the waveguide load cool. The third way to stress the tube is to shut off the first fan and let the temperature of the magnetron rise. The temperature rise was rather rapid with the high pulse rate of 1000 pulses per second.

The RF energy produced by the tube is delivered to a load at the end of the waveguide. By using a variable and movable load impedance attached to the waveguide, part of the RF energy can be reflected back to the magnetron, creating an abnormal operating situation. Changing the ratio of the reverse power to the forward power is the fourth way to stress the tube. The average RF output power of the magnetron was measured with a Hewlett-Packard Model 432A power meter. Two attenuators in series (each 10 dB) were used to connect the cable of the power meter to the waveguide when the power output was around 50 W, and one attenuator was used when the power output was around 5 W.

The normal pulse voltage is 15 kV and can be increased up to 21 kV. This is the fifth way to create abnormal operating parameters. However, this deviation may produce arcing among components inside the high-power supply cabinets. For this reason, it is used sparingly in this work. It was observed generally that when the tube was stressed, it produced abnormal current, and resulting abnormal RF pulses. The abnormal current pulse is captured with the masking feature of the oscilloscope, the data acquisition is stopped, and the magnitude of the FFT is calculated for the captured AE signal from the magnetron. The frequency spectrum of AE activity is expected to be less than 1 MHz. The sampling rate used was 250 MS/s. The displays on the oscilloscope screen were saved on a computer disc in bit map mode. This procedure was repeated many times with normal and varied abnormal functioning of the magnetron tube. The AE activity was monitored by the four previously described transducers in turn. The preamplifier gain and the post-amplifier gain were monitored and kept unchanged when two oscilloscope displays were considered for comparison.

7.5 Experimental results

The data in Figure 3 were obtained by using AE transducer C (see Table 1). Figure 3 shows the masking feature and mechanism provided by the oscilloscope for capturing a signal generated during the abnormal operation of the magnetron when it was stressed by changing one of its normal operating parameters. In both frames of Figure 3 the pulse rate was adjusted to 100 pulses per second and the oscillator filament primary voltage was set at the normal value of 115 volts. In both frames, oscilloscope channel 3 was connected to the output of the acoustic emission post- amplifier. Channel B was connected to the current sensor of the magnetron. It shows the zoom trace of the current pulse. The mask was set on channel D (zoom trace). Channel A shows the magnitude of the FFT of the acoustic emission signal from channel 3.

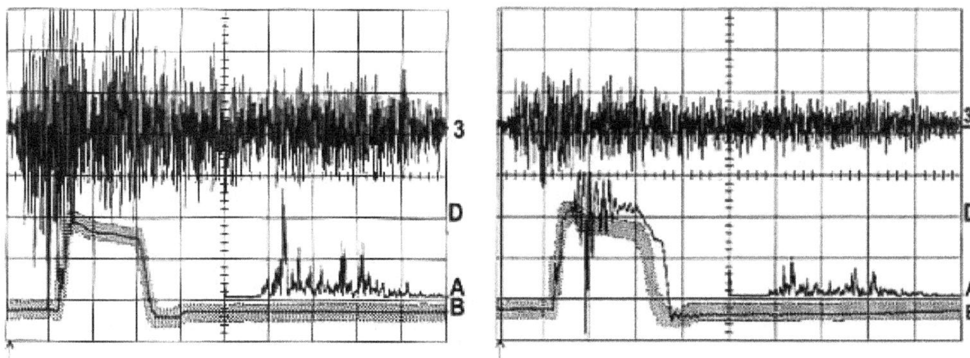

Fig. 3. Oscilloscope images of normal (pulse voltage: 15 kV) operation of the magnetron, showing AE signal on channel 3 for normal current pulse on channel B (top frame) and of abnormal (pulse voltage increased to 19.5 kV) operation showing AE signal of decreased amplitude due to abnormal current on channel B caught by masking feature (bottom frame).

In the normal operation, pulse voltage was set at the normal value of 15 kV. It then gradually increased to 19.5 kV when the abnormal current pulse occurred. When it was outside the mask value, the oscilloscope displayed the abnormal pulse. This display was then stored on the disc.

Let us compare the two acoustic emission signals. The preamplifier and the postamplifier gains were kept unchanged. In both frames of Figure 3 the vertical sensitivity on channels 3 and A were kept unchanged at 100 mV/division (millivolt/division) and 14.6 mV/division, respectively. Now compare the amplitudes of the acoustic emission signals in both frames on channel 3. When the abnormal pulse occurred, the acoustic emission signal amplitude was decreased considerably (right frame) in comparison with the one obtained during the normal operation (left frame). Also shown is the abnormal current pulse caught by the masking feature. Channel A shows the FFT of the AE signals.

Next, the data in Figure 4 were obtained using transducer A (see Table 1) coupled to the magnetron anode. The pre-amplifier gain at 20 dB and the postamplifier gain at 0 dB were held constant. The pulse rate was 1000 pulses per second and the pulse voltage was 15 kV. The pulsed microwave power of about 40 kW was sent forward to the load at the end of the waveguide and no power was reflected back. These normal parameters were held constant except for the oscillator's primary filament voltage. For the data in the left frame of Figure 4, it was set at the normal value of 115 volts, while for the data displayed in right frame of Figure 4, it was gradually decreased to the value of 95 volts when the abnormal current pulse occurred. The display from the oscilloscope screen was immediately recorded on the floppy disc. In records of both frames in Figures 4, channel 3 was connected to the output of the AE post- amplifier. Channel B in left frame of Figure 4 shows the zoom display of the normal current pulse while in right frame of Figure 4, it shows the zoom of the abnormal current pulse.

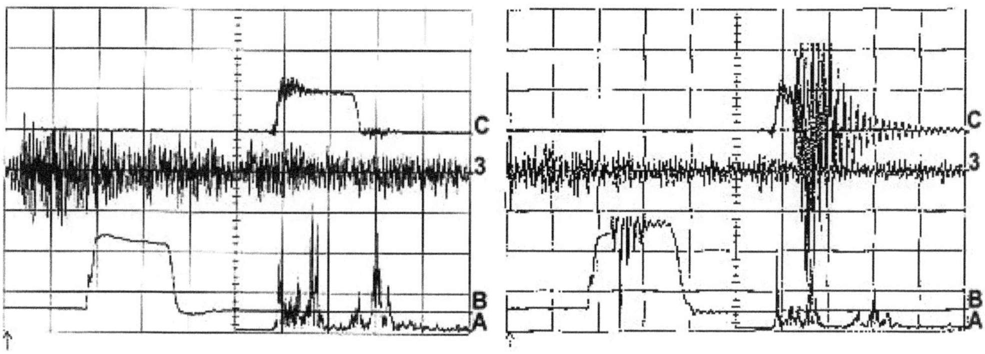

Fig. 4. Oscilloscope images of normal (filament transformer primary voltage 115 volts) operation of the magnetron, showing AE signal on channel 3 for normal current pulse on channel B (left frame) and of abnormal (filament primary voltage reduced to 95 volts) operation showing AE signal of decreased amplitude due to abnormal current on channel B (right frame). No mask is shown here. It is hidden.

Channel C in the left frame of Figure 4 shows the zoom display of the normal RF signal related to the RF power taken out from the waveguide, while in the right frame of Figure 4, it shows the zoom of the abnormal RF signal. AE signals are shown on channel 3, one of high amplitude in left frame due to normal operation and of decreased amplitude in the right frame due to abnormal operation. The magnetron was forced to behave abnormally by changing the value of the oscillator filament primary voltage. A similar decrease in AE signals was observed using different abnormal conditions such as load mismatch at waveguide, causing the power ratio (reflected power to incident power) to change. The details of different abnormal situations used and corresponding AE patterns obtained using different transducers are reported by Joshi et al. (2000, 2006 and 2007). Irrespective of the cause of the abnormal functioning, it produced a decrease in the amplitude of the burst type of AE signal. Channel A shows the FFT of the AE signals.

7.6 Discussion of magnetron experimental results

Under abnormal operation of a high-power magnetron radar tube induced by different stressing mechanisms, the AE signal consistently showed a measurable decrease in amplitude relative to the AE signal amplitude recorded during normal functioning of the tube. The phenomenon observed was independent of the AE sensors used to pick up the AE signal from the grounded cylindrical anode of the magnetron tube. The details of the mechanism of generation of acoustic emission due to the pulsating of the RF electromagnetic wave is under investigation. The pulsing electromagnetic energy generates a thermal shock wave in the cylindrical grounded anode that in turn produces the acoustic emission stress wave due to elastic properties of its material. Acoustic energy could also be generated by the interaction of

1-μs, 16-A current pulses with a strong magnetic field. The decrease in the pulsating electromagnetic energy in the abnormal functioning of the tube shows the related decrease in the energy of the acoustic emission signal. These result proved that one can use advanced AE technology for in-situ performance monitoring of radar tube like magnetron.

8. Application of AE technology to klystron high power microwave tube

8.1 Experiment on first klystron unit

The unit used in this work was a five-cavity klystron amplifier. Joshi et al. (2001, 2002, and 2007) and Ramirez et al. (2003) have published the details of the functioning of the klystron unit with auxiliary electronic circuits. The pulse repetition frequency (PRF) was chosen at 213 pulses per second. An adjacent pair of pulses was generated. The pulse width of the first pulse was 2 μs and that of the second pulse was 125 μs. The separation between two pulses was 125 μs. The RF frequency of the first pulse was 894.33 MHz. The second pulse was the chirp signal with the center frequency of 894.33 MHz and the variation of ±1/2 MHz. In other words, the chirp range was 1 MHz. The klystron had 48 channels between 851 and 942 MHz. The low band is made of channels 1 to 16. The middle band is made of channels 17 to 32. The high band is made of channels 33 to 48. The klystron was connected to the dummy load at the end of the waveguide. The dummy load was cooled by circulating water. In the normal operation, the cathode was held at -42 kV with respect to the grounded collector.

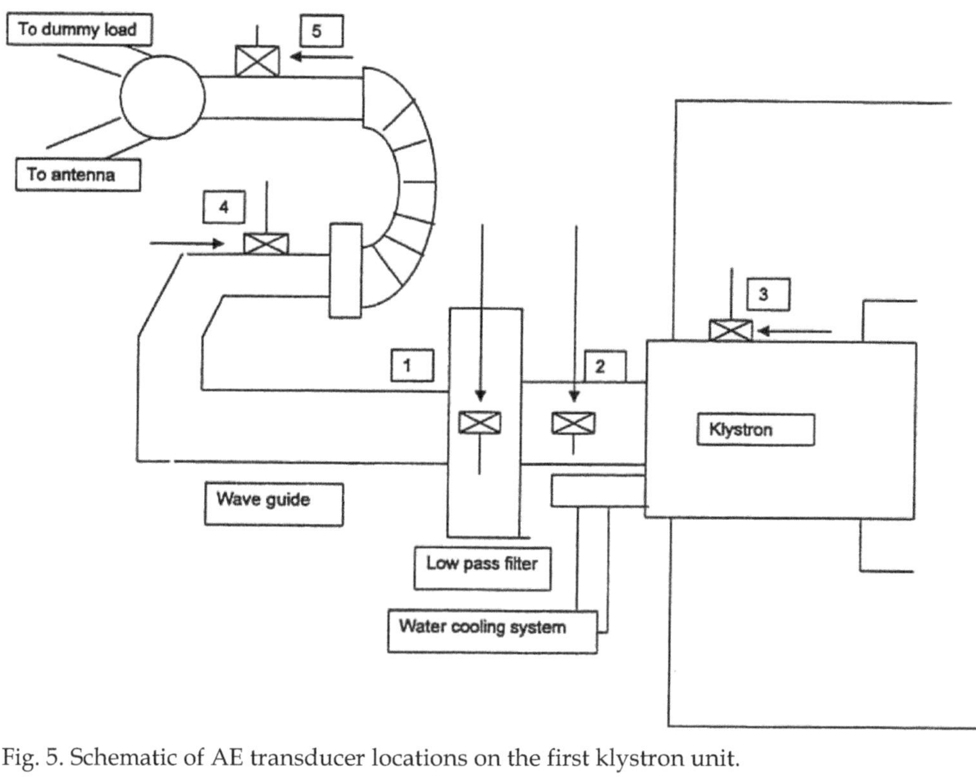

Fig. 5. Schematic of AE transducer locations on the first klystron unit.

8.2 Experimental setup

The acoustic emission system used included the general purpose transducer A (see Table 1), a preamplifier and a postamplifier as described earlier. Figure 5 shows the different locations on the klystron unit where the AE transducer was attached in turn to pick up emissions generated during the operation of the klystron. The pre-amplifier was wrapped in a bag of aluminum foil to prevent it from picking of extraneous signals. The appropriate coaxial cables (RG 223U) were used for all connections. The total gain of two amplifiers was set at 61 dB. The LeCroy 9354AM 500-MHz digital oscilloscope was used to search for abnormal acoustic emission signals generated and to store them on a disc.

8.3 Experimental results from first klystron unit

8.3.1 Effect of different locations of AE sensor

When all parameters such as gains of two AE amplifiers and transducer couplings were held constant, the strength of the AE signal differed from locations to locations during normal operating conditions of the klystron unit. The AE signals received from the location of 5 on the waveguide and from the location of 3 (Figure 5) on focusing coils inside the cabinet of the klystron unit were stronger than the one received from the location of 1 on a low pass filter.

8.3.2 Effect of increased peak power output

The klystron unit was well protected by various sensors to avoid abnormal variations of the operating parameters. The only way to stress the klystron unit was to change the peak power output, so it was decided to stress the unit by changing this single available parameter. The AE transducer was coupled to the location 3 (Figure 5) on the focusing coils of the klystron unit. In the left frame of Figure 6, four separate signals appear on the zoom trace A, which is the averaged acoustic AE signal from Channel 3. They corresponded with rise and fall of signals from beam current sensor connected to Channel 2 of the oscilloscope.

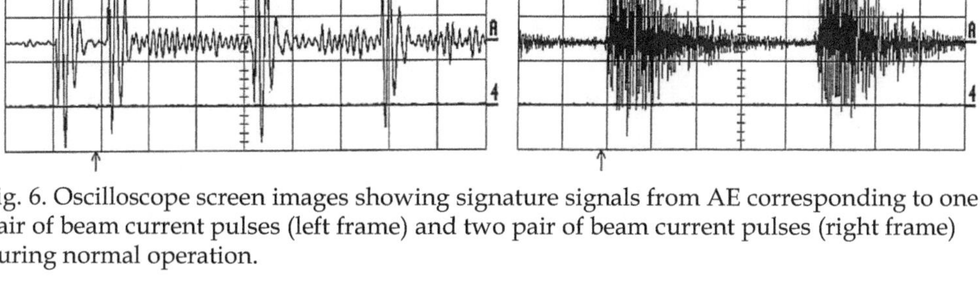

Fig. 6. Oscilloscope screen images showing signature signals from AE corresponding to one pair of beam current pulses (left frame) and two pair of beam current pulses (right frame) during normal operation.

Channel 1 shows two detected (negative 50 mV) RF drive signals. The first narrow signal 2-μs wide is around the second vertical division. The second signal, 125-μs wide and 50-mV deep, is immediately beyond fifth vertical division.

The pulse repetition rate selected was 213 pulses per second and the klystron unit was operated at a frequency of 851 MHz. In the right frame of Figure 6, the horizontal time scale was changed to 1 ms per division in order to see the next pair of pulses. The time interval between successive pair of pulses was 4.69 ms (milliseconds). Here, each pulse is a pair of pulses in rapid succession. Two pairs of beam current on channel 2 are separated by 4.7 ms (4.7 divisions x 1 ms per division). The same is true for RF drive signals (-50 mV) on channel 1. The operation of the klystron unit thus produced corresponding acoustic emission signals that appeared on channel 3. Their averaged version appears on zoom trace A in both frames of Figure 6. No AE signals appeared between the beam current pulses.

Since location 3 was giving strong AE signals, the transducer was set now set on location 4 (Figure 5). In both frames of Figure 7, channel 3 shows an AE signal. Channel 1 shows a RF drive signal and channel 2 shows a beam current sensor signal. The left frame of Figure 7 shows the normal picture when the klystron unit was operated under normal conditions. The normal operating range for the beam current was 0.9 to 1A. The peak power output was about 280 kW. To stress the klystron unit, the output power was increased by increasing the beam current slowly to the level of 1.1A, when suddenly an abnormal current pulse appeared and the threshold amplifier responded with a Transistor-Transitor Logic (TTL) high level output on channel 4 (as shown in right frame of Figure 7). The abnormal performance leads to a decrease in AE response (Channel 3) as expected. In Figure 7, vertical and horizontal sensitivities on channels 1 and 2 are the same. However, vertical sensitivity on channel 3 was 20 mV/division in the left frame of Figure 7 and was 50 mV/division in the right frame. When the vertical sensitivity on channel 3 of the right frame was changed to 20 mV/division, for comparison of two AE signals, the AE signal amplitudes in the right frame under abnormal operation would still be much smaller than the ones from the left frame of Figure 7. Thus, it was experimentally proven that the AE signal amplitudes decreased in abnormal performance of the klystron unit similar to tests on the magnetron.

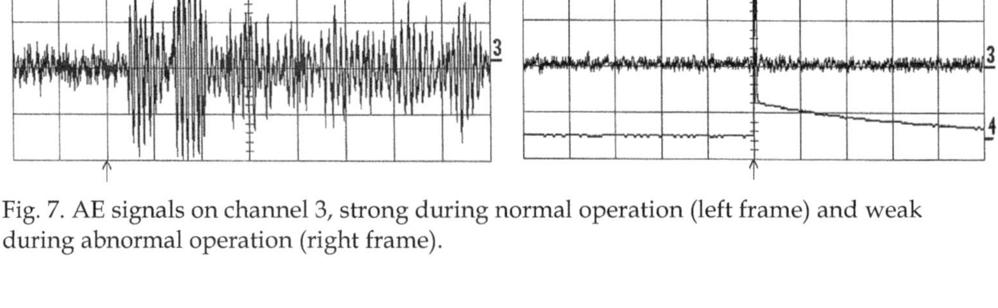

Fig. 7. AE signals on channel 3, strong during normal operation (left frame) and weak during abnormal operation (right frame).

8.4 Experiment on second klystron unit

Encouraged by the results of the acoustic emission nondestructive testing technique on the normal and abnormal operation of the first klystron unit, it was decided to repeat the experiments on the second klystron unit from a different location with different operating parameters. The second Klystron radar set had the instrumented range of 474 km (kilometers). The transmitter of the system amplifies the 2–mW (milliwatt) RF drive signal from the receiver to a 9.8- kW average power level. The amplification is performed in two stages by a driver, and the power klystron unit. The RF energy is routed to the antenna through the microwave group (coaxial cable and waveguide). The RF drive power rating was 30 W. The filament voltage requirement was 8 V. The focusing coil current required was 15 A. The frequency band called the mid-band extended from 881.333 to 910.222 MHz. The radar operates in frequency agile mode as well as in the fixed frequency mode. The fixed frequency used during this work was 902.999 MHz. Further details on both klystron units are available from Joshi et al. (2007).

8.4.1 Experimental setup for second klystron unit

The experimental equipment and arrangement was similar to that used with the first klystron unit (Figure 2). The radar was set into operation in the long pulse mode. The pulse repetition frequency (PRF) chosen was 254.4 Hz. The pulse repetition period was 3.93 ms. The duty cycle was 0.0377. The average RF power was 9.04 kW and average dBm (1 milliwatt reference) was 69.6. The short pulse width was 10 μs and the long pulse width was 137.5 μs. The pulse separation was 160 μs. The RF was 902.999 MHz. The RF signal was fed to a dummy load at the end of the waveguide.

8.4.2 Experimental results from second klystron unit

a. Collection of data during normal operation

Figure 8 identifies the various locations at which the AE transducer was mounted in turn to examine the AE signal patterns generated by the traveling RF pulse energy through the waveguide. To start with, the AE transducer B (see Table 1) was used to pick up AE signals. As before, the output of the transducer was connected to the input of the preamplifier set at the 40 dB gain. The output of the postamplifier set at 21 dB gain was connected to channel 3 of the oscilloscope. Channel 2 of the oscilloscope carried a T-shape BNC connector. Its one terminal was connected to klystron current (5A/V) terminal. The other terminal of the T was connected the high pass filter. The output of the high pass filter was connected to the detector. The output of the detector was connected to the low pass filter. The output of the low pass filter was connected to the threshold amplifier. The output of the threshold amplifier was connected to channel 4 of the oscilloscope. The zoom trace C was used to show the averaged AE signal from channel 3.

b. Effect on AE of different locations of AE sensor

To make sure that the equipment was not picking up extraneous signals from air, the transducer was held in air while the klystron current pulses were turned on. No AE signal was received on channel C of the oscilloscope. The AE transducer was then coupled with the viscous couplant to the top of the low pass filter at location 4. The AE signal in step with the

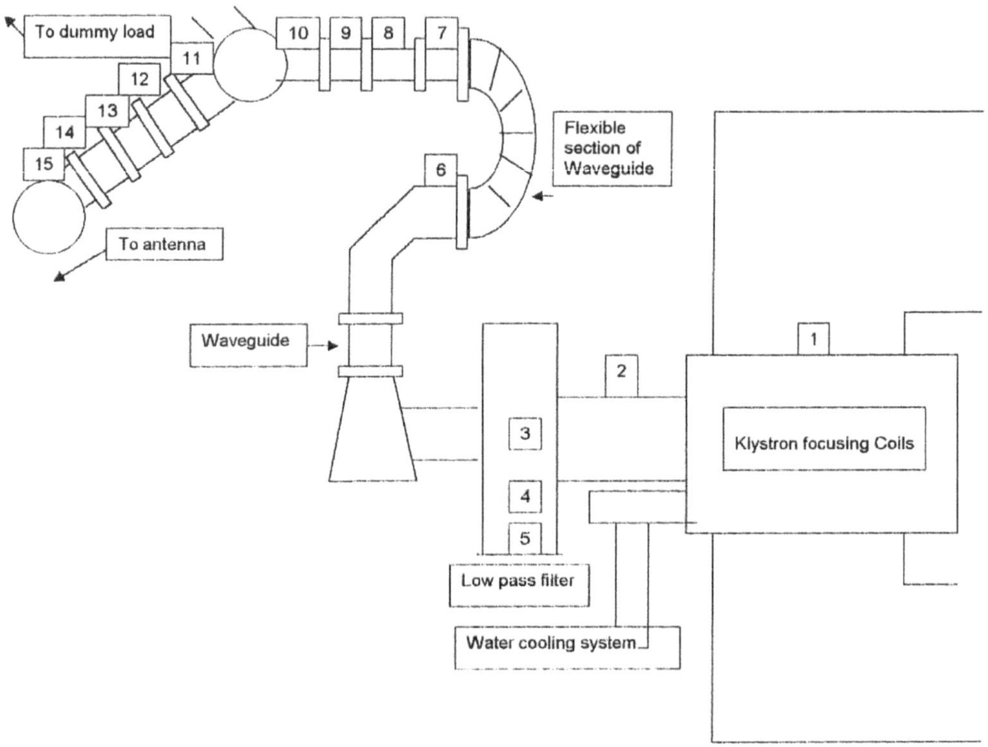

Fig. 8. Schematic of AE transducer locations on the second klystron unit.

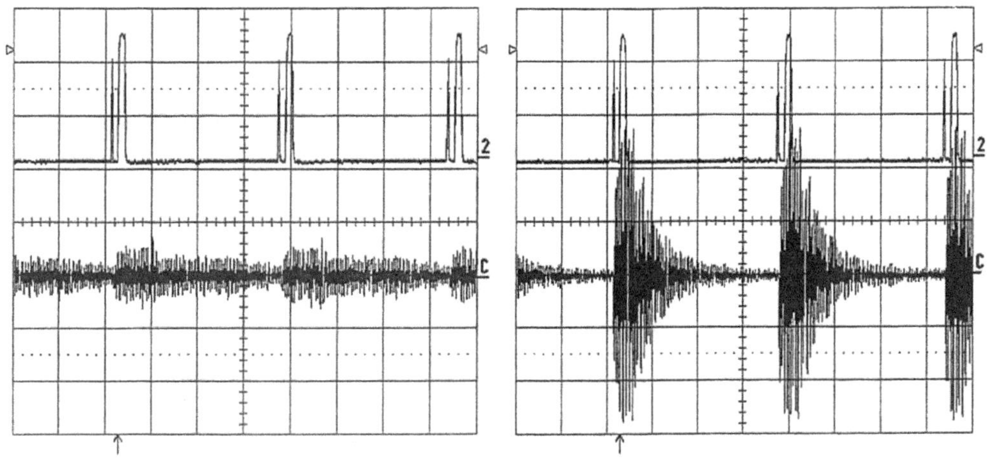

Fig. 9. AE signals in step with beam current pulses during normal operation. Signature signals from AE sensor at location 4 (left frame) and stronger signals from location 8 (right frame).

current pulses was recorded on channel C (averaged signal from channel 3) as shown in Figure 9. Channel 2 shows three pairs of the klystron beam current pulses separated by about 3.93 ms. The horizontal time scale on channel 2 was 1 ms/division. The acoustic emission activity of burst type appeared on channel C in step with beam current pulses.

The AE activity recorded was the strongest at location 8, on the isolator section of the waveguide, among all locations, as seen in the right frame of Figure 9. The vertical sensitivity on channel C was 50 mV/division in the left frame of Figure 9 while it was 0.50 V/division on channel C of the right frame. During collection of data from different locations on the waveguide, gains on both AE amplifiers were held constant so that AE signal data from different locations could be compared. The AE transducer was coupled properly at every new location so that no air bubbles were trapped in the couplant layer.

c. Data Collection during Stressed Operation of Klystron

The experimental setup was similar to the one used for experiment with the first klystron unit. The total gain of the AE preamplifier and the postamplifier was set at 61 dB. The AE transducer was set on the coaxial cable at the location 2 (Figure 8). The beam current was slightly less than 1 A. The AE activity was recorded during normal operation as shown in left frame of Figure 10. Channel 2 shows beam current pulses and channel C shows the averaged AE signal. Next, the high voltage was gradually increased beyond the normal value of 40 kV. The current showed a slight increase beyond 1 ampere and then the TTL output of the threshold amplifier was recorded on channel 4 as seen in the right frame of Figure 10. The recorded AE signal during abnormal operation is shown on channel 3 without being averaged. The averaging operation removes noise from the signal. The vertical sensitivity was set at 200 mV/division on channel C in the left frame (normal operation) while it was set at 50 mV/division on channel 3 in the right frame (abnormal operation). The comparison of strengths of two AE signals (200 mV/division x 4 divisions = 800 mV in the left frame while 50 mV/division x 4 divisions = 200 mV in the right frame) shows that during abnormal operation strength of AE signal decreased provided all other parameters were kept unchanged.

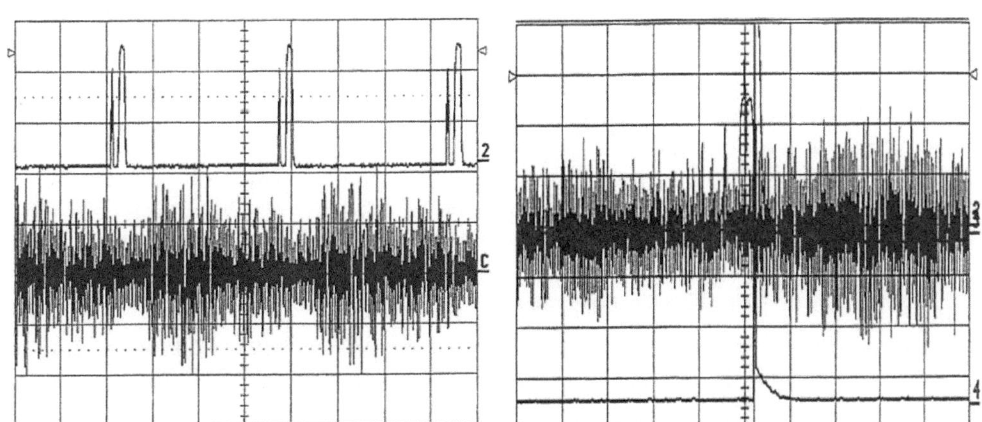

Fig. 10. Strong AE signal on channel C during normal current pulses on channel 2 (left frame) and weak signal due to abnormal current pulse on channel 2 (right frame).

In the right frame of Figure 11, the horizontal time scale was expanded to 0.5ms/division from 1ms/division so that the abnormal beam current pulse from the right frame of Figure 10 could be seen clearly. Signals from other channels were not displayed. In the right frame of Figure 11, the time scale was further expanded to 50 µs/division on the zoom trace A of channel 2 in order to see the fine structure of the abnormal pulse.

In the pair of consecutive pulses in the right frame of Figure 11, the first pulse is 10-µs wide, seen just before the first vertical line. The second pulse started just before the fourth vertical line. The separation between two consecutive pulses was 160 µs as expected. The second pulse of 137.5-µs width developed a dip in its structure as seen on the fifth vertical line. This abnormal structure in the second beam current pulse caused the threshold amplifier to trigger TTL outpu.

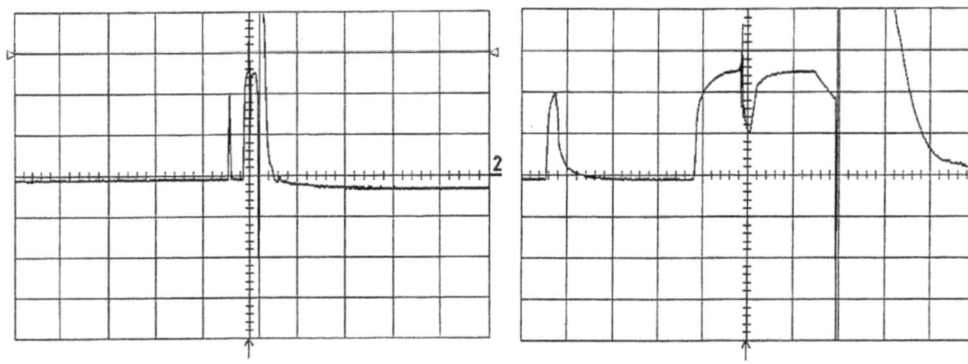

Fig. 11. Both left and right frames show expanded view of the same abnormal current pulses with horizontal time scale equal to 0.5 ms/division (left frame) and is equal to 50 µs/division.(right frame).

The normal or good pulse will not trigger the threshold amplifier output that was connected to channel 4. It takes its input from channel 2 (of the klystron beam current pulses) through the high pass filter, detector, and low pass filter. An abnormal pulse produced by stressing the klystron tube beyond its normal operating range has high-frequency components. The high pass filter (beyond 1.6 MHz) stops the regular pulse and let the abnormal pulse go through. The detector detects it and passes it to the low pass filter (below 15 MHz), which cleans up the signal. It was passed to the threshold amplifier, which responded with TTL output. Further details on the electronic circuits for detection of abnormal pulses can be obtained from Joshi et al. (2001 and 2002) and Ramirez et al. (2003).

8.4.3 Conclusions from klystron operations

The advanced nondestructive technique of acoustic emissions was applied for the first time in the current research project for in-situ performance monitoring of two high-power microwave tube klystron units. Positioning the AE transducer at different locations on the waveguides of the klystron units produced remarkably different AE patterns. The abnormal operations of tubes showed a decrease in strength of AE signals when compared with AE signals from a tube operating under normal conditions.

9. Application of AE technology to traveling wave tube

The TWT (Traveling Wave Tube) is an electron tube used for amplification at microwave frequencies. Microwave frequencies are generally identified as frequencies above 500 MHz. The operation of a TWT depends on the interaction of a beam of electrons with an electromagnetic wave. The most common traveling wave thermionic device is the linear beam TWT. Because of their very wide bandwidth, and high-power gain, traveling wave tubes are extensively used in radar, space communication, and electronic countermeasure (ECM) systems.

The high-power TWT is a microwave amplifier characterized by high gain, large bandwidth, relatively low efficiency, and high operating voltages. The electron gun assembly of a TWT consists of a filament and cathode, grid, anode, and collector. It is a tetrode tube. In addition, it has a helix structure for an RF signal to be amplified. Electrons leaving the electron gun are accelerated through the helix structure to the collector by the high voltages from the power supplies. The TWT uses a series of permanent magnets to provide a partial focus for the electron beam. The grid, anode, helix structure, and collector voltages 'fine tune' the beam focus so that a minimum of electrons strike the helix. Very few electrons strike the helix structure. An RF signal passes into a vacuum interior of the tube through a ceramic window at the RF input and then propagates through the slow-wave circuit (the helix) with a phase velocity approximately equal to the electron beam velocity. The beam and the propagating RF signal interact such that energy is transferred to the RF signal. Application of an RF signal to the input of the helix structure causes alternately accelerating and decelerating electric fields along the helix structure. Electrons passing through the helix will either speed up or slow down, depending on which part of the field they are located. Electrons that are slowed down meet with electrons that are accelerated by the preceding field. Thus, the input RF signal interacts with the electron beam, causing it to bunch up. This process grows as the beam travels through the helix and the amplitude of the RF wave associated with the bunched electrons increases. The amplified RF signal then passes out of the vacuum interior at the RF output, and the spent electron beam is absorbed by the collector. Heat energy is dissipated in the collector structure and heat sink, which are cooled by the high-efficiency air blower. The helix has a virtually constant phase velocity over a wide range of frequencies and has by far the largest bandwidth of any slow-wave structure. It possesses a high-interaction impedance and is thus able to produce a high gain in a short length.

9.1 Traveling Wave Tube Amplifier (TWT-A)

The TWT-A (Model A600/L-878A) unit used in this research work was designed and manufactured by Logimetrics. It is used to linearly amplify microwave signals over the operating bandwidth of 1 GHz (1 GHz to 2 GHz). The TWAT-A unit was self-contained. It obtained its power through 30-A/115-V thick cable from the three-prong wall socket. The intake air is filtered through a filter on the front panel. The exhaust fan is located in the rear of the unit. The front panel carried three meters, one for the collector current (range: 0 to 1 Ampere DC), one for the helix current (range: 0 to 100 mA DC), and one for power in watts (range 0 to 500 W forward and 0 to 100 W). The front panel also carried the amplifier RF gain control (a 24.5 turn potentiometer acting as a 0- to 20-dB attenuator). Five overload shutoff mechanisms protect the unit. The front panel carries five corresponding fault indicator lamps indicating TWT-A thermal overload, power supply thermal overload, cooling air supply faults, VSWR cutoff, and helix current faults. The unit has an AC power main breaker, a power-on switch, and a two-position Standby/RF on switch. The timer inside the unit waits 3 minutes after the main circuit beaker and 'power on' switch are turned on. When the standby indicator lamp is turned on, the switch can be changed to the 'RF on' position.

The TWT-A used in this work provides in excess of 200 W of CW (continuous wave) power over the operating range of 1 to 2 GHZ. The unit is protected by helix sensing current and voltage sensing circuitry. An internal RF system detects high VSWR and automatically turns the amplifier off while simultaneously lighting up a VSWR fault lamp on the front panel. The TWT-A operates safely into an impedance mismatch that will produce a VSWR of 2.5:1. The input and output impedance were 50 ohms. The input and output connectors were N-type. The small signal region of the input/output characteristics is characterized by linear operation, low intermodulation products, and low harmonic generation. The gain of TWT-A is usually 6 to 8 dB lower at saturation than the gain available at the small signal. The gain decreases further into the overdrive region. The details are available from Logimetrics (1998).

9.2 Experimental setup

In the normal operation of the TWT-A, the output is connected to the waveguide leading to an antenna. In the laboratory, it was connected through the high-power directional coupler to a dummy load of the ratings of 50 Ω and 225 W. To remove heat, the dummy load carried a fan that could be run directly by the regular power supply. The high-power directional coupler was rated at 200 W, with a frequency range of 960 to 1215 MHz, and a coupling of 30 dB. It carried four N-type connectors J1, J2, J3, and J4. The input connector J1 was connected through the N-type coaxial cable (Alpha wire 92194-RG 214/U) to the output terminal on the rear panel of TWT-A. The output connector J2 was connected through the N-type coaxial cable to the dummy load. The forward connector J3 was connected through the N-type coaxial cable to the 20-dB attenuator in series with the 10-dB attenuator. The other end of the 10-dB attenuator was connected to the input terminal (50 Ω and 9 MHz to 22 GHz) of the HP 8562A spectrum analyzer. The two attenuators had ratings of medium power of 20 W and a frequency range of DC to 11 GHz. They carried metal fins to dissipate heat. The details of the circuit diagram can be obtained from Joshi et al. (2008).

The "T" connector between the two attenuators carried an inline crystal detector with a N-type connector on one end and a BNC connector on the other end. It is a square law crystal detector made of a silicon microwave diode. It was connected through the regular BNC cable (Pomona Electronics 2249-C-48) to Channel 2 of the LeCroy 9354AM 500-MHz digital oscilloscope. The input signal to TWT-A was generated by a HP 8352A RF plug-in module of a HP 8350B sweep oscillator. The type-N output connector from the module was connected through the type-N coaxial cable to the type-N input connector on the rear panel of TWT-A.

The right cover of the rectangular box-shaped TWT-A unit was opened. The TWT tube enclosed in the metal box was detected under the air circulation duct. The AE transducer A (see Table 1) was then mounted on the side of the TWT metal casing using the couplant silicone compound (Mil-S-8660B, General Electric Company). It was held in position by padding of foam inserted between the back of the transducer and the frame of the TWT-A unit. The transducer is then connected to the pre-amplifier. The pre-amplifier output was connected to the post-amplifier. The output of the main amplifier is connected to channel 3 of the LeCroy digital oscilloscope. The sampling rate of the oscilloscope for single channel was 2 Gs/s with 1 million points. If four channels are used at the same time, the sampling rate would be 500 Ms/s with 250 kilopoints per channel.

9.3 Experimental procedure

The maximum input power recommended for the spectrum analyzer is 0 dBm (1 milliwatt reference). The input power from the 83525A RF plug-in can be changed in steps. The frequency of the CW input signal was initially chosen to be 1.2 GHz. The output of the TWT-A unit was connected to a 50-Ω dummy load through the high- power directional coupler. The RF gain control knob (20-dB attenuator) from the front panel was set all the way in the clockwise position, indicating that the selected full input power is applied to TWT-A for amplification. The front panel power indicator had two scales, one for the forward power in the up position of the switch and the second for the reverse power in the down position of the switch. This power was attenuated by 60 dB (30-dB directional coupler + 20-dB attenuator +10-dB attenuator) before it was fed to the input terminal of the spectrum analyzer.

9.4 Experiments and results

The purpose of this research project was to check the feasibility of using acoustic emission as the noninvasive tool to monitor performance changes in the TWA-A. Such application of AE technology for in-situ performance monitoring of high-power microwave radar tubes of a pulsed magnetron and a klystron unit were demonstrated successfully in the earlier part of this chapter. In the earlier experiments on two radar tubes of pulsed magnetron and klystron, the radar units made it easy to introduce variations in the normal working parameters of the tubes, thereby producing abnormal pulsed outputs. AE transducers immediately indicated differences between normal and abnormal functioning of tubes. In the case of TWT-A, the unit was protected with five sensors. Thus, the tube did not allow much variation in the normal working parameters. It was decided therefore to use the TWT-A unit in two different modes (CW mode and pulsed mode) and check the ability of acoustic emission to detect the difference.

9.5 Part I -CW mode with dummy load of 50 ohms

Here, the aim was to observe acoustic emission activity under the normal operation of TWT-A in CW mode when the output was connected to the 50-Ω dummy load through the 30-dB directional coupler. An AE transducer was mounted on the TWT near the output terminal of the helix. The normal noise band on channel 3 was covered with the mask (zoom trace D) of a height sufficient to avoid triggering on the noise signal. The trigger level was chosen slightly below the upper boundary of the mask. The trigger position was set at two divisions to the right of zero on the horizontal scale. If all the points of the signal remain within the mask, there will be no beep sound and no automatic dumping of the screen display on the

3.5-inch floppy disc in the disc drive of the oscilloscope. The arrangement ensured recording of the AE burst signals only, avoiding noise signals. To avoid damage to the oscilloscope screen, the mask on the zoom trace D is set and then turned off during prolonged operation of the equipment. Nine AE events were recorded on the disc in 3 hours. Changes in the detected RF level on channel 2 were expected, assuming that AE signals were produced due to the irregular functioning of TWT-A. In the left frame of Figure 12, the horizontal scale was set at 0.1 ms per division on channels 2 and 3. The vertical sensitivities were 10 mV per division on channel 2 and 50 mV per division on channel 3. Figure 12 (left frame) shows the typical AE signal on channel 3 and the detected RF level on channel 2 at -22 mV. No appreciable changes were detected in the AE signal or in RF level during the 3 hours. The sporadic presence of AE signals could be taken as an instantaneous variation in the normal parameters of the unit. Since the unit was protected by five sensors, the system gets restored instantaneously.

9.6 Part II (CW mode with no load-all power reflected back).

9.6.1 Effect of closing the front air inlet

It was then decided to stress the TWT-A operation by forcing it to operate at higher temperatures, which in turn could produce irregular RF amplified pulses and corresponding changes in AE signals. The front-panel air inlet was covered tight with aluminum foil sealed around the edges. The dummy load was disconnected and was replaced by a teflon (insulator) plug between the central pin and the surrounding metal cylinder of the N-type connector. This was done to reflect the RF power in order to stress the tube. Although AE signals were received at intervals, TWT-A got shut off within 8 hours due to thermal overload. The detected RF level on channel 2 did not change. No changes were detected in AE signals on channel 3 while operating the tube at a higher temperature. At this stage, the left-side and the top-side panels of the unit were opened. The five sensors protecting the unit were identified as a thermal sensor, an air flow sensor, a power supply sensor, a VSWR overload sensor, and a helix current sensor.

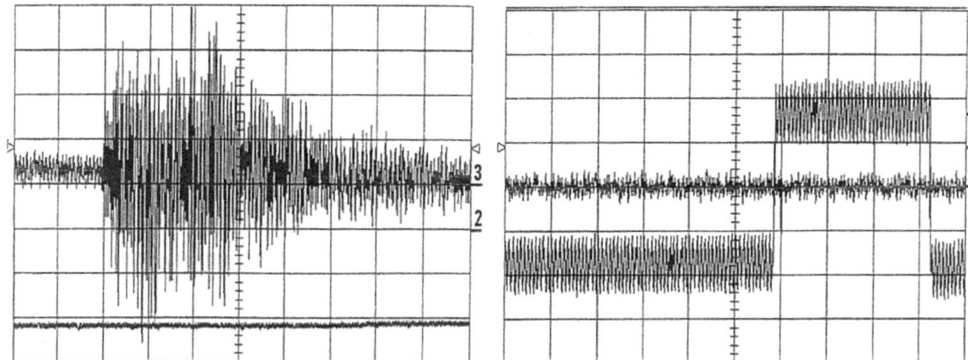

Fig. 12. Sporadic AE signal from the TWT-A in CW operation (left frame) and periodic AE signal on the sixth vertical division (compressed) from pulsing the TWT-A (right frame).

9.6.2 Effect of the different locations of the AE sensor

There was a need to find out whether the position of the AE transducer could make a difference in receiving the AE signals. For this purpose, the transducer was mounted on the metal coaxial cable running from helix output terminal to the output terminal on the rear panel. The new position of the transducer made no difference in recording AE activity. The AE activity was sporadic on and off, appearing intermittently as in the left frame of Figure 12. Next, it was decided to feed the pulsing RF input signal to TWT-A in place of the CW signal and observe the AE signatures.

9.7 Part III (Pulse mode with no load-all power reflected back)

The RF input frequency was chosen at 1.4770 GHz. The frequency span (Δf) was set on zero. The sweep trigger was on the line. The 'RF on' time was set at 2 seconds. The 'RF off' time was fixed internally at about 1 second. On the spectrum analyzer screen, the start frequency was 1 GHz at the left end. The stop frequency was 2 GHz at the right end of the screen. The resolution bandwidth was 300 kHz. The video bandwidth was 300 kHz. The sweep time was 50 ms. The input power is set 13.5 dBm on the RF plug-in. The ALC (automatic level control) mode was internal. The TWT-A was turned on. After a few minutes, the power meter on the front panel was showing a rise and fall in the output power in sync with the rise and fall of the RF signal on the spectrum analyzer screen. It was observed that the maximum power output was changing with the change of the input frequency. The gradual change in the input frequency showed that the output power peaked around 1.5040 GHz. The output power at this frequency rose to the level of 140 W. The AE transducer was mounted on the TWT-A tube near the helix output terminal. The gains of both amplifiers were kept constant. It was then observed that AE signals were generated in step with the oscillating output power of the TWT-A. They are not sporadic as before. The oscillating RF power might be producing induction heating. The subsequent contraction during the cooling period was possibly the cause of the generated acoustic emission energy. Figure 12 (right frame) shows the generation of AE signal on channel 3 at the stop of the detected 'RF on' segment and at the start of the 'RF off 'segment on channel 2. The AE signal appears to be a very narrow vertical line close to the sixth vertical division due to the choice of horizontal time scale equal to 0.2 second per division suited to show 'RF on' and 'RF off' segments, which are actually inverted in the figure. The vertical sensitivities were 10 mV per division on channel 2 and 50 mV per division on channel 3.

9.7.1 Effect of RF input duty cycle on the AE activity

Next, the effect of duty cycle on the generation of AE signals was examined. The 'RF on' interval was increased from 2 to 20 seconds at steps of 1 second. The 'RF off' interval was fixed internally at about 1 second. In every instance, the AE activity was observed to be in step with the changing RF output. Then the 'RF on' interval was reduced to 1 second, almost equal to the 'RF off' interval. The AE signals were still generated in step with changing RF output. The 'RF on' interval was further reduced to less than 1 second and AE activity started diminishing rapidly with further decrease in the 'RF on' time. It appears that the 'RF on' interval has to be greater than 'RF off' interval to generate the AE activity. It was decided, therefore, to set the 'RF on' interval at 2 seconds for future testing. Figure 13 (left frame) shows the AE signals generated with the pulse rate of 20 pulses per minute. In Figure 13 (right frame), the detected RF level is in the process of changing its value from that in left frame. For each pulse, the 'RF on' interval was 2 seconds and the 'RF off' interval was 1 second. In both figures, the horizontal time scale was set at 0.5 ms per division on channels 2 and 3, and the vertical sensitivities were 10 mV per division on channel 2 and 50 mV per division on channel 3. In the left frame of Figure 13, the zoom trace B of channel 2 shows moving detected RF voltage at a negative 22 mV level. In the right frame of Figure 13, the detected RF level is in the process of changing its value from that in the left frame. The AE signal on channel 3 in both frames of Figure 13 appear well expanded relative to the single vertical line that appears in the right frame of Figure 12. All power was reflected back since the dummy load was replaced with the teflon plug.

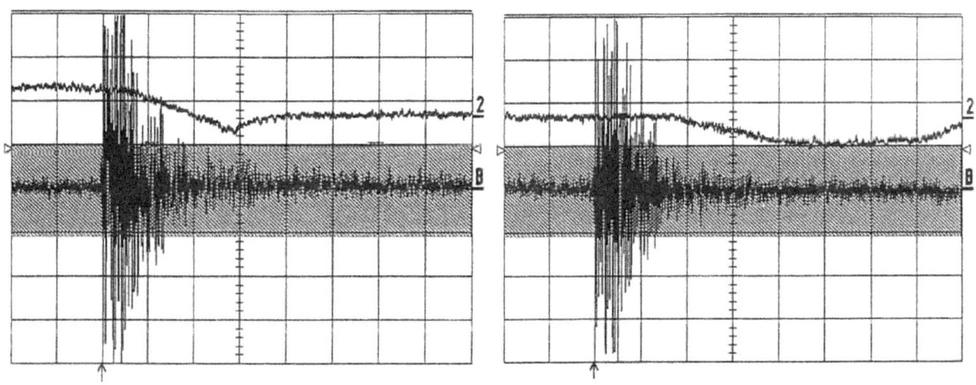

Fig. 13. Expanded view of signal from right frame of Figure 12.

To determine the 'RF off' interval precisely, the horizontal time scale was changed next, the screen displays were captured saved on the disc. In Figure 14, the horizontal time scale of 0.1 second/division was used. It was observed that the 'RF off' interval was 0.7 second (channel 2), slightly less than the assumed value of 1 second. The 'RF off' interval is internally fixed and hence cannot be changed. The AE signal from channel 3 is seen on the second vertical line in the left frame of Figure 14. Its expanded view was shown in the right frame.

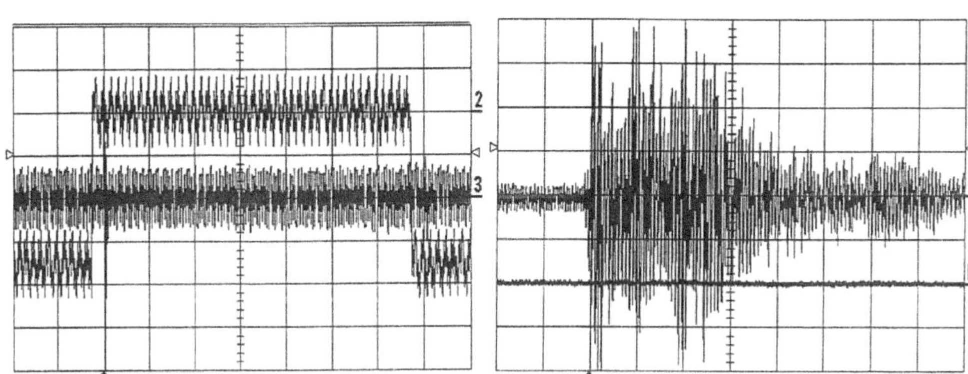

Fig. 14. Compressed AE signal (left frame) on the second vertical division in sync with the pulsating traveling wave tube amplifier and its expanded view in the right frame showing the expanded AE signal.

9.7.2 Effect of sudden 'cut off' of filament power on performance of TWT-A

The purpose of this experiment was to check the feasibility of application of advanced AE technology to detect abnormal functioning of TWT-A. Since the TWT-A unit used in the project was protected by five overload detection sensors, one has to use novel ways to induce the irregular performance of the unit. The power of the cathode/heater assembly suddenly and see its effect on the performance of the tube. Since the cathode/heater assembly was at a negative high voltage of about 2000 V with respect to the collector, the necessary high-voltage protection device was used during the sudden disconnection of the power supply to the filament. As the tube cooled off rapidly, many AE signals were generated in succession in about 1 minute before all the acoustic emissions stopped with the power output dropping rapidly from 140 W (forward power) to 0 W. It takes 15 seconds for the oscilloscope to store the screen display on its disc. It was impossible to save the AE signals on the floppy disc generated during the rapid cool-off period. However, a couple of screen displays were stored automatically during the cool-off period. The AE signals obtained during the simulated cold cathode experiment had relatively small amplitudes and longer ringing time.

9.7.3 Effect of cold start

The standby time between the 'power on' position of the switch and 'RF on' position is 3 minutes. The standby time was measured by closing both the air inlet and outlet windows. The times were about the same. Again, to produce some kind of irregularity in the functioning of the unit, the timing circuit governing the standby interval was bypassed by changes in the circuitry. The RF power input was set at 13.5 dBm, as before. The RF input power was pulsating with a 'RF on' interval of 2 seconds. No load was connected and all the power was reflected back. The main circuit beaker switch, the power-on switch, and the 'RF on' switch were turned on in succession to produce a cold start condition. The tube was not damaged. The only observable effect of the cold start was a little delay in receiving AE signals in step with pulsating RF power. The AE signals generated in the beginning were not strong. After about 10 minutes, strong AE signals started appearing in step with the pulsating input power and out powers.

9.8 PART IV (Pulse mode with dummy load of 50 Ohms)

The dummy load of 50 Ω was connected to the output cable. Then the RF input to TWT-A was set on CW mode as in the Part I experiment. The RF input power was 13.5 dBm. The gain setting was in the fully clockwise position, giving the maximum gain. There was no AE activity even after half an hour. Then, RF input was changed from the CW mode to the pulsating mode. The dummy load remained connected. Within a few minutes, AE signals were received just like in Figure 14. They were generated in step with pulsating RF output.

9.9 TWT amplifier operation conclusions

The CW mode operation did not produce AE activity when the output cable of the TWT-A unit was connected to the dummy load of 50 Ω (matched load), and when it was connected to the teflon plug (mismatched load), causing power to be reflected back. The pulsing RF input power generated the AE activity irrespective of the termination (matched or mismatched load) of the output cable. The AE activity is generated as long as the 'RF on' interval in the pulsing RF input power remains longer than the 'RF off' interval. Acoustic emission signals are generated at the end of 'RF on' interval and at the beginning of the 'RF off' interval of the output RF pulse. The cold start experiment showed that the strength of AE signals was low in the beginning but they picked up the intensity after the unit was in operation about 10 minutes with pulsating input power. The simulated cold cathode experiment showed that rapid changes in the filament power supply caused AE activity in succession and stopped with the diminishing power output. Overheating, the unit did not produce any new effects since the unit was protected by the thermal overload sensor. The maximum leveled output from the 83525A RF plug-in of the 8350B sweep oscillator was 17.5 dBm. However, the unit got shut off by the VSWR overload sensor when the input power exceeded 14 dBm.

The TWT-A used in this project was meant for CW mode operation. By operating it in the pulse mode, the AE activity was produced in step with the changing RF input power. During the operation in CW mode, it was observed that whenever there was change (increase or decrease) in the RF input power, AE activity was generated. In short, changes in the RF output power, either due to irregular changes or regular changes in the RF input power, caused the AE signals to appear. These experimental observations definitely indicate that AE activity will be produced whenever the RF output power of the unit changes, irrespective of the cause of the change external to the unit (RF input power) or internal to the unit (degradation of the tube performance). The results of this work show that acoustic emission technology can be used as the nondestructive testing tool for in-situ performance monitoring of normal and abnormal functioning of high-power microwave TWT amplifiers.

10. Overall conclusions

Advanced acoustic emission technology offers a simple and direct NDT technique for real-time monitoring of in-situ performance of high-power microwave devices such as the magnetron, klystron, and traveling wave tube. The technique has proven itself as a great advancement over the currently available data collection systems using as many as 11 electrical sensors measuring different current and voltages of tubes.

11.Acknowledgment

This research work was supported by American Society for Engineering Education (ASEE) and SPAWAR Systems Center San Diego.

12. References

ASNT Handbook on "Acoustic Emission Testing" Vol. 5, 1987.

Green Robert E., "Elastic Wave Analysis of Acoustic Emission "Mechanics of Nondestructive Testing, Editor: Stinchcomb W.W., Plenum, New York, 1980. pp55-76.

Leybovich Alexander and Ferdinand R. Mark, "Sputtering Target Arc Detection Using Acoustic Emission, "Materials Evaluation, Nov. 2001, pp. 1331-1335.

Sakoda T., Nieda H., and Ando K., "Characteristics of elastic waves caused by corona discharges in an oil-immersed pole transformers," IEEE Transactions on Dielectric and Electrical Insulation, vol.8,No. 2, April 2001, pp.276-283.

Boczar Tomasz, "Identification of Fundamental Forms of Partial Discharges Based on the Results of Frequency Analysis of Their Acoustic Emission," Journal of Acoustic Emission, Vol.17, 1999, pp.S7-S12.

Boczar Tomasz, "Results of frequency analysis of AE pulses generated by corona discharges," Technical University of Opole Poland, 2001 http://cmspro.fme.vutbr.cz/uk/odbory/vav/AE2001/contrib/boczar1.htm2001.

Vahaviolos S. J., and Saifi Mansoor, "Laser Spot Welding and Real-time Evaluation," IEEE Journal of Quantum Electronics, Vol. QE-12, No.2, 1976,p.870.

Tsukamoto O., and Iwasa Y., "Acoustic Emission diagnostic & monitoring techniques for superconducting magnets," Advanced Cryogenic Engineering, Vol.31, 1986, pp.259-268.

Theobald P., Rokowski R.T., Yan T., Jarvis D., Dowson S., and Jones B.E, "Reference Source for Calibration of Acoustic Emission Measurement," IEEE Instrumentation and Measurement, Technology Conference, Budapest, Hungary, May 21-23, 2001, pp.412-416.

Edwards T.C., "Introduction to Microwave Electronics," Publisher: Edward Arnold, 1984, p.32.

Connor F.R., "Wave Transmission," Publisher: Edward Arnold, 1972, p.7.

Love Wayne, "Magnetrons," Handbook of Microwave Technology, Volume 2, (Applications), Edited by T. Koryu Ishii, Academic Press, 1995, p.36.

Watson W.H. and Troy J.W., "Computer-controlled Performance Monitor for Microwave Tubes" NOSC Technical Document 1070, Naval Ocean Systems Center San Diego, CA 92152-5000, March 1987.

Joshi, N. R., Brock, D.W., Russell S.D., Kasa, S.D., and Garcia, G.A., "Acoustic Emission Technology for Radar Tubes," Technical Note 1810, published in August 2000 by SPAWAR Systems Center, San Diego (SSC-SD), California, pages 190.

Joshi, N.R., Russell, S.D., Ramirez, A.D., and Brock, D.W., "In-situ Performance Monitoring of High Power Microwave Tubes with Acoustic Emission," published in the paper summaries of ASNT Fall Conference in Houston, 23-27 October 2006, pp.7-11.

Joshi, N.R., Russell Stephen D., Ramirez Ayax D., Brock David W., and Lasher Markham E., "Acoustic Emission Technology for Performance Monitoring of High Power Microwave Tubes," Material Evaluation, Vol.65,No.4, April 2007, pp.411-416.

Joshi, N. R., Brock, D.W., Russell S.D., Lasher, M.E., and Kasa, S.D., "Built-in Test for High-Power Microwave Tubes Using Acoustic Emission Technology," Technical Note 1816, published in September 2001 by SPAWAR Systems Center, San Diego (SSC-SD), Ca 92152-5001, pages 99.

Joshi, N.R., Russell S. D., Ramirez A. D., and Brock D. W. "Monitoring of High-Power Microwave Tube Systems Using the Integrated Condition Assessment System (ICAS)," Technical Report 1885 published in July 2002 by SPAWAR Systems Center, San Diego (SSC-SD), Ca 92152-5001, pages 116 (53+A63).

Ramirez, A.D., Russell, S.D., Brock, D.W., and Joshi, N.R., "Automated Classification of Microwave Transmitter Failures Using Virtual Sensors, " published in the

Proceedings of 13th International Ship Control Systems Symposium, 7-9 April 2003, Orlando, Florida, USA, pages 10.

Joshi, N.R., Russell, S.D., Ramirez A.D., and Brock D.W., "Performance Monitoring of High Power Microwave Radar Klystron Units with Acoustic Emission Technology," Materials Evaluation, Vol.10, October 2007, pp.1048-1053.

Logimetrics. Traveling Wave Tube Amplifier (TWT-A), Technical Manual, Model A600/L878A, 50 Orville Drive, Bohemia, NY 11716, 1998 and TWT and TWTA Handbook, Hughes Aircraft Company, Electron Dynamics Division, P.O.BOX 299, Torrance, CA 90509-2999.

Joshi N.R., Russell S.D., Ramirez A.D., and Brock D.W., "Acoustic Emission as Noninvasive Tool for Performance Monitoring of High Power Microwave Radar Traveling Wave Tube," Materials Evaluation, Vol.66, No. 4, April 2008, pp.413-418.

4

Using Acoustic Emission to Evaluate Fracture Toughness Energy Release Rate (GI) at Mode I Delamination of Composite Materials

Amir Refahi Oskouei[1], Andrea Zucchelli[2],
Mehdi Ahmadi[3] and Giangiacomo Minak[4]
[1]Islamic Azad University, East Tehran Branch,
[2,4]University of Bologna,
[3]Amirkabir University of Technology (Tehran Polytechnic),
[1,3]Iran
[2,4]Italy

1. Introduction

Delamination is a critical damage mode in composite structures, not necessarily because it will cause the structure split into two or more pieces at the end of the damaging process, but because it can degrade the laminate strength to such a degree that it becomes useless in service. The design of composite structures to account for delamination and other forms of damage involves two fundamental considerations, namely damage resistance and damage tolerance. Knowledge of a laminated composite material's resistance to interlaminar fracture is useful for product development and material selection.

Damage resistance is the measure of the capability of a material or structure to resist the initial occurrence of damage. This aspect must be considered for designing and maintaining the integrity of load carrying structures (Pagano & Schoeppner, 2000). In this regard, we should also note that polymer matrix composites are given primary attention because of their importance and dominance in practical structural applications as well as the composite literature itself.

Mode I inter-laminar fracture has received the greatest attention from researchers. This is due to the fact that the delamination initiation energy is low compared to that of the shearing mode. Several studies were already conducted on the mode I fracture to determine the surface energies of composite materials by creating a crack propagation effect (Perrin et al., 2003). Researchers were thus able to consider the effect of ambient conditions and manufacturing variables on the strain energy release rate (G_I) (Velmurugan & Solaimurugan, 2007). The effect of fiber direction on delamination fracture toughness is a problem that was studied by (Kim & Mayer, 2003), and (Solaimurugan & Velmurugan, 2008). Literature results show that adjacent ply delamination fracture toughness decreased as the mismatch angle of fiber between the same plies increased (MIL-HDBK, 2002). The reason for fracture energy dependence on fiber mismatch angle is considered to be the stiffness of the plies thus fibers have the highest contribution to the stiffness of each ply. The effects of temperature on delamination growth in a carbon/epoxy composite under fatigue loading also shows that at elevated temperatures the strain energy release rate threshold values for delamination growth under fatigue loading are significantly lower than the critical energy release rates in static tests (ASTM, 1994; Sjogren & Asp, 2002). The effect of loading rate on fracture toughness of laminated composites was studied by (Kusaka et al., 1998) and (Hug, 2006). The results show a slight effect of loading rate on fracture toughness at rates up to 1.6 m/s, but beyond this rate threshold the fracture energy decreases (Benmedakhene et al., 1999). So far, many researchers have tried to improve delamination fracture toughness through different techniques. In some investigations, improvements in delamination fracture toughness were performed through 3D fiber architecture (stitching, knitting and braiding). It was observed that through thickness stitching is a promising reinforcing technique for improving interlaminar strength (Tsai & Chen, 2005).

Some analytical and theoretical models simulated via finite element method (FEM) are also used for predicting interlaminar delamination in composite structures (Chen, 1999). The computed results have been very similar and in close agreement with both the experimental results and solutions obtained from a corrected beam theory (Diaz, 2007).

Based on the extensive and thorough literature investigation it can be concluded that mode I delamination tests depend on many parameters such as opening rate, fiber direction, type of resin, condition of resin, temperature, etc. Crack propagation can be defined as the ultimate form of failure mechanism taking place in the interlaminar zone during delamination. As can be evaluated by direct observation or using a load-displacement plot, this is a macroscopic failure activity. Microscopic behavior of fiber matrix interface microdamage is more difficult to be investigated, although it represents a major part of the delamination process. For monitoring microscopic events, Acoustic Emission (AE) is an appropriate tool (Refahi Oskouei & Ahmadi, 2010; Ndiaye et al., 2000; R'Mili et al., 2008; Cesari et al., 2007; Huguet et al., 2002). However, the interpretation of the signals and, hence, the evaluation of the damage stages is a major problem of the AE method.

The combination of Double-cantilever beam (DCB) fracture-toughness tests with simultaneous AE monitoring provides useful information about microscopic and macroscopic aspects of the delamination process.

During the last 30 years, in few papers the AE has been considered as a tool to predict the delamination fracture toughness (Benmedakhene et al., 1999; Ndiaye et al., 2000; Niebergall et al., 1999; Bohse, 2000; Lorriot, Wu & Dzenis, 2003; Bohse & Brunner, 2008). Except the paper (Ndiaye et al., 2000), in all other works the mode I fracture toughness is estimated by conventional formulas provided by normative and the AE information are used to have a phenomenological description of the delamination progression. The paper presented by (Ndiaye et al., 2000) used the AE information to estimate the G_{Ic}, nevertheless their approach has some limitations concerning the theoretical approach and some details about that will be discussed later in the present paper.

From literature studies about delamination failure in composite laminate and its monitoring by AE three critical aspects can be listed: the first concerns how AE information is used to estimate the G_{Ic}, the second concerns the repeatability of the results and the application of the methodologies to different case studies, the third aspect concerns the comparison of the results obtained by using the AE information to the ones that can be obtained by the current standards, such as (ASTM, 1994). Regarding the first aspect, it can be noted that the classic way of using the AE information in order to estimate some of the relevant material performances (e.g. ultimate strength, fracture toughness, residual strength after damage) is based on empirical or semi-empirical relations where the mechanical and the AE information is used separately. In other words, in these studies the authors want to use the AE information as self-standing information to predict or estimate the mechanical performances. The results obtained from such studies are typically weak with respect to the repeatability and to the extension to case studies different from the one that is analyzed in the specific paper. The approach that is proposed and applied in the present study is based on a novel function that combines the mechanical and AE energies. This function is based on the idea that during the loading process a material is able to store strain energy and at the same time part of the stored energy is also released due to the internal failures. The AE events energy represents an important part of the released energy and can be used to weight the strain energy storing capability of the material. As greater is the damage due to internal failure as the cumulated AE increases and at the same time the strain energy storing attitude is reduced. The proposed function thus compares the strain energy stored in the material (E_s) to the released energy in the form of acoustic energy (E_a) by their ratio (E_s/E_a). It was proved that the best way to represent such a ratio is to consider its natural logarithm ($\ln(E_s/E_a)$), (Minak & Zucchelli, 2008). This function, called sentry function *f*, has been successfully applied to study the damage progression of some composite laminates (Minak & Zucchelli, 2008), to estimate the static and the fatigue residual strength of a plate subjected to indentation processes (Cesari et al., 2007; Minak & Zucchelli, 2008; Minak et al., 2009), and to estimate the residual torsional strength of a composite tube subjected to a lateral impact (Minak et al., 2009).

The purpose of the present study is to investigate the delamination mode I process of composite laminate in double cantilever beam (DCB) set-up by using the sentry function. Three types of specimens with different midplane layups, leading to different levels of inter-laminar performance, were therefore studied. The use of the sentry function enabled the analysis of a test phase during which it was not possible to identify any visible crack propagation. Such analysis is of particular interest to determine the crack length evolution and consequently the G_I variation during the test. In particular, it was possible to determine the test stage at which the crack propagation became visually detectable by means of the maximum value of the integral of the sentry function (Int(f)) that is a parameter used to estimate the material damage progression (Minak & Zucchelli, 2008; Minak et al., 2009). The Int(f) is defined as the integral of the sentry function over the displacement domain where acoustic emission has been detected:

$$\mathrm{Int}(f) = \int_{\Omega_{AE}} f(x)\,dx \tag{1}$$

Based on values of Int(f) it was possible to predict the test stage at which a delamination propagation becomes visible under an opening load condition. Furthermore, it was possible to highlight a bi-linear relation between the cumulative strain energy release rate (G_ICUM) and the integral of the sentry function over the displacement domain (Int(f)). The transition point in the bilinear relation enabled the estimation of the critical strain release rate (G_{Ic}) value. The G_{Ic} obtained by this approach was then compared to the values obtained by both ASTM D5528 standard test method (ASTM, 1994) and (Ndiaye et al., 2000) approaches. It was possible to obtain a very good agreement between the results from the new approach to the ones obtained by the ASTM D5528. At the contrary, a strong disagreement was noted between the results obtained by the Ndiaye approach with respect to the ones obtained by the new approach and the ASTM D5528 one.

2. Experimental procedure

2.1 Material and specimen preparation

The composite double cantilever beam (DCB) specimens used in this study are polyester resin reinforced with glass fiber. The properties of the polyester resin as a matrix material and fiber as a reinforcement are summarized in table 1. Three different interface types were used: woven/woven (T3), woven/unidirectional (T4) and unidirectional/unidirectional (T5). The laminates were prepared by hand lay-up. The starter crack was formed by inserting a Teflon film with a thickness of ~ 20 µm at mid-plane during molding as an initial crack for the mode I interlaminar fracture toughness test. Fiber orientations were measured with respect to beam axis along the specimen length. The angular orientation of woven lamina was represented by the orientation of warp directional fibers. The specimen dimensions were width b=20 mm, crack length a=70 mm, total length 250 mm, and thickness 2h=5mm.

Material	Density*	Ultimate tensile stress (MPa)	Elastic modulus (GPa)	Tensile failure strain (%)
Polyester resin	1.12	80	2.7	3
Unidirectional glass fiber	500	2150	74	2.6
Woven glass fiber	292	2150	74	2.6

*For polyester the unit of density is g/cm^3 and for glass fiber the areal or surface density is considered e.g. g/cm^2.

Table 1. Summary of matrix and fiber properties.

2.2 Testing procedure

The opening forces were applied to the DCB specimen with loading blocks (Fig. 1) bonded to one end of the specimen. The ends of the DCB were opened by controlling the crosshead movement. The tests were carried out with a displacement speed of 5 mm/min for each specimen, while the load and delamination length were recorded. Both edges of the specimens just ahead of the insert were coated with a thin layer of water-based typewriter correction fluid and marked in accordance with ASTM D5528 standard instruction to aid in visual detection of delamination onset. Five specimens for each type of sample were used for the DCB test.

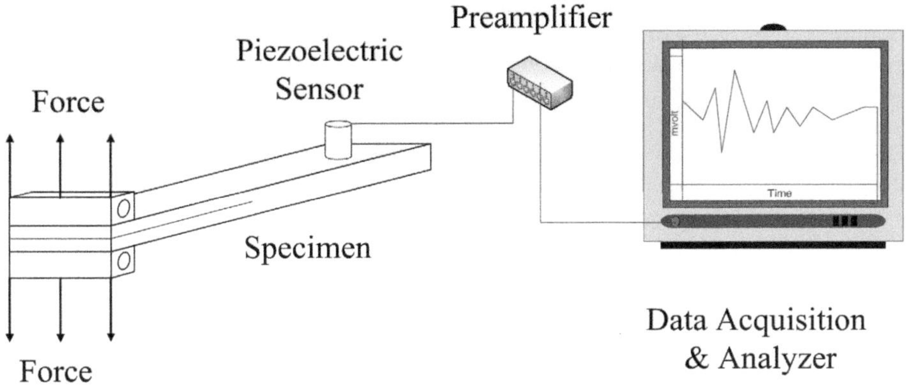

Fig. 1. View of the experimental device for AE monitoring of DCB specimen.

2.3 AE equipment

Acoustic emission software AEWin and a data acquisition system Physical Acoustics Corporation (PAC) PCI-2 with a maximum sampling rate of 40 MHz were used to record AE events. A broadband, resonant-type, single-crystal piezoelectric transducer from PAC, called PICO, was used as the AE sensor. The sensor had a resonance frequency of 513.28 kHz and an optimum operating range of 100–750 kHz. The surface of the sensor was covered with grease in order to provide good acoustic coupling between the specimen and the sensor. The signal was detected by the sensor and enhanced by a 2/4/6-AST pre-amplifier. The gain selector of the preamplifier was set to 40 dB. The test sampling rate was 1 MHz with 16 bits of resolution between 10 and 100 dB. Prior to the damage check, the data acquisition system was calibrated for each kind of specimen, according to a pencil lead break procedure. The pencil lead break procedure enables the generation of waves at the specimen surface that are used for the device calibration. At the same time, the velocity and attenuation of the AE waves were measured. The lead breakage operation was repeated several times and at different locations between the sensors. After the calibration step, AE signals were captured during mechanical testing. Signal descriptors, such as amplitude, duration, rise time, counts, and energy, were calculated by the AE software (AEWin).

2.4 Energy release rate

The beam theory expression for the strain energy release rate of a perfectly built-in (that is, clamped at the delamination front) double cantilever beam is as follows:

$$G_I = \frac{3P\delta}{2ba} \tag{2}$$

where P = Load, δ = load point displacement, b = specimen width and a = delamination length.

In practice, this expression will overestimate G_I because the beam is not perfectly built-in (that is, rotation may occur at the delamination front). One way to overcome this problem is to apply a correction factor to the modified beam theory (MB) for the strain energy release rate. The ASTM D5528 standard introduces the correction factors F and N in order to account for the presence of the end blocks and for the large opening displacements obtained as the crack length a increases (Equation (3)):

$$G_I = \frac{3P\delta}{2b\left(a+|\Delta|\right)}\frac{F}{N} \tag{3}$$

where Δ may be determined experimentally by generating a least-square plot of the cubic root of compliance, $C^{1/3}$, as a function of delamination length. The compliance, C, is the ratio of the load point displacement to the applied load, δ/P. The values used to generate this plot should be the load and displacements corresponding to the visually observed delamination onset on the edge and all the propagation values. As P reaches its critical value, P_c, the crack starts propagating. The corresponding G_I value is the critical energy release rate G_{Ic}. The latter is considered to be a material characteristic that represents the fracture toughness of the cracked interface. Based on an energy criterion, G_{Ic} yields an estimation of the amount of energy absorbed during crack growth until the delamination size reaches the critical length.

2.5 Sentry function

The sentry function is defined as the logarithm of the ratio between mechanical and acoustic energies and it can be formulated as follows:

$$\mathrm{f}(x) = Ln\left(\frac{E_s(x)}{E_a(x)}\right) \tag{4}$$

For the calculation of the two energies, two reference volumes were considered: the volume of the material where the strain energy is stored (V_1) and the volume where the delamination can propagate (V_2) and from which the AE events can take place. So, for the purpose of the analysis of the DCB tests, the strain energy, E_s, was normalized over the bended volume joined with the cracked volume (V_1+V_2) and the cumulative acoustic energy, E_a, was normalized over the cracked volume (V_2), see Fig. 2.

As described in previous studies (Cesari et al., 2007; Minak & Zucchelli, 2008; Minak et al., 2009; Minak et al., 2009), the function f is defined over displacement domain where the acoustic energy Ea is non zero (see Fig. 3, Ω_{AE}). Depending on the material damaging process, the resulting f can assume any combination of the five trends shown in Fig. 3. From the physical point of view, the parts of f characterized by an increasing trend, type I, represent the strain energy storing phases. When a significant internal material failure occurs, there is an instantaneous release of the stored energy that produces an AE event with high energy content. This fact is highlighted by the sudden drops of the function f that can be described by type II functions: PII(x). The constant behavior of f, described by PIII, is due to a progressive strain energy storing phase that is superimposed on an equivalent energy release due to material damage progression. The subsequent Bottom-Up (BU) trend indicates that a strengthening event induced an instantaneous energy storing capability in the material. Such an event can be related to hardening effects, self-healing effects or, as in the case of the present study, it can be related to fiber bridging effects. The decreasing behavior of *f*, type PIV, is related to the fact that the AE activity is greater then the material strain energy storing capability: the damage has reached a maximum and the material has no resources to sustain the load. The P_I and BU functions both have an increasing trend. Both functions, according to the definition of the sentry function represent events that occur when the strain energy storing capability exceeds the energy release (that is, in this procedure, measured by the acoustic energy). This means that if the material is able to store energy during the test, the sentry function can be described by a function of type I or a BU function type. In the case of moderate material damage, always at the beginning of the testing process, the strain energy storing capability exceeds the amount of energy that is released and the sentry function is characterized by an increasing function, but, due to the fact that the damage in the material is growing, the slope of this function progressively decreases. On the other hand, if a strong material configuration change occurs, such as the when fibers between two adjacent plies create a connection, also called fiber bridging, the strain energy storing capability increases instantaneously. In such a case the sentry function can be described by a linearly increasing function as the BU function is. So, comparing the PI and the BU functions, it can observed that the first one is characterized by a decreasing slope and the BU function has an almost linear slope. Moreover the BU function has a lower extension than the P_I one.

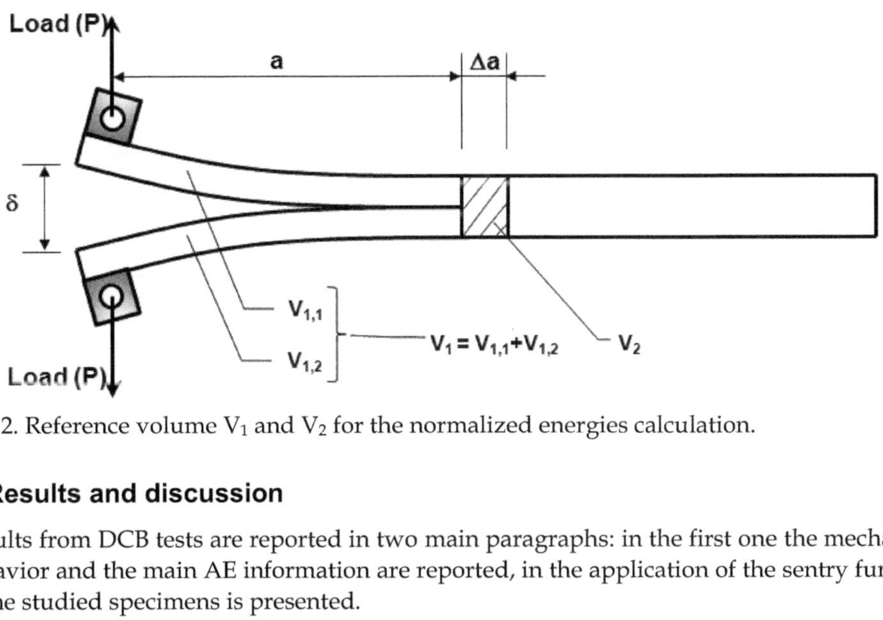

Fig. 2. Reference volume V_1 and V_2 for the normalized energies calculation.

3. Results and discussion

Results from DCB tests are reported in two main paragraphs: in the first one the mechanical behavior and the main AE information are reported, in the application of the sentry function to the studied specimens is presented.

3.1 Mechanical behavior and AE trends

Fig. 4 shows an example of the DCB specimen mechanical behavior in terms of load-displacement and crack length-displacement diagrams for each mid-plane interface type (the load was normalized by means of the specimen thickness)

Referring to Fig. 4, it can be noted that at the beginning of loading, the load and displacement relation is almost linear until the maximum load. Beyond the maximum load, the nonlinearity is considerable. Also, the crack propagation initiates near the maximum load. Note that there is no visible crack propagation at the beginning of loading. Referring now to the crack length diagrams, the comparison of three plots in Fig. 4 shows that the specimen T3 has stable-like crack propagation and there is a smooth increasing curve during the loading after the maximum load. For specimens T4 and T5, the crack length curve development has a completely different trend. This is due to fiber bridging during the delamination process that affects the load-displacement curve. In some cases, before maximum load, fiber bridging has occurred and the kick points (sudden drop off in load) are highlighted in the load-displacement curve; however, such a physical phenomenon does not seem to have any effect on the crack length development. Moreover, the other conditions like crack jumping from one layer to another one and the cutting condition (cutting edge) are factors leading to kick point existence too.

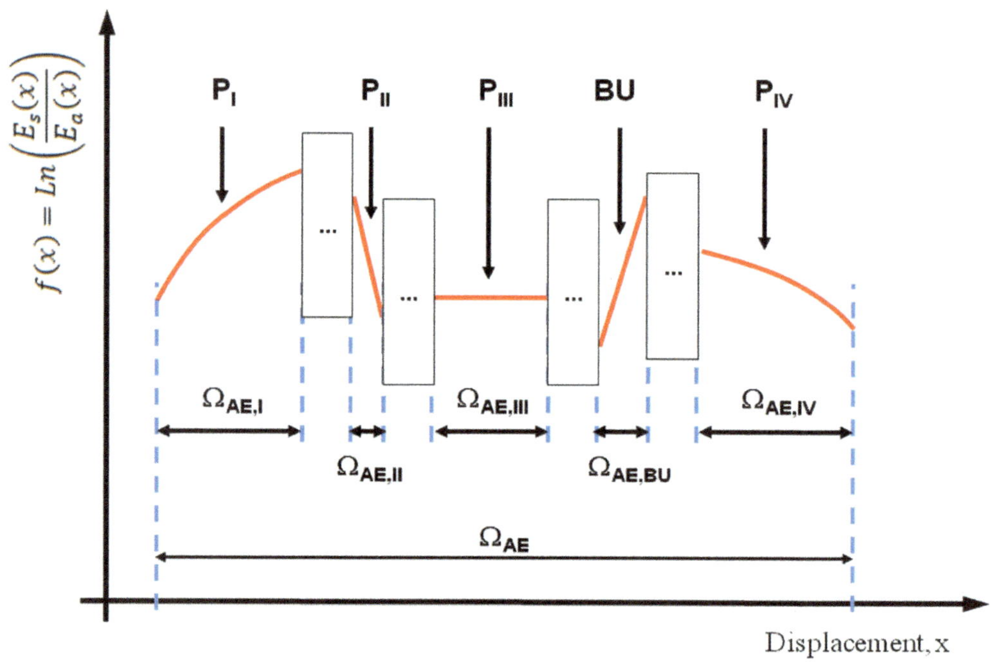

Fig. 3. The basic functions P_I, P_{II}, P_{III} and P_{IV}, used to describe the function f.

Diagrams in Fig. 5 show the trend of the G_I calculated by the formula (2) from the ASTM D5528 standard. From such diagrams it can be observed that up to maximum load the strain energy release rate is increasing, but beyond maximum load there is a slight decrease in the G_I trend. Such reduced values of the strain energy release rate are related to the drop off of the load shown in the load-displacement diagram. The load drop off is due to the crack propagation process based on a sudden extension and a subsequent stop (the "stop and go" phenomenon).

Fig. 6 summarizes some of the most relevant information extracted from the mechanical behavior. In particular, the load values which were considered are the maximum load and the load at which the first crack extension was visibly observed. Both values are useful for the subsequent calculation of the G_I. The displacement values at the maximum load and at the first observed crack extension are both useful to characterize the different behavior of the mid-plane interface quality with respect to mode I of delamination.

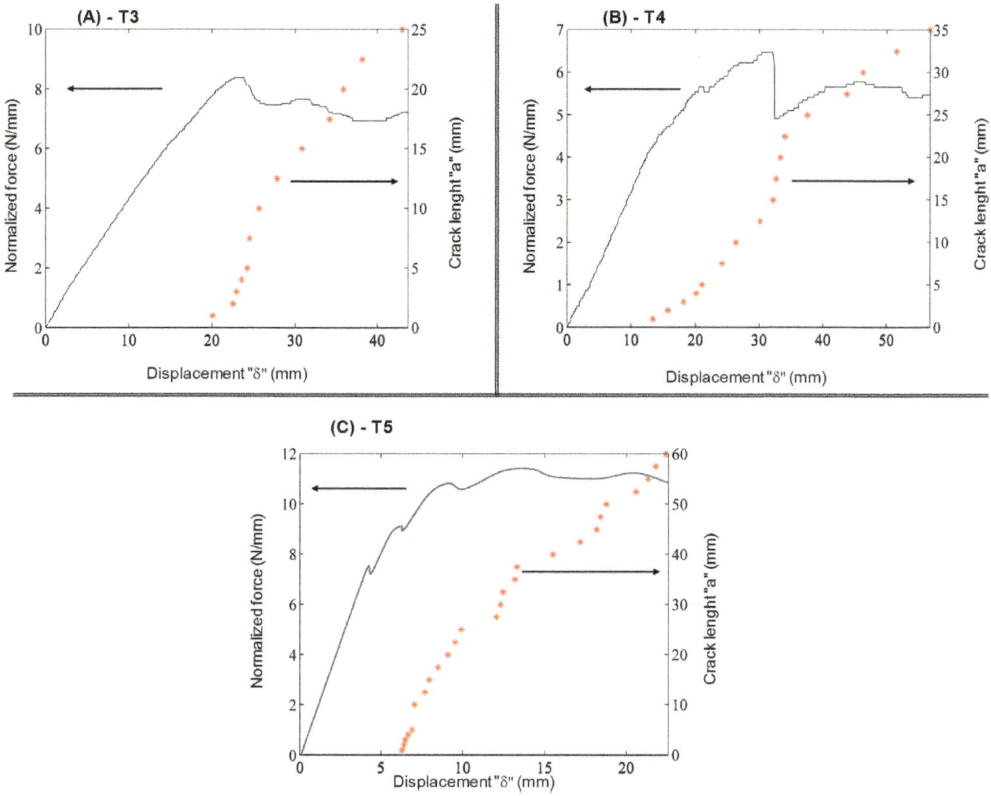

Fig. 4. Typical load and crack propagation length versus displacement for the three types of specimen mid-plane interfaces.

Fig. 7 shows the acoustic emission events behavior during loading. Different AE behaviors were noted between the different interface types and in particular great differences are highlighted where AE energies were first detected at the early stage of loading with alternating appearances up to an arbitrary delaminated length. Variations in AE event energy also reflect different damage mechanisms. The early stage of loading is free of acoustic events, but near the maximum load the first AE events are detectable. During the crack propagation process, the number of cumulative AE events and the cumulative AE event energy increase (Fig. 8). In particular, it can be seen that depending on the interface type the trend of cumulative events and of cumulative AE energy versus opening displacement change. It can thus be observed that at a fixed opening displacement in the case of the tougher interface, T5, the detected events are fewer than with the brittle one, T4, and the AE energy can vary because of different dominant damage mechanisms in each specimen.

A more detailed description of AE events can be obtained from Fig. 8 where the cumulative counts and AE event energy are plotted versus the displacement (in Fig. 8, the load and the G_I are also reported in each diagram and all diagrams have the same horizontal scale). In particular, it can be noted that increasing the toughness of the interface, from T4 up to T5, the number of cumulative counts recorded at the visible crack propagation point and at the maximum load decreases.

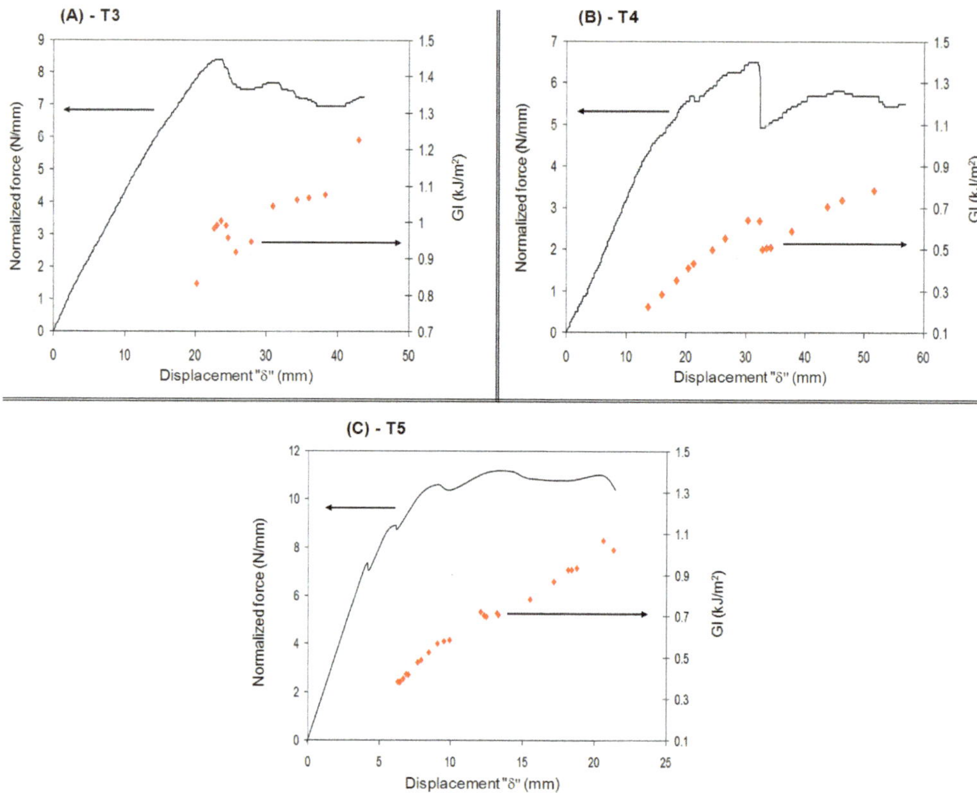

Fig. 5. Typical trends of load and G_I versus displacement for the three mid-plane interfaces.

In particular, it is of interest to consider the number of cumulative counts (Count CUM) and the cumulative AE event energy (Ea CUM) at the first visible crack extension. These values are summarized in Table 2 in terms of mean value (M.V.) and coefficient of variation (C.V.) of cumulative counts and cumulative AE energy.

The data in Table 2 show that the Count CUM or the cumulative AE event energy are not useful variables to predict the test stage at which the first visible crack extension occurs because the coefficients of variation are too high.

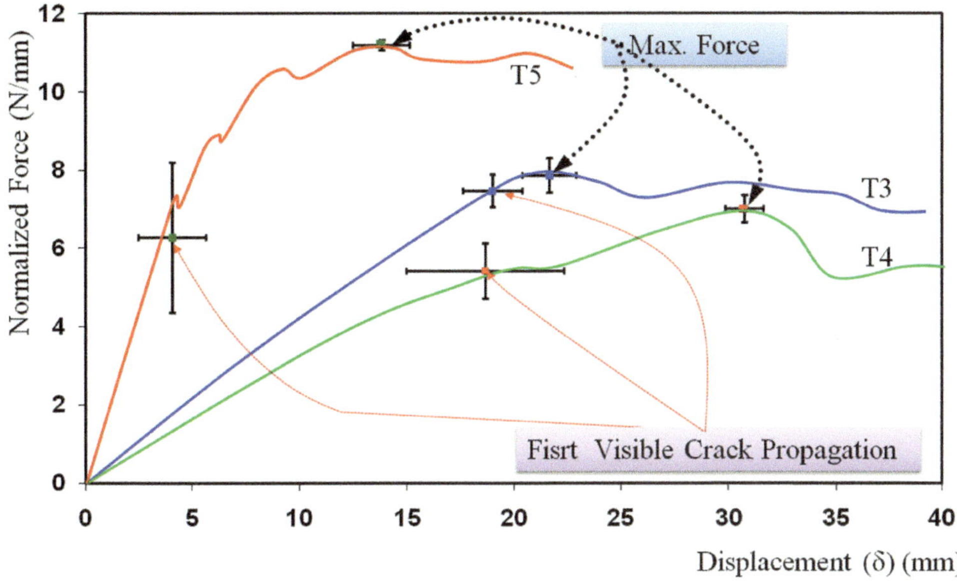

Fig. 6. Comparison of the normalized force vs. displacement diagrams for each of the three interface types with the points where there is the maximum of the force and where crack propagation begins (the confidence bars are based on the single standard deviation).

	Count CUM		Ea CUM (J)	
	M.V.	C.V.	M.V.	C.V.
T3	10920	110%	1.3E-13	171%
T4	244	49%	3.8E-16	97%
T5	195	59%	2.5E-17	42%

Table 2. Count CUM and Ea CUM at the first visible crack extension.

3.2 The sentry function application and results discussion

The sentry function is the logarithm of the ratio between the normalized Strain Energy and the normalized Cumulative Acoustic Energy as in Equation 4. The Strain energy is calculated by using the load-displacement information, Fig. 4, and the cumulative AE energy is calculated as the summation of the AE event energy, Fig. 8. Both energies are normalized by the proper reference volume as previously described, Fig. 2: the reference volume to normalize the strain energy is made by the bended volume joined with the volume where the delamination grows (V1+V2), the reference volume for the acoustic energy is calculated where the delamination grows (V2) considering each measurement of crack extension, see Fig. 4.

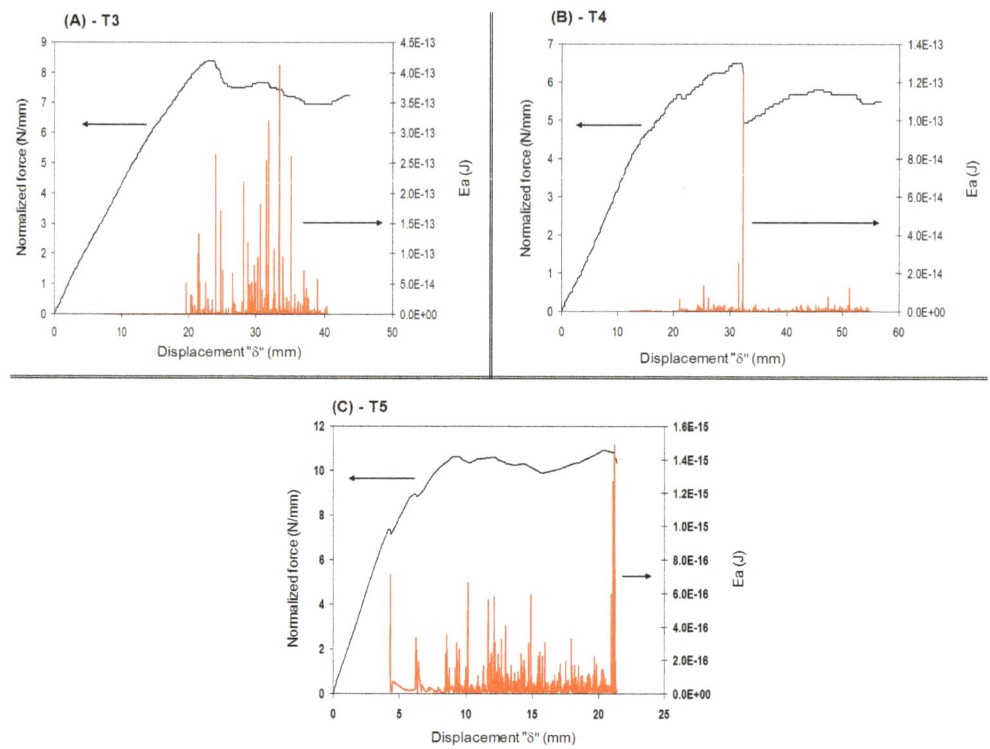

Fig. 7. Normalized load vs. displacement and acoustic energy (Ea) vs. displacement diagrams for the three types of interfaces.

In diagrams (A), (C) and (E) of Fig. 9 the single sentry function trends are highlighted according to the scheme reported in Fig. 3. The ascription of the sentry function trends to the experimental diagram of *f* is based on the basic discussion reported in the paragraph 2.5. In particular, it is interesting to note that in the case of T3 interface types (diagram A) no BU trends are highlighted; on the contrary, this trend can be seen in the case of the other two interface types. This can be explained by the fact that in the case of T4 and T5 interface types the fiber bridging effects were experimentally observed. It can also be observed that for all interface types, the sentry function has many drops (P_{II} trend) but one of them is prevalent with respect to the others (the "big fall"). In the particular case of T3 and T4 interface types, the big fall occurs when the load undergoes a big drop. For the T3 and T4 interfaces, this behavior, see also diagrams (B) and (D) in Fig. 9, can be related to the abrupt change in the energy release rate G_I and can be physically described as the test stage at which the delamination growth reaches the instability condition. In the case of the T5 interface, the big fall of the sentry function is not related to a significant load drop, but it is related to the second small load drop. So, unlike the case of the T3 and T4 interface types, the big drop of the T5 sentry function can not be related to the instable delamination growth.

Two other interesting analyses on the sentry functions can be performed by considering respectively its trends before and after the big drops and its shape with respect to the first

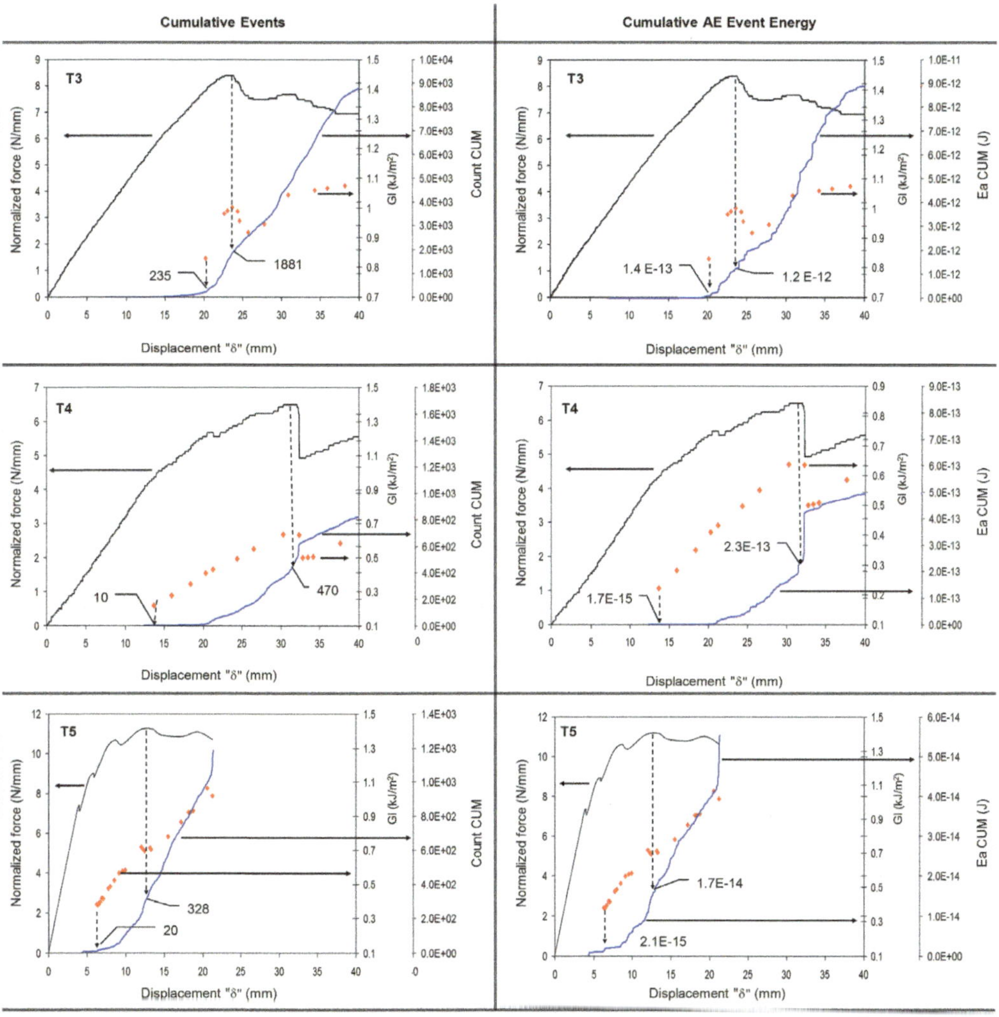

Fig. 8. Typical AE cumulative event (column 1) and cumulative AE event energy (column 2) during the delamination process for each type of interface.

visible crack propagation. Regarding the first aspect, it can be noted that, before the big fall of *f*, in the case of T3 and T4 specimens the sentry function trends consist of combinations of P_{II}, P_{III} and P_{IV} trends. These compositions indicate the fact that the AE released energy tends to overcome the strain energy storing capability of the material. In the case of the T4 specimen, it also has to be noted that before the big fall of the sentry function some BU trends are present. As stated previously, the BU trends highlight the presence of some fiber bridging due to the presence of a unidirectional ply in the T4 interface structure. In the case of T3 and T4 interface types, after the big drop the sentry function consists of a waved composition of P_I, P_{III} and P_{IV} trends. These compositions for the sentry function indicate a renewed strain energy storing capability in the material that is, in any case, partial and

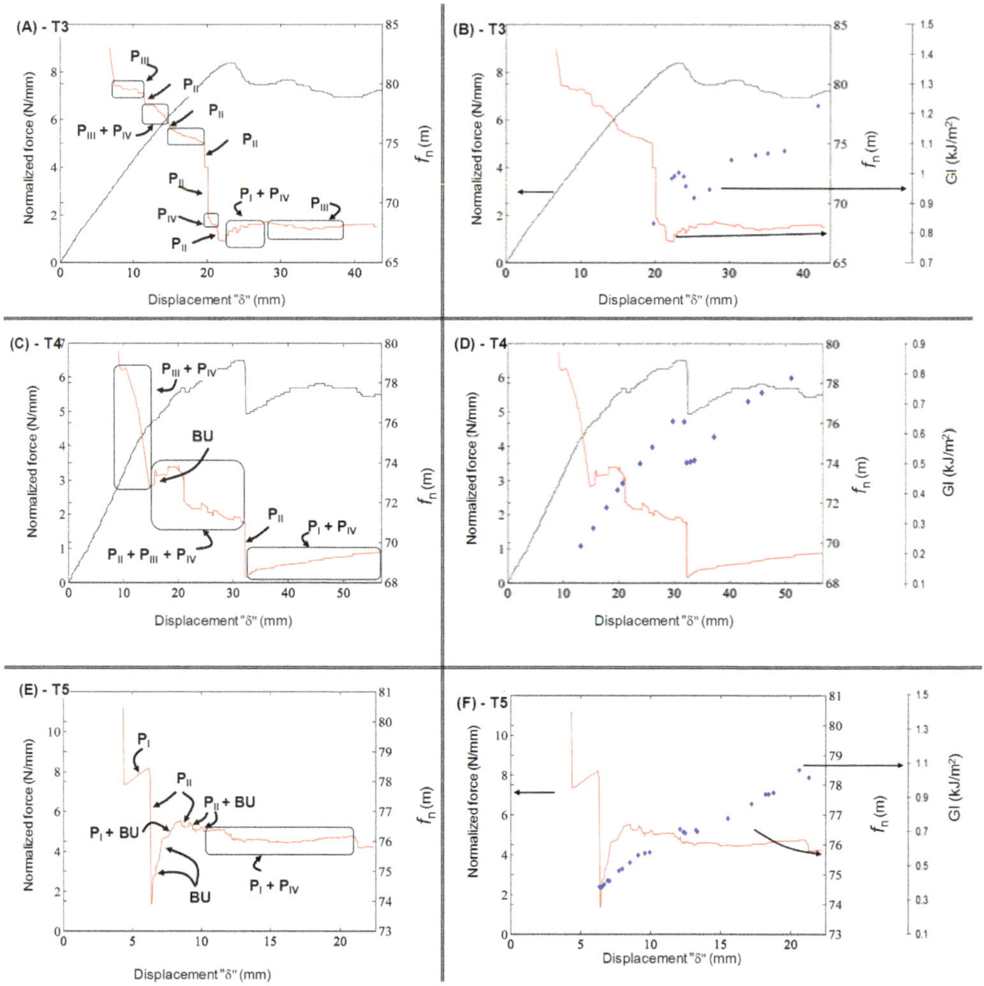

Fig. 9. Examples of sentry function trends obtained by the normalized energies in the case of woven-woven (A-T3), woven-unidirectional (B-T4) and unidirectional-unidirectional (C-T5) interface types (the load-displacement and the GI-displacement are also reported in each diagram).

discontinuous due to continuous "stop and go" progression of the post instability delamination growth process. In the case of the T5 interface type, before the big fall, the sentry function is characterized by trends of types P_I and P_{II}. These trends indicate the strain energy storing capability and the abrupt failure at the DCB specimen interface level. After the big fall, the sentry function trend of T5 specimens is characterized by the presence of P_I, P_{II}, P_{IV} and BU trends. This behavior indicates the fact that, due to the fiber bridging effect, the specimen at the interface zone has a residual strain energy storing capability. In particular, this behavior is sustained until the instability growth of the delamination is reached at about 13 mm of displacement. After the instability of the delamination process, the sentry function consists of a composition of P_I, P_{III} and P_{IV} that is waved in shape due to the continuous "stop and go" progression of the crack front.

The second aspect concerns the relationship between the sentry function and the first visible crack propagation. The discussion is based on the fact that the sentry function simultaneously takes into account the strain energy storing capability (E_s) and the release of the internal energy (E_a) due to damage. So, when the energy storing capability decreases this indicates that the material damage is increased, and, at the same time, the tendency of the material to release the internal energy will decrease as well. Thus, when the material damage increases, the ratio between the energy storing capability and the cumulated released energy (E_s/E_a), that is the sentry function, will decrease. The sentry function can thus be used to describe the history of the material damage. In particular, considering the integral of the sentry function over the AE displacement domain we obtain a value that is proportional to the cumulated material damage. This integral (Int(f)) can be used as a material damage indicator processes (Cesari et al., 2007; Minak & Zucchelli, 2008; Minak et al., 2009). In fact, in previous studies, it was shown how the integral of the sentry function is related to the material damage and in particular how it can be used to predict the residual strength of composite laminate after an accident (such as an indentation or a low velocity impact phenomenon) processes (Cesari et al., 2007; Minak & Zucchelli, 2008). Based on that observation and taking advantage of previous experiences, the integral of the sentry function in the present study is related to the test stage at which the crack propagation became visible. The aim of this analysis is to find a novel way to predict when the crack propagation became visibly detectable (a similar analysis was previously performed considering the cumulative counts and the cumulative AE energy without any success). Implementing this idea, the integral of the sentry function was calculated between the displacement at which the AE was revealed for the first time and the displacement at which the crack progression became visible. Obviously the first crack propagation was accurately estimated by a continuous crack tip monitoring at both specimens edges.

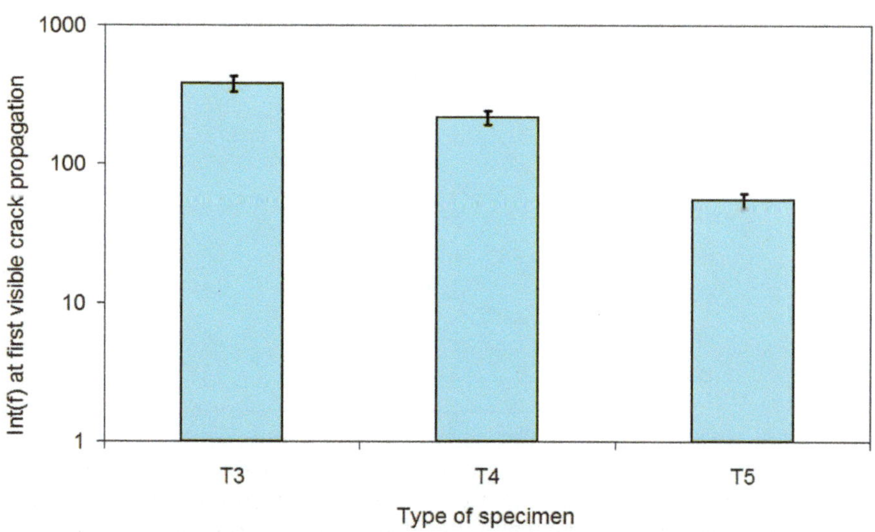

Fig. 10. Mean values and standard deviation of the integral of the sentry function between the beginning of the AE and the first visible crack propagation (value in log-scale).

From Fig. 10 it can be seen that the value of the integral of the sentry function between the beginning of the AE and the first visible crack propagation has a very low deviation from the mean value for each type of specimen (the coefficient of variations was close to 13%). These results indicate that for a specific type of interface the value of the integral of the sentry function can be used as an alarm, indicating when the crack propagation became visible. From the practical point of view, this way of using the Int(f) could be usefully applied to detect, in a real size structure, when a delamination crack, under a mode I loading, starts propagating.

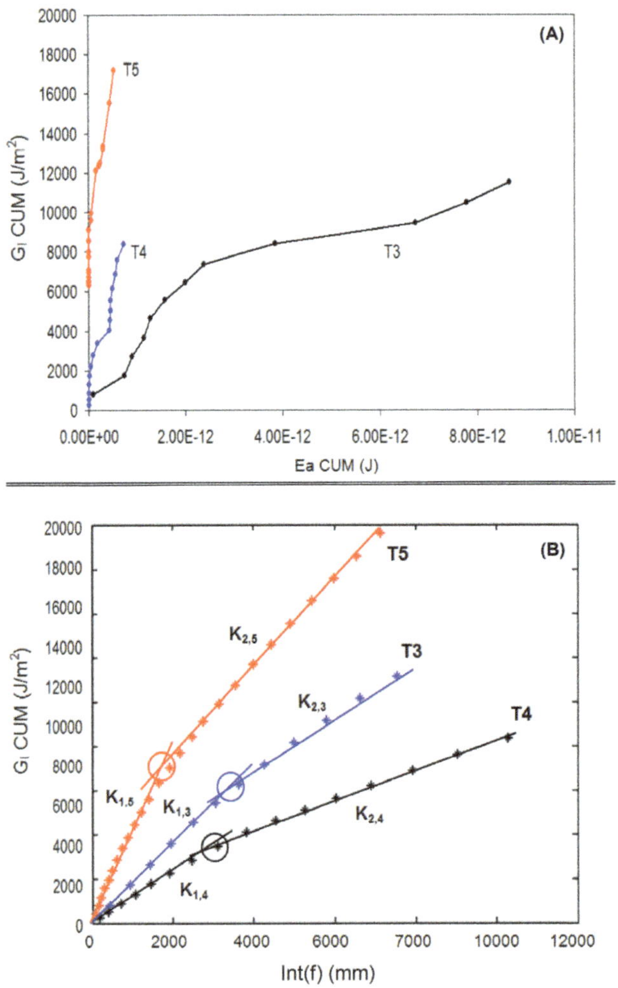

Fig. 11. (A) plot of GI CUM versus Ea CUM as proposed in Ref. (Ndiaye et al., 2000); (B) plot of cumulative G_I versus integration of *f* function, in this case the bilinear functions for each specimen type are visually characterized by means of the transition points highlighted by the circles.

A second analysis that was performed using the sentry function concerns the relationship between the integral of the sentry function itself and the strain energy release rate. This analysis is inspired by the work of (Ndiaye et al., 2000) where the cumulative of G_I is related to the cumulative of the AE event energy, see Fig. 11 (A); a different approach is nevertheless proposed considering the plot of cumulative G_I (G_ICUM) versus the integral of the sentry function, Int(f), see Fig. 11 (B).

The plots in Fig. 11 (A) do not highlight any particular trend; however, this plot has been proposed as a basis to determine the critical stain energy release rate by means of a linear fitting of the data. On the other hand, for each plot in Fig. 11 (B), a bi-linear relationship can be found between the G_ICUM and Int(f) where a transition point separates the two linear trends. Our idea is that using the plots in Fig. 11(B) it is possible to estimate the critical crack length and consequently the G_{Ic} and in the following some arguments are developed following this aim.

A preliminary observation from Fig. 11(B) concern the fact that the two variables, Int(f) and G_ICUM, exhibit a proportionality. This fact is supported by the fact that the Int(f) is related to the material damage (Minak & Zucchelli, 2008; Minak et al., 2009) and the G_ICUM is related to the ability of the material to yield the stored strain energy to create new cracked surfaces. So, both variables are related to the material damage and this support their proportionality: cumulating the material damage (Int(f)), the cumulated strain energy release rate (G_ICUM) increases as well. The slope of the fitting line indicates how fast the cumulated material damage influences the strain energy release rate variation during the test. So, considering the slope of the linear $K_{1,i}$ (i=3,4,5), Fig. 11(B), related to the specimens T3, T4 and T5 respectively, it can be observed that even if T5 interface type is supposed to be the toughest one, it is also more susceptible to the material damage than the other. This mean that T5 is more susceptible to the crack length extension then the other type of specimens: when and if the delamination crack is extended in the specimen, the T5 has a faster attitude to cumulate the damage and to increase the strain energy release rate than the others.

The second observation regard to the second slope of the bilinear trends shown in Fig. 11(B). In fact the analysis of the bilinear relations shown in Fig. 11(B) reveal the fact that the slope of the linear $K_{1,i}$ (i=3,4,5) before the transition point is always higher than the slope $K_{2,i}$ (i=3,4,5) after the transition point. The slope variation is due to a change of the material strength behavior respect to the delamination crack propagation. After the transition points the material exhibits reduced crack propagation strength and the slope decreases. This fact indicate that when the cumulated strain energy release rate (G_ICUM) increases of a small amount this cause a considerable increment in the cumulated material damage (Int(f)).

Combining the first and the second observations it is possible to note that before the transition point the linear relation between Int(f) and GICUM refers to the stable crack propagation, while after the transition point the linear relation refers to the unsteady state crack propagation. Furthermore it can be also noted that for all specimens the transition point usually occurred near the maximum load in the load-displacement plot.

It can be concluded that Int(f)-G_ICUM plot (Fig. 11 (B)) can distinguish the specimens with different resistance to the delamination process in mode I.

Based on the previously developed arguments the transition point can be used to calculate the G_{Ic}.

The critical strain energy release rate can be calculated by different techniques has proposed in standards and the literature (Ndiaye et al., 2000). Based on the ASTM D5528 standard, there are three different definitions for an initiation value of G_{Ic} that can be evaluated during DCB testing. These include G_{Ic} values determined using the load and deflection measured respectively (1) at the point of deviation from linearity in the load-displacement curve (NL), (2) at the point where delamination is visually observed on the edge (VIS) measured with a microscope and (3) at the point where the compliance has increased by 5% or where the load has reached a maximum value. In the present work the ASTM D5528 method (1) and (3) will be considered to calculate reference values of the G_{Ic} that will be compared to the ones obtaind following the method proposed in (Ndiaye et al., 2000) and the new method based on the sentry function (Fig. 12).

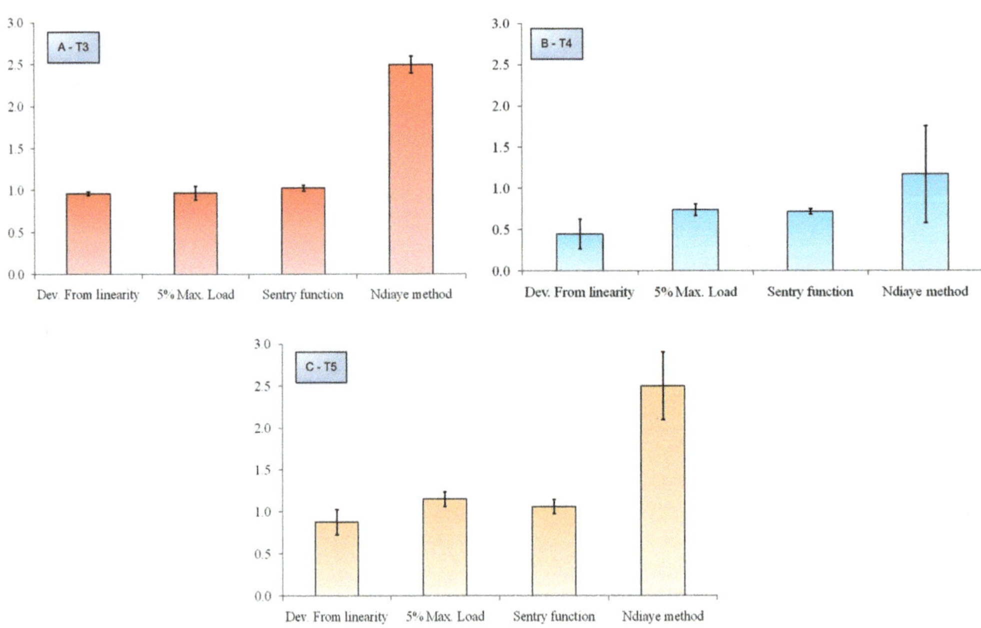

Fig. 12. Comparison between the results from the different methods for G_{Ic} calculation: (A) woven-woven interface, (B) woven-unidirectional interface, (C) unidirectional-unidirectional interface.

The calculated results are obtained by finding the desired points based on the applied method. These points include propagated length of crack, displacement and the load values. For the sentry function method, G_{Ic} is calculated by finding the knee point taking place in the G_I cumulative versus Int(f) plot. The knee point in this plot is in relation to the propagated crack length that can be found from the load-displacement plot. With regard to the crack length, the displacement point and load value of the critical strain energy release rate are calculated. Comparison between the results shows that the Ndiaye method (Ndiaye et al., 2000) has estimated G_{Ic} values about 2 or 3 times more than the ASTM D5528 results. But G_{Ic} calculated from the sentry function method is in good agreement with the results obtained from the ASTM D5528 method (5% max. Load). Moreover it can be also noted that the standard deviation related to the results obtained by the new method are smaller than the ones obtained applying the ASTM D5528 methods.

It can thus be concluded that the sentry function is another alternative method for calculating G_{Ic} directly from acoustic emission data.

4. Conclusions

In the present study the acoustic emission technique has been used in an unconventional way to monitor and study mode I delamination behavior of different composite laminate interfaces. The study is based on a novel function that combines mechanical information, the strain energy, and the acoustical information, the acoustic emission event energy. This function, called sentry function, is used to study the mode I delamination behavior of the three considered interfaces, and in particular it was used to identify damage progression for each interface type as well as to identify the fiber bridging events. More than this, the sentry function was also used to obtain two considerable results: the first one concerns the identification of the test stage at which the delamination propagation became visible, while the second result is a new way to estimate the G_{Ic}. The proposed approach, based on the sentry function, gives values for the G_{Ic} of each composite interface that are in good agreement with the values estimated by means of the ASTM D5528 standard procedures. It can also be observed that the proposed method enables an estimation of the G_{Ic} with a lower standard deviation from the mean value if compared to the results obtained by the other procedures. The main advantage of the proposed approach with respect to those in the literature is, therefore, the simple and objective identification of the test stage at which the crack delamination length became critical, thus enabling a robust calculation of the G_{Ic} value.

5. Acknowledgements

The author is especially indebted to working groups in the engineering faculty and laboratory of Bologna University for valuable discussions of the problems to which the above results pertain.

6. References

ASTM Standard D5528. (1994). Standard Test Method for Mode I Interlaminar Fracture Toughness of Unidirectional Fiber-Reinforced Polymer Matrix Composites.

Benmedakhene, S.; Kenane, M. & Benzeggagh, M. L. (1999). Initiation and Growth of Delamination in Glass/Epoxy Composites Subjected to Static and Dynamic Loading by Acoustic Emission Monitoring. *Composites Science and Technology*, Vol.59, No.2, (February 1999), pp. 201-208, ISSN 0266-3538

Bohse, J. & Brunner, A. J. (2008). Acoustic Emission in Delamination Investigation, In: Delamination Behaviour of Composites. S. Sridharan, (Ed.), 217-277, CRC Press, ISBN 978-1-4200-7967-8, New York

Bohse, J. (2000). Acoustic Emission Characteristics of Micro-Failure Processes in Polymer Blends and Composites. *Composites Science and Technology*, Vol.60, No.8, (June 2000), pp. 1213-1226, ISSN 0266-3538

Cesari, F.; Dal Re, V.; Minak, G. & Zucchelli, A. (2007). Damage and Residual Strength of Laminated Carbon–Epoxy Composite Circular Plates Loaded at the Centre. *Composites Part A: Applied Science and Manufacturing*, Vol.38, No.4, (April 2007), pp. 1163-1173, ISSN 1359-835X

Chen, J.; Crisfield, M.; Kinloch, A. J.; Busso, E. P.; Matthews, F. L. & Qui, Y. (1999). Predicting Progressive Delamination of Composite Material Specimens via Interface Elements. *Mechanics of Composite Materials Structures*, Vol.6, No.4, (October 1999), pp. 301–317, ISSN 1075-9417

Diaz, A. D.; Caron, J. F. & Ehrlacher, A. (2007). Analytical Determination of the Modes I, II and III Energy Release Rates in a Delaminated Laminate and Validation of a Delamination Criterion. *Composite Structures*, Vol.78, No.3, (May 2007), pp. 424–432, ISSN 0263-8223

Hug G, Thevenet, P.; Fitoussi, J. & Baptiste, D. (2006). Effect of the Loading Rate on Mode I Interlaminar Fracture Toughness of Laminated Composites. *Engineering Fracture Mechanics*, Vol.73, No.16, (November 2006), pp. 2456–2462, ISSN 0013-7944

Huguet, S.; Godin, N.; Gaertner, R.; Salmon, L. & Villard, D. (2002). Use of Acoustic Emission to Identify Damage Modes in Glass Fibre Reinforced Polyester. *Composites Science and Technology*, Vol.62, No.10-11, (August 2002), pp. 1433–1444, ISSN 0266-3538

Kim, B. W. & Mayer A. H. (2003). Influence of Fiber Direction and Mixed-Mode Ratio on Delamination Fracture Toughness of Carbon/Epoxy Laminates. *Composites Science and Technology*, Vol.63, No.5, (April 2003), pp. 695–713, ISSN 0266-3538

Kusaka, T.; Hojo, M; Mai, Y. W.; Kurokawa, T.; Nojima, T. & Ochiaib S. (1998). Rate Dependence of Mode I Fracture Behaviour in Carbon-Fibre/Epoxy Composite Laminates, *Composites Science and Technology*, Vol.58, No.3-4, (March-April 1998), pp. 591-602, ISSN 0266-3538

Lorriot, T. h.; Marion, G.; Harry, R. & Wargnier, H. (2003). Onset of Free-Edge Delamination in Composite Laminates under Tensile Loading. *Composites Part B: Engineering*, Vol.34, No.5, (July 2003), pp. 459–471, ISSN 1359-8368

MIL-HDBK-17-3F, (2002). Department of Defense Handbook, Composite Materials Handbook, Volume 3. Polymer Matrix Composites Materials Usage, Design and Analysis, Chapter 5, Paragraph 5.4.5, ed. 17, (June 2002).

Minak, G. & Zucchelli, A. (2008). Damage Evaluation and Residual Strength Prediction of CFRP Laminates by Means of Acoustic Emission Techniques. In: Composite Materials Research Progress, L.P. Durand, (Ed.), 165-207, Nova Science Publishers, ISBN 978-1-60692-496-9, New York

Minak, G.; Abrate, S.; Ghelli, D.; Panciroli, R. & Zucchelli, A. (2010). Residual Torsional Strength after Impact of CFRP Tubes, *Composites Part B: Engineering,* Vol.41, No.8, (December 2010), pp. 637-645, ISSN 1359-8368

Minak, G.; Morelli, P. & Zucchelli, A. (2009). Fatigue Residual Strength of Circular Laminate Graphite–Epoxy Composite Plates Damaged by Transverse Load, *Composites Science and Technology,* Vol.69, No.9, (July 2009), pp. 1358-1363, ISSN 0266-3538

Ndiaye, I; Maslouhi, A. & Denault, J. (2000). Characterization of Interfacial Properties of Composite Materials by Acoustic Emission. *Polymer Composites,* Vol.21, No.4, (August 2000), pp. 595-604, ISSN 0272-8397

Niebergall, U.; Bohse, J.; Seidler, S.; Grellmann, W. & Schürmann, B. L. (1999). Relationship of Fracture Behavior and Morphology in Polyolefin Blends. *Polymer Engineering & Science,* Vol.39, No.6, (June 1999), pp. 1109-1118, ISSN 1548-2634

Pagano, N. J. & Schoeppner, G. A. (2000). Delamination of Polymer Matrix Composites: Problems and Assessment. In: Comprehensive Composite Materials. A. Kelly, C. Zweben, (Ed.), 433-528, Elsevier Science, ISBN 0080437206, Amesterdam

Perrin, F.; Bureau, M. N.; Denault, J. & Dickson, J. I. (2003). Mode I interlaminar Crack Propagation in Continuous Glass Fiber/Polypropylene Composites: Temperature and Molding Condition Dependence. *Composites Science and Technology,* Vol. 63 No.5, (April 2003), pp. 597-607, ISSN 0266-3538

R'Mili, M.; Moevus, M. & Godin, N. (2008). Statistical Fracture of E-Glass Fibres Using a Bundle Tensile Test and Acoustic Emission Monitoring. *Composites Science and Technology,* Vol.68, No.7-8, (June 2008), pp. 1800-1808, ISSN 0266-3538

Refahi Oskouei, A. & Ahmadi, M. (2010). Acoustic Emission Characteristics of Mode I Delamination in Glass/Polyester Composites. *Composite Materials,* Vol.44, No.7, (April 2010), pp. 793-807, ISSN 0021-9983

Sjogren, A. & Asp, L. E. (2002). Effects of Temperature on Delamination Growth in a Carbon/Epoxy Composite under Fatigue Loading. *International Journal of Fatigue,* Vol.24, No.2-4, (February-April 2002), pp. 179–184, ISSN 0142-1123

Solaimurugan, S. & Velmurugan, R. (2008). Influence of In-Plane Fibre Orientation on Mode I Interlaminar Fracture Toughness of Stitched Glass/Polyester Composites. *Composites Science and Technology,* Vol.68, No.7-8, (June 2008), pp. 1742–1752, ISSN 0266-3538

Tsai, G. C. & Chen, J. W. (2005). Effect of Stitching on Mode I Strain Energy Release Rate. *Composite Structures,* Vol.69, No.1, (June 2005), pp. 1–9, ISSN 0263-8223

Velmurugan, R. & Solaimurugan, S. (2007). Improvements in Mode I Interlaminar Fracture Toughness and In-Plane Mechanical Properties of Stitched Glass/Polyester Composites. *Composites Science and Technology,* Vol. 67, No.1, (January 2007), pp. 61-69. ISSN 0266-3538

Wu, X. F. & Dzenis, Y. A. (2005). Experimental Determination of Probabilistic Edge-Delamination Strength of a Graphite-Fiber/Epoxy Composite. *Composite Structures*, Vol.70, No.1, (August 2005), pp. 100–108, ISSN 0263-8223

5

Detection, Recognition and Location of Partial Discharge Sources Using Acoustic Emission Method

Wojciech Sikorski[1] and Waldemar Ziomek[2]
[1]*Poznan University of Technology,*
[2]*University of Manitoba,*
[1]*Poland,*
[2]*Canada*

1. Introduction

The acoustic emission (AE) is a group of phenomena involving generation of transient elastic (acoustic or vibro-acoustic) waves, resulting from liberation of intermolecular bond energy (deformation, cracking, phase transitions). This fact contributed to elaboration of the acoustic emission method, which is widely used in: (i) diagnostics of cutting tools and machine cutting process, (ii) machinery and civil structures condition assessment, (iii) fatigue and fracture materials research, (iv) detection of material defects, (v) orogenic and earthquake prediction research, (vi) rockbursts prediction in deep mines and (vii) in chemical reactions and phase transitions research. This method is also applied in electrical power engineering for detection and location of partial discharge (PD) sources in insulation system of large power transformers and GIS/GIL (Gas Insulated Substations/Gas Insulated transmission Lines).

According to the International Electrotechnical Commission IEC-60270 standard, the partial discharge is defined as localized electrical discharge that only partially bridges the insulation between conductors and which can or cannot occur adjacent to a conductor. Partial discharges are in general a consequence of local electrical stress concentrations in the insulation or on the surface of the insulation. During a partial discharge, there occurs impulse conversion of some part of electric energy (about 1-5%) to mechanical energy, which is an acoustic emission wave. The remaining part includes electric, thermal and chemical energy. The most defects of the oil/paper transformer insulation system are characterized by partial discharges. It is a phenomenon, which can, but does not have to, lead to damage of a high-voltage device in short time. Some defects can develop for years as well. The phenomenon can also have a passing character and depend on different factors. Therefore, the assessment of dynamics changes of PD intensity in time is essential and it can be realized by monitoring system perfectly (Feser, 2003; Gulski et al., 1999; Gulski & Smitt, 2007; Sikorski et al., 2007c). At present, there are a few methods, which allow to monitor the development of partial discharges, such as: DGA - *Dissolved Gas in oil Analysis*, UHF - *Ultra High Frequency* and AE - *Acoustic Emission*. Each method is characterized by certain advantages and disadvantages, which were summarized in Table 1. From the listed methods, the most applicable and least expensive seems to be the AE method (Sikorski & Walczak, 2010).

Possibilities	Method		
	DGA	UHF	AE
Detection	Yes	Yes	Yes
Detection sensitivity	High (~10pC, depending on time of PD activity)	High (~1-10pC, depending on distance and location of the PD source)	Moderate (> 300 pC, depending on location of the PD source)
Intensity measurement	No	Limited	Limited
Identification	No	Yes	Yes
Location	Limited	Limited	Yes
Installation difficulty	Moderate (transformer must be turned off)	Moderate (through oil valve, transformer under load / through dielectric window, transformer must be turned off and opened)	Low (transformer under load)
Sensors	1 (due to high costs)	limited by number of oil valves or dielectric windows	open structure, typically 1-16

Table 1. The comparison of the PD monitoring methods for large power transformers

The main advantages of the AE method are: (i) tests can be taken when the transformer is on-line (during applied voltage or induced voltage tests), (ii) simple test procedure, (iii) susceptibility to external electrical disturbances is relatively low, and (iv) possibility of partial discharge recognition, location and on-line monitoring (Institute of Electrical and Electronics Engineers [IEEE], 2007; Kuffel et al., 2000; Lundgaard, 1992a, 1992b; Markalous et al., 2003).

In order to develop and optimize the acoustic emission for power transformer diagnostics, the author followed the approach outlined below:

- selection of optimal AE signal processing methods and parameters for efficient PD detection and recognition,
- laboratory research of PD's parameters generated in properly prepared models of oil/paper insulating systems (imitating real PD sources in power transformers),
- completion and developing of PD-patterns database based on laboratory results and selected parameters,
- verification tests on real power transformers.

2. AE measurement system for partial discharge detection

A typical measurement system for partial discharges detection based on acoustic emission method is composed of: (i) piezoelectric sensors, (ii) preamplifiers, (iii) signal conditioning unit, (iv) signal acquisition unit and (v) specialized software for digital signal processing. The most advanced systems dedicated for power transformer on-line monitoring are additionally adjusted to work in difficult weather conditions (embedded climate control unit, waterproof chassis/enclosure) and they are equipped with an expert system software for automatic PD detection and recognition. The system for power transformer online monitoring, which was in detail described in (Sikorski & Walczak, 2010), was presented in Fig 1.

Fig. 1. System for large power transformer on-line monitoring based on AE method

The choice of the optimal type of acoustic emission sensor for PD detection is difficult, because it involves the need for finding a compromise between the required sensitivity and bandwidth. Some experts recommend the use of narrowband sensors, emphasizing their high sensitivity in the lower frequencies (20-100 kHz) (Bengtsson & Jönsson, 1997). These sensors provide high efficiency of detection of surface discharges (generating acoustic emission signals with frequencies from 20 to 110 kHz), which generally have high energy and can cause accelerated degradation of the insulation system. In turn, other experts suggest the choice of broadband transducers (up to 1 MHz) with the resonance frequency in the range from 60 kHz to 150 kHz (Boczar, 2001; Lundgaard, 1992a, 1992b; Markalous et al., 2003; Sakoda et al., 1999; Sikorski & Siodla, 2005; Varlow et al., 1999). Theoretically, they enable the detection of all types of partial discharges that may occur in transformers with oil/paper insulation. The only disadvantage of this solution is smaller sensitivity of surface discharge detection.

A fundamental importance in the detection of partial discharges in a power transformer is a proper acoustic coupling between the sensor and the surface of the tank (American Society for Testing and Materials [ASTM], 2007). For this purpose one can use silicone grease or gel dedicated for ultrasonic applications. A lack of direct contact of acoustic sensor with the tank causes a strong attenuation of the AE amplitude signal, and thus a strong decrease in sensitivity of partial discharges detection. A practical solution facilitating the mounting of AE sensor to the tank wall with a constant force is the use of special handles equipped with permanent magnets (Fig. 2).

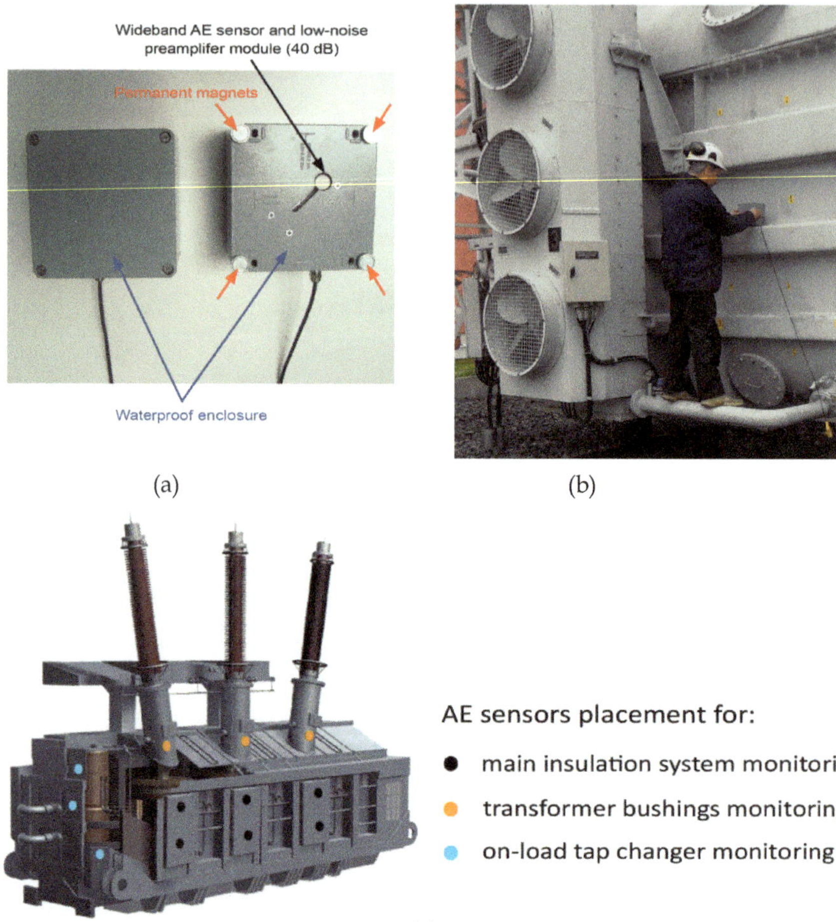

Fig. 2. General view of the AE sensor's waterproof enclosure with permanent magnets (a); mounting of the sensor on transformer tank (b); typical AE sensors placement on transformer tank (c).

3. Possibilities and limits of acoustic emission in power transformer diagnostics

In case of large power transformers the main problem is an environmental heterogeneity between partial discharge source and AE sensor (Sikorski et al., 2007c). The acoustic field inside the tank is very complex due to wave reflection and diffraction in different materials (pressboard, copper, pressboard, paper, oil). The acoustic wave propagates in materials, from which the transformer is made, with different velocity: 1413 m/s in transformer oil, 1500 m/s in impregnated pressboard, 3570 m/s in copper and 5100 m/s in steel. The velocity of the AE wave propagation in oil strongly depends on its temperature. When the oil temperature equals 20°C the velocity of wave propagation amounts to 1413 m/s. For higher oil temperatures the velocity falls and equals: 1300 m/s by 50°C, 1200 m/s by 80°C and 1100 m/s by 110°C. The velocity in metal is greater than in oil, therefore very often the wave, which most of its way travels in tank wall (structure-borne path), arrives at the sensor first. Afterwards the sensor registers the wave, which propagated in oil slowly (direct acoustic path). The problem of structure-borne waves may be minimized by placing the acoustic sensor inside the tank. In order to achieve this goal, hydrophones or optical fibre sensors are used (IEEE, 2007; Lundgaard, 1992a, 1992b). The model of AE waves propagation in power transformer tank and its influence on signal acquisition outcomes was presented in Fig. 3.

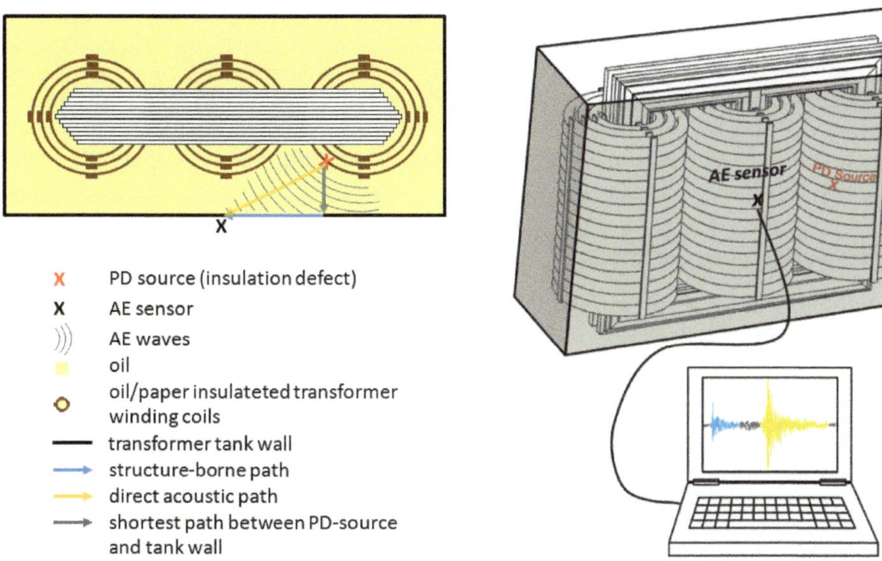

Fig. 3. The propagation of AE waves in power transformer tank and its influence on signal acquisition outcomes.

The correct interpretation of measurement results may be made difficult by disturbances during on-site PD detection. The main sources of the disturbances are:

- switching of on-load tap changer,
- thermal faults of transformer's active part,
- high-voltage switchgear operations near the investigated transformer,
- environmental noises (thunderstorms, rain, wind),
- core magnetostriction noise (Barkhausen effect),
- loose shielding connection in transformer tank.

In order to interpret the results of the AE method correctly, mostly the comparative analysis with additional parameters of the transformer during the test, such as: load, voltage, oil/windings temperature, information on time and number of switching operations of on-load tap changer and cooling fans etc., is conducted. However the best results are achieved by simultaneous use of the AE and electrical method. The conventional electrical PD detectors conformable with IEC-60270 standard are not suitable for on-site measurement (due to corona discharges and other electrical disturbances), but it is possible to use the UHF probe inside the transformer tank installed through a drain valve or Rogowski coil (HFCT/RFCT sensors as well) installed on the transformer tank ground wire (IEEE, 2007).

4. Selected methods of AE signals processing

The first stage of the research was related to the selection of signal processing methods which would support both the detection and recognition of PDs. The parameters used for software selection included immunity to disturbances and wide- and narrow-band noises, minimum computing time, and overall efficiency (Sikorski & Siodla, 2005; Sikorski et al., 2007a).

Based on results of numerous computer simulations and a widespread literature study, *high-resolution spectral analysis* (HRSA) and *joint time-frequency analysis* (JTFA) were chosen as the optimum methods for establishing the AE-based PD-patterns (Antoniadis & Oppenheim, 1995; Holschneider, 1998; Kia et al., 2007; Lobos et al., 2000, 2001; Ma et al., 2002; Shim et al., 2000; Yang & Judd, 2003; Zhang et al., 2003).

4.1 High-resolution spectral analysis (HRSA)

High-resolution spectral analysis methods, also known as *subspace methods*, generate the frequency component estimates for a signal based on eigenanalysis or eigendecomposition of the correlation matrix (Hayes, 1996; Marple, 1987). These methods exploit the noise subspace to estimate unknown frequency parameters of a random process and are very effective in the detection of sinusoids (a PD signal can be approximated as an exponentially attenuated sinusoidal oscillation) buried in noise, especially when the signal-to-noise ratios (SNR) are low. This property makes the analysis method particularly attractive for efficient detection of PD signals recorded under conditions of strong external disturbances. Due to this fact, the detection of internal discharges occurring in power transformer in which the acoustic signal is damped by pressboard barriers is also possible. It is worth mentioning that such efficient detection with the use of classic Fast Fourier Transform (FFT) is not possible. Among available subspace methods **Mu**ltiple **Si**gnal **C**lassification (MUSIC) was chosen (Schmidt, 1986). The MUSIC method very precisely identifies harmonic components (Besson & Stoica, 1996; Hayes, 1996; Lobos et al., 2000, 2001; Marple, 1987). This property is illustrated below by the results of simulation in which two waveforms were used (Fig. 4).

The first waveform analysed (Fig. 4a) was registered in an electrode system, which enabled the generation of different types of multi-source discharges. In this case PD in an internal gas void and surface discharges were generated simultaneously (at ~30 mV noise level). The second waveform analysed (Fig. 4b) has the same PD signal, but it is buried in additive white Gaussian noise. One can see that the MUSIC method, independent of the noise level, efficiently detects all the dominant harmonic components (in this case ~100 kHz and ~280 kHz). Fast Fourier Transform allows detection of PD signals only in the absence of other noise sources, because all the noise components are also visible in the spectrum. Therefore, in a case of measurements under industrial conditions (e.g. in power substation), it is advantageous to replace the classic FFT analysis with the MUSIC algorithm, or similar high-resolution spectral method. This results in enhanced PD detection sensitivity and fault recognition efficiency.

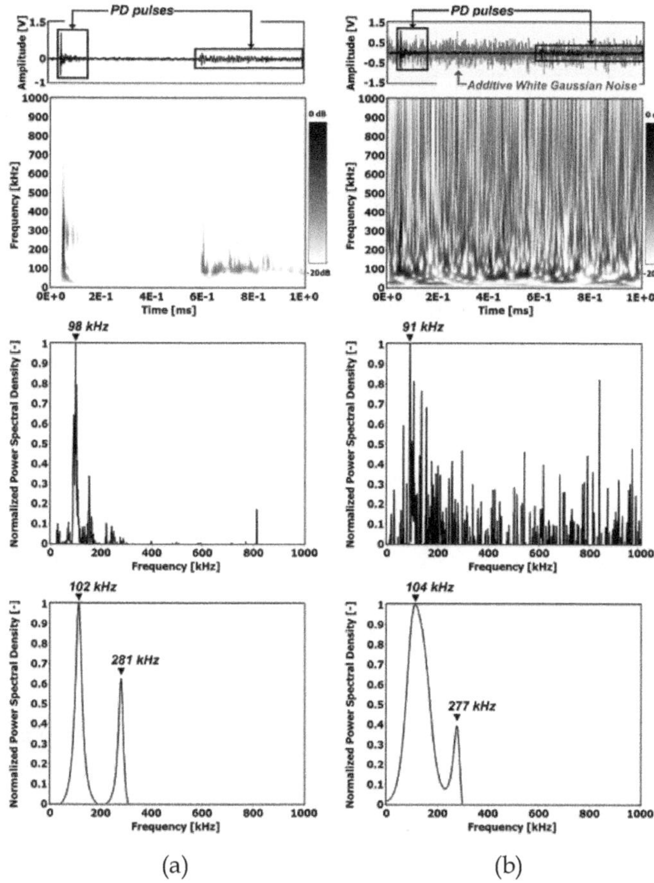

Fig. 4. Time waveform, Continuous Wavelet Transform (CWT) spectrogram, normalized FFT spectrum and normalized MUSIC spectrum determined for: a) PD pulses, b) PD pulses buried in additive white Gaussian noise.

In the case of subspaces methods, the signal model $y(x)$ is defined as the sum of p complex sinusoids (of varied amplitude A_k, frequency F_k and phase φ_k) recorded in additive white noise $n(x)$ (Besson & Stoica, 1996; Hayes, 1996 ; Marple, 1987; Schmidt, 1986):

$$y(x) = s(x) + n(x) = \sum_{k=1}^{p} A_k e^{j(2\pi F_k x + \varphi_k)} + n(x) \tag{1}$$

For the recorded N-sample vector $y = [y(x),...,y(x+N-1)]$ of the noisy signal, equation (1) can be written as:

$$y = Sa + n \tag{2}$$

where n is a zero mean Gaussian white noise vector with variance σ_n^2 and

$$S = [s_1, s_2, ..., s_p] \tag{3}$$

$$s_k = [1, e^{j2\pi F_k}, ..., e^{j2\pi F_k (N-1)}]^T \tag{4}$$

$$a = [A_1, A_2, ..., A_p]^T \tag{5}$$

The first step of the MUSIC algorithm is the calculation of the value of the autocorrelation function

$$R_{yy} = E[yy^H] \tag{6}$$

for the registered signal y, where the exponent H denotes the Hermitian transpose and E denotes the expectation. Next, from the calculated autocorrelation function a $N \times N$ matrix is built. Using matrix notation this operation can be expressed as:

$$R_{yy} = R_{ss} + R_{nn} = SPS^H + \sigma_n^2 I \tag{7}$$

where R_{ss} and R_{nn} are the autocorrelation matrices of the signal and noise respectively, I is a square diagonal identity matrix of dimension $N \times N$ and $P = aa^H = diag[A_1^2, A_2^2, ..., A_p^2]$ is the diagonal matrix.

In the next step of the MUSIC algorithm an eigendecomposition of the autocorrelation matrix R_{yy} is performed. Because the autocorrelation matrix R_{ss} of p complex sinusoids has only p non-zero eigenvalues, one can also express the diagonal matrix of the noise in terms of the eigenvectors of R_{ss}. Then eigendecomposition of the autocorrelation matrix R_{yy} can be written as:

$$\begin{aligned} R_{yy} &= R_{ss} + R_{nn} = \sum_{k=1}^{p} \lambda_k v_k v_k^H + \sigma_n^2 I \\ &= \sum_{k=1}^{p} \lambda_k v_k v_k^H + \sigma_n^2 \sum_{k=1}^{N} v_k v_k^H \\ &= \sum_{k=1}^{p} (\lambda_k + \sigma_n^2) v_k v_k^H + \sigma_n^2 \sum_{k=p+1}^{N} v_k v_k^H \end{aligned} \tag{8}$$

where v_k are the eigenvectors and λ_k are the eigenvalues of the matrix R_{ss} respectively and the expression $(\lambda_k + \sigma_n^2)$ complies the eigenvalues of the matrix R_{yy}.

In this way the calculated eigenvectors and eigenvalues are partitioned into two subsets. The first set of eigenvectors, associated with the p largest eigenvalues, span the *signal subspace* and the second subset of eigenvectors span the *noise subspace* and have σ_n^2 as their eigenvalues. The signal subspace (signal eigenvectors) and the noise subspace (noise eigenvectors) are orthogonal, therefore all p harmonic signal vectors s_i $(s_1, s_2, .., s_p)$ are orthogonal to the noise subspace:

$$s_i^H(f) v_k = \sum_{x=0}^{N-1} v_k(x) e^{j2\pi F_i x} = 0 \qquad k=p+1,..,N \tag{9}$$

Then, the MUSIC spectrum is defined as:

$$P_{MUSIC}(f) = \frac{1}{\sum_{k=p+1}^{N} \left| s^H(f) v_k \right|^2} = \frac{1}{s^H(f) V V^H s(f)} \tag{10}$$

where $s(f) = [1, e^{j2\pi f}, ..., e^{j2\pi(N-1)f}]^T$ is the complex sinusoidal vector and $V = [v_{p+1}, v_{p+2}, ..., v_N]$ is the matrix of eigenvectors of the noise subspace.

4.2 Joint Time-Frequency Analysis (JTFA)

Joint Time-Frequency Analysis is also an effective tool for acoustic emission signal processing (Boczar & Zmarzly, 2004). The application of JTFA allows one not only to observe the changes in amplitude spectrum, but also makes possible detecting in investigated signal narrow-band noises and disturbances. A given type of PD can be described as a three dimensional image (*time-frequency-amplitude*). For joint time-frequency analysis of acoustic emission signals originating from partial discharges, the Continuous Wavelet Transform (CWT) was chosen.

Simplifying, the operational rule of this transform can be defined as an operation of comparison of investigated signal $y(x)$ to elementary components $\gamma_{\tau,a}(x)$ obtained by scaling and shifting of basic wavelet $\gamma(x)$ called *mother wavelet*. The scaling result $\gamma(x)$ of wavelet with the scale coefficient $a > 0$ can be expressed as follows (Antoniadis & Oppenheim, 1995; Holschneider, 1998):

$$\gamma_a(x) = \frac{1}{\sqrt{a}} \gamma\left(\frac{x}{a}\right) \tag{11}$$

Meanwhile the shifting operation of rescaled wavelet $\gamma_a(x)$ and the shifting coefficient is written as follows:

$$\gamma_{\tau,a}(x) = \gamma_a(x - \tau) = \frac{1}{\sqrt{a}} \gamma\left(\frac{x-\tau}{a}\right), \text{ for a} > 0 \tag{12}$$

Then the operation of comparison of calculated basic components $\gamma_{\tau,a}(x)$ with investigated signal $y(x)$ can be defined as:

$$CWT(\tau, a) = \int_{-\infty}^{\infty} y(x) \gamma_{\tau,a}^*(x) dx = \frac{1}{\sqrt{a}} \int_{-\infty}^{\infty} y(x) \gamma^*\left(\frac{x-\tau}{a}\right) dx \tag{13}$$

were $\gamma\left(\frac{x-\tau}{a}\right)$ is scaled with coefficient a and shifted in time by τ, wavelet $\gamma(x)$. CWT value in (τ, a) point expresses similarity (correlation) of signal $y(x)$ to scaled and shifted wavelet (the higher value, the greater similarity), whereas the $*$ symbol denotes the complex conjugate.

Presentation of continuous wavelet transform as a picture of *time-frequency-amplitudes* called a scalogram, and its values are calculated as a square of the module $CWT(\tau, a)$:

$$SCAL(\tau,a)=\left|CWT(\tau,a)\right|^2 \tag{14}$$

After calculating the scale coefficient a to corresponding frequency values the *time-frequency-amplitude* image (spectrogram) is obtained.

5. Partial discharges recorded in oil-paper insulation models

Properly planed and performed investigations conducted on models are important sources of information on electrical withstand of parts of transformer insulation, such as coil-to-coil insulation, clearance to core, space at the bushings etc. But if the conducted investigations should have a cognitive and scientific value, it is necessary to fulfil some basic postulates, such as:

- the construction of models should have the same material structure, shape and proportional geometrical dimensions as the part of the insulation that is modelled; it should reproduce the same mechanism of initiation and development of partial discharges,
- the construction of models should reproduce real distribution of electric field,
- the oil in the model should have similar composition as the real one - content of water, dissolved gases and solid impurities.

The model investigations performed with utilizing the above mentioned conditions, allowed to register the waveforms of AE signals generated by partial discharges typical for oil/paper insulation system of a power transformer (Bengtsson & Jönsson, 1997; Boczar, 2002; Lundgaard, 2000; Elborki et al., 2002; Harrold, 1975; Massingue et al., 2006) (Table 2).

PD1	Surface discharges in uniform electric field, where the normal component of field strength vector is insignificant
PD2	Surface discharges in moderate non-uniform electric field with small normal component of field strength vector
PD3	Surface discharges in non-uniform electric field with large normal component of field strength vector
PD4	Partial discharges in internal gas void
PD5	Partial discharges in gas bubbles in oil
PD6	Pressboard penetrating discharges (puncture)
PD7	Partial discharges from a needle in oil (point-to-plane electrode system)
PD8	Partial discharges from a needle on free potential in oil
PD9	Partial discharges in oil wedge at the oil-pressboard-electrode triple junction
PD10	Turn-to-turn insulation discharges (interturn)

Table 2. The investigated types of partial discharges

During the laboratory research the influence of some factors on AE signals frequency parameters was investigated. The factors, among other things, were: (i) moisture content, (ii) aging degree of pressboard/paper samples, (iii) test voltage polarity, (iv) electrode and insulation samples geometry, as well as (v) type of applied AE sensor (narrow/wideband) (Sikorski, 2006). Furthermore, the pressboard barriers influence on AE attenuation effects and their frequency parameters was examined (Sikorski et al., 2007b).

The presentation of model research outcomes was limited to four types of partial discharges (PD1÷PD4), recognized in power transformer, which test results were shown in the further part of the paper (section 7). Schematic diagrams of electrode systems, in which the discussed kinds of PD's were generated, are presented in the figures 5-8.

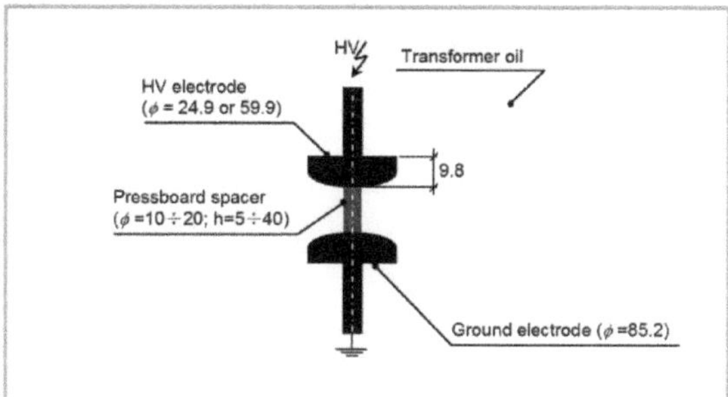

Fig. 5. Schematic diagrams of electrode system, in which were generated surface discharges in uniform electric field, where the normal component of field strength vector is insignificant (PD1).

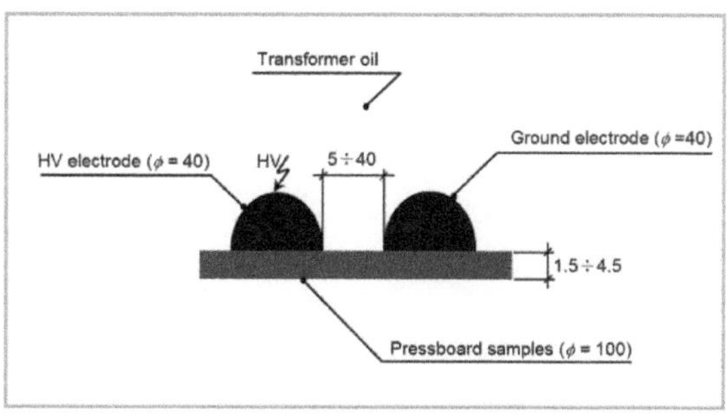

Fig. 6. Schematic diagrams of electrode system, in which were generated surface discharges in moderate non-uniform electric field with small normal component of field strength vector (PD2).

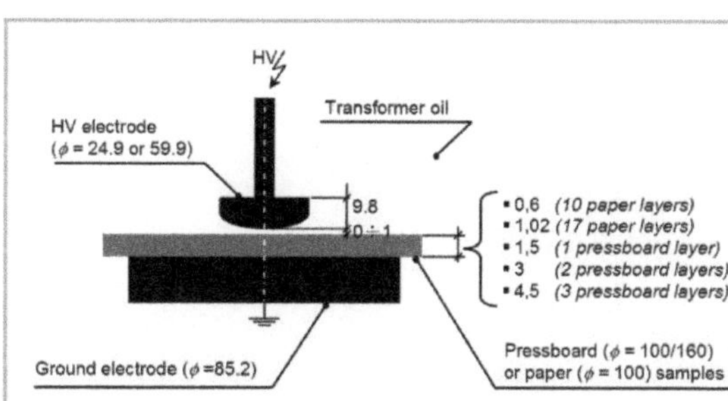

Fig. 7. Schematic diagrams of electrode system, in which were generated surface discharges in non-uniform electric field with large normal component of field strength vector (PD3).

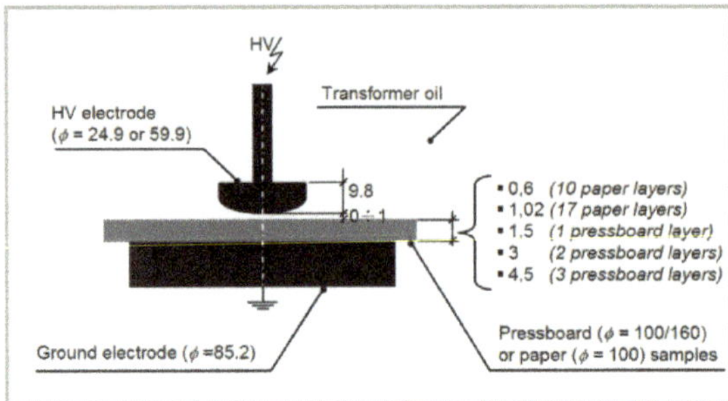

Fig. 8. Schematic diagrams of electrode system, in which were generated partial discharges in internal gas void (PD4).

The laboratory investigation was conducted in a specially designed tank model in which the electrode systems with oil/paper insulation samples were placed (Fig. 9). The model casing of dimension equal to 600x300x500 *mm* was made of stainless steel and the thickness of sheet metal was 3 *mm*. Additionally, the side walls were modified for installation of UHF probes inside the tank.

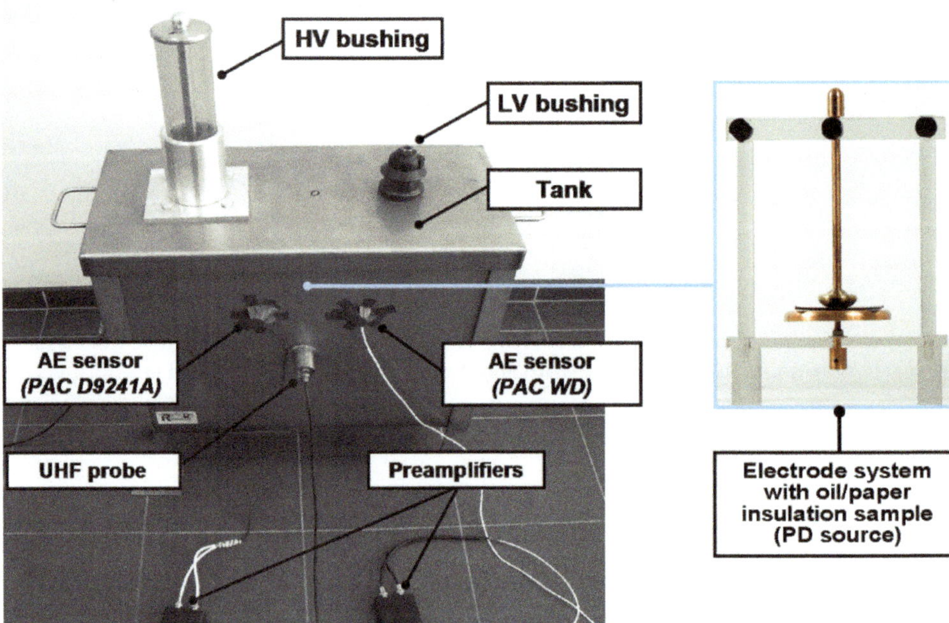

Fig. 9. The model of the transformer tank applied in laboratory investigation.

Registration and analysis of AE pulses generated by partial discharges were carried out in measuring system that consists of the following components: wideband (PAC WD) and narrowband (PAC D9241A) acoustic emission sensors, preamplifiers (40dB gain), signal conditioning unit with high-pass filters and amplifiers (signal bandwidth from 20 kHz to

1000 kHz), digital oscilloscope and portable industrial computer with ultra-high speed (20 MS/s) data acquisition card (Adlink PCI-9812). The frequency response characteristics of the applied sensors were shown in the Fig. 10.

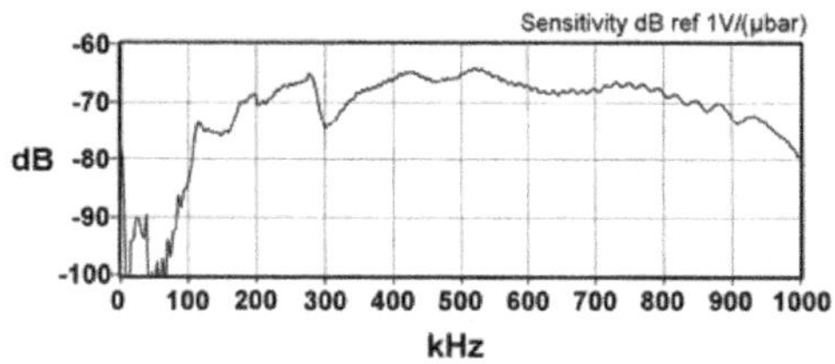

(a)

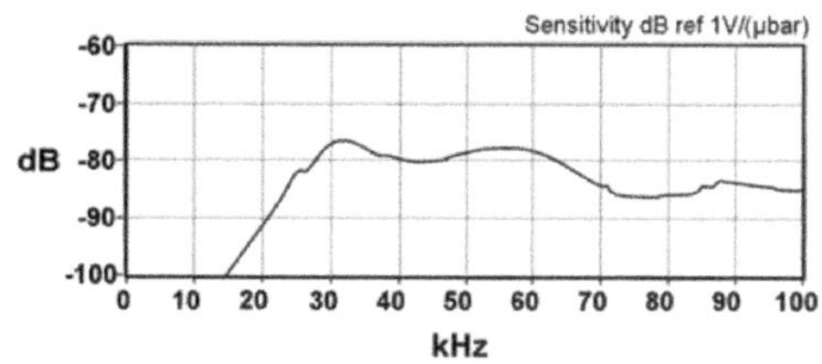

(b)

Fig. 10. Frequency response of the: a) wideband sensor type WD and b) narrowband sensor type D9241A.

To verify the investigation results obtained with the use of AE method, the registration of PD was simultaneously conducted with AE sensors, UHF probe and wideband PD detector according to IEC 60270.

6. Partial discharge patterns (PD-Patterns)

For all the AE pulses registered, the high-resolution MUSIC spectra were calculated. Their analysis showed that each investigated PD type has different dominant frequency components characteristic for itself only. This property allowed establishing the PD patterns.

The database prepared with the PD-patterns in the form of high-resolution MUSIC spectra was implemented to specialized computer system (*PD-Identifier*) realizing the procedures of automatic detection and recognition of discharges occurring in oil-paper insulation (Sikorski et al., 2007a).

Efficiency improvement of PDs' detection and identification process can be achieved by conducting additional analysis in joint time-frequency domain. To attain this goal for the investigated kinds of PDs, processed with CWT algorithm, the *time-frequency-amplitude* images were convoluted.

The patterns of four investigated PD types (PD1÷PD4), which were registered simultaneously with the WD wideband sensor and D9241A narrowband sensor (especially dedicated for power transformer monitoring), are presented in the Figures 11-14.

The analysis results of the surface discharges in uniform electric field, where the normal component of field strength vector is insignificant (PD1) are presented in the Figure 11. Comparing time waveforms, one can notice that the amplitude the signal registered with the D9241A has slightly higher amplitude, although oscillations of the signal registered with the wideband sensor are more regular. The joint time-frequency analysis showed that both sensors had registered the AE signal in the range of ~25÷40 kHz. In next step, the high-resolution spectral analysis with the use of the MUSIC method allowed to define the main harmonic component frequency more precisely (as 28 kHz). The author (Boczar, 2002) obtained exactly the same results while registering this type of discharges with the use of other sensors (Brüel&Kjær 8312, ENDEVCO 752-10 Isotron, ENDEVCO 7259A-10 Isotron).

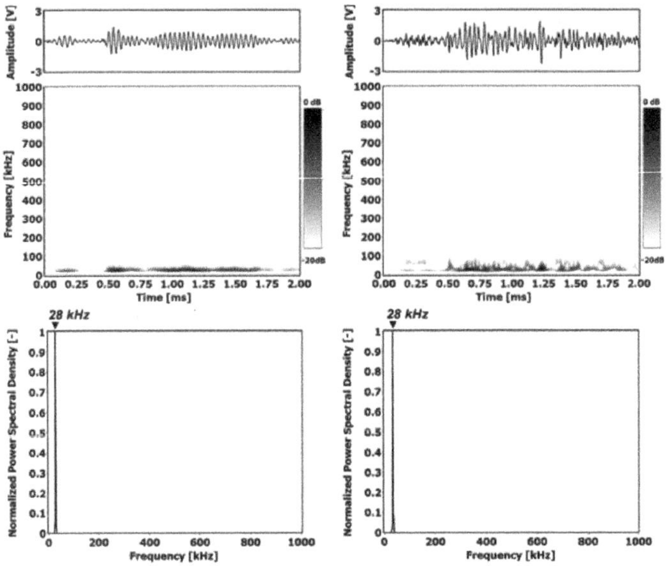

(a) (b)

Fig. 11. Time waveform, continuous wavelet transform (CWT) spectrogram and normalized MUSIC spectrum prepared for acoustic emission signals generated by surface discharges in uniform electric field, where the normal component of field strength vector is insignificant (PD1), which were registered with wideband sensor type WD (a) and narrowband sensor type D9241A (b).

The model research proved that the narrowband sensor is more sensitive in detection of surface discharges generated in moderate non-uniform electric field with small normal component of field strength vector (PD2) or in non-uniform electric field with large normal component of field strength vector (PD3). On the CWT spectrograms one can notice that in case of the WD sensor, the presence of normal component of field strength vector results in occurrence of harmonic components in the range of ~90÷110 kHz (Fig. 12a and Fig 13a). In turn, in case of narrowband sensor the additional harmonic components are transferred in the range of ~50÷100 kHz (Fig. 12b and Fig. 13b).

This divergence in the obtained PD-patterns results from different frequency response of both sensors. As one can see on frequency response characteristics (Fig. 10), the WD wideband sensor, compared to narrowband sensor, is not capable to transfer the acoustic signal in the range of ~50÷90 kHz.

One may subsequently state that the parameters characterizing an AE sensor modify the signal and resulting CWT pattern. The measured results confirm that the PD pattern strongly depends on the type of applied acoustic sensor.

Although the narrowband sensors are typically characterized by high sensitivity, they are less universal than wideband sensors due to limited bandwidth. Wideband sensors have advantage in case of detecting certain types of PD e.g. discharges in oil (from sharp point), discharges in gas bubbles in oil and discharges in internal gas voids (voids within delaminations).

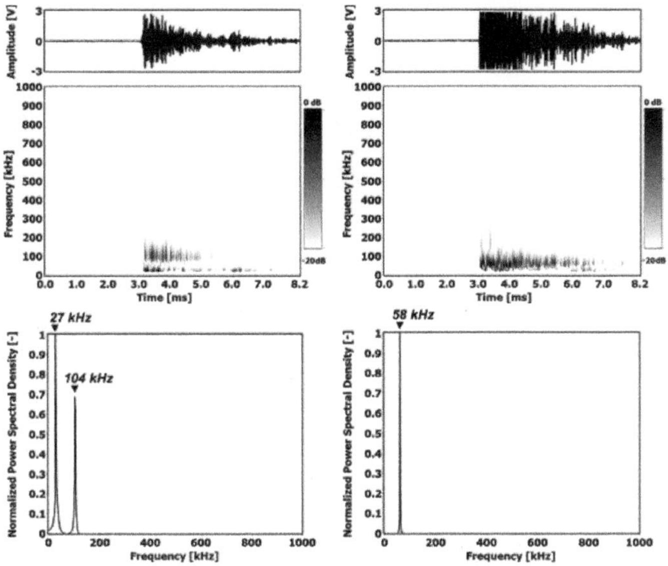

(a) (b)

Fig. 12. Time waveform, continuous wavelet transform (CWT) spectrogram and normalized MUSIC spectrum prepared for acoustic emission signals generated by surface discharges in moderate non-uniform electric field with small normal component of field strength vector (PD2), which were registered with wideband sensor type WD (a) and narrowband sensor type D9241A (b)

The analysis of AE pulses generated by PD in internal gas void (PD4) was shown in the Figure 14. During the laboratory research the AE signals of that type were registered only by WD wideband sensors (compare time waveforms in the Fig. 14a and the Fig. 14b), because the D9241A sensor does not register the signals of frequency higher than 200 kHz. Time duration of this burst is very short (< 0.15 ms) and the dominating frequency band is in the range of ~270÷380 kHz.

7. Partial discharge detection, recognition and location in transformer

This section presents the example of power transformer diagnostics with the use of the AE method. During the research the sensitivity of PD detection and the efficiency of their recognition and location were verified.

7.1 Investigation object – mobile transformer

The mobile transformer designated to operate with two reconnectable voltages on HV side and two voltages at LV side (see Table 3 for rated data) was built with series-parallel (S/P) switches (see Fig. 15).

The unit was subjected to standard dielectric testing per CSA Standard C88-M90. The dielectric tests included: (i) lightning impulse test at both series and parallel connections;

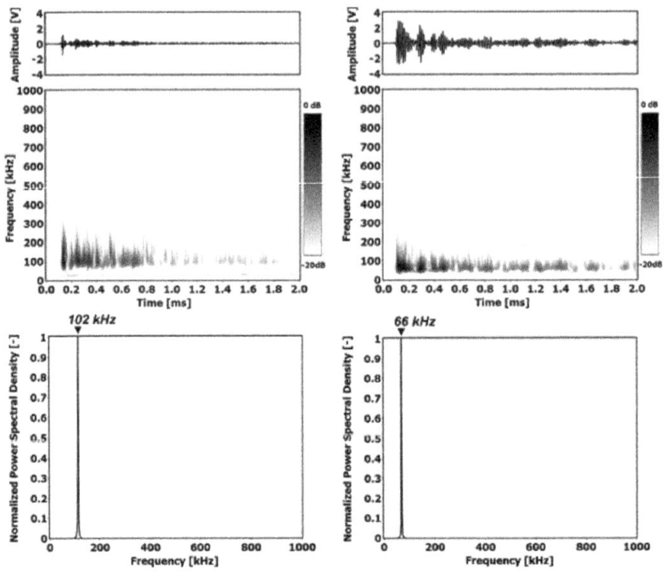

(a) (b)

Fig. 13. Time waveform, continuous wavelet transform (CWT) spectrogram and normalized MUSIC spectrum prepared for acoustic emission signals generated by surface discharges in non-uniform electric field with large normal component of field strength vector (PD3), which were registered with wideband sensor type WD (a) and narrowband sensor type D9241A (b)

(ii) applied voltage test, (iii) induced voltage test with simultaneous measurement of partial discharge level at parallel connection of HV (voltage to ground: 71.3kV for 7200 cycles and 62.9 kV for 1 hour) and at series connection (HV voltage to ground: 142.5 kV for 7200 cycles and 125.7 kV for 1 hour).

The transformer experienced partial discharge problems during the induced voltage test in a series connection. The PD level (the apparent charge value) during the test in a series connection (i.e. with a nominal voltage U_n=66kV) at terminal H1 was high at 60% of U_n already, reaching 800÷1000 pC, while phases 2 and 3 were showing low levels, below 500 pC at 150% of U_n. It created opportunity to employ the AE measurements for localization and recognition of PD source.

Rated voltage	132 x 66 – 25Y x 12.5Y [kV]
BIL	550 x 350-150 x 125 [kV LI]
Power rating	18 MVA
Cooling method	ODAF
Temperature rises	oil: 65°C; windings: average 75°C/ hotspot 80°C
Power frequency	60 Hz
DETC	+/- 10% HV

Table 3. Rated data of mobile transformer under investigation

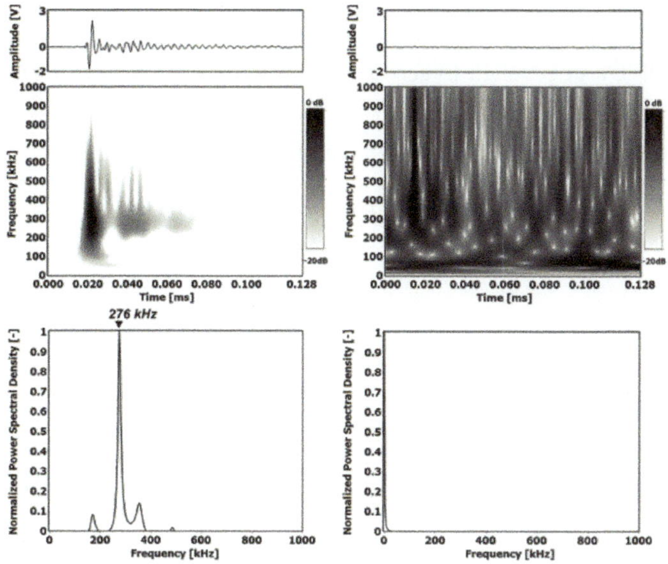

(a) (b)

Fig. 14. Time waveform, continuous wavelet transform (CWT) spectrogram and normalized MUSIC power spectrum density prepared for acoustic emission signals generated by partial discharges in internal gas void (PD4), which were registered with wideband sensor type WD (a) and narrowband sensor type D9241A (b)

(a) (b)

Fig. 15. The investigated mobile transformer: a) the LV series-parallel switch visible in front (3 phase switch), b) the HV series-parallel switch (with 2 poles per phase) in front of the winding assemblies

7.2 Partial discharge detection and location – conventional electrical method

The investigation tests were performed with 4 combinations of HV and LV connections: (i) HV in series, LV in parallel, (ii) HV in series, LV in series, (iii) HV in parallel, LV in series, (iv) HV in parallel, LV in parallel. Every time when HV side was in series connection, the PD level was elevated on the terminal H1.

The electric PD pattern (see Fig. 16) with maximum value of PD pulses appearing close to maximum of the voltage waveshape where implying the location of a PD source close to the HV 1 line, HV 1 terminal, or metal parts of HV 1 circuit, physically in proximity of a two-pole series-parallel switch on phase HV 1 (see Fig. 17).

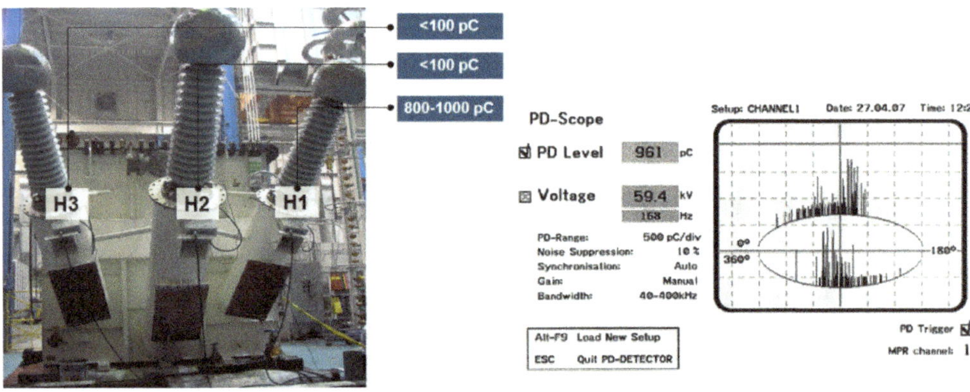

Fig. 16. The results of the PD detection with the use of conventional electric method and typical electric PD pattern registered at terminal H1 (Q_{max}=961 pC).

Fig. 17. Approximate location of PD source at the S/P switch on phase HV 1 (red rectangle).

7.3 Partial discharge detection, recognition and location – acoustic emission method

The location of partial discharges with the use of acoustic emission method may be conducted by: (i) measurement of AE pulses amplitude in different distances from PD source (auscultatory method), or (ii) AE pulses arrival timings at sensors fixed in various places of the transformer tank (triangulation method) (Markalous et al., 2008).

The auscultatory method applied in this investigation is the easiest way to locate the PD sources (IEEE, 2007). Despite the fact that its accuracy is lower than that of triangulation method's, it may be very efficient and useful in cases, when (i) the measurement time is limited by the time of voltage tests, (ii) during the voltage test the phase with PD source has been detected with the use of wideband electrical method, (iii) acoustic measurement is conducted with one-channel-setup, (iv) the detail construction of the investigated transformer is known, (v) the investigated transformer is relatively small.

To avoid the measurement errors and increase the PD amplitude measurement accuracy, the measurement channel calibration was conducted each time after fixing the sensor on the tank wall.

Among the 8 points forming the measurement grid, the highest AE pulse amplitudes due to internal PDs in a transformer were registered in point #5 and #3 (Fig. 18b).

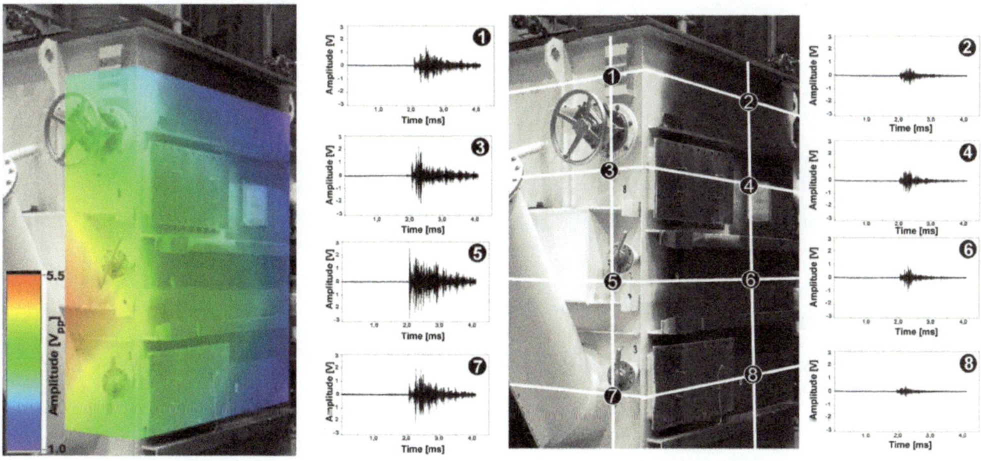

(a) (b)

Fig. 18. The results of PD sources location with the use of auscultatory method: acoustic map (a) and AE waveforms registered in selected places of power transformer tank (b)

Three different types of AE signals were registered in the investigated mobile transformer (Fig. 19). The comparative analysis with PD-patterns obtained in model research (Fig. 11-14) suggested that following types of discharges could have occurred in a transformer:

- surface discharges in uniform electric field, where the normal component of field strength vector is insignificant (Fig. 19a),
- surface discharges in moderate non-uniform electric field with small normal component of field strength vector (Fig. 19b),

- partial discharges in internal gas void (Fig. 19c).

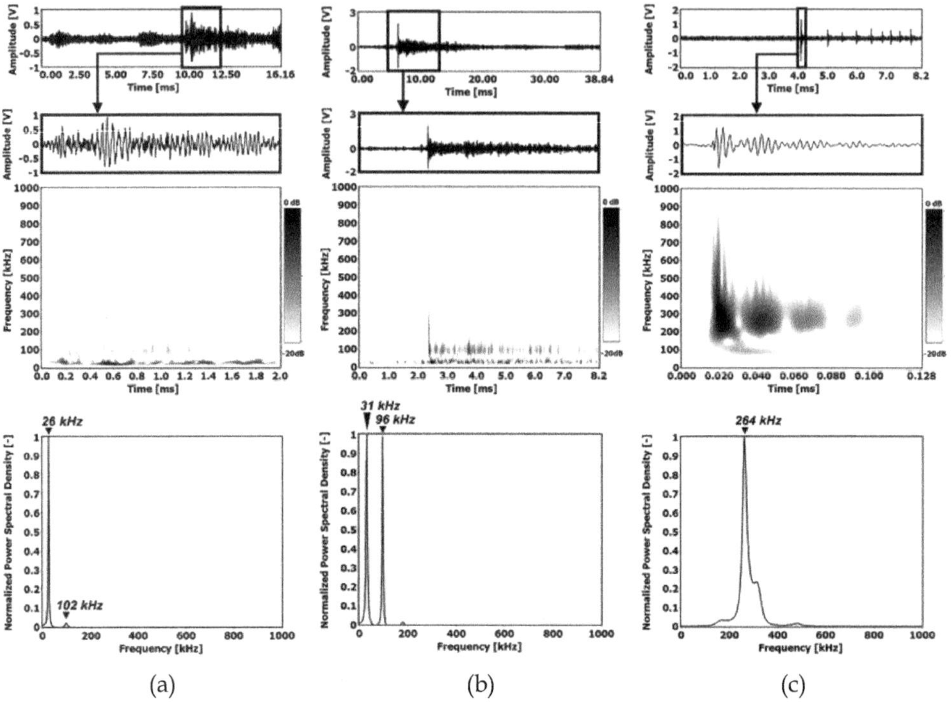

Fig. 19. The analysis of AE pulses registered in the investigated transformer (using PAC WD wideband sensor), which were recognized as partial discharges of the PD1 (a), PD2 (b) and PD4 type (c).

7.4 Disassembly and findings

The disassembly of the transformer resulted in finding two possible sources of the PD problem:

1. the shaft of DETC switch (laminated tube made of phenolic resin) had delamination (voids) with internal PD traces (Fig. 20),
2. there were tracking traces along the edge of supporting beam (made of phenolic resin board) of a S/P switch (Fig. 21).

For these both cases of identified PD sources the electric field analysis was performed at the affected areas, i.e. at DETC and S/P switches. The electric stress at the switches was low, insulating parts passed both creep and strike criteria: the field intensity was not high enough for PD inception. Also, the point stresses were found to be low, within acceptable limits: E_{max}=7.26 *kV/mm* for a DETC switch (Fig. 22a) and E_{max}=5.05 *kV/mm* for a S/P switch (Fig. 22b), both values calculated at 150% U_n, while the PD problem was already active at the voltage as low as 60% of U_n. It indicated that the design is correct and the problem may be related to a material defect, or imperfection.

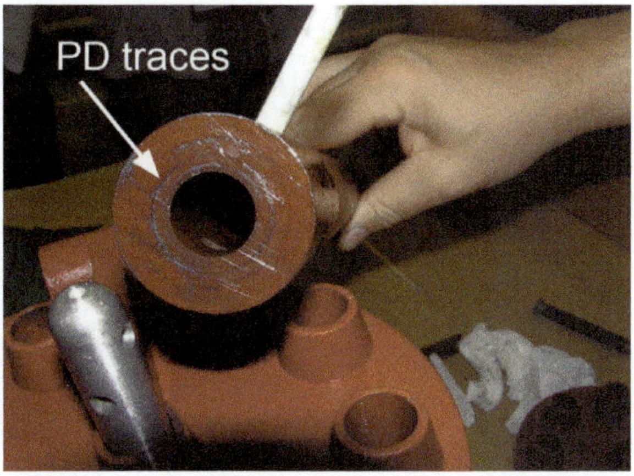

Fig. 20. The defected (delamination) shaft of DETC switch with internal PD traces.

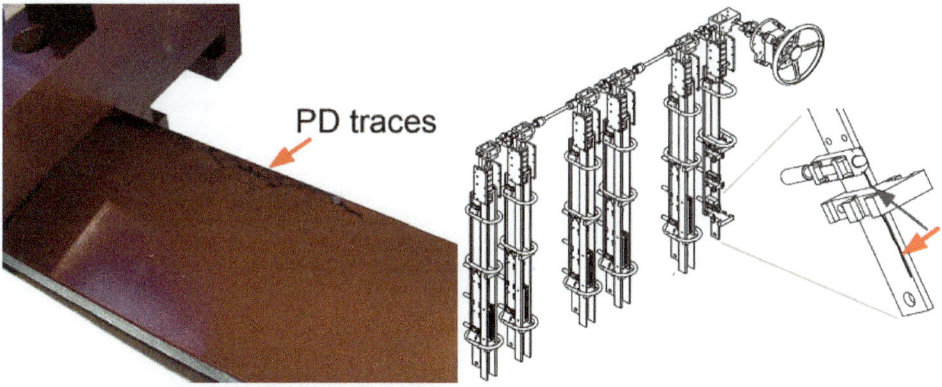

Fig. 21. The supporting beam of S/P switch with visible PD traces.

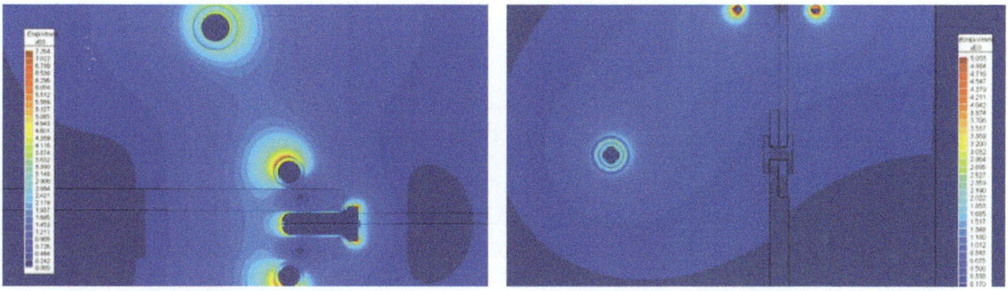

(a) (b)

Fig. 22. Electric filed stress distribution: a) E_{max}=7.26 *kV/mm* for a DETC switch, b) E_{max}=5.05 *kV/mm* for a S/P switch.

Therefore, the S/P switch was disassembled and the tracking traces were found on its pressboard support. The pressboard support was subjected to X-rays and a metallic contaminant was found, buried at the edge of the board where PD tracking was originating from (Fig. 23).

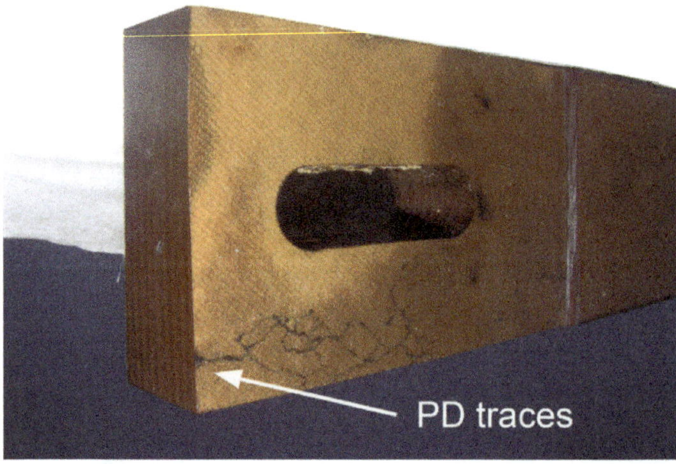

Fig. 23. The location of embedded metal particle under the surface of pressboard beam of S/P switch (the PD tracking originated from this area).

Based on recorded test results and physical evidence, the following scenario for the PD problem was determined:

1. in a series connection a high electric stress was developed at the S/P switch - between the pressboard support and the tank;
2. this electric stress led to initiation of PD at the edge of metallic contaminant embedded in the board;
3. next, the surface discharge developed along the edge of pressboard beam;
4. the progressing surface discharges resulted in degradation of dielectric properties of mineral oil, giving rise to PD in adjacent insulating elements of S/P switch as well as DETC switch.

8. Conclusions

The example of efficient detection and recognition of partial discharge sources using acoustic emission method with advanced signal processing was presented in the chapter.

High-resolution Multiple Signal Classification (MUSIC) method, chosen for the frequency analysis, effectively detects in registered AE signal (also buried in noise) frequency components corresponding to partial discharges. In turn, joint time-frequency analysis using Continuous Wavelet Transform (CWT) allows to detect multi-source discharges of different types, as well as narrowband noises.

On the basis of selected digital signal processing methods and complex model research, the database with PD-Patterns was prepared. An important conclusion of this model research is that each of the investigated insulation defects (PD sources) generates repeatable and characteristic (unique) acoustic emission signals. Moreover, it was shown that the shape of PD-Patterns strongly depends on the type of the applied sensor.

The increase of PD detection and recognition efficiency can be achieved by simultaneous application of narrow- (20÷100 kHz) and wideband (20÷500 kHz) sensors. Due to that, a high sensitivity of measurement with a possibility of detecting all PD types is combined.

The measurements performed in the power transformer factory showed that PD-patterns acquired during a laboratory research can be successfully employed in detection and recognition of discharges occurring in real power transformers.

9. References

Antoniadis, A. & Oppenheim, G. (1995). *Wavelets and Statistics,* Springer-Verlag, ISBN 9780387945644, New York, USA

ASTM E650-97 (2007). *Standard Guide for Mounting Piezoelectric Acoustic Emission Sensors,* American Society for Testing and Materials, 2007

Bengtsson, T. & Jönsson, B. (1997). Transformer PD diagnosis using acoustic emission technique, *10th International Symposium on High Voltage Engineering,* Montreal, Canada, 25-29 August 1997

Besson, O. & Stoica, P. (1996). Analysis of MUSIC and ESPRIT Frequency Estimates for Sinusoidal Signals with Lowpass Envelopes, *IEEE Transactions in Signal Processing,* Vol.44, No.9, pp. 2359-2364, ISSN 1053-587X

Boczar, T. (2001). Identification of a specific type of PD from acoustic emission frequency spectra, *IEEE Transactions on Dielectrics and Electrical Insulation,* Vol.8, No.4, pp. 598-606, ISSN 1070-9878

Boczar, T. (2002). Acoustic emission of multisource surface partial discharges generated in insulating oil, *6th Symposium on High Voltage Engineering IW-2002,* pp. 14-17, Poznan, Poland, May 2002

Boczar, T. & Zmarzly, D. (2004). Application of wavelet analysis to acoustic emission pulses generated by partial discharges, *IEEE Transactions on Dielectrics and Electrical Insulation,* Vol.11, No.3, pp. 433-449, ISSN 1070-9878

Elburki, M.A., Crossley, P.A.; Wang, Z.D.; Darwin, A. & Edwards, G. (2002). Detection and Characterization of Partial Discharges in Transformer Defect Models, *IEEE Power Engineering Society Summer Meeting,* pp. 405-410, July 2002

Feser, K. (2003). Developments and challenges in high voltage engineering, *13th International Symposium on High Voltage Engineering,* p. 3, ISBN 978-9077017791, Delft, Netherlands, August 2003

Gulski, E.; Smith, J.J.; Brooks, R. & Turner, M. (1999). Experiences with Digital Analysis of Discharges in High Voltage Components, *IEEE Electrical Insulation Magazine,* Vol.15, No.3, (May/June 1992), pp. 15-24, ISSN 0883-7554

Gulski, E. & Smitt, J.J. (2007). Condition Assessment of Transmission Network Infrastructures (April 2007), *Power Tech, 2007 IEEE Lausanne,* pp. 2189–2194, Lausanne, Switzerland, July 2007

Harrold, R.T. (1975). Ultrasonic spectrum signatures of under-oil corona sources, *IEEE Transactions on Electrical Insulation*, Vol.10, No.4, pp. 109-112, ISSN 0018-9367

Hayes, M.H. (1996). *Statistical Digital Signal Processing and Modeling*, John Wiley & Sons Inc., ISBN 978-0-471-59431-4, San Francisco, USA

Holschneider, M. (1998). *Wavelets: an analysis tool*, Clarendon Press, ISBN 0-19-850521-3, Oxford, UK

IEEE Guide for the Detection and Location of Acoustic Emissions from Partial Discharges in Oil-Immersed Power Transformers and Reactors, C57.127-2007, Institute of Electrical and Electronics Engineers, August 2007

Kia, S.H.; Hena, H. & Capolino, G.A. (2007). High-Resolution Frequency Estimation Method for Three-Phase Induction Machine Fault Detection, *IEEE Transactions on Industrial Electronics*, Vol.54, No.4, (August 2007), pp. 2305-2314, ISSN 0278-0046

Kuffel, E.; Zaengl, W.S. & Kuffel, J. (2000). *High Voltage Engineering Fundamentals (Second edition)*, Butterworth-Heineman Publication, ISBN 978-0-7506-3634-6, Oxford, UK

Lobos, T.; Kozin, T. & Leonowicz, Z. (2000). High Resolution Spectrum Estimation Methods for Signal Analysis in Power Electronics and Systems, *2000 IEEE International Symposium on Circuits and Systems*, pp. 553-556, Geneva, Switzerland, May 2000

Lobos, T.; Leonowicz, Z. & Rezmer, J. (2003). Advanced Spectrum Estimation Methods for Signal Analysis in Power Electronics, *IEEE Transactions on Industrial Electronics*, Vol.50, No.3, pp. 514-519, June 2003, ISSN 0278-0046

Lundgaard, L.E. (1992a). Partial Discharge - Part XIII: Acoustic Partial Discharge Detection - Fundamental Considerations, *IEEE Electrical Insulation Magazine*, Vol.8, No.4, (July/August 1992), pp. 25-31, ISSN 0883-7554

Lundgaard, L.E. (1992b). Partial Discharge - Part XIII: Acoustic Partial Discharge Detection - Fundamental Considerations, *IEEE Electrical Insulation Magazine*, Vol.8, No.5, (September/October 1992), pp. 34-43, ISSN 0883-7554

Lundgaard, L.E. (2000). Partial discharges in transformer insulation, CIGRE Task Force 15.01.04, Paper 15-302, Paris, France, 2000

Ma, X.; Zhou, C. & Kemp, I.J. (2002). Interpretation of wavelet analysis and its application in partial discharge detection, *IEEE Transactions on Dielectrics and Electrical Insulation*, Vol.9, No.3, (June 2002), pp. 446-457, ISSN 1070-9878

Markalous, S. ; Grossmann, E. & Feser, K. (2003). Online Acoustic PD-Measurements of Oil/Paper-Insulated Transformers - Methods and Results, *13th International Symposium on High Voltage Engineering*, p. 324, ISBN 978-9077017791, Delft, Netherlands, August 2003

Markalous, S.; Tenbohlen, S. & Feser, K. (2008). Detection and location of partial discharges in power transformers using acoustic and electromagnetic signals, *IEEE Transactions on Dielectrics and Electrical Insulation*, Vol.15, No.6, pp. 1576–1583, ISSN 1070-9878

Marple, S.L. (1987). *Digital Spectral Analysis with Applications*, Prentice Hall Inc., ISBN 0-132-14149-3, New Jersey, USA

Massingue, F.; Meijer, S.; Agoris, D.; Smit, J.J. & Lopez-Roldan, J. (2006). Partial Discharge Pattern Analysis of Modeled Insulation Defects in Transformer Insulation, *Conference Record of the 2006 IEEE International Symposium on Electrical Insulation,* (June 2006), pp. 542-545, ISSN 1089-084X

Sakoda, T.; Arita, T.; Nieda, H. & Ando K. (1999). Analysis of Acoustic Emissions Caused by the Partial Discharge in the Insulation Oil, *Proceedings of 13th International Conference on Dielectric Liquids (ICDL '99),* ISBN 0-7803-4759-5, Nara, Japan, July 20-25, 1999

Schmidt, R.O. (1986). Multiple Emitter Location and Signal Parameter Estimation, *IEEE Trans. Antennas Propagation,* Vol.34, No.3, pp. 276-280, ISSN: 0018-926X

Shim, I.; Soraghan, J.J. & Siew, W.H. (2000). Digital signal processing applied to the detection of partial discharge: an overview, *IEEE Electrical Insulation Magazine,* Vol.16, No.3, (May/June 2000), pp. 6-12, ISSN 0883-7554

Sikorski, W. & Siodla, K. (2005). Identification of partial discharge sources in high voltage insulating systems using acoustic emission method, *14th International Symposium on High Voltage Engineering*, Beijing, China, 2005

Sikorski, W. (2006). Defects recognition in chosen insulation systems based on acoustic emission signal analysis, Ph.D. dissertation, Institute of Electric Power Engineering, Poznan University of Technology, Poznan, Poland, May 2006

Sikorski, W.; Staniek, P. & Siodla, K. (2006). The influence of pressboard barriers on signal suppression of acoustic emission generated by partial discharges, *13th International Workshop on High Voltage Engineering,* p. 7, Kleinwalsertal, Austria, September 2006

Sikorski, W. ; Siodla, K. & Staniek, P. (2007a). Identification of partial discharge sources in high voltage insulating systems using acoustic emission method, *15th International Symposium on High Voltage Engineering*, Ljubljana, Slovenia, 2007

Sikorski, W.; Staniek, P. & Siodla K. (2007b). The influence of pressboard barriers on acoustic emission signal generated by partial discharges in a power transformer, *Poznan University of Technology Academic Journals. Electrical Engineering,* No. 56, 2007, ISSN 1897-0737

Sikorski, W.; Staniek, P. & Walczak K. (2007c). Joint UHF/AE Method of Partial Discharges Detection and Identification, International Conference on Advances in Processing, Testing and Application of Dielectric Materials APTADM'2007, Wroclaw, 2007

Sikorski, W. & Walczak, K. (2010). Online Condition Monitoring and Expert System for Power Transformers, *ARWtr2010 - International Advanced Research Workshop On Transformers,* Santiago de Compostela, Spain, 3-6 October, 2010

Varlow, B.R.; Auckland, D.W.; Smith, C.D. & Zhao J. (1999). Acoustic emission analysis of high voltage insulation, *IEE Proceedings in Science, Measurement and Technology,* Vol.146, No.5, pp. 260-263, ISSN 1350-2344

Yang, L. & Judd, M.D. (2003). Recognizing multiple partial discharge sources in power transformers by wavelet analysis of UHF signals, *IEE Proceedings - Science, Measurement and Technology,* Vol.150, No.3, pp. 119-127, May 2003

Zhang, X.; Zhang, L.; Yue, B. & Xie, H. (2003). Investigation on Aging of Stator Bar Insulation based on JTFA of PD Pulse Waveforms, *Proceedings of the 7th International Conference on Properties and Applications of Dielectric Materials*, pp. 270-273, Nagoya, Japan, June 2003

6

Identifying and Monitoring Evolving AE Sources

Rúnar Unnþórsson
University of Iceland
Iceland

1. Introduction

Fatigue is a stochastic process influenced by several random factors such as material, manufacturing and in-service variations. Due to the uncertainties involved, the true system cannot be accurately represented by a mathematical fatigue model. By combining condition monitoring techniques with fatigue modelling, critical material degradation processes can be identified, failure predicted, and preventive actions planned. This approach is known as condition-based maintenance. However, by using effective condition based maintenance, instead of corrective (after failure) or preventive (calendar-based) maintenance, substantial savings can be made. The savings will be in the form of extended part life, reduced number of unexpected breakdowns, lower risk of secondary failure, and increased safety.

Recent years have seen an increasing interest in the use of carbon fibres for applications in the automotive, aerospace and biomedical industries. Their popularity stems largely from their advantageous material properties such as high strength-to-weight ratio, fatigue strength, corrosion and heat resistance. Despite these properties, damage in Carbon Fibre-Reinforced Polymer (CFRP) composites develops early in service (Degrieck & Paepegem, 2001; Dzenis & Qian, 2001; Halverson et al., 1997) and continues to accumulate throughout the service life. The fatigue tolerance can be attributed to resistance to inhomogeneous damage growth, which is a property of highly inhomogeneous materials (Halverson et al., 1997). The high damage tolerance of composites means that composites are able to meet their in-service requirements for a prolonged period of time while damages accumulate and grow. Consequently, there is a definite advantage in being able to detect, monitor and evaluate individual damage mechanisms before the failure of a composite.

Acoustic Emission can be used to detect delamination, matrix cracking, debonding, fibre cracking and fibre pull-out (Giordano et al., 1998; Green, 1998; Nayeb-Hashemi et al., 1999; Tsamtsakis et al., 1998; Wevers, 1997). However, the high sensitivity of the AE technique means that the measured AE signal may contain a high number of AE transients from sources located both in the composite and the environment. The sources in the composite include damage growth, rubbing of crack surfaces and friction between the fibres and the matrix due to their different material properties. The varying material properties will result in an anisotropic speed of propagation (Duesing, 1989). In addition, reflection and attenuation of the AE waves add to the complexity. Attenuation can be caused by geometric spreading, dispersion, internal friction and scattering (Prosser, 1996). Furthermore, the AE waves from damage growth can be buried in the AE generated by the friction and rubbing of crack surfaces (Mouritz, 2003). As a result, multiple AE transients with varying amplitude, duration, and frequency can be emitted simultaneously. Furthermore, the values of the AE signal features from cumulated damage usually fall in the same range as those that result from damage growth (Dzenis & Qian, 2001; Tsamtsakis et al., 1998).

An intuitive approach to condition monitoring using AE signals is to keep track of one or more waveform parameters, which characterize the AE from the source of interest. Each parameter, however, will follow a probability distribution, which changes when the source (damage) changes. Because of the parameter fluctuations and the high rate of AE with similar parameter values, waveform parameters alone are not sufficient for distinguishing between sources. Additional indication is therefore needed. For an example, if an AE source always emits an AE at the same load level, then the load level of AE occurrence is sufficient to distinguish between the sources. However, when a source evolves, e.g. a growing delamination crack, the load level of AE occurrence changes.

This chapter presents a methodology for processing, presenting, and quantifying AE data. It was designed for the purpose of identifying and tracking locations of multiple evolving AE sources in CRFP subjected to cyclic loading using a cyclic reference signal. The methodology can also be used for normal AE health monitoring of materials/machinery and is neither limited to AE signals nor to periodic reference signals. The chapter also presents an AE hit pattern feature which is made by fusing and coding AE hit-based features with timing information, i.e. the inter-spike intervals (ISI). The methodology was presented in (Unnthorsson et al., 2007a) and the feature was studied in (Unnthorsson et al., 2007b; 2008b).

The remaining of the chapter is organized as follows. Sections 2 and 3 introduce and explain, respectively, the methodology and the AE hit pattern feature. Then, an experimental case study is presented in section 4 and in section 5 both the methodology and the AE hit pattern feature are illustrated using the experimental data. Finally, section 6 concludes this chapter and suggests future research topics.

2. Monitoring evolving AE sources

In order to identify evolving AE sources in a signal which contains high AE activity from many different sources, e.g. evolving damages, non-evolving damages, friction, machinery and more, a set of procedures and signal processing methods need to be combined. In this section a methodology for this purpose is introduced. By using this methodology one can identify and locate interesting AE signals for further study, or for tracking, which otherwise would be difficult to accomplish due to the overwhelming number of sources with similar characteristics. Figure 1 shows a schematic overview of the methodology. It is divided into four steps which will be described and discussed in this section. The four steps are: segmentation, band-pass filtering, feature extraction and visualization.

2.1 First step: segmentation

Most professional AE systems offer the possibility of acquiring external parameters such as pressure, strain, load and displacement. If these parameters, or reference signals, are acquired simultaneously with the AE then any of them can be used as reference signals. By segmenting the AE signal so that all segments are exactly one cycle, of equal length and start at the same

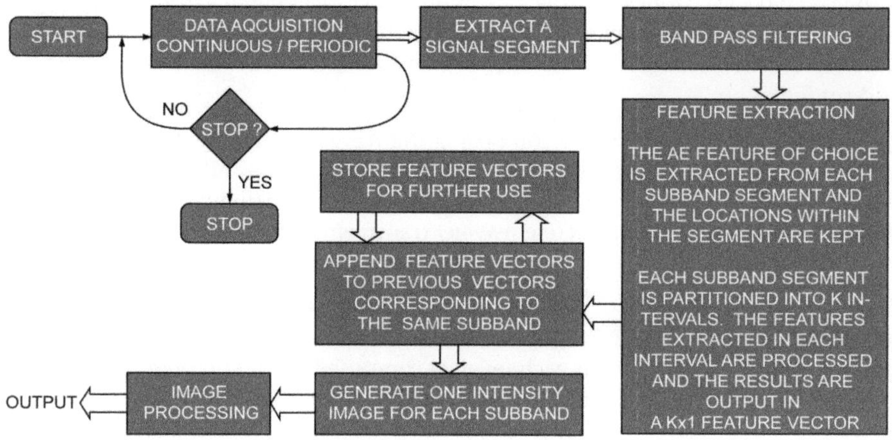

Fig. 1. Schematic overview of the methodology.

reference point of the reference signal then it is possible to compare segments and identify and monitor evolving AE sources. The segmentation, illustrated on the left side of in Fig. 2, can be applied to both periodic and aperiodic signals.

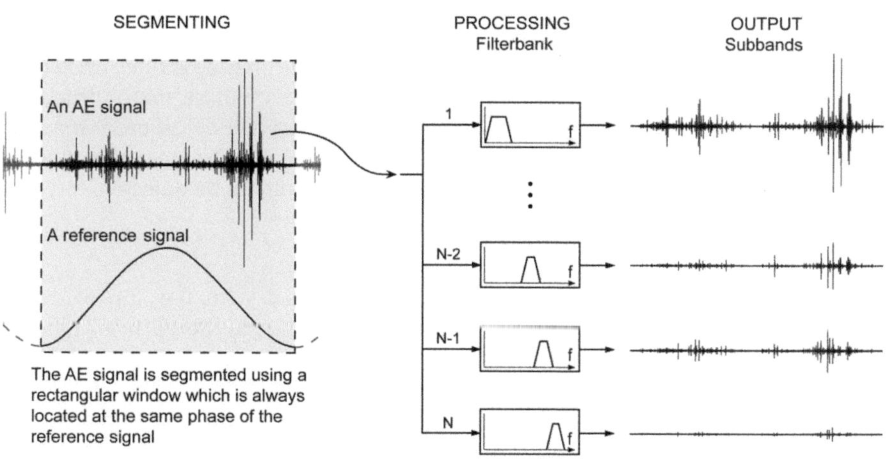

Fig. 2. Illustration of how the AE signal is segmented using a reference signal and then split into N subands. All segments start at the same phase of the reference signal and are exactly one cycle.

2.2 Second step: Band-pass filtering

Multi-resolution analysis of AE signals is a valuable tool for identifying band-limited AE sources. This is because AE from different damage mechanisms resides in frequency bands (Bohse, 2000; De Groot et al., 1995; Iwamoto et al., 1999; Kamala et al., 2001). Hence, the subband filtering can enable detection of sources which otherwise could be masked out and left undetected. The decomposition of each segment, s, into N subbands is illustrated on the right side of Fig. 2.

Several methods are available for decomposing the AE signal into subbands. The selection of the method, the type of filters used, the number of subbands (N) and the bandwidth of each subband, depends on the user and his preferences. Two methods are mentioned here; a filter bank representation of the discrete wavelet transform and a method known as phaseless filtering (Mercer, 2001).

The filter bank representation of the discrete wavelet transform (DWT) offers a convenient method for decomposing the AE signal into subbands. In the filter bank, the conventional discrete wavelet decomposition, at each level, is computed by filtering the input signal with low and high pass filters, producing two sequences, and keeping only their even numbered coefficieFnts. All subsequences of the original signal are called wavelet packets, and together they form a wavelet packet tree. Each packet retains the necessary information in order to reconstruct the signal in the corresponding subband. This means that packets from different levels can be used to fully reconstruct the original signal if, together, they span the full bandwidth.

Phaseless filtering can also be used to split the signal into subbands. The filtering is made phaseless by filtering twice. After the first filtering, the signal is reversed, filtered and reversed once again. The type of filter used depends on the preferences of the user. The use of phaseless filtering is important for identifying and monitoring evolution of AE sources. The fixed reference point enables the detection of trends in the signal. If phaseless filtering is not used then the identification and monitoring of evolving AE sources will be more difficult. This is because a phase lag will be caused by the filters and the lag will be different for different frequencies. For an evolving AE source which emits AE earlier and changes the frequency the phase lag could possibly delay the signal so that the change will not be detected.

2.3 Third step: Feature extraction

Over the years many research projects have been conducted with the aim of extracting useful information from AE signals. The extracted information is stored in n-dimensional data structures, known as features. Commonly used features for monitoring AE activity are number of AE hits and the signal's energy (Carlos & Vallen, 2005; Hellier, 2001). Conventional features extracted from AE hits include amplitude, duration, energy, number of peaks above certain threshold (ring-down count) and rise time (Carlos & Vallen, 2005). One of the advantages of extracting and using features is that they provide a huge data reduction compared with a full waveform acquisition, but there are several practical disadvantages. First, the fact that the same type of damage can emit AE signals with different amplitudes (Prosser et al., 1997) makes it difficult to set the threshold for hit detection and to interpret. Second, AE signals in composites suffer from high attenuation, which means that emissions close to the transducer are stronger and more likely to be detected than those generated further away. Third, the rugged frequency response of resonance AE transducers means that frequencies which are spaced only a few tens of kilohertz apart, will be magnified differently. The magnification can differ by several decibels. Finally, if only the AE features are stored then the analysis is limited to these features, e.g. it is impossible to study different results using other threshold settings.

The procedure for extracting features from each subband segment, s_{sb}, is a two step procedure. In the first step the AE feature of interest is extracted from the segment and both the feature values and their locations within the segment are logged. In the second step, the segment is partitioned into K intervals and the features extracted within each interval are processed, e.g. occurrences counted or the maximum value picked. The user selects the number of intervals, K. The results from the processing are output in a feature vector ($K \times 1$), i.e. one vector for each feature. Depending on the processing in the second step, the first step can in some cases be omitted, e.g. when the energy in each interval or the maximum amplitude is computed. Figure 3 illustrates this procedure by showing how a a feature vector is generated when the maximum amplitude in each interval is used. Each subband segment is first rectified and partitioned into K intervals and then the maximum amplitude within the interval is found, i.e. a piecewise constant envelope is generated. The envelope is then down-sampled by a factor L/K, where L is the length of the subband segment. The resulting feature vector contains one sample from each interval of the envelope. The amplitude-filtering and down-sampling process extracts the amplitude of the strongest transient in each interval. Hence, the tracking capability is limited by this filtering. However, the filtering is performed in all the subbands, which consequently improves the tracking ability because the AE energy from different sources often resides in different subbands.

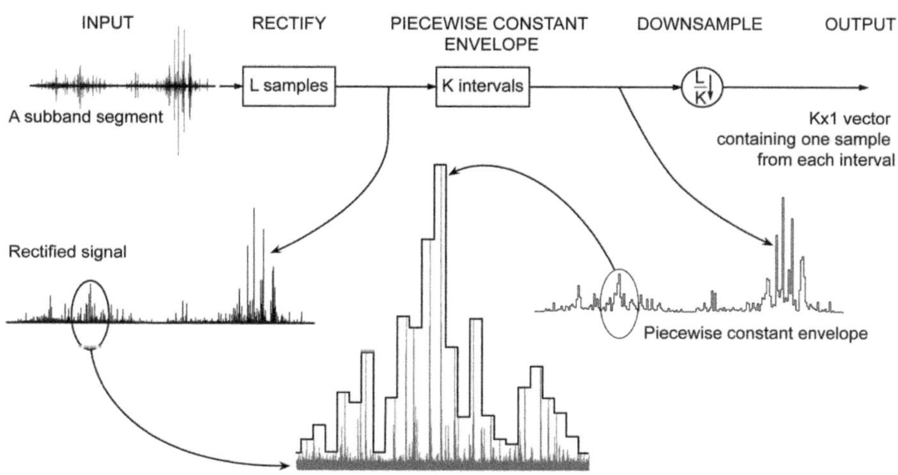

Fig. 3. For each subband segment, the new feature vector is computed by first rectifying the signal, then computing a piecewise constant envelope, and finally down-sampling the envelope.

2.4 Fourth step: Visualization

Visualization of measurement data is a powerful tool for detecting trends in complex data. In fact most AE analysis is done graphically using histograms or by plotting the feature of choice either as a function of time or as a function of another feature. The methodology presented here is designed for detecting temporal trends in feature values and in the positions of emissions relative to a reference point in the reference signal. If one feature value is studied at a time, then the data becomes 3-dimensional. A convenient way for studying 3D data is by using a 3D surface. A 3D surface representation of the data is generated for each subband by appending each new feature vector to previous vectors from the same subband. The first step in Fig. 4 illustrates this procedure where the duration of one cycle (other parameters can be used) is on the x-axis, the segment number is on the y-axis and the feature value is on the z-axis.

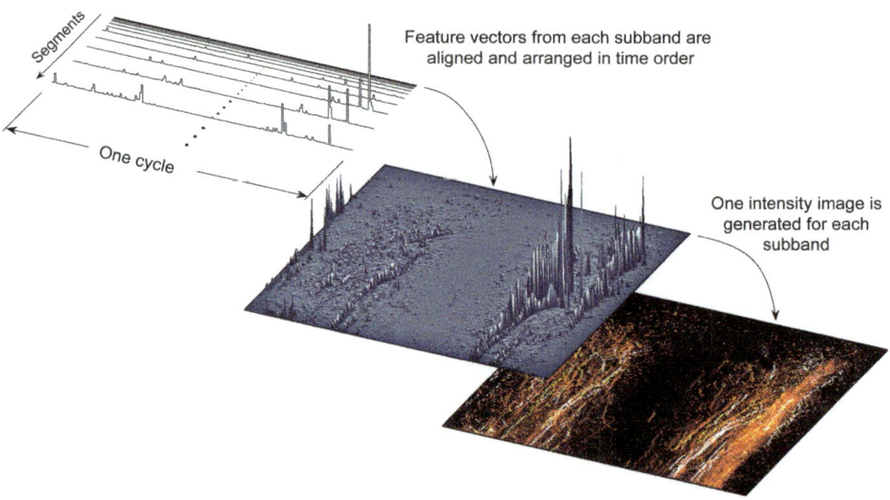

Fig. 4. For each subband, new feature vectors are appended to previous vectors and a 3D surface generated. The 3D surfaces can be converted into 2D intensity images for visualization and further processing.

Trends can be detected by visually inspecting the 3D surfaces, e.g. few trends can be detected from the 3D surface in Fig. 4. Any change in either the feature value or the relative position indicates an evolving AE source. AE from growing cracks will be emitted at different load/time and the corresponding feature value will shift on the x-axis. This type of a change will be seen as a curved ridge on the surface. Stationary AE sources will be represented by straight vertical ridges. Ascending and descending ridges need a special attention because the energy of the corresponding AE source may be shifting to another subband.

Another useful way to visualize the AE data is to use intensity images. Intensity images can be generated from 3D surfaces by transforming and colour coding the z-axis, i.e. the feature values. Logarithmic transformation, e.g. base-10, can be a good choice as it changes the dynamic range of the signal by enhancing low values, while compressing high values. Monochrome colour coding is often used for intensity images. By using monochrome coding the transformed feature values are represented in shades of one colour and the higher the value, the brighter the image pixel. A bright pixel shows the location of an AE source with a high feature value. The horizontal axis is the same as the x-axis in the 3D surfaces representation and shows the location of an source relatively to the phase of the reference signal. The segments are arranged on the vertical axis with numbers increasing from top to bottom, i.e. the first segment is at the top of the image. The illustration in Fig. 4 shows the results of transforming a 3D surface into a 2D intensity image. The ridges on the surface appear as paths in the image and random AE sources appear as random points. Ascending and descending ridges can be identified by the changing pixel intensity in a path. The paths can be used to identify and monitor evolving AE sources and also to isolate the signal for further analysis. Advanced image processing can be applied to the images in order to enhance the images and to make the paths more prominent.

2.5 Summary

The methodology presented in this section was designed to facilitate trend detection and monitoring of evolving AE sources when the AE activity is high. Identification of evolving AE sources is performed by visual inspection of 3D surfaces or intensity images. These two visual representations are valuable for interpreting the AE behaviour and for gaining a deeper understanding of the active damage mechanisms. Trends can be relatively easy to identify from the intensity images but, the resolution and quality depend on the colour coding quantization and the coding range (the max/min values). Due to the limited number of quantized values a decreasing feature value may be detected later than when performing visual inspection of 3D surfaces. For a thorough inspection of a 3D surface, the user needs to have several tools at his disposal, such as rotate, zoom and section.

3. AE hit pattern feature

One of the fundamental elements of music is rhythm. Rhythm can be determined by the relation between note accents (attack) and the rests between notes (Olson, 1967). The term can be used to refer to either a repetitive pulse, or a beat, which is repeated throughout the music or a temporal pattern of pulses. The modelling and the interpretation of temporal patterns are of interest to people working in different disciplines. In the field of music information retrieval rhythm-based features have been extracted from audio and used to classify music styles (Dixon et al., 2003), e.g. blues, disco, polka, etc. In the field of computational neuroscience, patterns in spike trains are studied in order to understand the "language of the brain" (Rieke et al., 1997). For this reason, the author has formulated an AE feature to investigate whether patterns exist in AE signals and if they can be used to extract valuable information from the AE signal. This AE feature will now be introduced.

The first step in working with temporal patterns is to determine the pulses. The detection of pulses usually begins with the processing of the signal in order to make the detection more accurate. The resulting signal is called a detection function. An example of a detection function is one made using a model called a rhythm track (Sethares et al., 2005). The rhythm-track model is based on the assumption that the audio can be considered to be a random signal and the signal's energy increases significantly when a pulse occurs. The resulting rhythm track contains the locations of all the pulses. These pulses are known as AE hits when working with AE signals. An intuitive rhythm track is a vector, of the same length as the signal, containing ones where the AE hits start and zeros elsewhere.

Interpreting patterns in the rhythm-track is a challenging task because similar patterns can be generated differently. For example, closely spaced transients can be attributed to factors such as rapid AE release, reflections, and simultaneous emissions from multiple sources. The results from an investigation into fusing AE features (Unnthorsson et al., 2005) provided an impetus to investigate the fusion of AE features with the rhythm track information. As a result, a methodology for fusing AE features, and for finding and locating patterns within the fused data representation, has been elaborated. The additional information provided by the waveform-based features comes at higher computational cost, but may help to distinguish and interpret rhythm track patterns which are generated differently, or at different locations.

The fused and coded AE features are collected in a vector, called a coding vector, where each hit is represented by a subvector. The length of the subvector, is the same for all hits but, depends on both the number and type of features used. In order to limit the number of patterns found in the coding vector the features are first processed and then their values are quantized to a relatively small set of integers. The quantization suggested here is to transform the processed features logarithmically ($\log_{10}$), then shift, scale, and round the results so that they are represented by integers ranging from 1 to $N_{FEATURE}$, where $N_{FEATURE}$ is an integer number chosen by the user.

The scaling operation requires the user to know the extreme values of the features. If the extreme values can only be approximated then the coded elements will be occasionally out of range. In order to solve this, hard limits can be used, i.e. if the values go out of range then they will be set to the allowable maximum. However, the limits do not have to be critical; if the coded elements are allowed to exceed the limit then relatively few new patterns will be added. Henceforth, the processing and quantization will be referred to as coding. The coding of each element depends on the corresponding feature.

Figure 5 illustrates how the coding vector is generated when two element subvectors are used. The elements of the subvectors, shown in the figure, are the coded maximum amplitude of the hit and the coded inter-spike interval (ISI) between the peaks of the current hit and the previous hit. The coding of the ISI is made by first logarithmically transforming ($\log_{10}$) the time between the peak amplitudes of successive hits, measured in milliseconds. The results are then quantized so that each ISI is represented by integer values ranging from 1 to N_{ISI}. The coding of the peak amplitudes is performed in a similar way. The amplitudes are logarithmically transformed ($\log_{10}$) and quantized to integer values between 1 and N_{AMP}.

The procedure for searching for hit patterns in the coding vector is illustrated in Fig. 6. The locations where a pattern, H_P, is observed are stored in an observations vector, O_P, where $P = 1, \ldots, N_P$ and N_P is the number of hit patterns. Each observation vector is of same length as the original AE signal and initially contains only zeros. The locations where a pattern, H_P, is observed within the AE signal are indicated by ones in the corresponding observation vector. The ones are placed where the patterns start.

3.1 Summary

This section introduced and formulated an AE feature which contains information about the timing between hits and a methodology for fusing it and coding with other AE features. The section also developed a procedure for searching for AE-hit patterns in the coded representation and formulating the search results. The experimental results from using this feature will be demonstrated in section 5.

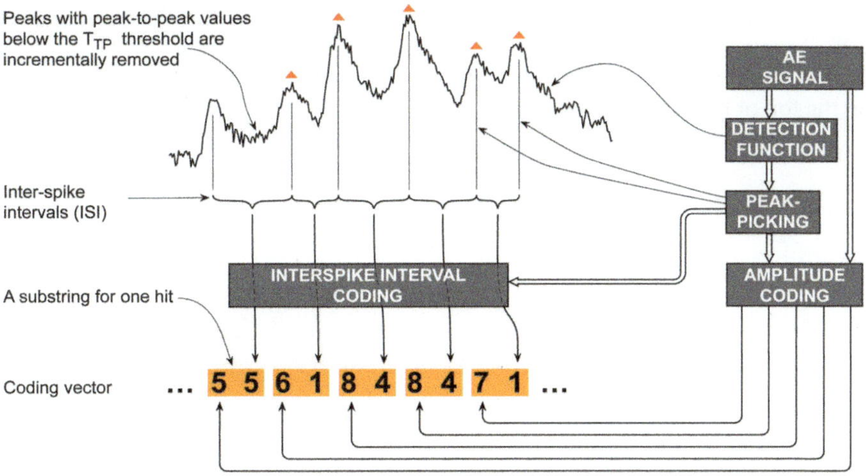

Fig. 5. The generation of the coding vector illustrated using two element subvectors for each hit. The two elements are coded maximum amplitude and coded ISI respectively.

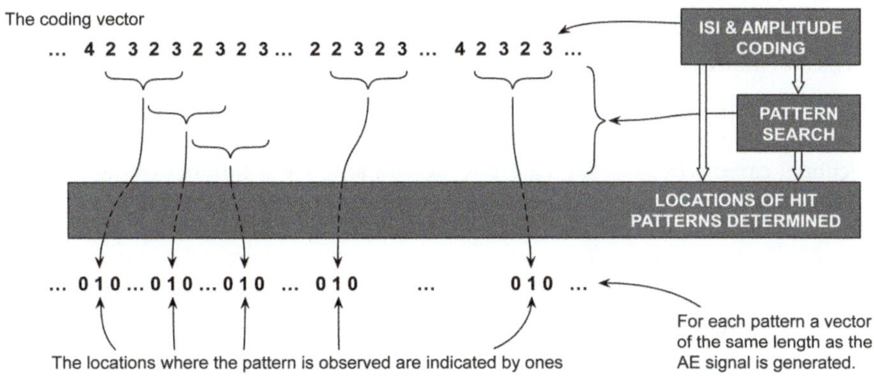

Fig. 6. The procedure for finding hit patterns in the coded representation.

4. Experimental setup

Experimental data is used for demonstrating both the methodology and the AE Hit pattern feature. An overview of the experimental setup is provided in this section. The section describes test specimen, the experimental procedure, and the equipment used. For detailed information about the experimental setup the reader is referred to author's PhD thesis (Unnthorsson, 2008).

4.1 The test specimen

The test specimen used in this study is an assembled prosthetic foot of the type Vari-Flex, made by Össur hf. The Vari-Flex consists of three CFRP composite components: a dual part heel and split-toe foot. In addition, a shock-absorbing crepe is glued under the forefoot. The foot assembly is illustrated in Fig. 7. All components are curved with both varying thickness and width. Figure 7a depicts the varying thickness and Fig. 7b shows the split toe. A male pyramid, for fastening the foot to endoskeletal pylon components and the test machine, is bolted to the top of the foot component.

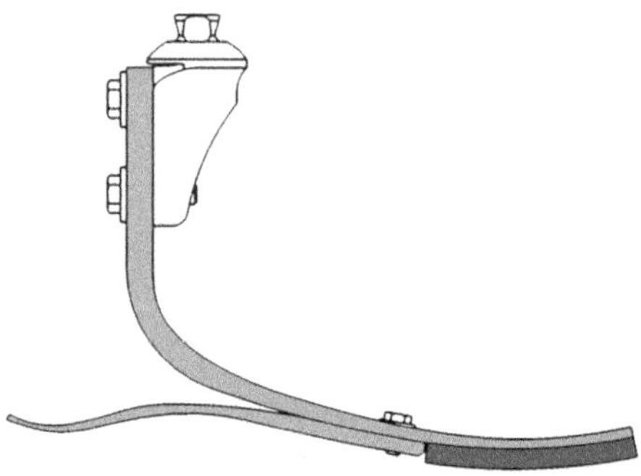

(a) A side view of the Vari-Flex. The diagram shows the varying thicknesses of the components.

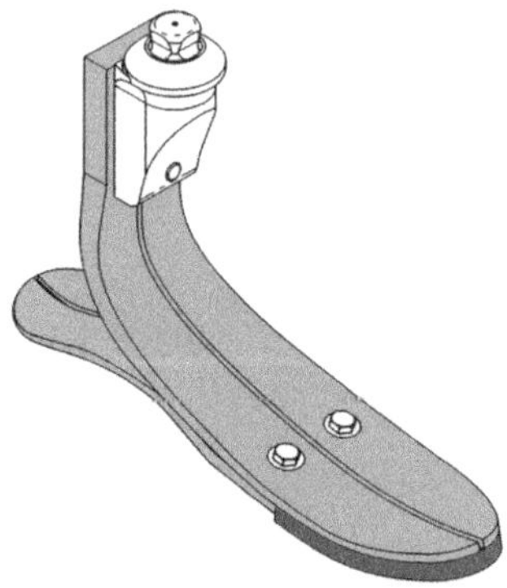

(b) An isometric view of the Vari-Flex.

Fig. 7. The assembled Vari-Flex. The Vari-Flex is made up of two heel parts and one toe part.

The layer orientation and sequence are similar for all the components. The four outermost layers, on each side, are woven carbon/epoxy prepregs, laid at 45°. Between the outer layers unidirectional carbon/epoxy prepreg tapes are laid at 0°. The number of the unidirectional tapes differs between components and their lengths are varied in order to obtain varying stiffness, resulting in a tapered thickness. Around the holes, at the top of the foot, extra unidirectional carbon/epoxy prepreg tapes are laid at 90° for added strength.

4.2 Test setup & procedure

The fatigue tests were performed in an ISO 10328 Foot/Limb test machine run under PID closed loop control. In each test, a foot was placed in the test machine where two actuators were used to flex the foot using 90° phased sinusoidal loading. Figure 8 shows a schematic representation of the experimental setup. One actuator loaded the forefoot and the other loaded the heel. The actuator that flexed the forefoot was rotated 20° from the vertical and the one that flexed the heel was rotated −15° from the vertical. Furthermore, the foot was rotated 7° out of the vertical plane defined by the movement of the actuators.

The maximum loading used was based on the stiffness category of the toe unit and the minimum loading was set to 50 N. In order to accelerate the tests the maximum loading was set 50% higher than is used for testing at Össur. In addition a 2° plastic wedge was placed between the heel and the toe components. The wedge is used by amputees in order to stiffen the foot. The increased load and the use of the wedge result in considerably shorter fatigue tests.

During one test, both the minimum and maximum loads were held constant, with allowable variation of ±2%. The operating frequency was 1.0 Hz. It took the test machine approximately

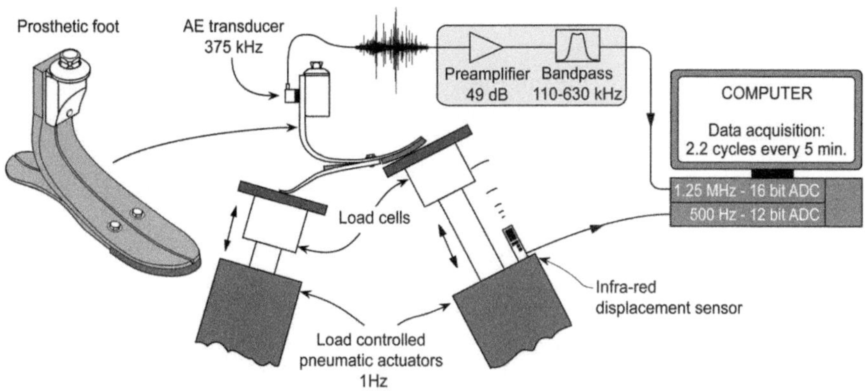

Fig. 8. Schematic representation of the experimental setup for both the AE and the position measurements.

1000–2000 cycles to reach the specified maximum and minimum load values. After this run-in period, the limits of the failure criterion were defined. The failure criterion is a heuristic criterion used in-house at Össur. It defines a failure when a 10% change in the displacement of either actuator, with respect to initial value, is observed. All fatigue tests were run until the failure criterion was met.

4.3 Data acquisition and processing

The additional parameter acquired simultaneously with the AE was the displacement of the forefoot actuator. Due to the geometry and structure of the Vari-Flex, the forefoot's actuator always meets the 10% displacement criterion before the heel's actuator. For this reason, only the position of the forefoot's actuator was measured. A L-Gage Q50A infra-red displacement sensor from Banner Engineering Corporation was used for measuring the position of the actuator. The sensor's location is illustrated in the lower left corner of Fig. 8. The analogue signal from the displacement sensor was digitally converted by a PCI-6024E 12 bit A/D converter (manufactured by National Instruments Corp.).

For acquiring the AE, the VS375-M AE transducer and the AEP3 preamplifier from Vallen Systeme GmbH were used. The transducer was located between the two bolts used to fasten the foot to the pyramid. This was the only part of the foot for where a full face contact with the transducer could be ensured. The transducer was held by a plastic c-clamp and a heavy duty high vacuum silicon grease from Wacker Chemie GmbH was used as coupling medium.

The gain of the preamplifier was set to 49 dB. The gain settings of the preamplifier are based on 28 VDC supply. The actual amplification, however, depends on the input voltage to the AEP3 preamplifier. With less than 28 V the saturation point is earlier and the gain goes down slightly. After consultation and discussion with the manufacturer 29.9 VDC was used. and powered using Velleman PS3003 – a linearly regulated laboratory power supply. A linear regulated power supply was recommended because switching power supplies may add unwanted noise. Table 1 lists the equipment along with the settings used.

AE acquisition	
VS375-M	An AE transducer with resonance at 375 kHz
AEP3	Preamplifier,
	49 dB gain
	110 kHz high pass filter (54 dB/octave)
	630 kHz low pass filter (30 dB/octave)
DCPL1	Decoupling box
PS3003	Linearly regulated laboratory power supply
	29.9 VDC
PCI-6250	16 bit A/D converter
	1250 kHz sampling rate on one channel
Position measurements	
L-Gage Q50A	Infrared displacement sensor
	Fast response mode used
PCI-6024E	12 bit A/D converter
	500 kHz sampling rate on one channel

Table 1. The measurement equipment used for acquiring AE and the position of the forefoot's actuator during fatigue testing.

The position of the forefoot's actuator and the AE signal were measured simultaneously at 500 Hz and 1.25 MHz sampling rates respectively. In order synchronize the two A/D converters their Real-Time System Integration (RTSI) buses were connected. The data was acquired automatically, for 2.2 seconds every 5 minutes. After the data acquisition, the data was trimmed so that it represented exactly one fatigue cycle, starting at the lowest position of the forefoot's actuator. The data was also high-pass filtered in order to remove DC and other low frequency disturbances. A fifth-order elliptic filter with 1 dB passband ripple and corner frequency of 80 kHz was used. The stopband attenuation was set to 30 dB at 50 kHz. No attempts were made in order to correct the acquired signal, i.e. remove the amplification made by the preamplifier and the transducer.

4.4 Summary

This section presented the test specimen and the test equipment used. An accelerated fatigue testing procedure was also presented. Frequently the damage mechanics change when the stress level changes, however, preparatory tests showed that the damage mechanisms leading to final failure were the same as observed under normal fatigue testing conditions.

5. Experimental results

Experimental data was acquired from fatigue testing of 75 nominally identical samples of the prosthetic foot. This section presents a case study of one foot and is split into three subsections. In the first subsection the evolution of few AE features throughout the fatigue life of the foot are studied and compared against their average evolution. The second and the third subsections present an in-depth analysis of the experimental data using conventional AE features and AE Hit patterns respectively. Both subsections show how the results provided by methodology can be used to facilitate early damage diagnosis and failure.

5.1 Evolution of AE features

The evolution of few AE features throughout the fatigue life of one prosthetic foot will now be studied. Each feature is overlaid onto its corresponding mean and standard deviation. Both are computed by from the evolution curves of all the other feet. The gray area in Fig. 9 and Fig. 11 represents all values which lie within one standard deviation from the mean. The position measurements of the forefoot's actuator are converted to cycle time and used as a reference signal for segmenting. The segments start at the lowest position of the actuator, or at 0° phase angle, and end when the actuator returns back to this position. Because the actuator moves at a constant angular velocity the segments are all of equal length. The features are extracted from the whole segment and without subband filtering or, in the terms of the methodology presented in 2, the features are extracted by using one interval ($K = 1$) and the signal's full bandwidth ($N = 1$).

Based on the results of visual and acoustic inspection performed by the author, i.e. watching and listening, during the cyclic testing of the foot, the temporal behaviour of the AE features are interpreted. Detection and determination of AE hits is made using the methodology presented in (Unnthorsson et al., 2008a) and (Unnthorsson, 2008). The STFT detection function is used, with segment size of $k = 128$ samples and $d = 120$ sample overlapping. The hits are located by setting the trough-to-peak threshold (T_{tp}) at 304 dB V-s and determined by setting the determination threshold (T_{AE}) at 3 mV.

Shortly after the cyclic test was initiated, or after 6k cycles, small splinters started to form on both sides of the split-toe foot. During the next 4.2k cycles the splinters grew larger and rubbed against the sides. As can be observed in Fig. 9, this resulted in a very steep increase in the AE energy, but only a slight increase in the number of AE hits. Two spikes can be observed in the AE energy when 12.3k and 13.2k cycles have elapsed. At this time a medium-sized splinter formed on the right side of the split-toe foot. The two measurements corresponding to the spikes in AE energy were taken at points when the splinters were growing. This means that the readings are composed of AE from both rubbing and damage growth. Thus, spikes are observed rather than a permanent increase. The reader is reminded that measurements are made at 5-minute intervals, or every 300 cycles; hence, the probability of recording AE from damage growth is low.

In the interval between cycles 15k to 18.6k cycles, a large splinter on the right side of the split-toe foot had formed. The splinter, depicted in Fig. 10a, caused both the upward and downward bending stiffnesses to drop. As a consequence, the displacement of the forefoot's actuator increased abruptly. The abrupt displacement increase will be referred to as step (or event) A. The formation of the splinter was accompanied by an increase in the AE energy and an abrupt jump in the AE hit count.

Within a few cycles after the formation of the splinter, or when 21.6k fatigue cycles had been applied, the outer woven layers on the left side delaminated from the unidirectional layers. The delamination crack initiated from the splinter crack and grew for undetermined number of cycles. The number of cycles is undetermined because it was not possible to establish if and when the crack growth stopped using visual inspection. In each cycle, when the crack opened an audible AE was produced. Due to this and also because the crack grew over time, the AE energy increases. From 21.6k to 28.5k cycles the AE energy increased at a relatively

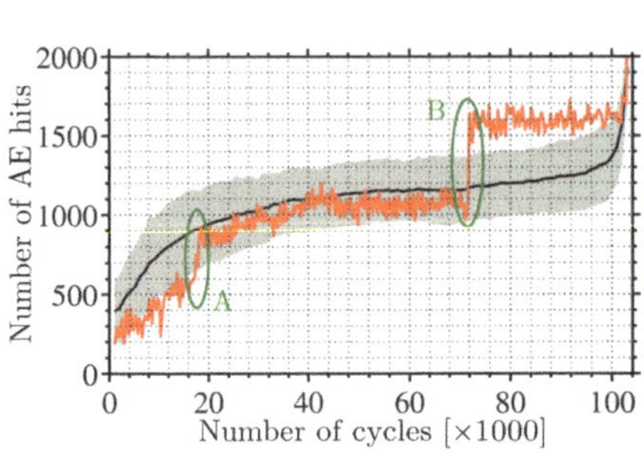

(a) The evolution of the AE hit count.

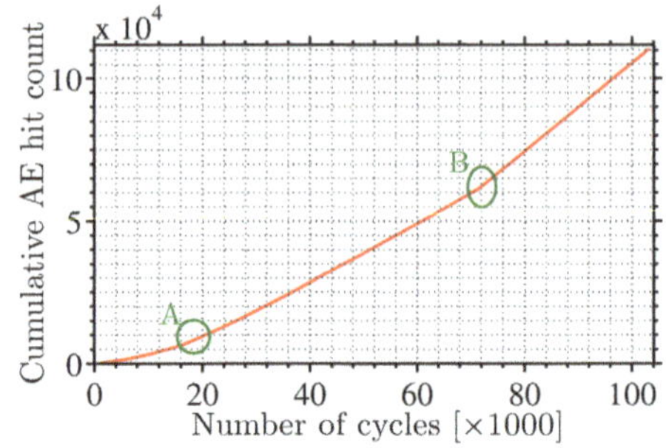

(b) The evolution of the cumulative AE hit count.

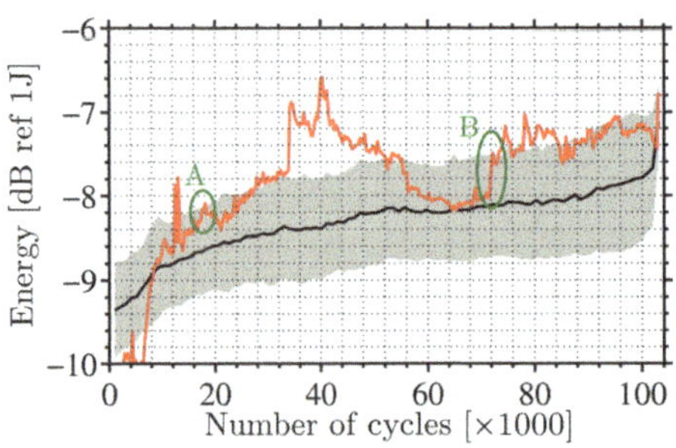

(c) The evolution of the energy

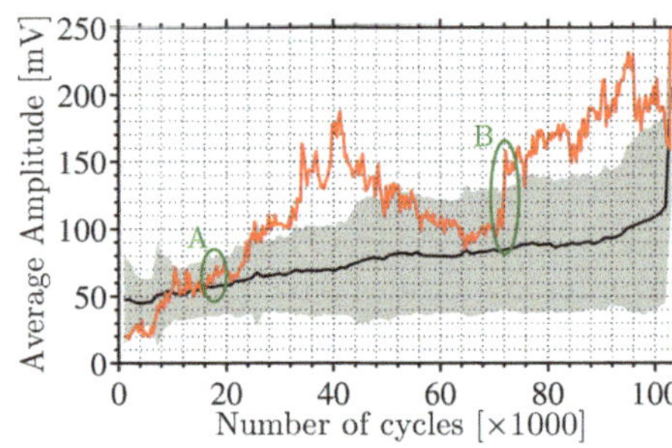

(d) The evolution of the average amplitude

Fig. 9. The evolution of selected AE features throughout one fatigue test. The grey area for each feature represents all values which lie within one standard deviation from the mean (black curve). The mean is computed by averaging the evolution curves from all other feet.

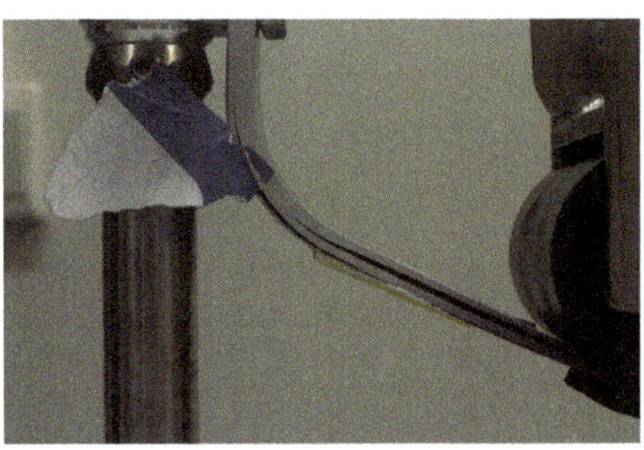

(a) Right side of the split-toe foot under static loading.

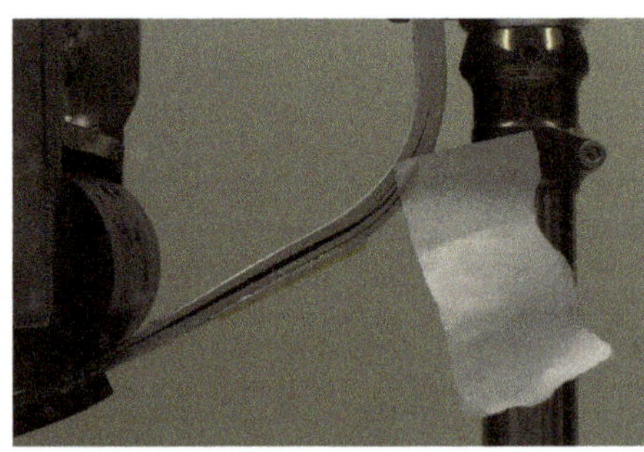

(b) Left side of the split-toe foot under static loading.

Fig. 10. The left and right sides of the split-toe foot after cyclic testing.

steady high rate, but then it became constant. At 28.5k cycles a medium-sized splinter formed on the left side. The splinter, shown in Fig. 10b, did not affect the bending stiffnesses and was not detectable in the evolution of any of the AE features. At 33.9k cycles there was an abrupt jump in the AE energy and over the next 4.1k cycles the energy fell back by 50% of the jump. Shortly after that, or at 40.2k cycles, another abrupt jump in the AE energy occured. From this point the energy started to decline, initially at a high rate, but the rate decreased temporarily at 41.7k cycles, and then resumed at 55.2k cycles for 900 cycles. After this, the energy reduced at a low rate until the reduction ceased, at 64.8k cycles The two abrupt jumps in the energy, at 33.9k and 40.2k cycles, were not accompanied by any changes in the bending stiffnesses. Subtle changes, however, could be detected in the slope of the AE hit count, and duration at 40.2k cycles.

At 72k cycles, the left half of the split-toe foot delaminated. This resulted in a drop in the bending stiffnesses. The corresponding abrupt displacement increase will be referred to as event B. Both events are shown in Fig. 9 and Fig. 11. Abrupt jumps in the AE energy and AE hit count were also observed.

The cumulative AE hit count, presented in Fig. 9b, is a commonly used parameter for the study of acoustic emission. Although the cumulative sum contains the same information as the AE hit counts, the information provided by subtle slope changes, small jumps, and fluctuations is harder to detect. One can observe from the figure that the curve can be divided into three segments, each with a different slope. The slope changes occur at events A and B. However, the AE hit count, shown in Fig. 9a, is the slope of the curve in Fig. 9b at every measurement. The AE hit count is absolute in that it does not depend on prior values, but each value of the cumulative curve depends on all prior values. This means that missing or late measurements do not affect the results when monitoring the AE hit count, but the slope of the cumulative curve is a function of the frequency of the measurements, i.e. different curves are obtained by summing up the AE hit counts from measurements made at different or irregular intervals. Hence, the cumulative AE hit count works best when the interval between the measurements can be fixed.

The evolution of the average amplitude, depicted in Fig. 9d, is strongly correlated with the AE energy on a decibel scale, shown in Fig. 9d. The Pearson and Spearman correlation coefficients for the two curves are 0.88 and 0.92 respectively.

The AE hit pattern feature used is computed by first determining the time from a trough to a peak (and also from a peak to a trough) for each AE hit, then coding the results and then counting the occurrences of all patterns found in the coded representation. A trough-to-peak interval is a variant of the Inter-spike Interval (ISI). It can also be recognized to include variants of two commonly used AE hit-based features, namely the rise time and the fall time.

The methodology described in Sect. 2 is used to count the pattern occurrences. The AE hits are located using the procedure described above, but no determination is performed, i.e. the determination threshold (T_{AE}) is set to 0 mV. The trough-to-peak intervals, in microsecond, are quantized by using a natural logarithm and rounding the result to the nearest integer.

Figure 11 shows the evolution of the total number of observations for four selected trough-to-peak patterns of length 2. These patterns contain the coded values of the fall time of one hit and the rise time of the next adjacent hit. The evolution curves of the trough-to-peak patterns have a slope change at around 60% of the lifetime. The slope then remains constant until failure. This slope change is not present in the evolution curves for the other AE features. This suggests that these trough-to-peak pattern features are capturing important information from the AE signal, e.g. the slope change may be caused by the formation of a damage which grows until failure. Intuitively, the salient slope change can be used to provide an early warning about the health of the composite.

In (Unnthorsson et al., 2008b) the trough-to-peak patterns were the only features studied which probability distributions at 50% and 95% of the lifetime could be reasonably well separated using a Bayes optimal decision boundary. The Bayes optimal decision thresholds are shown in Fig. 11. Patterns made using different coding, length, features, or a combination of features, can possibly be used to obtain more information. The additional information can be combined into a feature vector which can be used as an input into classifier system for more accurate warnings.

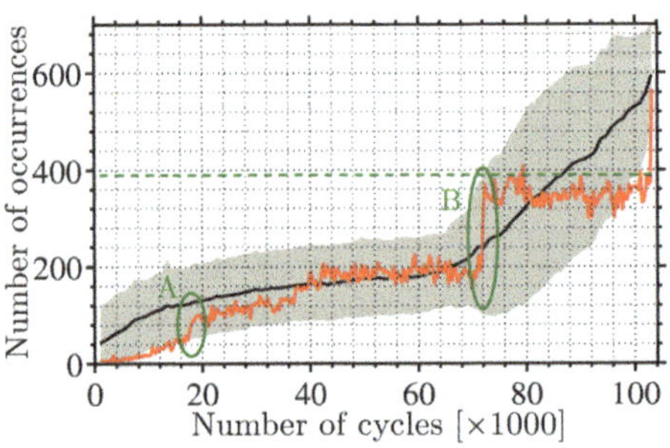

(a) The evolution of the number of occurrences of trough-to-peak pattern no. 87.

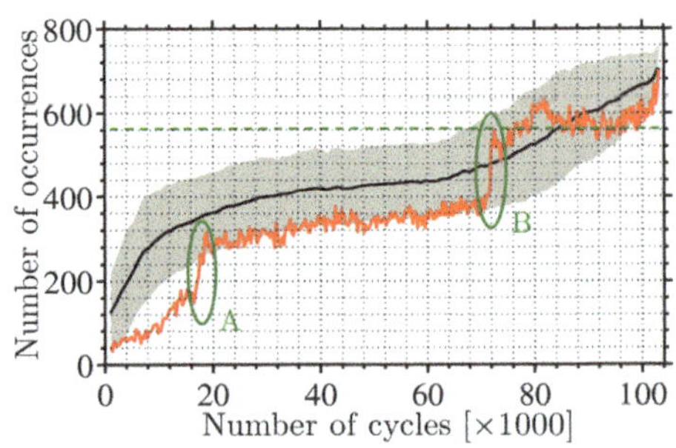

(b) The evolution of the number of occurrences of trough-to-peak pattern no. 94.

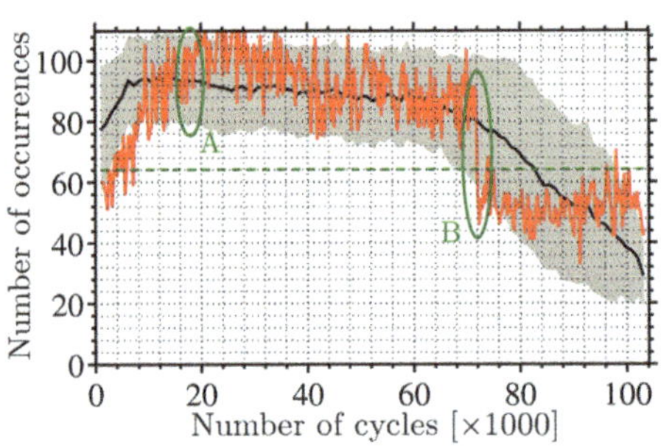

(c) The evolution of the number of occurrences of trough-to-peak pattern no. 95.

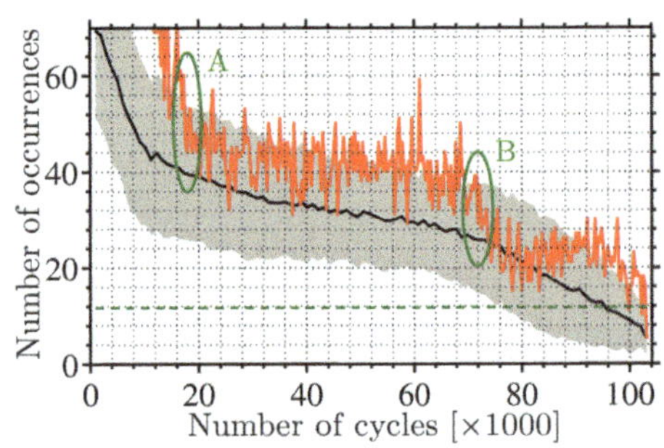

(d) The evolution of the number of occurrences of trough-to-peak pattern no. 102.

Fig. 11. The evolution of the number of occurrences for four trough-to-peak (ISI) patterns computed from the AE segments. The grey area for each pattern represents all values which lie within one standard deviation from the mean (black curve). The mean is computed by averaging the evolution curves from all other feet.

5.2 AE feature tracking

The results presented above were obtained by using one interval ($K = 1$) and full bandwidth ($N = 1$). In this subsection each subband segment is divided into $K = 200$ equally-sized intervals and phaseless filtering is used to bandpass filter the AE signal into $N = 19$ subbands, each with 33 kHz bandwidth. By dividing the segments into shorter intervals, the dimension of time within a segment is added to the analysis. Furthermore, by bandpass filtering the signal and studying each subband separately the results are divided into N different subresults, one for each subband. For visualizing the results 2D intensity images will be used.

Figure 12 shows the resulting intensity image for the 133–166 kHz subband using the maximum amplitude of the rectified signal in each interval to generate the feature vector. The procedure is explained in Fig. 3. Also depicted in the figure is the evolution of the AE energy and the AE hit count. The AE energy and the hit count are computed the same way as was done in last subsection, i.e. by setting $K = 1$ (one interval) and $N = 1$ (full bandwidth).

Figure 13 shows four more intensity images corresponding to the 266–300 kHz, 366–400 kHz, 466–500 kHz and 566–600 kHz subbands. By comparing the evolution of the AE energy and the AE hit count in Fig. 12 against the intensity images depicted in Fig. 12 and Fig. 13, one can see that the intensity images facilitate a better understanding of the changes which occur in the material than the evolution of the AE features in last subsection.

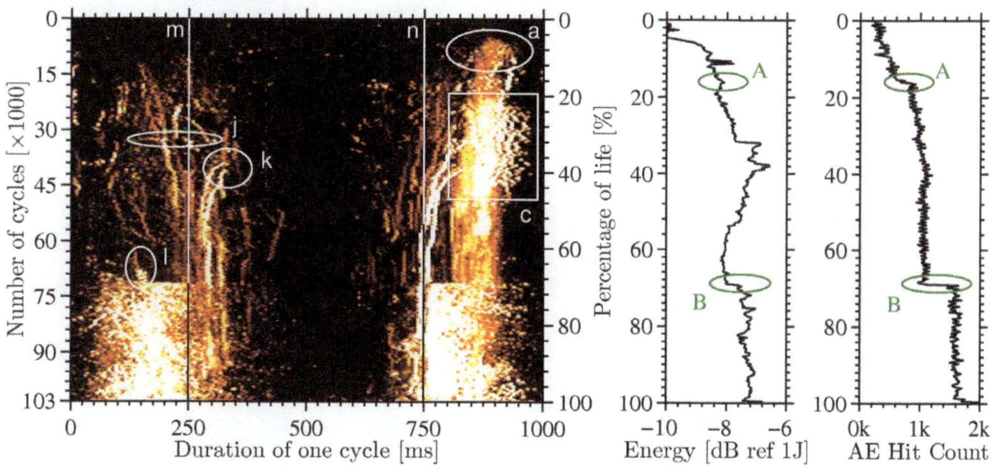

Fig. 12. On the left is the resulting intensity image for the 133–166 kHz subband. Each value in the image is the maximum amplitude in the corresponding interval. On the right is the evolution of the total AE energy and AE hit count from each segment (computed using the signal's full bandwidth).

The steep increase in the AE energy early in the fatigue test is attributed to formation of small splinters on the side of the split-toe foot. In each fatigue cycle these splinters rub against the toe-foot component and occasionally grow. The AE from these splinters is presented as a growing cluster in the upper left corner of the intensity image in Fig. 12. The cluster is circled and labelled a. At 12.3k and 13.2k cycles two spikes are observed in the evolution of the AE energy. These spikes are due to the formation of a medium-sized splinter on the right side of the split-toe foot. Although the AE energy only shows two spikes, one can see, from the intensity images, two new paths initiate at this time from within the cluster labelled a (Fig. 12). The paths are circled and labelled b (Fig. 13d).

A large splinter on the left side of the split-toe foot forms during the period from 15k to 18.6k cycles. The splinter causes a drop in both bending stiffnesses and can be detected as a subtle change in the evolution of the AE energy, but as an abrupt jump in the AE hit count (labelled A in Fig. 12. In the intensity images, this event can be detected by the formation of new paths and a sudden change in the left path of the two circled and labelled b (Fig. 13d). The left path changes its course, shifts to the left and disappears.

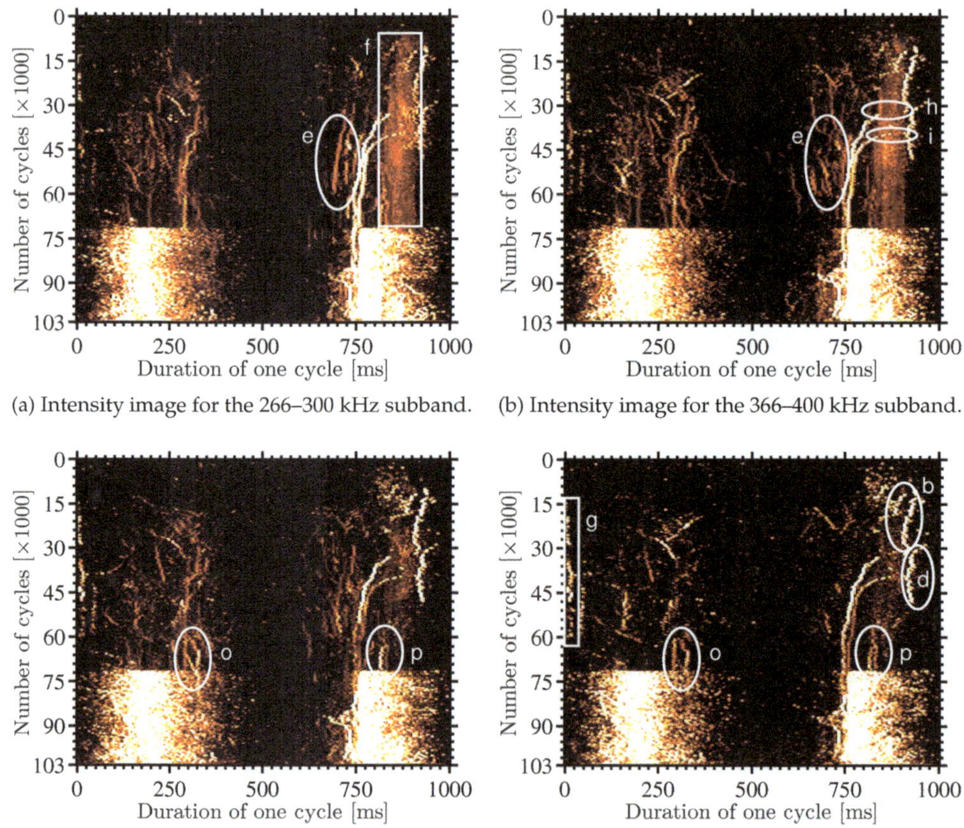

(a) Intensity image for the 266–300 kHz subband. (b) Intensity image for the 366–400 kHz subband.

(c) Intensity image for the 466–500 kHz subband. (d) Intensity image for the 566–600 kHz subband.

Fig. 13. The resulting intensity images for four selected subbands. The brightness of each pixel is based on the amplitude in the corresponding interval.

Within a few hundred cycles after the formation of the splinter, the woven layers at the left bottom of the split-toe foot delaminate from the unidirectional layers. The delamination crack opens every time that the heel's actuator is nearly fully loaded and an audible AE is generated. The crack grows and the AE energy increases at a steady high rate. The AE from the delamination crack resides in the lower frequency bands and can be observed in the boxed area labelled c (Fig. 12). In the 133–166 kHz subband, the AE from the delamination crack masks out other AE. However, the AE is bandlimited; hence, different frequency bands can be used to monitor some of the masked-out damages, e.g. the circled paths labelled b and d (Fig. 13d).

At 28.5k cycles a medium-sized splinter forms on the left side of the split-toe foot. The formation cannot be detected from the the evolution of the AE energy, AE hit count nor the bending stiffnesses. The AE emitted from the splinter is masked out in the 133–166 kHz subband, but can be detected and monitored in intensity images for the higher frequency bands. The corresponding evolution path is circled and labelled d in Fig. 13d.

By studying different subbands, one can in some cases detect and distinguish between damages that emit bandlimited AE signals at the same time, but are evolving in different directions, as can be seen by comparing the circled paths labelled e in Fig. 13b and Fig. 13c. Most of the energy from the frictional AE caused by the rubbing of the splinters is located in the lower frequencies. The boxed region labelled f (Fig. 13a) shows where the frictional AE is located within the fatigue cycles. In addition to the frictional AE, the splinters also, for a limited time, produce strong AE at the end of their push-out movement (the circled paths labelled b and d in Fig. 13d) and also when they snap back in (the boxed area labelled g in Fig. 13d).

The two abrupt jumps in the AE energy and amplitude at 33.9k and 40.2k cycles are accompanied only by subtle changes in the slope of the other AE features, but no changes in the bending stiffnesses. The intensity images, however, show the initiation of two paths originating from within the area where the frictional AE from the splinters is located. The area of initiations are circled and labelled h and i in Fig. 13b.

In the first half of the segment in which the first path starts (labelled h in Fig. 13b), a high amplitude AE can be observed in the intensity image for the 133–166 kHz subband (Fig. 12). This portion is circled and labelled j. Furthermore, two evolving high amplitude paths end in this portion. Simultaneously with the initiation of the second path, labelled i (Fig. 13b), another high amplitude AE path initiates in the first half of the loading cycle. The path can be seen in all subbands. The beginning of the path is circled and labelled k in Fig. 12. The two paths, labelled h and i (Fig. 13b), evolve asymptotically towards a line close to and parallel to the 750 ms line, labelled n in Fig. 12. Initially the paths evolve at a high rate but as they approach the line the rate decreases. The vertical asymptote line is located where the tensional, compressional and shear stresses change their signs. This location in the fatigue cycle acts as an attractor for most evolving AE sources in the left half of the intensity images, i.e. during the downward movement of the forefoot's actuator. Conversely, during the upward movement of the actuator, the so-called attractor is around 250 ms, shown by a line labelled m in Fig. 12. As can be observed in the figure, the attractor is offset to the right. This is because the the loading is not symmetrical around the 500 ms.

The left half of the split-toe foot delaminates at 72k cycles. This event (labelled B in Fig. 12) can be observed in the evolution of all AE features and also in the downward bending stiffness. After the delamination, a high amplitude AE is generated in a large portion of each cycle and in all subbands. During the 12k cycles leading up to the delamination the AE activity, as indicated by the evolution of the AE features in Fig. 9, remains at a relatively steady level. However, the intensity images show few changes which can be interpreted as warnings. First, several new high-amplitude paths start during this period. These paths are circled and labelled l, o and p in Figs. 12, 13c and 13d. Second, during this period one can observe that the path which began in the circled area labelled h moves one step closer to the 750 ms line.

(a) Intensity image computed using the ring-down counts of the hit with the maximum amplitude in each interval.

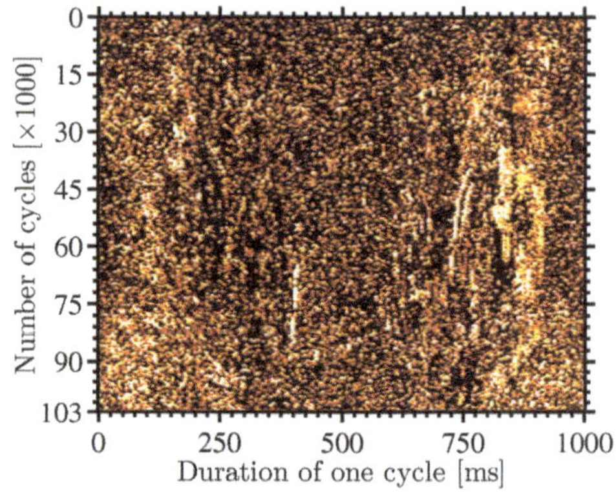

(b) Intensity image computed using the duration of the hit with the maximum amplitude in each interval.

Fig. 14. Two intensity images, for the 133–166 kHz subband, made using threshold-based AE features. The brightness of each pixel is based on the value of the AE feature in the corresponding interval.

The AE energy feature produces nearly identical results to those presented here. Figure 14a and Fig.14b show results computed using the ring-down counts and the duration of the hit with the maximum amplitude in each interval respectively. In order to compute the features, the determination threshold, T_{AE}, is set to 3 mV. These images do not have as well defined paths as the images presented above, but different features can reveal new patterns (e.g. the circled path labelled q in Fig. 14a). The main challenge in using this approach for threshold-based features is the task of adjusting for each subband the trough-to-peak threshold, T_{tp}, used for locating hits from the detection function, and the determination threshold, T_{AE}, used both for determining hits and for extracting the threshold-based features.

5.3 AE pattern tracking

The results presented above were obtained using the maximum amplitude in an interval. Also demonstrated were results from using both the ring down count and the duration of the hit with the largest amplitude in an interval. Here the results of using AE-hit patterns will be presented. The segmentation and subband filtering as before, i.e. $K = 200$ and $N = 19$.

The algorithm for detecting hits is adjusted for each subband so that the average number of detected hits in the first 5 measurements is around 10.000. In other words, the average pulse duration in each subband is 0.1 ms. By using these settings small pulsations in the AE signal's amplitude are detected as hits but, the hit determination is the same as before, i.e. $T_{AE} = 3$ mV.

Two different coding representations are studied: one using only ISI information, and another using both ISI and the peak amplitude of the hits. The values of N_{ISI} and N_{AMP} are both set to 10 when computing the feature vectors. The resulting feature vector for each pattern contains the total number of observations in each interval. In order to fight the curse of dimensionality only two pattern lengths are used for both coding representations: $L = 2$ and $L = 4$.

5.3.1 ISI coding

Approximately 60 patterns are found in each subband when using ISI coding. However, only a handful of them produce intensity images with detectable paths. Figure 15 shows an example of two intensity images which have no detectable paths.

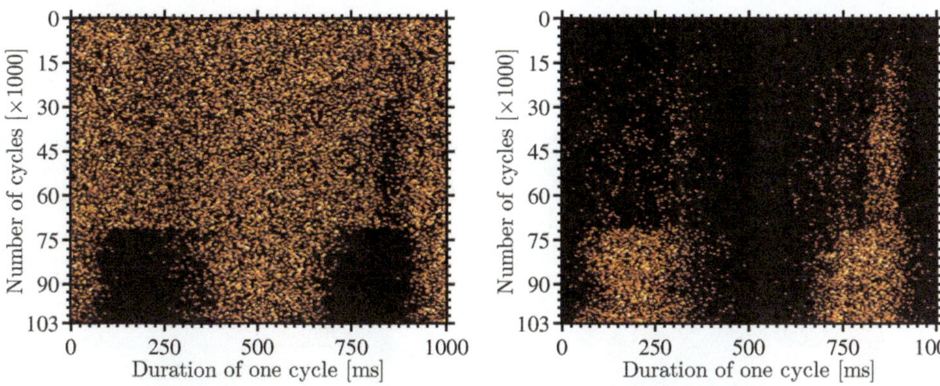

(a) An intensity image for an ISI pattern which does not show any paths.

(b) An intensity image for an ISI pattern which does not show any paths.

Fig. 15. Two intensity images which have no detectable paths. The brightness of each pixel is based on how often a pattern is observed in the corresponding interval.

Figure 16 shows intensity images for four handpicked patterns. As one can observe, the circled paths labelled 1, 2, 3 and 6 in the figure are the same paths as those depicted in Fig. 12 and Fig. 13. However, further comparison of the images, in these three figures, shows that only small portions of the circled paths labelled 4 and 5 in Fig. 16 can be detected in the other two figures. Consequently, by using AE patterns based on the ISI, more defined and more visually detectable paths can be detected. These paths can be used to locate AE sources for tracking and for detailed analysis of their AE signals.

Figures 16b and 16c show that different patterns may be used to monitor the same AE source. This is because the ISI, as used here, is the based on the time between the small pulsations in the AE amplitude and the AE emitted from a source may contain several patterns.

The intensity images for some patterns do not have any prominent paths, but instead clusters. An example of such an image is shown in Fig. 16d. The high values in this image are clustered where the frictional AE is emitted, indicating that some patterns may be used to monitor frictional AE.

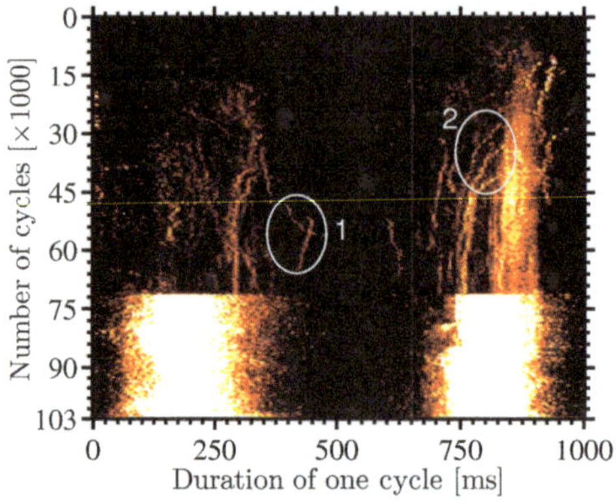

(a) Intensity image for a pattern in the 333–366 kHz subband.

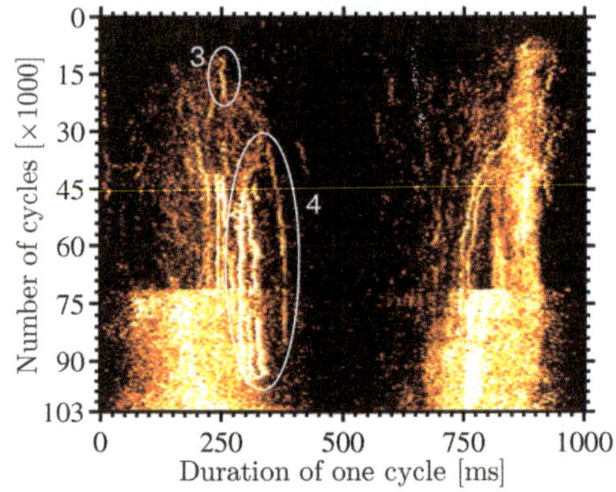

(b) Intensity image for a pattern in the 133–166 kHz subband.

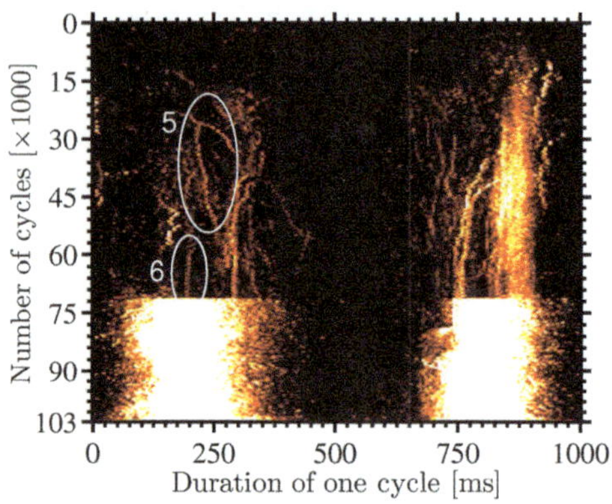

(c) Intensity image for a pattern in the 366–400 kHz subband.

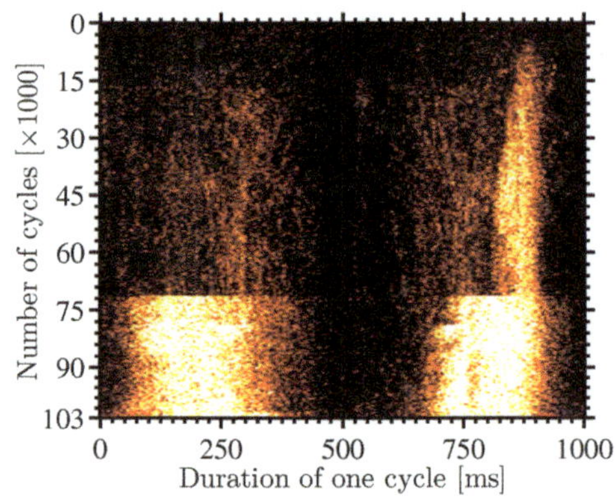

(d) Intensity image for a pattern in the 366–400 kHz subband.

Fig. 16. Intensity images corresponding to four selected ISI patterns. The brightness of each pixel is based on how often a pattern is observed in the corresponding interval.

5.3.2 ISI/peak amplitude coding

By combining the ISI coding with the peak amplitude, the total number of patterns increases up to approximately 600 patterns for each subband. The increase is a function of the number of quantization levels used for the amplitude,N_{AMP}.

Figure 17 shows intensity images corresponding to four handpicked patterns. The circled paths labelled I, III and IV can also be detected in the images obtained using only ISI coding (see Fig. 16b). However, these paths are more prominent in Fig. 17. This is because the addition of the peak amplitude works like a filter. The observations of patterns with certain ISI coding are divided between patterns with the same ISI coding, but different amplitude coding. As a result, the number of AE patterns is higher. This filtering helps with detecting patterns which would otherwise pass undetected, e.g. the circled paths labelled II in Fig. 17b.

By visually inspecting the intensity images for different patterns and extracting prominent paths, e.g. those circled in Fig. 17, a composite image can be made by piecing together the individual paths. Figure 18 shows a composite image, made by picking out, overlaying and enhancing paths extracted from 32 handpicked intensity images. The images are from all subbands. On the right side the composite image is the evolution of the AE energy in each segment.

A comparison of the intensity images in Fig. 16 with the intensity image in Fig. 18 reveals that the addition of the peak amplitude makes it possible to track the evolution of several AE sources after the left half of the split-toe foot delaminates at 72k cycles. The tracking improvement can also be observed by comparing the 750-1000 ms region of the images (where the frictional AE due to the rubbing of the splinters is located). In this region of the fatigue cycle, the initiation of 3 paths can be observed (indicated by arrows). Furthermore, the first path, which starts at event A, can now be tracked until shortly before the delamination of the left half. Based on these results, it can be deduced that the splinters initiate damages which grow until delamination occurs.

5.4 Discussion and summary

In this section the methodology was demonstrated and used for two different purposes. One is for general health monitoring where the evolution of a feature, extracted from each segment, is monitored in order to detect changes and issuing timely warning signs. The other is for detailed health monitoring where individual AE sources are identified, their evolution monitored and their nature determined.

The results show that none of the AE features studied here is suitable for general health monitoring. None of the features could be used to give timely warnings about the delamination which occurs at step B when the method of extracting one AE feature from each segment and studying its evolution is used. The abrupt jumps in the AE energy (and the average amplitude) at 33.9k and 40.2k cycles may perhaps be considered as early warning signs. However, because the AE energy decreases rapidly after the jumps and no significant changes are observed in the evolution of the other AE features, nor in the bending stiffness, nor from the visual tests, the potential warning signs cannot be interpreted or verified. Although these results are discouraging, there are also positive results. The average evolution curve of the four selected hit patterns has a slope change at around 70% of the lifetime. This is different from the other features studied and indicates that pattern features are able to capture important information from the AE signal.

The results show that the methodology is a valuable tool for detailed health monitoring. By using a conventional AE feature such as the maximum amplitude in an interval, valuable information about the location and the evolution of multiple AE sources was obtained. The results obtained using other hit-based features indicate that they are not good for this task. The sources are identified and monitored using the paths in the resulting intensity images. The nature of sources can also be determined by studying the paths (evolving or stationary source) or clusters (rubbing) in the images. Furthermore, by bandpass filtering the AE signal and studying each subband separately, band-limited AE sources can be identified and monitored. These sources may otherwise pass unnoticed.

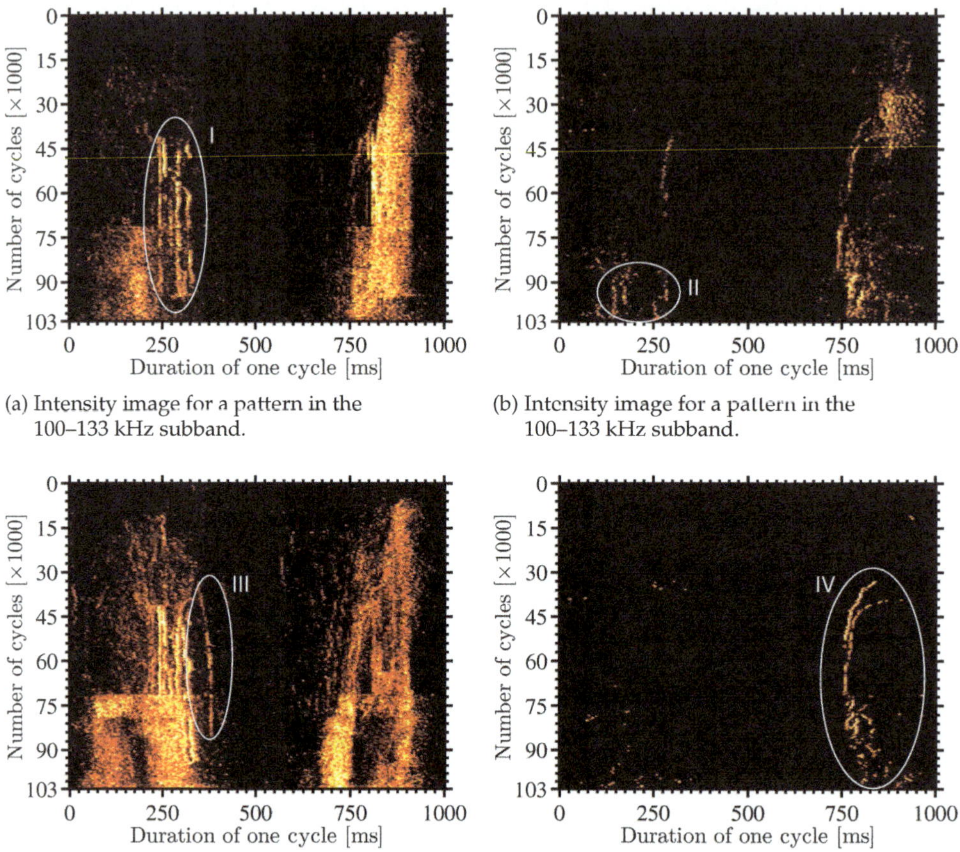

(a) Intensity image for a pattern in the 100–133 kHz subband.

(b) Intensity image for a pattern in the 100–133 kHz subband.

(c) Intensity image for a pattern in the 133–166 kHz subband.

(d) Intensity image for a pattern in the 133–166 kHz subband.

Fig. 17. Intensity images corresponding to four handpicked patterns. The brightness of each pixel is based on how often a pattern is observed in the corresponding interval.

From the resulting intensity images, the initiation of two AE sources was identified and tracked until delamination of the foot. Visual inspection of the intensity images revealed that the corresponding paths initiated where the AE from the splinters was located. This suggested that the corresponding damages were caused by the splinters. Hence, the paths can be used to significantly improve our understanding of the changes occurring in the material, and lead to more focused analysis of the AE signal. The analysis becomes more focused and detailed because it is possible to locate the AE from one particular AE source in the subband of interest. This means that the detection of paths is important for all further analysis and interpretation of the AE signal. The important task is therefore to find all existing patterns.

The results show that the AE patterns made from inter-spike intervals (ISI) can be used to track the locations of AE sources. The paths obtained using the patterns are nearly the same

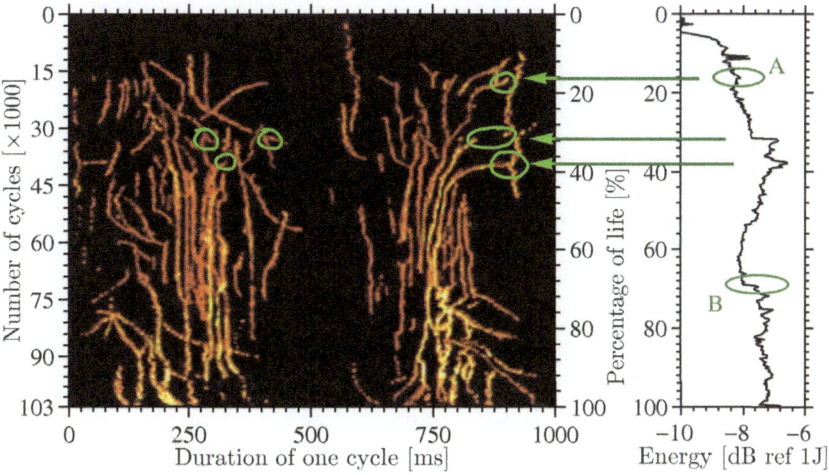

Fig. 18. On the left is a composite image made by overlaying and enhancing the results from 32 patterns. On the right is the evolution of the AE energy.

as the those obtained using the AE energy. A significant improvement was accomplished by combining the ISI information with the peak amplitude of each hit. This is because the addition of the amplitude acts like a filter on the results obtained using ISI patterns. Hence, by adding the peak amplitude it is possible to track evolving sources when they are buried in AE from delamination or rubbing. It can also be used to do the opposite: to filter out and monitor the evolution of a rubbing AE.

6. Conclusion and future research

This chapter presented a methodology for identifying and monitoring evolving AE sources and demonstrated it using experimental data acquired during fatigue testing of assembled CFRP prosthetic feet. By using this methodology, both frictional AE and AE from evolving damage growth could be identified, located and tracked. Furthermore, because AE from specific AE sources can be identified and isolated, further analysis of the AE can be made more effective. The methodology can be used to study changes, or artifacts, in AE or other signals using either periodic or aperiodic reference signals. Examples of signals which can be studied include AE from relays and valves (aperiodic), vibrational data, charging/discharging of batteries, etc.

The chapter also presented a procedure for extracting/generating an AE hit pattern feature. The AE hit pattern feature is a technique for: a) fusing the features extracted from AE hits, b) combining the fused data, and c) finding and locating patterns which appear within the fused data representation. The favourable results show that the AE hit pattern features can be used to extract valuable information from the acquired AE signal.

Despite the promising results, there is a lot of room for improvement. All analysis of the AE signal strongly depends on the detection of clusters or evolving paths in the intensity images. For this reason, different signal and image processing techniques need to be investigated at each step, from subband filtering to feature extraction, and from feature extraction to the generation of the 3D surfaces and/or intensity images. The investigation should also include a further in-depth study of AE features, e.g. the effect of using different coding for the hit pattern feature or different pattern lengths.

Furthermore, automation is needed. The identification and selection of hit patterns for AE tracking was done manually through visual inspection of intensity images. It is unlikely that the same patterns will be useful for other test specimens. Furthermore, evolving paths only appear in certain intensity images, depending both on the type of feature and on the subband. Hence, in order to make both the method and the AE hit pattern feature attractable an automatic detection of paths and patterns in the intensity images is needed.

Finally, if one wishes to use a reference signal other than relative time, e.g. load or displacement, then the methodology must be augmented to handle reference signals with both a variable rate and changing dynamic range.

7. Acknowledgements

The author wishes to thank Össur hf. for both providing prosthetic feet for testing and access to their testing facilities. Furthermore, the author wants to acknowledge the financial support from the University of Iceland Research Fund, the Icelandic Research Council Research Fund, the Icelandic Research Council Graduate Research Fund and Landsvirkjun's Energy Research Fund.

8. References

Bohse, J. (2000). Acoustic emission characteristics of micro-failure processes in polymer blends and composites, *Composites Science and Technology* 60(8): 1213–1226.

Carlos, M. F. & Vallen, H. (2005). Acoustic Emission Signal Processing, *in* P. O. Moore (ed.), *Acoustic Emission Testing*, 3rd edn, Vol. 6 of *Nondestructive testing handbook*, American Society for Nondestructive Testing, Inc., Columbus, pp. 153–154.

De Groot, P. J., Wijnen, P. A. M. & Janssen, R. B. F. (1995). Real-time frequency determination of acoustic emission for different fracture mechanisms in carbon/epoxy composites, *Composites Science and Technology* 55(4): 405–412.

Degrieck, J. & Paepegem, W. V. (2001). Fatigue damage modeling of fibre-reinforced composite materials: Review, *Applied Mechanics Reviews* 54(4): 279–300.

Dixon, S., Pampalk, E. & Widmer, G. (2003). Classification of dance music by periodicity pattern, *International Symposium on Music Information Retrieval*, ISMIR.

Duesing, L. A. (1989). Acoustic emission testing of composite materials, *Reliability and Maintainability Symposium, 1989. Proceedings., Annual* pp. 128–134.

Dzenis, Y. A. & Qian, J. (2001). Analysis of microdamage evolution histories in composites, *International Journal of Solids and Structures* 38(10-13): 1831–1854.

Giordano, M., Calabro, A., Esposito, C., D'Amore, A. & Nicolais, L. (1998). An acoustic-emission characterization of the failure modes in polymer-composite materials, *Composites Science and Technology* 58(12): 1923–1928.

Green, E. R. (1998). Acoustic emission in composite laminates, *Journal of Nondestructive Evaluation* 17(3): 117–127.

Halverson, H. G., Curtin, W. A. & Reifsnider, K. L. (1997). Fatigue life of individual composite specimens based on intrinsic fatigue behavior, *International Journal of Fatigue* 19(5): 369–377.

Hellier, C. J. (2001). *Handbook of Nondestructive Evaluation*, McGraw-Hill, New York.

Iwamoto, M., Ni, Q.-Q., Fujiwara, T. & Kurashiki, K. (1999). Intralaminar fracture mechanism in unidirectional CFRP composites: Part I: Intralaminar toughness and AE characteristics, *Engineering Fracture Mechanics* 64(6): 721–745.

Kamala, G., Hashemi, J. & Barhorst, A. A. (2001). Discrete-Wavelet Analysis of Acoustic Emissions During Fatigue Loading of Carbon Fiber Reinforced Composites, *Journal of Reinforced Plastics and Composites* 20(3): 222–238.

Mercer, C. A. (2001). Removing phase delay using phaseless filtering - a phaseless filtering technique that eliminates time lag. http://www.prosig.com/signal-processing/PhaselessFiltering.html.

Mouritz, A. P. (2003). Non-destructive evalutation of damage accumulation, *in* B. Harris (ed.), *Fatigue in Composites*, Woodhead Publishing Ltd., Cambridge, pp. 242–266.

Nayeb-Hashemi, H., Kasomino, P. & Saniei, N. (1999). Nondestructive evaluation of fiberglass reinforced plastic subjected to combined localized heat damage and fatigue damage using acoustic emission, *Journal of Nondestructive Evaluation* 18(4): 127–137.

Olson, H. F. (1967). *Music, Physics and Engineering*, 2nd ed. edn, Dover Publications, New York.

Prosser, W. H. (1996). Applications of Advanced Waveform-Based AE Techniques for Testing Composite Materials, *Proceedings of the SPIE conference on Nondestructive Evaluation Techniques for Aging Infrastructure and Manufacturing: Materials and Composites*, Scottsdale, Arizona, pp. 146–153.

Prosser, W. H., Jackson, K. E., Kellas, S., Smith, B. T., McKeon, J. & Friedman, A. (1997). Advanced waveform-based acoustic emission detection of matrix cracking in composites, *NDT and E International* 30(2): 108.

Rieke, F., Warland, D., de Ruyter van Steveninck, R. & Bialek, W. (1997). *Spikes: Exploring the Neural Code*, Computational Neuroscience, The MIT Press.

Sethares, W. A., Morris, R. D. & Sethares, J. C. (2005). Beat tracking of musical performances using low-level audio features, *Speech and Audio Processing, IEEE Transactions on* 13(2): 275–285.

Tsamtsakis, D., Wevers, M. & De Meester, P. (1998). Acoustic Emission from CFRP Laminates During Fatigue Loading, *Journal of Reinforced Plastics and Composites* 17(13): 1185–1201.

Unnthorsson, R. (2008). *Acoustic Emission Monitoring of CFRP Laminated Composites Subjected to Multi-axial Cyclic Loading*, Ph.D. dissertation, University of Iceland.

Unnthorsson, R., Pontoppidan, N. H. & Jonsson, M. T. (2005). Extracting Information from Conventional AE Features for Fatigue onset Damage Detection in Carbon Fiber Composites, *The 59th meeting of the Society for Machinery Failure Prevention Technology*, Society for Machinery Failure Prevention Technology, Virginia Beach, USA, pp. 293–302.

Unnthorsson, R., Runarsson, T. P. & Jonsson, M. T. (2007a). Monitoring The Evolution of Individual AE Sources In Cyclically Loaded FRP Composites, *Journal of Acoustic Emission* 25(January-December): 253–259.

Unnthorsson, R., Runarsson, T. P. & Jonsson, M. T. (2007b). On Using AE Hit Patterns for Monitoring Cyclically Loaded CFRP, *Journal of Acoustic Emission* 25(January-December): 260–266.

Unnthorsson, R., Runarsson, T. P. & Jonsson, M. T. (2008a). Acoustic Emission Based Fatigue Failure Criterion for CFRP, *International Journal of Fatigue* 30(1): 11–20. doi:10.1016/j.ijfatigue.2007.02.024.

Unnthorsson, R., Runarsson, T. P. & Jonsson, M. T. (2008b). AE Feature for Early Failure Warning of CFRP Composites Subjected to Cyclic Fatigue, *Journal of Acoustic Emission* 26(January-December): 229–239.

Wevers, M. (1997). Listening to the sound of materials: acoustic emission for the analysis of material behaviour, *NDT and E International* 30(2): 99–106.

7

New In-Situ Characterization Technique of Active Materials in Batteries: Electrochemical Acoustic Emission Method

Hiroshi Inoue
Osaka Prefecture University
Japan

1. Introduction

Active materials, reversible reactants at the positive and negative electrodes, in secondary batteries like lithium ion batteries and nickel-metalhydride (Ni-MH) batteries etc. expand and shrink their volume in charging and discharging processes. For example, in Ni-MH batteries, hydrogen storage alloy negative electrodes initially form hydrogen atoms (H_{ad}) on the alloy surface during the charging process (eq. 1).

$$H_2O + e^- \text{-->} H_{ad} + OH^- \qquad (1)$$

The resultant hydrogen atoms are absorbed into the alloy particles (eq. 2) while in the final stage of the charging process hydrogen evolution also occurs on the alloy surface (eq. 3).

$$H_{ad} \text{-->} H_{ab} \qquad (2)$$

$$2H_{ad} \text{-->} H_2 \qquad (3)$$

Here, H_{ab} is a hydrogen atom absorbed into alloy particles. When hydrogen atoms are absorbed into the alloy particles, their lattice volume expands. When the alloys absorb hydrogen atoms more and more, they eventually crack due to excessive strain. Consequently, fresh alloy surfaces appear, leading to improvement in reactivity for hydrogen absorption and discharge capacity, but the progress of the cracking causes the degradation of the alloys, leading to the decrease in discharge capacity. Therefore, characterization of the cracking, i.e. when and how the alloys crack, is important for the design of new alloys with higher performance. However, thus far, cracking of the alloys during charging has only been confirmed *ex situ* from scanning electron microscopy of the alloy surface.

Some *in situ* characterization techniques for the change in lattice volume of hydrogen storage alloys in charging have been developed. For example, *in situ* X-ray and neutron diffractometry clarified the relationship between lattice volume and state of charge, and phase transformation during charging in detail. On the other hand, *in situ* scanning tunneling microscopy demonstrated reversible volume expansion of the secondary phase in a V-Ti-Ni-based multiphase alloy. However, these techniques are so skilled to get satisfactory data that simpler *in situ* techniques are desirable.

The acoustic emission (AE) method is a nondestructive and sensitive technique for detecting elastic waves generated by the cracking of materials. In addition, the AE method is a real time *in situ* analytical technique, and it's a low-priced method compared to X-ray and neutron diffractometry and STM. Physical phenomena such as corrosion, oxide formation, gas evolution, thin film rupture, cracking etc. can be discriminated by the AE method. In this chapter, I would like to introduce some examples in which the AE method was applied to the direct analysis of physical phenomena linked to electrochemical processes occurring at some hydrogen storage alloy negative electrodes during charging, including cracking of alloy particles and hydrogen evolution.

2. Experimental details

Ingots of $MmNi_{3.6}Mn_{0.4}Al_{0.3}Co_{0.7}$ (Mm: Mischmetal) and $TiCr_{0.3}V_{1.8}Ni_{0.3}$ alloys were prepared by an arc melting method in an Ar atmosphere. The mischmetal is a mixture of rare earth elements (La, Ce, Pr, Nd etc.). Each alloy ingot was ground into powder, and sieved into powders with particle sizes of 106-125 μm and less than 25 μm for the $MmNi_{3.6}Mn_{0.4}Al_{0.3}Co_{0.7}$ alloy and 75-106 μm and 25-53 μm for the $TiCr_{0.3}V_{1.8}Ni_{0.3}$ alloy. Each alloy powder was mixed with Cu powder with a mass ratio of 1:3, and the mixture was subjected to a pressure of 200 MPa at room temperature for 1 min to yield pellets for a negative electrode.

A typical experimental cell assembly for AE monitoring is shown in Fig. 1. The pellet-type negative electrode, a sulfonated-polypropylene separator and a $Ni(OH)_2$/NiOOH positive electrode were stacked as shown in Fig. 1. A nickel sheet was used as a current collector for the negative electrode and protected the AE sensor from a 6M KOH aqueous solution which was used as the electrolyte. Before fabricating the cell assembly, the separator was thoroughly soaked in a 6M KOH aqueous solution. An Hg/HgO electrode was used as the reference electrode.

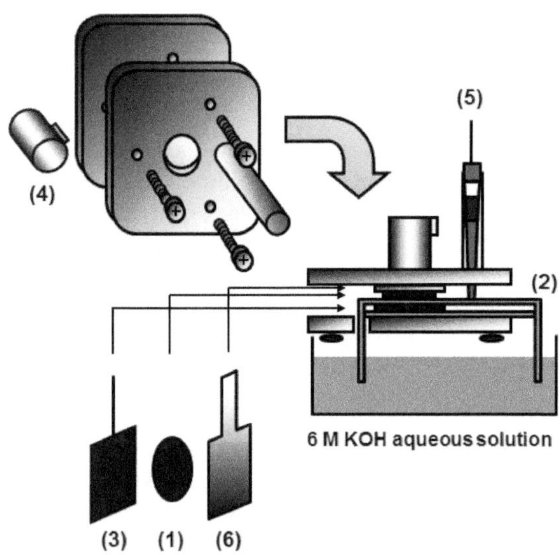

Fig. 1. A typical experimental cell assembly for monitoring AE signals. (1) Negative electrode, (2) Separator, (3) Positive electrode, (4) AE sensor, (5) Reference electrode, (6) Ni current collector.

The $MmNi_{3.6}Mn_{0.4}Al_{0.3}Co_{0.7}$ electrodes were charged at 100 mA g(alloy)$^{-1}$ for 3 h and discharged at 50 mA g(alloy)$^{-1}$ to -0.65 V vs. Hg/HgO, while the $TiV_{1.8}Cr_{0.3}Ni_{0.3}$ electrodes were charged at 100 mA g(alloy)$^{-1}$ for 6 h and discharged at 50 mA g(alloy)$^{-1}$ to -0.75 V vs. Hg/HgO. The unit "g(alloy)" means mass of alloy. After every charging the circuit was opened for 10 min. A setup for AE monitoring is shown in Fig. 2. An automatic AE monitoring system (NF Electronic Instruments, 7600/0710) was used for AE monitoring with an AE transducer (NF electronic Instruments, AE-900S-WB) and a preamplifier (30 dB gain for $MmNi_{3.6}Mn_{0.4}Al_{0.3}Co_{0.7}$ and 40 dB gain for $TiV_{1.8}Cr_{0.3}Ni_{0.3}$). The elastic waves generated by various phenomena such as cracking were detected by the AE transducer and transformed into AE signals. The AE signals were amplified by the preamplifier and stored in the automatic AE monitoring system. The stored data were displayed as time histories, power spectra and AE waveforms of the AE signals. The power spectra and AE waveforms express the distribution of frequencies for the AE signals and duration of the AE signals, respectively. All experiments were carried out at room temperature.

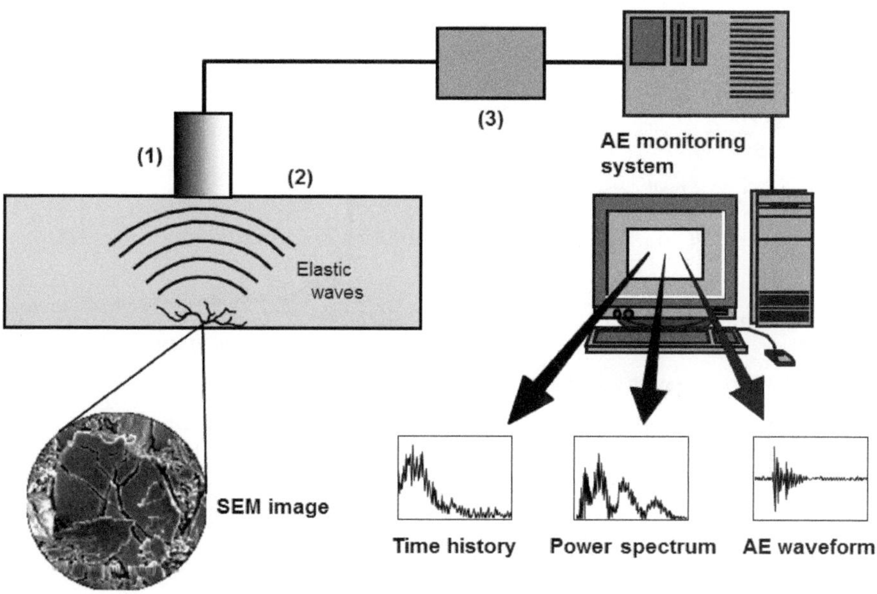

Fig. 2. A setup for AE monitoring. (1) AE transducer, (2) Sample, (3) Preamplifier.

3. Acoustic emission for hydrogen evolution on a copper electrode

During charging, hydrogen evolution occurs on the surface of the hydrogen storage alloy negative electrodes as a side reaction of the hydrogen absorption into the alloy. Not only the cracking of alloy particles but also the release of hydrogen bubbles on the electrode surface can be detected as AE signals. Figure 3 shows typical AE waveforms and power spectra for hydrogen evolution on a Cu electrode at -1.0 and -1.1 V vs. Hg/HgO. The AE waveforms at both applied potentials had long durations of over 0.15 ms. The AE waveform at –1.1 V had a larger maximum amplitude and longer duration than that at -1.0 V, suggesting hydrogen evolution at -1.1 V was more vigorous. As for power spectra, both had a relatively narrow frequency distribution of approximately 0.6 MHz and maximum amplitude at about 0.1

MHz. The AE waveforms and power spectra in Fig. 3 did not change with time. Sawai et al. have also obtained an AE waveform and power spectrum for hydrogen evolution when a sodium hydrogencarbonate aqueous solution was neutralized by dilute sulfuric acid. The present data for electrochemical hydrogen evolution were quite similar to their data. Therefore, the electrochemical hydrogen evolution is characterized by the AE waveform with a long duration of over 0.15 ms and a power spectrum with a relatively narrow frequency distribution of around 0.6 MHz and the maximum amplitude at about 0.1 MHz.

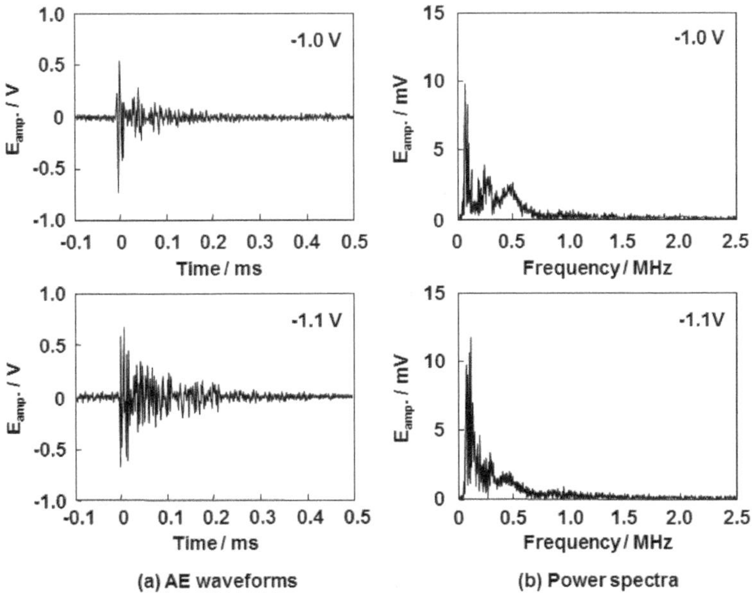

Fig. 3. (a) AE waveforms and (b) power spectra for hydrogen evolution on a Cu electrode at -1.0 and -1.1 V vs. Hg/HgO in a 6 M KOH aqueous solution. (Reprinted with permission from *Electrochem. Solid-State Lett.*, 9, A504 (2006). Copyright 2006 The Electrochemical Society.)

4. Acoustic emission from hydrogen storage alloy negative electrodes while charging

4.1 $MmNi_{3.6}Mn_{0.4}Al_{0.3}Co_{0.7}$ electrode

AE waveforms and power spectra for a $MmNi_{3.6}Mn_{0.4}Al_{0.3}Co_{0.7}$ (106-125 μm) negative electrode were measured every 1 h during the 1st charging process and summarized in Fig. 4. AE waveforms and power spectra observed for the initial 1 h were quite different from those identified as the hydrogen evolution. In the former, AE waveforms have a sharp spike with duration shorter than 0.1 ms and larger amplitude, and power spectra have wide frequency distribution, particularly some peaks at frequencies more than 0.6 MHz, which is not characteristic of hydrogen evolution. The AE waveforms and power spectra, however, changed to those due to hydrogen evolution between 1 h and 2 h after start of the charging process. During the first half of the 1st charging process, a phenomenon other than hydrogen evolution, probably cracking of individual alloy particles, must have occurred. From the comparison of scanning electron micrographs (see Fig. 7) for the surface of

$MmNi_{3.6}Mn_{0.4}Al_{0.3}Co_{0.7}$ particles before and after the 1st charge-discharge cycle, a number of large and small cracks were observed on the alloy surface after the charge-discharge cycle. Therefore, it is inferred that the alloy particles cracked in the first half of the charging process. Thus the cracking of alloy particles is characterized by a burst-type AE waveform with duration shorter than 0.1 ms and large amplitude, and a power spectrum with wide frequency distribution, particularly with some peaks at frequencies more than 0.6 MHz.

In the 2nd charging process, AE waveforms and power spectra for the $MmNi_{3.6}Mn_{0.4}Al_{0.3}Co_{0.7}$ electrode measured every 1 h are summarized in Fig. 5. The AE waveforms after 1 h and 2 h had a sharp spike with duration shorter than 0.1 ms, but their amplitude was much smaller than that in the 1st charging process. In addition, power spectra after 1 h and 2h had wide frequency distribution. These results clearly exhibit that the cracking of alloy particles occurs at 1 h and 2 h after start of the 2nd charging process. On the other hand, the AE waveforms and power spectra at the beginning and 3 h suggested hydrogen evolution. Thus, in the 2nd charging process the cracking occurred preferentially after 1 h and 2 h after start of charging while hydrogen evolution proceeded throughout charging, but it was not vigorous compared to the 1st charging process.

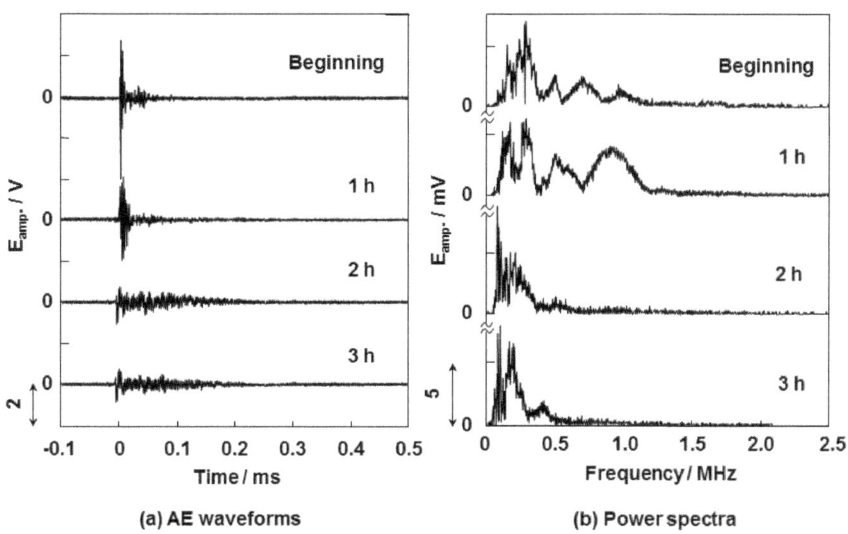

Fig. 4. (a) AE waveforms and (b) power spectra for electrochemical phenomena on a $MmNi_{3.6}Mn_{0.4}Al_{0.3}Co_{0.7}$ (106-125 μm) negative electrode measured every 1 h during the 1st charging process. (Reprinted with permission from *Electrochem. Solid-State Lett.*, 9, A504 (2006). Copyright 2006 The Electrochemical Society.)

Figure 6 shows initial activation behavior for a $MmNi_{3.6}Mn_{0.4}Al_{0.3}Co_{0.7}$ electrode and time history of frequency of AE signals and time course of electrode potential in the 1st, 2nd or 6th charging process. In each time history, the frequency of AE signals is defined as the number of AE signals with amplitude over a threshold of 0.5 V for one minute. In the initial activation behavior, discharge capacity was ca. 140 mAh $g(alloy)^{-1}$ at 1st cycle, and then it increased with cycle number and approximately reached a constant value (ca. 220 mAh $g(alloy)^{-1}$) at the 4th cycle, indicating that the initial activation was completed for four charge-discharge cycles.

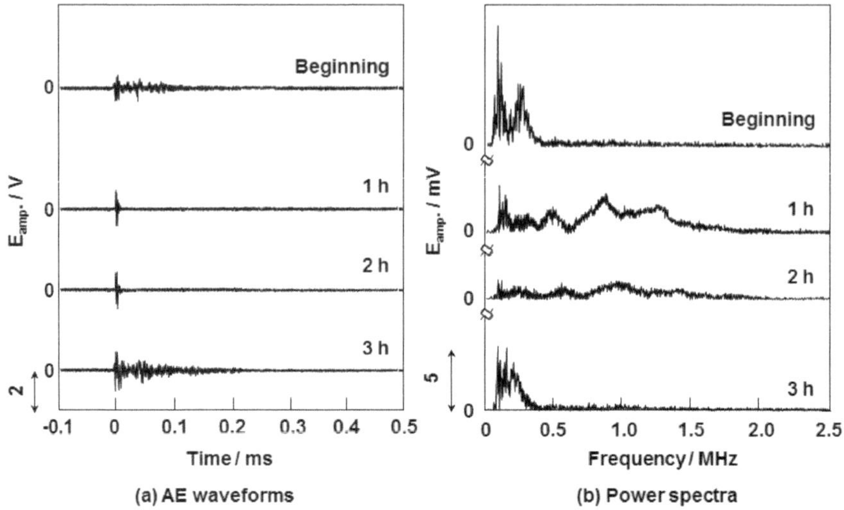

Fig. 5. (a) AE waveforms and (b) power spectra for electrochemical phenomena on a $MmNi_{3.6}Mn_{0.4}Al_{0.3}Co_{0.7}$ (106-125 μm) negative electrode measured every 1 h during the 2nd charging process. (Reprinted with permission from *J. Alloys Compd.*, 446-447, 681 (2007). Copyright 2007 Elsevier.)

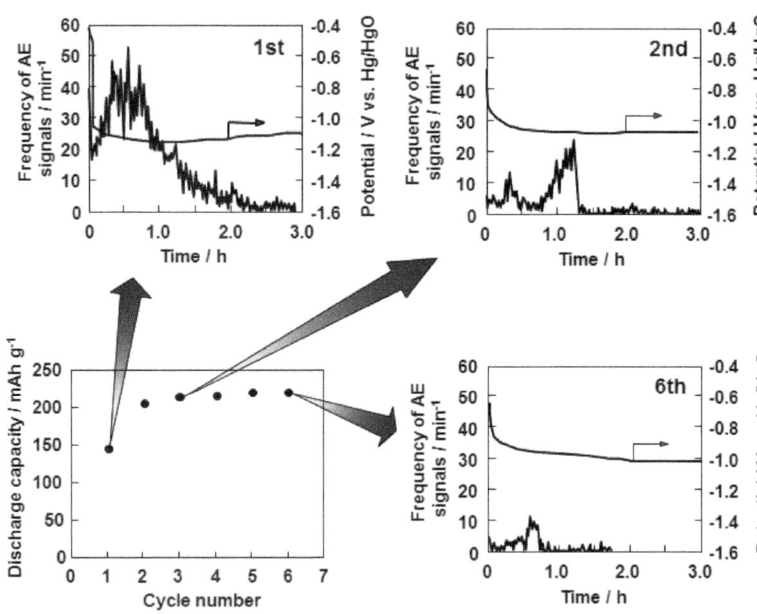

Fig. 6. Initial activation behavior for a $MmNi_{3.6}Mn_{0.4}Al_{0.3}Co_{0.7}$ (106-125 μm) negative electrode and its time history of frequency of AE signals and electrode potential in 1st, 2nd and 6th charging processes. (Reprinted with permission from *J. Alloys Compd.*, 446-447, 681 (2007). Copyright 2007 Elsevier.)

In the first half of the 1st charging process, a large number of AE signals were observed. This comes from the cracking based on lattice expansion due to hydrogen absorption, which has also been confirmed by the AE wave forms and power spectra in the same period (Fig. 4). The electrode potential reached ca. -1.1 V immediately after start of charging and then it hardly changed during the charging. This suggests that hydrogen evolution proceeds throughout the charging. Figure 7 shows scanning electron micrographs of the $MmNi_{3.6}Mn_{0.4}Al_{0.3}Co_{0.7}$ electrode surface before charging and after the 1st, 2nd and 6th cycles. As described above, a number of large and small cracks appeared on the alloy surface after the 1st charge-discharge cycle, supporting a large number of AE signals observed in Fig. 6. In the 2nd charging process, the frequency of AE signals greatly decreased compared to the 1st charging process as shown in Fig. 6, and the scanning electron micrograph after 2nd charging showed that smaller cracks increased compared to after 1st charging. These results indicate that the fine cracking of alloy particles occur in the 2nd charging process. After 6th cycle, the number of AE signals significantly decreased and discharge capacity showed a constant value as shown in Fig. 6. A lot of finer cracks were observed after 6th cycle as shown in Fig. 7. Thus, *in situ* analysis by the AE technique confirms the well-known conclusion that during initial activation the $MmNi_{3.6}Mn_{0.4}Al_{0.3}Co_{0.7}$ particles frequently cracks and the increase in the cracks with cycle number leads to the increase in discharge capacity.

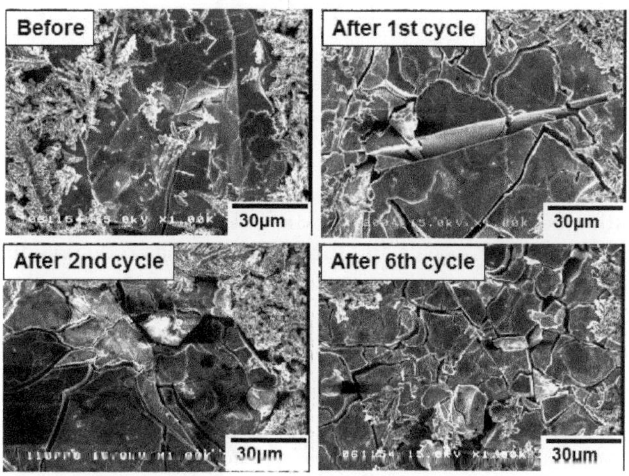

Fig. 7. Scanning electron micrographs of a $MmNi_{3.6}Mn_{0.4}Al_{0.3}Co_{0.7}$ (106-125 μm) negative electrode surface before charging and after 1st, 2nd and 6th charging processes. (Reprinted with permission from *J. Alloys Compd.*, 446-447, 681 (2007). Copyright 2007 Elsevier.)

The pattern of cracking during charging may change with size of alloy particles. The cracking behavior of small alloy particles was also characterized by the AE technique. Figure 8 shows initial activation behavior for a $MmNi_{3.6}Mn_{0.4}Al_{0.3}Co_{0.7}$ (less than 25 μm) electrode and time history of frequency of AE signals and time course of electrode potential in the 1st, 2nd or 6th charging process. In this case, discharge capacity was ca. 220 mAh $g(alloy)^{-1}$ even at the 1st cycle, and then it was almost constant. This indicates that the small alloy particles do not have to be activated. In the time history at the 1st cycle, there were very few AE signals compared to the time history for the larger alloy particles (see Fig. 6), suggesting that the cracking of the small alloy particles poorly occurred probably because the lattice stress of the small particles with lattice expansion was not so large as larger particles. The similar results were observed in the 2nd and 6th cycles. Figure 9 shows scanning electron micrographs of the $MmNi_{3.6}Mn_{0.4}Al_{0.3}Co_{0.7}$ (less than 25 μm) electrode surface before charging and after 1st, 2nd and 6th cycles. There were very few cracks even after the 1st cycle, which proves poor cracking of the small ally particles. Thus the small particles with sizes of less than 25 μm are hard to crack even though they absorb a great amount of hydrogen in charging.

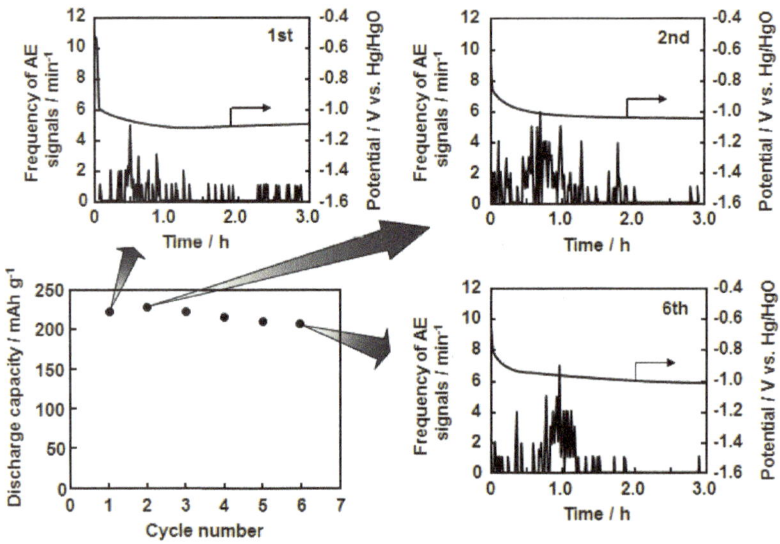

Fig. 8. Initial activation behavior for a $MmNi_{3.6}Mn_{0.4}Al_{0.3}Co_{0.7}$ (less than 25 μm) electrode and time history of frequency of AE signals and time course of electrode potential in the 1st, 2nd or 6th charging process.

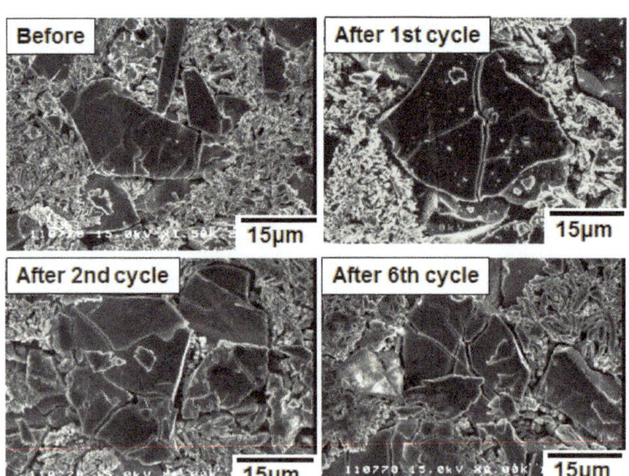

Fig. 9. Scanning electron micrographs of a $MmNi_{3.6}Mn_{0.4}Al_{0.3}Co_{0.7}$ (less than 25 μm) negative electrode surface before charging and after 1st, 2nd and 6th charging processes.

Figure 10 shows AE waveforms and power spectra for the $MmNi_{3.6}Mn_{0.4}Al_{0.3}Co_{0.7}$ (less than 25 μm) negative electrode measured every 1 h during the 1st charging process. AE waveforms and power spectra at 1 h after start of charging suggested the cracking of particles, but otherwise hydrogen evolution was dominant. Moreover, in the 1st charging process electrode potential reached ca. -1.1 V immediately after start of charging and then it hardly changed during the charging (Fig. 8), suggesting that hydrogen evolution proceeded throughout the charging.

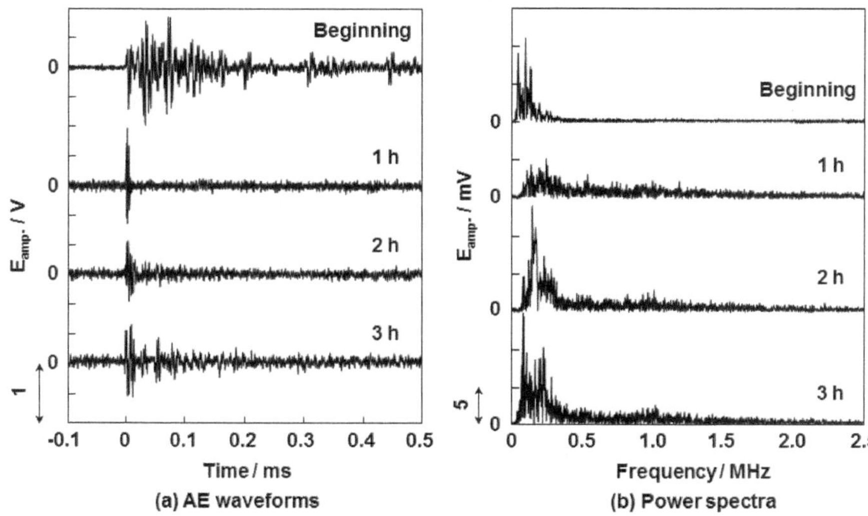

Fig. 10. (a) AE waveforms and (b) power spectra for electrochemical phenomena on a $MmNi_{3.6}Mn_{0.4}Al_{0.3}Co_{0.7}$ (less than 25 μm) negative electrode measured every 1 h during the 1st charging process.

4.2 $TiV_{1.8}Cr_{0.3}Ni_{0.3}$ electrode

V-based hydrogen storage alloys with a body-centered cubic (BCC) structure as a main phase are potential negative electrode active materials for nickel-metal hydride batteries and hydrogen reservoirs for fuel cells because of high hydrogen storage capacity per volume. A $TiV_{2.1}Ni_{0.3}$ negative electrode, one of V-based alloys prepared by the Author's laboratory, had high discharge capacity of 460 mAh g^{-1} and poor cycle durability. To improve cycle durability, it is effective to substitute Cr for the V constituent of $TiV_{2.1}Ni_{0.3}$ because the oxidative dissolution of V into an electrolyte solution is suppressed. In the preliminary experiments, a $TiV_{1.8}Cr_{0.3}Ni_{0.3}$ electrode significantly improved cycle durability. So the charging behavior of the $TiCr_{0.3}V_{1.8}Ni_{0.3}$ electrode was analyzed by the AE technique.

Time history of frequency of AE signals and electrode potential in the 1st, 2nd and 6th charging processes and initial activation behavior for the $TiV_{1.8}Cr_{0.3}Ni_{0.3}$ (75-106 μm) electrode were summarized in Fig. 11. In the 1st charging process, a large number of AE signals were observed for initial 2 h. Since electrode potential was ca. -1.1 V in this term, hydrogen evolution is also expected to proceed.

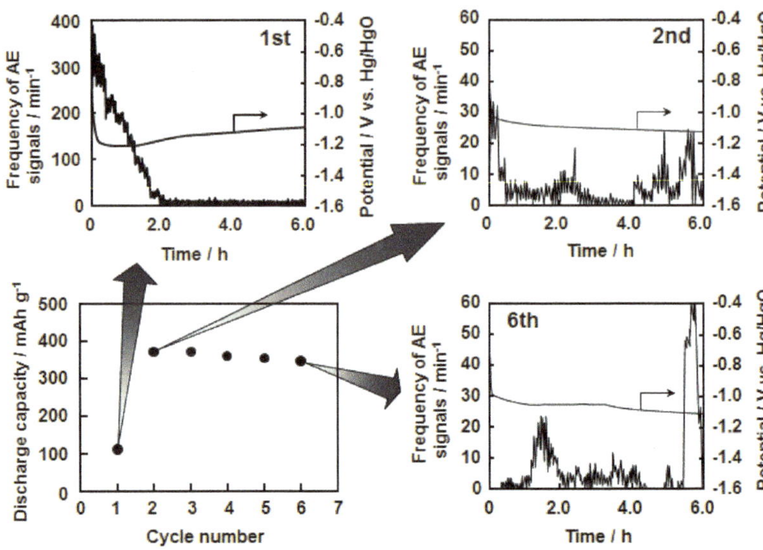

Fig. 11. Initial activation behavior for a $TiV_{1.8}Cr_{0.3}Ni_{0.3}$ (75-106 μm) electrode and time history of frequency of AE signals and time course of electrode potential in the 1st, 2nd or 6th charging process. (Reprinted with permission from *J. Alloys Compd.*, 446-447, 681 (2007). Copyright 2007 Elsevier.)

Figure 12 and Fig. 13 show AE waveforms and power spectra for the $TiV_{1.8}Cr_{0.3}Ni_{0.3}$ (75-106 μm) electrode measured every 1 h during the 1st charging process. Both the AE waveform and power spectrum at 1 h exhibit that hydrogen evolution proceeds, supporting the inference from Fig. 11. In further charging, the AE waveform and power spectrum due to

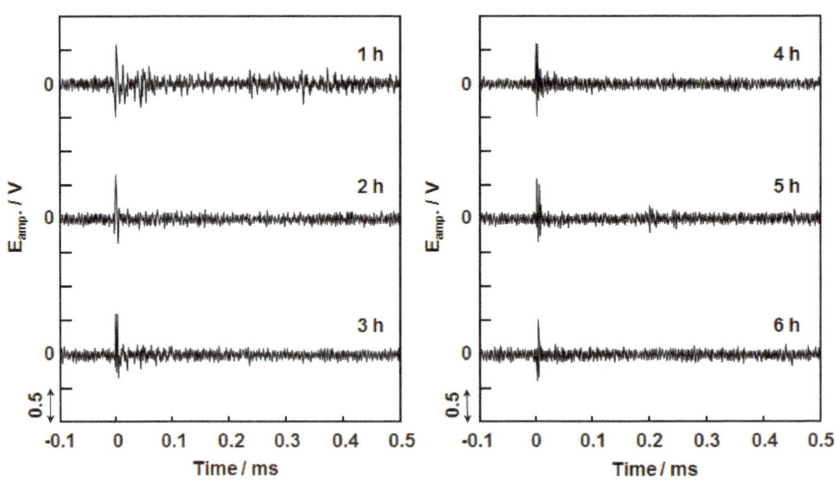

Fig. 12. AE waveforms for electrochemical phenomena on a $TiV_{1.8}Cr_{0.3}Ni_{0.3}$ (75-106 μm) negative electrode measured every 1 h during the 1st charging process. (Reprinted with permission from *J. Alloys Compd.*, 446-447, 681 (2007). Copyright 2007 Elsevier.)

the cracking were mainly observed, but judging from the time history of the frequency of AE signals, the cracking was sporadic. The $TiV_{1.8}Cr_{0.3}Ni_{0.3}$ electrode began to crack later than the $MmNi_{3.6}Mn_{0.4}Al_{0.3}Co_{0.7}$ electrode, suggesting that the former was more patient for cracking than the latter. The maximum amplitude of AE waveforms was just ca. one forth of that for the $MmNi_{3.6}Mn_{0.4}Al_{0.3}Co_{0.7}$ electrode, suggesting that relatively small cracks were formed on the $TiV_{1.8}Cr_{0.3}Ni_{0.3}$ alloy surface. Figure 14 shows scanning electron micrographs of the $TiV_{1.8}Cr_{0.3}Ni_{0.3}$ (75-106 μm) electrode surface before charging and after the 1st, 2nd and 6th cycles. The scanning electron micrograph after 1st cycle exhibits that a few and relatively small cracks were observed on the $TiV_{1.8}Cr_{0.3}Ni_{0.3}$ particles, which agreed with the suggestion from the AE measurements.

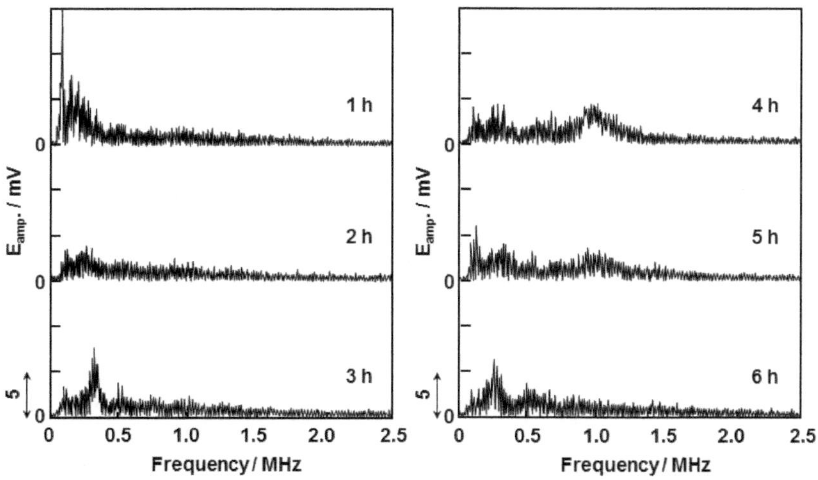

Fig. 13. Power spectra for electrochemical phenomena on a $TiV_{1.8}Cr_{0.3}Ni_{0.3}$ (75-106 μm) negative electrode measured every 1 h during the 1st charging process. (Reprinted with permission from *J. Alloys Compd.*, 446-447, 681 (2007). Copyright 2007 Elsevier.)

In Fig. 11, frequency of AE signals decreased greatly with repeating charge-discharge cycles. The AE waveforms (Fig. 15) and power spectra (Fig. 16) in the 2nd charging process showed that the cracking based on hydrogen absorption occurred from the beginning. As can be seen from charge-discharge cycle performance in Fig. 11, discharge capacity greatly increases for the first 2 cycles due to initial activation. In the initial activation the surface oxides inhibiting hydrogen absorption into alloy seemed to be removed by sporadic cracking of alloy particles during the 1st cycle to expose fresh surface, leading to hydrogen absorption from the beginning of the 2nd charging and the increase in discharge capacity. Moreover, Fig. 14 shows that the $TiV_{1.8}Cr_{0.3}Ni_{0.3}$ particles are not frequently pulverized like the $MmNi_{3.6}Mn_{0.4}Al_{0.3}Co_{0.7}$ particles in initial activation. Nevertheless the former has higher discharge capacity than the latter and is comparable to the $TiV_{2.1}Ni_{0.3}$ electrode. The $TiV_{1.8}Cr_{0.3}Ni_{0.3}$ is also composed of two phases like $TiV_{2.1}Ni_{0.3}$. The main phase has a Ti-V solid solution BCC structure and serves as the main hydrogen reservoir. On the other hand, the secondary phase has a TiNi-type structure and serves as a catalyst for generating hydrogen atoms and transferring them into the main phase. The partial substitution with the Cr component effectively inhibits corrosion and does not bring any serious problems on hydrogen diffusion in alloy and electron transfer at electrode surface.

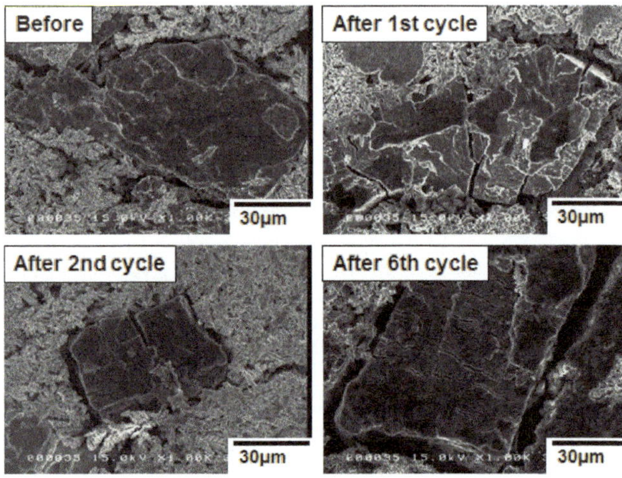

Fig. 14. Scanning electron micrographs of a $TiV_{1.8}Cr_{0.3}Ni_{0.3}$ (75-106 μm) negative electrode surface before charging and after 1st, 2nd and 6th charging processes. (Reprinted with permission from *J. Alloys Compd.*, 446-447, 681 (2007). Copyright 2007 Elsevier.)

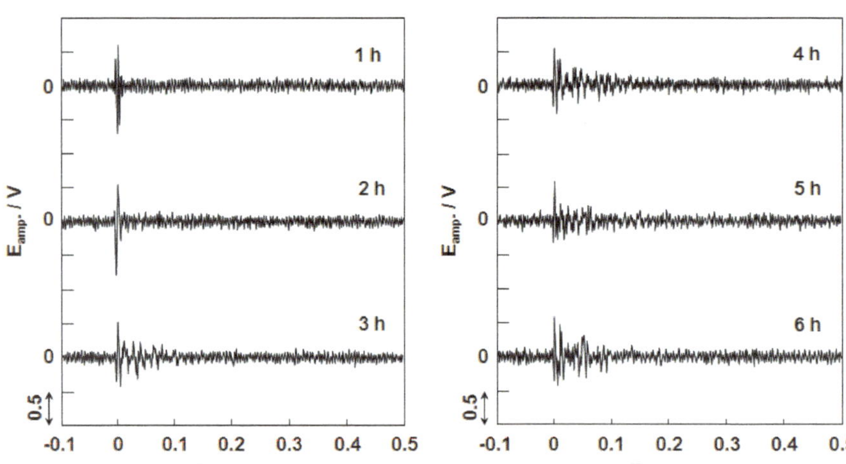

Fig. 15. AE waveforms for electrochemical phenomena on a $TiV_{1.8}Cr_{0.3}Ni_{0.3}$ (75-106 μm) negative electrode measured every 1 h during the 2nd charging process.

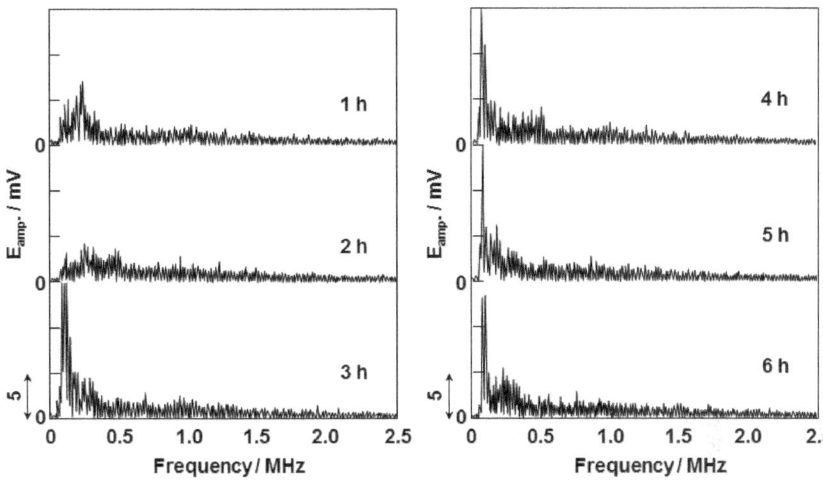

Fig. 16. Power spectra for electrochemical phenomena on a $TiV_{1.8}Cr_{0.3}Ni_{0.3}$ (75-106 μm) negative electrode measured every 1 h during the 2nd charging process.

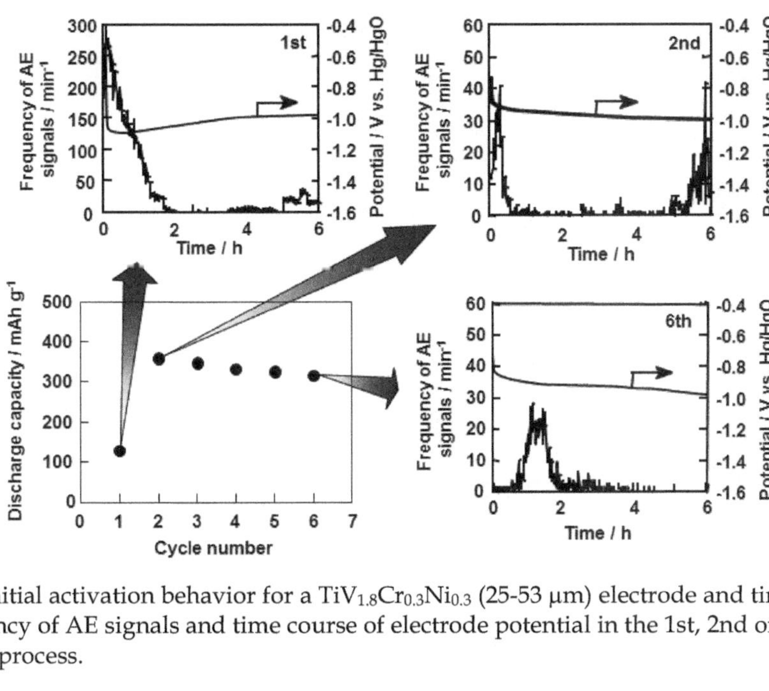

Fig. 17. Initial activation behavior for a $TiV_{1.8}Cr_{0.3}Ni_{0.3}$ (25-53 μm) electrode and time history of frequency of AE signals and time course of electrode potential in the 1st, 2nd or 6th charging process.

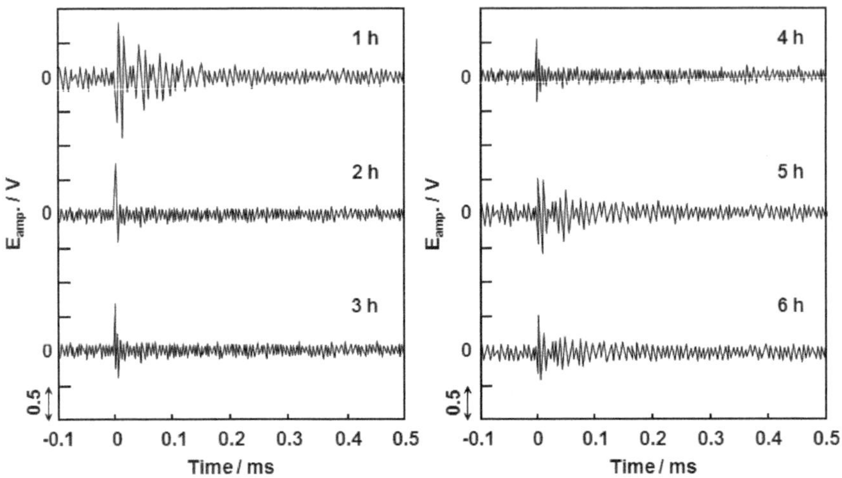

Fig. 18. AE waveforms for electrochemical phenomena on a $TiV_{1.8}Cr_{0.3}Ni_{0.3}$ (25-53 μm) negative electrode measured every 1 h during the 1st charging process.

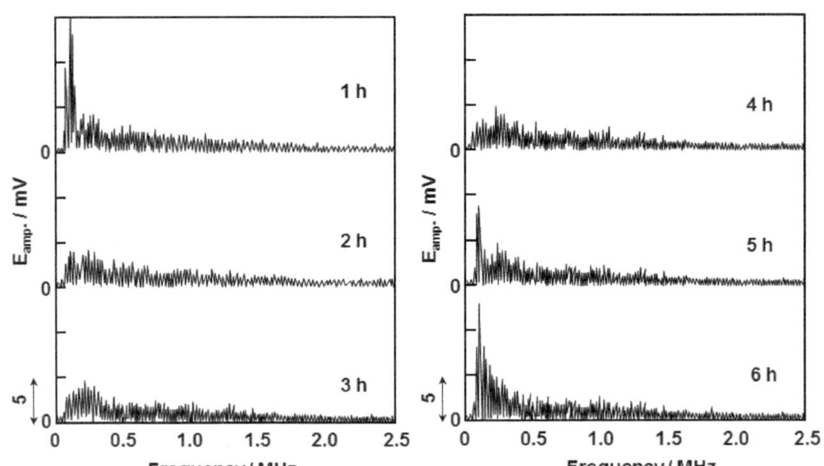

Fig. 19. Power spectra for electrochemical phenomena on a $TiV_{1.8}Cr_{0.3}Ni_{0.3}$ (25-53 μm) negative electrode measured every 1 h during the 1st charging process.

Figure 17 shows initial activation behavior for a $TiV_{1.8}Cr_{0.3}Ni_{0.3}$ (25-53 μm) electrode and time history of frequency of AE signals and time course of electrode potential in the 1st, 2nd or 6th charging process. From Fig. 11 and Fig. 17, charge-discharge cycle performance is independent of particle size of the alloy. As for time history, in the 1st charging process, a large number of AE signals were observed for initial 2 h, which is similar to the alloy particles with sizes of 75-106 μm. The frequency of AE signals significantly decreased with repeating charge-discharge cycling. Electrode potential in this term, however, was -1.0 V which was more positive compared to the $TiV_{1.8}Cr_{0.3}Ni_{0.3}$ (75-106 μm) electrode. In this case hydrogen evolution is expected to proceed but it seems to be poor. This is also suggested from Fig. 18 and Fig. 19. Both AE waveforms and power spectra in the 1st charging process show hydrogen evolution occurs for initial 2 h, and then cracking also occurs. But the term for cracking was shorter than that for the $TiV_{1.8}Cr_{0.3}Ni_{0.3}$ (75-106 μm) electrode, suggesting that the cracking was poorer for smaller alloy particles. Figure 20 shows scanning electron micrographs of the $TiV_{1.8}Cr_{0.3}Ni_{0.3}$ (25-53 μm) electrode surface before charging and after the 1st, 2nd and 6th cycles. The scanning electron micrograph after 1st cycle exhibits that very few cracks were observed on the $TiV_{1.8}Cr_{0.3}Ni_{0.3}$ particles, indicating that the number of cracks decreased with particle size. The number of cracks slightly increased with cycle number.

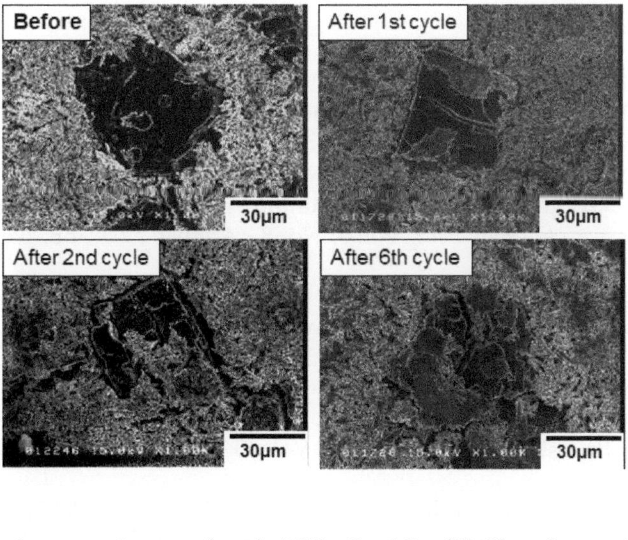

Fig. 20. Scanning electron micrographs of a $TiV_{1.8}Cr_{0.3}Ni_{0.3}$ (25-53 μm) negative electrode surface before charging and after 1st, 2nd and 6th charging processes.

5. Conclusion

In charging of hydrogen storage alloy negative electrodes, $MmNi_{3.6}Mn_{0.4}Al_{0.3}Co_{0.7}$ and $TiV_{1.8}Cr_{0.3}Ni_{0.3}$ electrodes, cracking based on hydrogen absorption into alloy and hydrogen evolution were clearly discriminated using an AE technique in terms of waveforms, power spectra and time history of the AE signals. The $MmNi_{3.6}Mn_{0.4}Al_{0.3}Co_{0.7}$ (106-125 μm) particles cracked intensively in the first half of the 1st charging process and were broken down with many large and small cracks during initial activation. In contrast, the $MmNi_{3.6}Mn_{0.4}Al_{0.3}Co_{0.7}$ particles (less than 25 μm) had high discharge capacity even in the 1st charging process, but it poorly cracked, suggesting that the particle size was enough small to diffuse throughout the particles without cracking. As for the $TiCr_{0.3}V_{1.8}Ni_{0.3}$ particles, irrespective of particle size the cracking sporadically occurred after hydrogen evolution in the 1st charging process and the AE signals for the cracking were very weak compared to those for the $MmNi_{3.6}Mn_{0.4}Al_{0.3}Co_{0.7}$ particles, indicating that the $TiCr_{0.3}V_{1.8}Ni_{0.3}$ particles were hard to crack. Moreover, the smaller particles (25-53 μm) were more resistant to cracking than the larger ones (75-106 μm).

Lithium ion battery (LIB) is the primary candidate for plug-in hybrid electric vehicles and electric vehicles because of their high energy and power density compared to Ni-MH battery. At both electrodes in LIB, lithium intercalation/deintercalation reactions occur. Recently, to improve energy and power density more and more, researches on new negative electrode active materials with high capacity like Si, Sn etc. have been actively performed. Such active material particles are easy to crack with repeated lithium intercalation/deintercalation, leading to capacity loss in a LIB or short lifetime. So the *in situ* real analysis of stress and cracking are important for clarifying the mechanism of capacity degradation, and the electrochemical AE technique must be a powerful tool.

6. References

Bernard, J.; Boinet, M.; Chatenet, M. & Dalard, F. (2005). Contribution of the Acoustic Emission Technique to Study Aluminum Behavior in Aqueous Alkaline Solution. *Electrochem. Solid-State Lett.*, Vol. 8, No. 7, (July 2005), pp. E53-E55, ISSN 0013-4651

Darowicki, K.; Orlikowski, J.; Arutunow, A. & Jurczak, W. (2005). Passive Layer Cracking Studies Performed on A95056 Aluminum Alloy by DEIS and Acoustic Emission. *Electrochem. Solid-State Lett.*, Vol. 8, No. 8, (August 2005), pp. B55-B57, ISSN 0013-4651

Etiemble, A.; Idrissi, H. & Roue, L. (2011). On the Decrepitation Mechanism of MgNi and $LaNi_5$-Based Electrodes Studied by in Situ Acoustic Emission. *J. Power Sources*, Vol. 196, No. 11, (June 2011), pp. 5168-5173, ISSN 0378-7753

Fregonese, M.; Idrissi, H.; Mazille, H.; Renaud, L. & Cetre, Y. (2001). Initiation and Propagation Steps in Pitting Corrosion of Austenitic Stainless Steels: Monitoring by Acoustic Emission. *Corrosion Science*, Vol. 43, No. 4, (April 2001), pp. 627-641, ISSN 0010-928X

Inoue, H.; Miyauchi, R.; Shin-ya, R.; Choi, W. -K. & Iwakura, C. (2002). Charge–Discharge Characteristics of $TiV_{2.1}Ni_{0.3}$ Alloy Surface-Modified by Ball-Milling with Ni or Raney Ni. *J. Alloys Compd.*, Vol. 330-332 (January 2002) pp. 597-600, ISSN 0925-8388

Inoue, H.; Tsuzuki, R.; Nohara, S. & Iwakura, C. (2006). *In Situ* Monitoring of Hydrogen Storage Alloy Negative Electrode during Charging by an Acoustic Emission Technique. *Electrochem. Solid-State Lett.*, Vol. 9, No. 11, (November 2006), pp. A504-A506, ISSN 0013-4651

Inoue, H.; Tsuzuki, R.; Nohara, S. & Iwakura, C. (2007). Characterization of Initial Activation Behavior for Hydrogen Storage Alloys by Acoustic Emission Technique. *J. Alloys Compd.*, Vol. 446-447, (October 2007), pp. 681-686, ISSN 0925-8388

Iwakura, C.; Choi, W. –K.; Miyauchi, R. & Inoue, H. (2000). Electrochemical and Structural Characterization of Ti-V-Ni Hydrogen Storage Alloys with BCC Structure. *J. Electrochem. Soc.*, Vol. 147, No. 7, (July 2000). pp. 2503-2506, ISSN 0013-4651

Iwakura, C.; Inoue, H. & Nohara, S. (2001). Hydrogen–Metal Systems: Electrochemical Reactions (Fundamentals and Applications), In: *Encyclopedia of Materials: Science and Technology*, Buschow, K.H.J.; Cahn, R.; Flemings M.; Ilschner B.; Kramer E.; Mahajan S. & Veyssiere, P., pp. 3929-3941, Elsevier Science, ISBN 10: 0-08-043152-6, Amsterdam

Kalisvaart, W. P.; Latroche, M.; Cuevas, F. & Notten, P. H. L. (2008). In Situ Neutron Diffraction Study on Pd-doped $Mg_{0.65}Sc_{0.35}$ Electrode Material. *J. Solid State Chem.*, Vol. 181, No. 5, (May 2008), pp. 1141-1148, ISSN 0022-4596

Kalnaus, S., Rhodes, K. & Daniel, C. (2011). A Study of Lithium Ion Intercalation Induced Fracture of Silicon Particles Used as Anode Material in Li-ion Battery. *J. Power Sources*, Vol. 196, No. 19, (October 2011), pp. 8116-8124, ISSN 0378-7753

Kuriyama, N.; Chartouni, D.; Tsukahara, M.; Takahashi, K.; Takeshita, H. T.; Tanaka, H.; Schlapbach, L.; Sakai, T. & Uehara, I. (1998). Scanning Tunnelling Microscopy In Situ Observation of Phase-Selective Cathodic Hydrogenation of a V-Ti-Ni-Based Multiphase Alloy Electrode. *Electrochem. Solid-State Lett.*, Vol. 1, No. 1, (January 1998), pp. 37-38, ISSN 0013-4651

Notten, P. H. L.; Daams, J. L. C. & Einerhand, R. E. F. (1994). On the Nature of the Electrochemical Cycling of Non-stoichiometric $LaNi_5$-based Hydride-forming Compounds Part II. In Situ X-ray Diffractiometry. *J. Alloys Compd.*, Vol. 210, No. 1-2, (August 1994), pp. 233-241, ISSN 0925-8388

Ohzuku, T.; Tomura, H. & Sawai, K. (1997). Monitoring of Particle Fracture by Acoustic Emission during Charge and Discharge of Li/MnO_2 Cells. *J. Electrochem. Soc.*, Vol. 144, No. 10, (October 1997), pp. 3496-3500, ISSN 0013-4651

Ramesh, R.; Mukhopadhyay, C. K.; Jayakumar, T. & Raj, B. (1998). Characterization of Acoustic Emission generated during Electrochemical Charging and Discharging of Hydrogen in Palladium. *Scr. Met.*, Vol. 38, No. 4, (January 1998), pp. 661-665, ISSN 1359-6462

Rhodes, K.; Dudney, N.; Lara-Curzio, E. & Daniel, C. (2010). Understanding the Degradation of Silicon Electrodes for Lithium-Ion Batteries Using Acoustic Emission. *J. Electrochem. Soc.*, Vol. 157, No. 12, (December 2010), pp. A1354-A1360, ISSN 0013-4651

Sawai, K.; Tomura,H. & Ohzuku T. (1998). Acoustic Emission Histometry for Battery Material Research. *Denki Kagaku*, Vol. 66, No. 3, (March 1998), pp. 301-307, ISSN 1344-3542

Tsai, S. T. & Shih,H. C. (1998). Correlation Between Acoustic Emission Signals and Hydrogen Permeation in High Strength, Low Alloy Steel Cracking in Wet H_2S. *J. Electrochem. Soc.*, Vol. 145, No. 6, (June 1998), pp. 1968-1976, ISSN 0013-4651

8

Acoustic Emission and Electromagnetic Effects in Loaded Rocks

Alexander Zakupin[1], Leonid Bogomolov[2], Virginia Mubassarova[1], Galina Kachesova[1] and Boris Borovsky[1]
[1]Research Station RAS in Bishkek
[2]Institute of Marine Geology and Geophysics, FEB RAS
[1]Kyrgyzstan
[2]Russia

1. Introduction

At the turn of the XX century, the progress of science and technology has pointed out a new possible way to approach the problem of reducing the hazard of strong earthquakes. This unexpected way is related to powerful physical fields which may influence the structures in terrestrial crust to induce unloading of tectonic overstress and, thus, to reduce seismic hazard. Initially, the effects allowing to control deformation processes in seismogenerating zones manifested themselves as induced seismicity, which resulted from underground nuclear explosions, fluid industrial waste injection to borehole located in seismic area, variation of water level in large water storage, or from mining operations etc. Thereafter it was revealed that dynamical actions may redistribute the seismicity in the following manner. They are to decrease the number of major events due to growth of energy released by week earthquakes. There are some ways which are the most acceptable from geoenvironmental consideration for man-caused relaxation of tectonic stress in terrestrial crust. One of them is electromagnetic actions by electric current flashing. Pioneer results of the effect of power electromagnetic pulses produced by magnetohydrodynamic (MHD) generators to test the seismic activity in regions of Bishkek and Garm testing fields were obtained in Russia and Kyrgyzstan (UIPE, OIVTRAN) (Tarasov & Tarasova, 2004). It is very important that such external impacts have always triggered the seismic events of minor magnitude (M<5). Such power actions are adequate for civil and scientific purposes only. At the present time the results of Tarasov have been confirmed by other works which involve a new approach to the experiment with the MHD generator (Chelidze et al, 2006) as well as the fresh data on powerful electromagnetic soundings (Bogomolov et al., 2005; Sychev et al, 2008). It is evident that the similar effect must occur in loaded rock specimens when the pulses of electromagnetic fields act on them in addition to the main mechanical load. The method of acoustic emission (AE) measurements which has recommended itself as a good tool for geophysical applications (Vinogradov, 1989; Paparo et al., 2002; Sobolev & Ponomarev, 2003) allows to get information about the aforementioned effect during laboratory experiments. This work is devoted to the simulation of phenomenon of weak seismicity triggering by actions of impulsive physical fields (so-called energy influence, EI). Concerning terminology it should be noted that the signatures of electromagnetic influence on defects accumulation in loaded geological medium are described as seismic, seismic-acoustic or acoustic emission response (AE), depending on the scale. In our previous experiments (Bogomolov et al., 2004, 2004a; Zakupin et al., 2006, 2009; Bogomolov & Zakupin, 2008) we have revealed the effect of AE activity increment stimulated by external electromagnetic fields, which indicates the effect of medium's factors on crack formation. The AE activity has proved to be a rather informative parameter reflecting both the process of structure defects accumulation in load-carrying medium and the variations of its rate under the effect of external fields. But not only easy building and visualization of activity's temporal plot during AE data processing have made it the main object of investigations in the above mentioned publications. The similarity between AE response to the physical fields effect and induced seismicity initiated with electromagnetic pulses of natural and man-made origin has been revealed just by activity variations, and this circumstance is working. The responses to the actions of nonstationary physical fields, detected on various scales: from the laboratory length of 1-10 cm, to the natural one of kilometers-length (Tarasov & Tarasova, 2004; Sychev et al, 2008; Zakupin, 2010), to be complement each other. At the same time, the problem of mechanism (or mechanisms) of generation of acoustic emission signals' responses (destructure indicators) to electromagnetic effect has not been fully ascertained yet. A theoretical complicacy is related mostly to the fact that a very wide range of effects, realizing on the various structural-scaled levels, are able to stimulation of AE activity, for example: inverse seismic-electric effect and piezoelectric effect occurred in the separate impurities, relaxation of electrical polarization, electroplastic and magnetoplastic, effects of spreading electro-magnetoelastic waves etc. The main problem is connected with a significant (by many orders of magnitude) difference between typical wavelength of external perturbations and defects' sizes. Nevertheless, it has been supposed that currently the understanding of electromagnetic effect on the base of physical model is possible in a case of tested specimens rather than natural, full-scale phenomena during the crust straining. Previously all relevant experiments to study electromagnetic influence on a specimen have been conducted in the conditions of quasi-static compressive loading. In addition, in such experiments tensometric measurements either have not conducted, or they have been too coarse (only AE). In this work we have tried, as far as possible, to expand the ideas about responses to EI by the arrangement of experiment in new conditions. Measuring of tensometric parameters, such as compressing load, longitudinal and lateral deformations, is an integral part of laboratory experiments, since specified parameters characterize deflected mode of a test specimen. Note that the system of AE data gathering has passed upgrade both by quality of record signals, and by registration speed (several hundreds of acts per second) and sensitivity of system. Using the new system of loading (lever press) and material (marble versus granitoids has apparent semibrittle properties) is a logical development of the experiments we have conducted earlier. The experiments have been held at Bishkek Geodynamic Research Center - RS RAS.

2. Measuring equipment

While forming the experimental complex, the significant role is played by the choice of a press with technological characteristics which provide a solution of methodical tasks. Primarily this is the securing of automated loading mode with ability of continuous monitoring of load changes and absence of instrumental noise – the false acoustic signals. At the present time, the tests in the RS RAS are conducted with lever press PLT-L. This press has been created in RS RAS based on the spring rheological plant for long-term tests (PLT) with maximal press tonnage of 100 tons (Stavrogin & Protosenja, 1979). PLT-L provided a load up to 35 tons and noiseless conditions of experiment carrying out, including experimental sessions with constant uniaxial compressive loading and with loading, increasing by addition of weight on the long lever arm. The photos of the press and the specimen are shown on the fig.1.

Fig. 1. Specimen with sensors and the lever press: a specimen with AE sensor, LVDT sensors of axial and lateral strain and electrode (the left frame); general view (the right frame)

This testing method (uniaxial compression) has been chosen in connection with the fact that the stressed state, initiated during the process, in a specimen with prevalence of compression stresses in the single direction is similar, in a certain extent, to a mountain mass, where tectonic stresses prevail over a lithostatic one (Kropotkin et al, 1987; Heidbach et al, 2008). From the viewpoint of continuum mechanics, the uniaxial compression is considered as elementary, and all the investigations of rocks always begin with it. Moreover, this method of tests is technically easy realized. Specimen material is more compliant then a material the pressing thrust journals are made of; therefore a specimen compression may be believed to occur between absolutely firm, rigid planes. Side face of a specimen is stress-free. And on its bearing faces (butts) through which pressure has been propagated to a specimen, conditions are determined with available or absence of a grease on contacting surfaces of a specimen and press. In our case when grease is absent, direct contact of a specimen bearing surfaces and press platters is realized. On these surfaces the strong bearing friction appears; the possibility of couple sliding of contacting surfaces' elementary parts is practically ruled out. In a specimen aggregate (heterogeneous) the stress state is created. Specimen volume elements located near the center of butts are in condition close to hydrostatic compression. On butts' ends the zone of stresses concentration with maximal difference of normal compressive stresses along different directions is formed. Therefore the maximal tangential shearing stresses developed here and the cleavage cracks are possible. In average horizontal cross-section of a specimen close to its center we observe irregular uniform compression with prevalence of vertical component that brings stretching deformations in horizontal direction to a specimen. It should bring to both cleavage and tensile cracks initiation. During the research of rock for uniaxial compression, the main methodical troubles appear by deformation in the area which is above ultimate strength. These troubles are connected with rather high brittleness of rocks under condition of uniaxial compression. That is why in the new series of experiments marble has been chosen rather than granite. In the work (Gzovsky, 1975) it is shown that it is impossible to fulfill all the similarity conditions when modeling the seismic process simultaneously in a single trial, even if all the process parameters and properties of a material at a depth, as well deformations and stresses were known. The work presents (Vinogradov, 1989) different similarity factors for deformation processes in the Earth crust and modeling specimens, derived by using various models of deformation. We use the method of acoustic emission in experiments, and this is the best way for investigation of different rocks of magmatic and sedimentary types with strongly expressed acoustic-emissive properties. Choosing solid materials of these rocks, we do not provide similarity by coefficient of viscosity but perform similarity by compressional speed and Poisson ratio, which allows talking about deflected mode similarity. In the experiments, AE signals have been registered with piezosensors in the frequency range from 80 kHz to 2 MHz. According to the experiment tasks and the methodology of acoustic emission measurements the hardware and software complex for wideband AE measurements has been worked out and used in the RS RAS (Bogomolov et at., 2004 a). The hardware of the complex allows to record signals coming from the AE sensor located on the tested object (rock specimen) and converted them to electrical signals. The specially designed program for data acquisition with the USB3000 modulus allows to fulfill signal processing and calculation of the parameters pointing out the changes of tested specimen deflected mode. It should be noted that due to calibration of used SE2MEG sensors (DECI, 2009) the output signal from AE measuring channel is proportional to the oscillating pressure on sensor-to-specimen contacting surface. The data registration and processing programs are developed on the C++ programming language in WINDOWS environment. For preprocessing of the gathered information, the software including viewers and correcting of received information programs is used. To calculate the activity of acoustic emission, we use the program of batch processing and forming of single time series of activity over all the data series received from one specimen. In the program, the algorithms of numerical differentiation and smoothing with Laplace running window are used. The equipment has worked in a standby mode, the launch has been executed by the set threshold value exceeding of a signal on the output of measuring AE channel. Signals have been digitized by a high-performance eight-channel unit of ADC USB 3000 (capacity is 14 bit, maximal frequency is 3MHz) and written onto the hard disk of a personal computer in automatic mode. As sensors for deformation (longitudinal and lateral) and load registration, we use the linear-variable differential transformers LVDT designed for linear displacements measurements. Operation of these sensors is based on the electromagnetic induction phenomena. Their output voltage is in proportion to position of a moving magnetic core. Primary coil is excited with a source of alternating voltage inducing in secondary coils voltages, which change with shift of a magnetic core within a set. Secondary coils are opposite winded, and when a core is in the center, voltages on coils are equal in value and opposite in sign, and output resulting voltage is zero. When core shifts from the center, voltage in secondary coil (in side of which core shifts) increased. As a result, output differential voltage lineary changes depending on a core's position. Usually a core has thread within to ease fastening of non-magnetic rod, which in his turn is fixed to an object, whose displacement will be measured. These sensors have high accuracy, linearity, sensitivity and definition, and also provide work without friction with high rigidity. In the

RS RAS the two types of LVDT sensors of Lucas Schaewitz company are used: MHR 010 and MHR 050, they have the following ranges of registered linear displacements: ±0,01"≈ ± 0,254 mm and ±0,05"≈ ± 1,27 mm respectively (Pallas-Areny & Webster, 1991). Load value is evaluated by circumstantial way with use of special construction, located in the bottom bearing of the test unit. Its principle of operation is based on measurement of the rigid membrane deflection which appears under action of applied mechanical force. Experimental specimen is placed between the dynamometer membrane and the top bearing. Inductance coil of the sensor is located in the bottom bearing of the test construction. Sensor core is rigidly tied to the membrane. When the membrane deflects under the action of loading, the core position is changed within sensor. Accordingly, the differential electrical voltage from secondary coils of the sensor inductive transformer is changed as well. Deformation is defined immediately by the value of core shift within a sensor. To measure longitudinal deformation, the sensor is located between the bottom and top plates of the test unit together with a specimen. For lateral deformations measurement, the special surrounding case has been made, so the shift of specimen surface gives the deformation in respect to case with sensors. The unit of conversion and gaining is a part of the system of registration of tested rocks tensometric parameters. In this system, converted and amplified signal comes to ADC of the LA-I24USB registrator, whence quantized voltage values are transferred to the computer. Conversion efficiencies of the deformation conversion scheme are the following: for the MHR050 – k=0,5034 mV/μm, for the MHR010 – k=0,9469 mV/μm. Conversion efficiency for load conversion scheme for MHR 050 sensor is k=64,64 mV/t. Based on multiloop feedback scheme, the Butterwort low-pass filter of 6th order with cutoff frequency 0,25 Hz, is used for data processsing. Cutoff frequency is chosen according to the Kotelnikov theorem, because the digitizing rate is 1 Hz.

3. Experiment methodology

For the tests, the marble specimens of parallelepiped-shape with 100x50x25 mm^3 sizes have been assorted. Marble as a model material fits nice for tests, first of all as pseudoplastic material which exposes to semibrittle fracture. Experiments have been conducted with indoor temperature and humidity. Electromagnetic pulses to produce an effect have been supplied by graphite electrodes maintained on lateral surfaces of a specimen. It is essential that the platen contacting to specimen butt ends are electrically isolated one-of-another with the aids of a dielectric spacer placed between the average traverse of the press and AE sensor, with preamplifiers and ball joint. At the same time it allows to avoid external noise induction in a closed loop and, thereby, provide purity of an experiment. Experiments have been conducted both with only mechanical loading, and with additional action of external source of EM field. The following model sources of EI have been used: the rectangular pulse generator G5-54, the capacitor disharger (CD), creating electric pulses with steepness of edge about 1 μs and peak voltage about 1kV, and also the inductance coil for magnetic field induction. In the case of CD the scheme of 10 dischargers following in 15 second intervals (2,5 minutes in tote) has been used. Amplitude of pulses has been chosen 800V stably. In the case of G5-54 generator the following parameters have been set: positive polarity, frequency 2 kHz, duration 60μs, amplitude 50V. Duration of generator operation has been chosen 1 hour. During a session with combined action 10 capacitor discharges have been supplied 2 times, each 10 discharges series have followed by half-an-hour period of G5-54 generator action. Choice of the generator G5-54 as a source of quasi-periodic electric pulses is stipulated also for the following reasons. A laboratory experiment with a source of electric pulses of such a form is identified with the natural experiments, which actually had taken place during the period of using geophysical MHD-generators on the Harm and Bishkek polygons (Tarasov et al., 1999; Tarasov & Tarasova, 2004). The scheme of the experiment with crossed electromagnetic field (Cr.EMF) is described in work (Bogomolov & Zakupin, 2008). It involves both: stages of constant load (CL), and that of growing load with the rate of 300 kg per hour. AE signals represented a flow of random pulsed train of an experimental specimen surface's acoustic oscillation. (Greshikov & Drobot, 1976; Stantchitz & Tomilin, 1984). The spectrum and dynamic range of these signals are wide enough, that is why it is necessary to have wideband sensors and low-noise measuring amplifiers for their registration. It is also important to note the importance of registration and visual control of AE signals form for rejection of man-made noise and interference in AE events flow, which appear as a result of scattered electromagnetic fields effect when switching of current pulses simulating an electromagnetic effect on the geological medium. For the rejection, great work in viewing of wave form of tens of thousands AE events has been carried out, that has provided validity of the data passing the further stages of mathematic processing and interpretation. Experimental data interpretation has been realized on the basis of analysis of diagrams of AE activity time dependence. Such a representation form reflects changes of specimen deformation mode most informative. The processing by moving window has been fulfilled to calculate the AE activity by experimental data (files of registered signals). The width of the window is determined by average time to accumulate the certain number of events. As a result we have received the time dependences reflecting quantity of AE acts per second (this value mathematically identified with the derivative dN/dt). The meaning of the AE activity estimated by such manner is correspondent with the standard parameter "activity of the AE events". This is presented (with amplitude and duration) in the most of hardware for technology applications of AE (Greshikov & Drobot, 1976). The analysis of AE activity means determination of a reaction (response) of a loading specimen to the electromagnetic field effect. This reaction is expressed in variations of AE activity, which amplitude exceed the root-mean-square deviation of the AE's temporal dependence, the RMS deviation being defined by the prehistory (for times before the EI source power up). The experiments with effect for the specimens series have been conducted with loads more than 50% from fracture (in our previous works it has been shown that the medium sensitivity to EM field is observed with loads more than 75-80% from fracture load) with load incremental step over 100 kg.

4. Results

In this chapter we will describe results of the experiments with marble specimens. It consists of six experiments. One of them was performed without any external EM actions and other with several sources of EM field or their combinations.

4.1 The experiment without electromagnetic field effect

Let's begin with the experiment where only mechanical loading has effected on the specimen. During the experiment three periods have been realized: CL and with load increase of 300kg per hour. In the fig.2 the diagrams reflecting load and extension changes during all the experiment are shown.

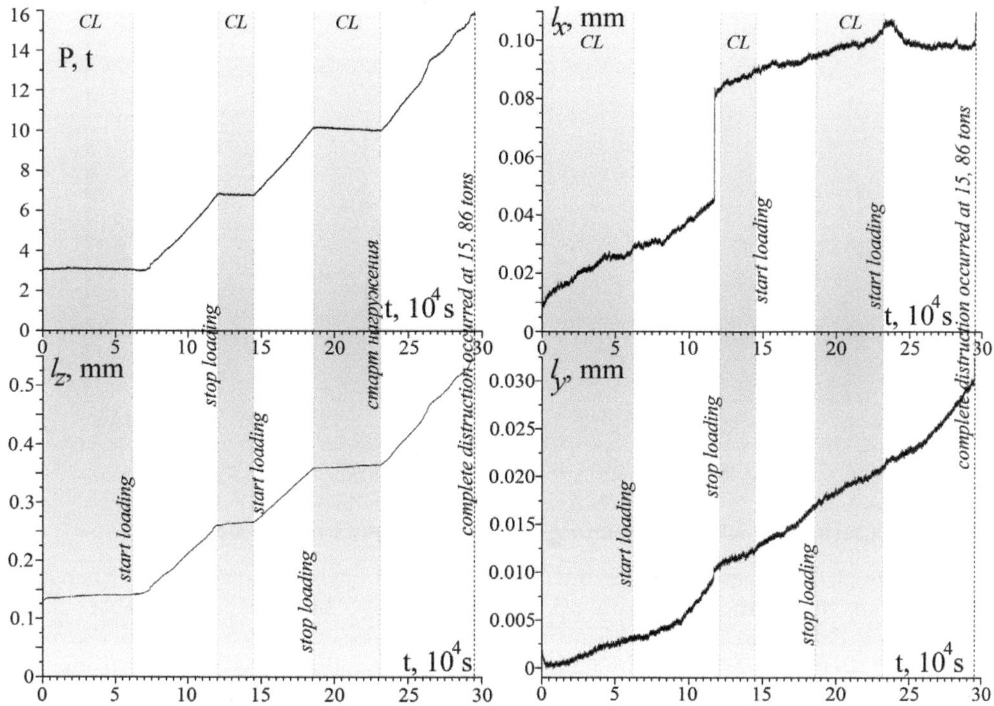

Fig. 2. Change of specimen's load and extensions (shortenings) in three directions (x, y, z).

The experiment has lasted for 81.86 hours in total. As one can see from the diagram the load increase arouses the adequate increase of longitudinal deformation. Lateral dimensions of the specimen have smoothly changed also demonstrating the growth. However in the two cases, we note sharp changes of specimen size in the X direction. By Y component the displacement increase has evenly progressed, going a bit faster during transition from CL into loading mode. Displacement by X has reached the maximum value of 0.11 mm, whereas by Y component it has been only 0.03 mm. Longitudinal shortening at the end of the experiment has been marked at the level of 0.54 mm. On the diagram of AE activity (fig.3) one can see variations after 7.4 tons. Activity slowly grows and about 11 tons sharply increases up to 80 events per second. This scattered activization ends with sharp decrease and during transition in the CL mode the activity goes to zero. During increasing load, the AE activates fast enough opposite to previous time, reaches a certain level, and however doesn't exceed 0.15 events per second. After transition in the CL mode, the character decrease is observed. Before the third period of increasing load its value has made up a little bit more than 10 tons, and AE activity has been close to zero.

The load beginning to rise once again has excited the activity about 0.2 events per second quickly enough, and then short-term spike up to 0.6 events per second. After 5 hours a fracture has come when maximum quantity of AE signals has been 131 events per second.

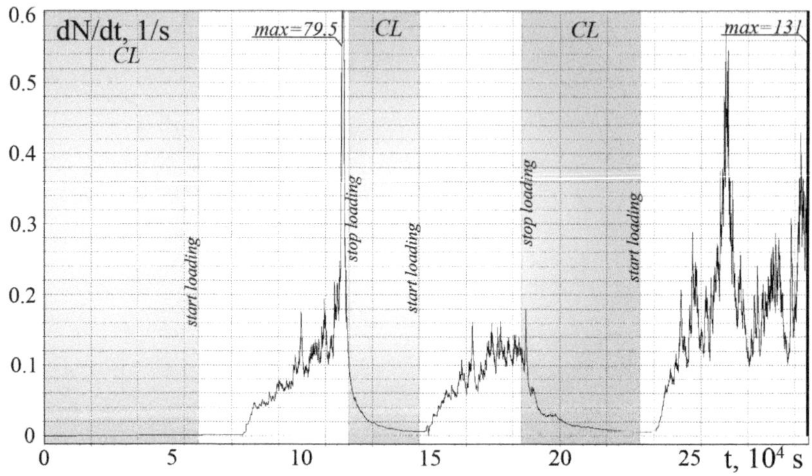

Fig. 3. AE activity during deformation in the CL and load constant increase modes.

4.2 The experiments with the discharges and G5-54 generator effect

The results of the experiment with the G5-54 generator have not revealed any changes in deformations during the generator work, and AE responses have been comparable with the responses registered in the experiments on the spring presses (on granite, salt and gabbro). Note only that, as earlier, the medium reaction has followed with some (sometimes prolonged) delay. The specimen has failed with most big loading, which has been 19.4 tons. We won't stop on this experiment in detail, because in the early works (Bogomolov et al., 2004; Zakupin et al., 2006) the main data volume has been received just by using this source of influence. In the experiment with the G5-54 on the marble specimen its high bearing value has become the general outstanding feature, it will be considered in detail when comparing the data in the following chapter (results discussion). Let's proceed to the experiments where effect to a specimen has been carried out by the CD. The two following experiments have foreseen EI to a specimen by the capacitor discharges with sessions according to the scheme: a) 10 discharges have been supplied during two minutes with a pause of half an hour, b) 10 discharges have been supplied during two minutes for one time per hour. In the first case, six such sessions have been conducted, and the three of them in the CL mode. The fracture loading has amounted 11.78 tons. The experiment has lasted for 47.7 hours in total. Changes of the specimen lateral dimensions in the course of the experiment are inessential. The single significant change of the X component (increase for 0.066 mm) has occurred because of sharp increase of load for 150 kg. Note that at the initial stage of the loading the lever press has one particular point, when transition to the ball bearing in the junction of third and second levers occurs. At such point sharp leaps of load (not always) are possible, which cause sharp shifts of a specimen faces, in this case such variations have taken place by Z component. In this experiment some sessions have been conducted in the CL mode and it is interesting that some changes of the longitudinal deformation l_z may be tied with influence of discharges series (fig.4). On the diagram it is seen that the longitudinal deformation has slightly decreased in the CL period, and three times sharply increased, two times in the moments of dischargers supply in the CL mode at its final part.

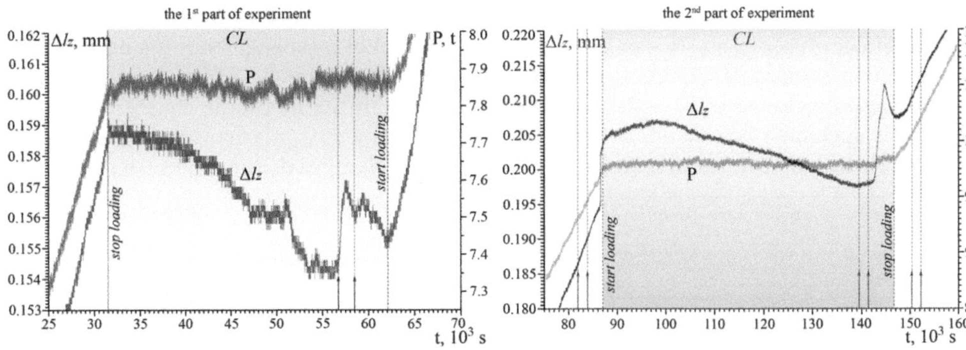

Fig. 4. Change of the load and longitudinal deformation, the supply moments of capacitor discharges of 800V are marked with arrows

The first time it was without a delay with increase for 0.003 mm and second time it was with a delay of 23 minutes and increase for 0.016 mm. Finally one more time l_z has increased in the mode of growing load in 40 minutes after discharges supplying (increased for 0.011 mm). Note that supply of discharges at the initial stage of the CL has not effected to the level of longitudinal deformation. Also the pulses in the mode of increasing load have not exerted influence on l_z. The experiments don't reveal significant influence on the components of deformations, when only 10 discharges (rather than 20) occurred. Note that in this experiment we have not conducted the sessions in the CL mode. At the same time at the late loading stages AE responses have been registered clear enough both in this and another case (fig.5).

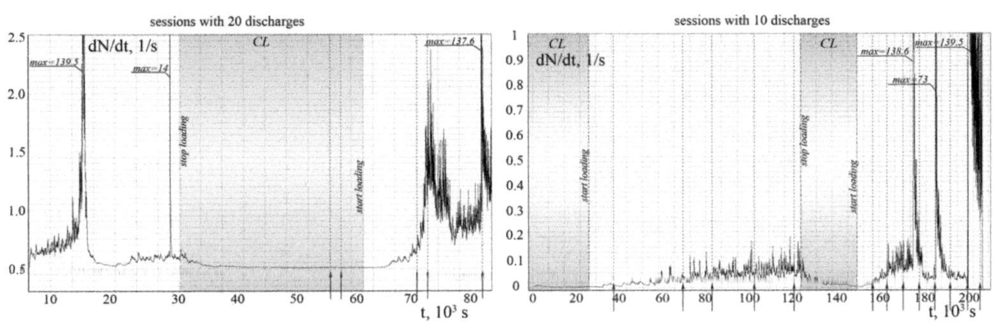

Fig. 5. The acoustic emission activity

Note, that in three sessions (fig.5) AE peaks have been observed immediately after supply of the discharges. It is important that maximal response activity has been by three orders greater than background level.

4.3 The experiment with combined effect of discharges and G5-54 generator

In the previous experiments we have revealed AE responses to effect of discharges. They have been observed both in the mode of 10 discharges per session, and in the mode of 20 discharges. The responses to the effect of series of light pulses from G5-54 generator have been weaker and with a delay. Note that during sessions, the deformations have not been changed. It has been interesting to conduct an experiment, where in a session the G5-54 generator effect precedes a series of powerful discharges. Let us speak about the experiment with combined variant of EI. So, as mentioned above, for EI the G5-54 generator and CD have been used. Scheme of EI order has been described in the methodic chapter. Fracture load for this specimen has made up 14.9 tons. The experiment has lasted for 81.49 hours in total. Fig.6 shows diagrams reflecting load and extension changes during all the experiment.

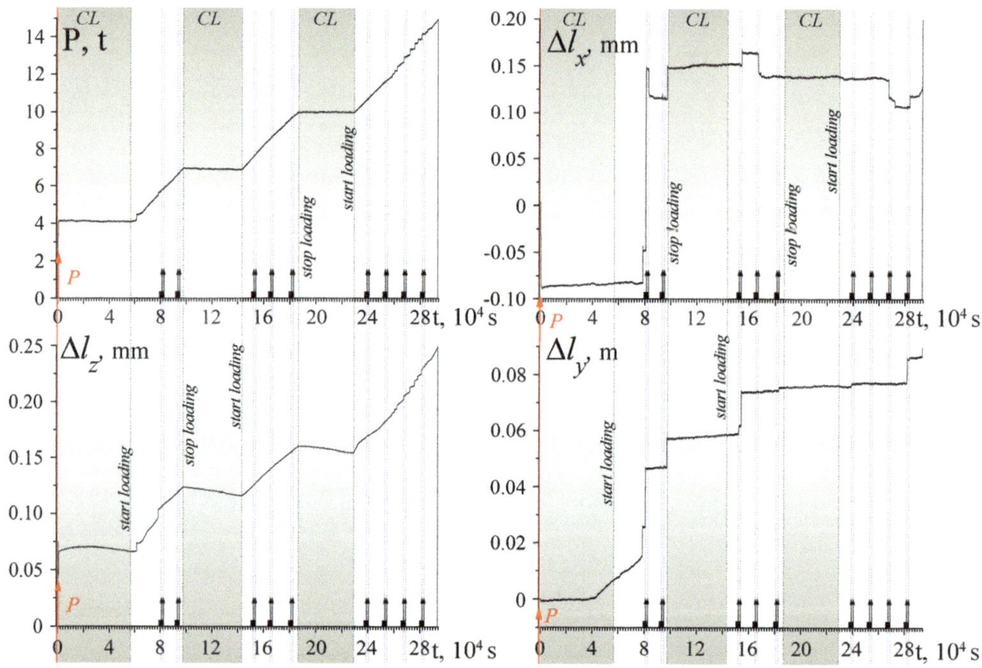

Fig. 6. Change of specimen's load and extensions (shortenings) in three directions (x, y, z).

Note that in the specimen some decrease of longitudinal shortening value has been observed in the CL mode by practically constant load and lateral sizes. The fact of strengthening may be related to influence on the specimen, because at the first CL such effect has not been observed. The longitudinal deformation has also grown during load increasing. All the nine sessions with the effect has been conducted in the mode of increasing load and have not exerted an influence on the longitudinal deformation. At the same time very interesting data have been received for the displacements in the lateral directions. Opposite to the first experiment, the change of these parameters is characterized with sharp spikes. It is notable that all significant variations accurately concur with the first or second series of discharges. In the fig.7a the results of tensometric measurements for lateral components of deformation (displacement) are shown. For example, in the first session lx on first series of discharges has increased only for 0.03 mm, and after half an hour on second series responded with decrease for 0.02 mm, and l_y has only increased. Second series of discharges has effected on synchronous increase of l_y and l_x for 0.01 and 0.02 respectively and so on. In the third session the responses by both components have taken place, and l_y have responded to both discharges series with a value increase, and l_x has responded only to the last discharges and also by increasing. In the fourth session the first discharges series has induced decreasing of the shift value by l_x, and on l_y it has not influenced at all. The fifth and sixth sessions are similar to each other. One could see the responses of l_y values. The difference is that in the fifth session a response has been induced with the first series of discharges and in the sixth session it has been the second series. Just one session №7 in the experiment, not to provoke some changes in the displacement values, has been conducted by the load of 11.535 tons. In the eighth session l_x has decreased for 0.012mm, and l_y has not changed. In the last ninth session only the value of l_y has increased to the first series of discharges, to the second series only l_x, and both for the value about 0.01mm. Using the results of earlier experiments with the action of the capacitor discharges only, one can distinguish the role of the periodical action of G5-54 generator pulses. It seemingly consists in some material preparation before effect of powerful discharges. The results of AE processing involve nine sessions of the EI. As in the case with the specimen which has not been exposed to the EI effect, there is first phase of AE acts accumulation, which ends by 6 tons up to 64 acts per second. The difference is that little variations are present after the spike and they are attached to the first two sessions of the effect. The two following stages involve a lot of spikes, and again after EI. The fig 7b demonstrates the results of last part of the experiment, which involves 4 sessions.

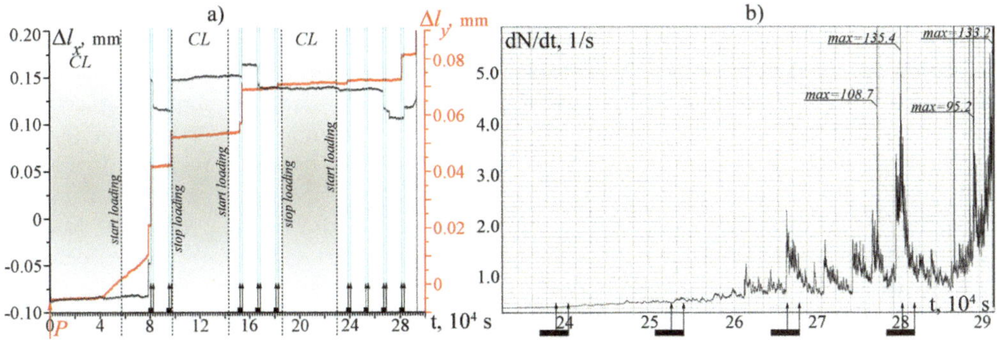

Fig. 7. a) Change of extension (shortening) of the specimen in the y, x directions; b) AE activity

Influence of the last two sessions has been effective, while the medium response has been clearly distinguishable on the background of AE fluctuation (at loads more than 80% from fracture). And if in the next to last session the activity increases times to the first discharges series, in the last session it has already occurred at the stage of G5-54 action, and discharges series has led to the second, but very powerful release. As in the many earlier results, a trigger character of the effect is evident here, because energy released is limited by its reserves and doesn't occur at the last ten discharges (in both cases) after its activation decrease (response). In tote the results by this experiment show the character of response to EI both in AE, and in deformation measurements.

4.4 The experiment with the crossed electromagnetic field effect

For the experiment with crossed electromagnetic field no less interesting results have been received. Fracture load for this specimen has made up 11.85 tons. The experiment has lasted for 70.1 hours in tote. As it follows from the results, the effect of crossed field has not practically affected on the deformation components change. Only the fifth session has a little variation in the longitudinal direction (growth has sharply slowed down) and during the sixth session in the y direction a little shortening has occurred (fig.8a).

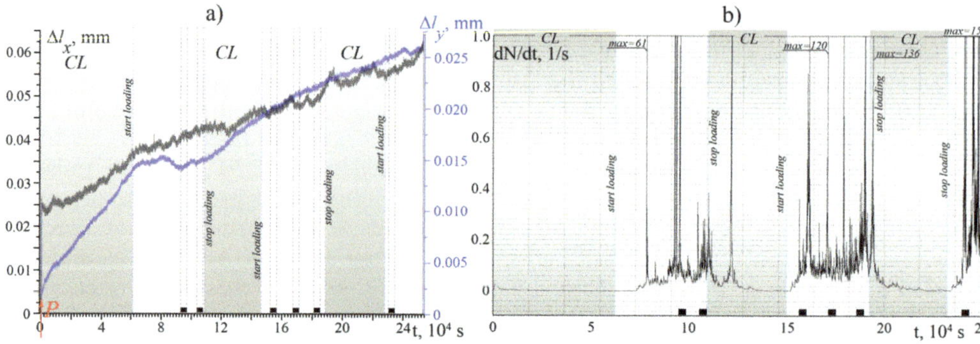

Fig. 8. a) Change of extension (shortening) of the specimen in the y, x directions; b) AE activity

In tote, the deformations smoothly increased during all the experiment. It is worth to note that in comparison to the previous specimens the lowest value of longitudinal shortening has been observed in this case. On average by the series of tested specimen to the moment, when load of 12 tons (about 100MPa) acts to a specimen, longitudinal shortening amounts more than 0.35 mm, but in this case it has not exceeded 0.09 mm. It is very notable that the general result by the medium reaction to the crossed field has been revealed in the AE activity (fig.8b). On the fig.8b all the six sessions of this experiment are shown. In five sessions (except the first one) the sharp peaks of AE activity during the period of field effect are noted. And it is not difficult to single out sharp and very big (by amplitude) peaks of activity (responses to EI), although in the time of session the beginning activity demonstrates some growth typical for the increasing compressive load as it is. So in the third and the sixth sessions the activity growth has reached 100-150 acts per second on a peak. Also, a durable aftereffect may be called a typical one. In our experiment this effect has been noted in some other form. In principle, in the earlier experiments we may not see it, because we have conducted the sessions only in the mode of constant compressive stress. In the new experiments it has been noted that after deformation transition from loading mode to the CL (especially at low loads) the AE activity either immediately ceases, or goes down to small values quickly enough. At the same time, as it is seen from the fig.8 that the second and the fifth sessions have been conducted exactly before the loading cutoff, but AE activation has being continued in the CL as a result of the effect. After the response to the second session it has been even the repeated peak. The relevant and possibly the most interesting results have been received in the tests on the lever press in the semipermanent deformation mode on the specimen of granite under action of crossed fields (Zakupin et al., 2009).

4.5 Comparison of the results

Comparing the results of deformations measurements and acoustic emission one can reveal the following features of the specimens' deformation in the CL and constant increase of load modes. In the CL mode the load is constant, and the longitudinal deformation may have tendencies both to little increase, and decrease. For example, in the case of the test of the specimen №1 (without EI) l_z in the CL has slightly increased, however the lateral sizes in this mode have grown as well. On the contrary, the lateral sizes of the specimen №2, which has exposed to the effect of discharges of the capacity discharger, by some decrease of l_z have remained invariable. Acoustic emission activity in the CL mode, as in the early experiments, is characterized with the Omory's law (Utsu et al., 1995). The relaxation release process of the energy, obtained by a specimen during tightening weight, revealing in the form of powerful AE flow, is a natural analogue of the aftershock activity known in seismology. In the both cases these processes have a damping character. The classic Omory's law for the time dependence of aftershock process can be described by means of the simple formula $n=k/(t_0+t)$. Here n is a number of aftershocks per time unit, the constant k depend on the main shock magnitude (or, in our case, on the main compressive stress and its increase), t_0 - the value, approximately equal to thirty six hours. In the generalization [Sobolev &Ponomarev, 2003] it is noted, that behavior of the AE events flow after mechanical tightening weight conforms to the Omory's law. It is shown there that the AE activity change on the drooping line after a tightening weight can be described using the activity value instead n, new parameters, determined the initial AE activity and its relaxation intensity after a tightening weight, instead k and t_0 coefficients. Depending on the load value t_0 will slightly change, however in the general case it has not to exceed 1-2 hours (the time, after which fast decrease of the activity ceases). Our dependencies have also been close to this law with fair accuracy. As it is seen from the diagrams 5-7, it is typical both for low and for high levels of load acting on the specimen. In the experiments on the spring press (PLT) we have observed some spontaneous activity on the decaying trend after a tightening weight, however in those experiments the loading has been more rigid. Under the rigid loading we mean a tightening weight for 2-3 tons during the period of 2-3 minutes. By the tests on the lever press a soft mode of loading practically approaches the received dependencies to the Omory's law without applying of approximation. Interesting result obtained in the experiment with the the Cr.EMF action is that the AE activity, which had begun after the pulses of Cr.EMF, is continuing even after the loading stop. And as it is seen from the figure 8, the activity peaks in the CL mode (aftereffect) have occurred higher than the major responses. It is worth to remark that EI by prolonged CL close to its end doesn't lead to responses such as described in the original works (Bogomolov & Zakupin, 2008) as well. Starting from 3 and more tons in the specimen, the slow increase of AE activity begins (during 10 hours), which then rapidly changes with a phase of fast release of acoustic emission, when the activity increases in hundreds times. The fast phase lasts about 25 minutes, and upon its completion the activity goes to zero at that (it occurs by load about 6 tons). This process may be considered as the first phase of fracture and the EI is ineffective here, although there may be exceptions. Then the similar phase of the avalanche cracking is observed in the specimens, but by one of the sources of EI it significantly differs from simple loading. Responses to EI reduce average level of AE intensity, and quantity of powerful spikes increases. Let's turn to the characteristics of the described effect of the EM field with respect to stress-strain properties of the specimen and its fracture. In the table 1 the general characteristics are shown (Explanations: P - loading without EI, for σ and ε the maximum values are given).

EI	σ, MPa	ε_x 10^{-3}	ε_y 10^{-3}	ε_z 10^{-3}	AE counts	AE response	Def. response
CD (20)	142	0,06	4,4	6,5	11977	+	+
CD (10)	93	0,02	0,2	2,5	55195	+	-
G5-54+CD(20)	119	3,4	3,2	2,5	36458	+	+
P	122	2	1,25	5,5	18129	n/a	n/a
Cr.EMF	95	1,1	0,91	0,9	16887	+	+
G5-54	155	1,5	2	4,6	17456	+	-

Table 1. Macrodeformative parameters of the tested marble specimens

From the results represented in the table, it can be concluded that maximal number of the AE acts fall to the specimen suffered minimal lateral deformations. Also in this case the minimal fracture load has been registered. Maximal lateral deformations have been observed for the third specimen (G5-54+CD). Only in this experiment the value of longitudinal deformation has been even less than lateral components, but it is necessary to note, that only here the hardening effect (decrease of longitudinal component by a constant load) has been observed. And finally, only in this experiment the nice responses to EI both on the deformation and on the acoustics have been received. As it is seen, the character of deformation and fracture of the specimens under the EI effect is varied and points to simple laws. That is why let us note only the most evident. The case with fracture of the specimen, which has not undergone to EI, can be supposed as classic. We have observed the prevalent axial component and moderate lateral components, which are less but comparable to axial one. By the acoustics in this case we observe a slight growth of the activity and minimal number of the sharp peaks (only two). All the rest cases essentially differ from the "classic" in one or another aspect. It is well-known (Lyakhovsky et al., 1997; Ben-Zion and Lyakhovsky, 2006) that the mechanical aspects of existing damage in a solid are modeled by generalizing the strain energy function as follows:

$$U = 1/\rho \cdot (\tfrac{\lambda}{2} I_1^2 + \mu I_1 - \gamma I_1 \sqrt{I_2}) \cdot \tag{1}$$

where $I_1 = \varepsilon_{kk}$ and $I_2 = \varepsilon_{ij}\,\varepsilon_{ij}$ are the first and second invariants of the elastic strain tensor ε_{ij}, ρ is the mass density, λ and μ are the Lame´ parameters, and γ is a third modulus of a damaged solid, introduced in (Lyakhovsky et al., 1997b). The first two terms of (1) give the classical strain potential for the isotropic elastic media. But the presence of damages (microcracks with some dominant orientation) allows "non-isotropic" third term in (1). This term was derived in (Lyakhovsky et al., 1997) by using the effective medium theory with noninteracting cracks or by expanding the strain energy potential as a general second-order function of I_1 and I_2 and eliminating nonphysical terms (Ben-Zion & Lyakhovsky, 2006). So, the third term in (1) describes the seeming "contribution" of damages to the energy function and thus the interrelation between AE signals flow and changes in parameters of a specimen straining (tensometry). The following stress-strain relation is correspondent with the above energy function (1):

$$\sigma_{ij} = \rho\, \partial U/\partial \varepsilon_{ij} = (\lambda - \gamma \sqrt{I_2}/I_1) \cdot I_1 \delta_{ij} + (2\mu - \gamma I_1/\sqrt{I_2}) \cdot \varepsilon_{ij} \tag{2}$$

Expression (2) describes the reason why damage accumulation (growth of factor γ) recorded by AE data may entail seeming change in elastic module. One can see that the signatures of damage growth in (2) may produce an impression of "strengthening" (due to bracket with λ) when the increment of γ takes place in a short time, without considerable variations of macro-parameters σ, and ε. This is correspondent with the enhanced value of averaged angular coefficient for σ-ε plot in case with Cr.EMF. But the stress-strain diagram for Cr.EMF case does not mean physical strengthening, because subsequent re-distribution of local deformation inside specimen is to result in great "delayed" growth of the main deformation (axial shortening). Such behavior (similar to pseudo-plasticity) of specimens under action of linearly growing compression and crossed EM field was described by (Urusovkaya, et al., 2000). Experiment of (Urusovkaya, et al., 2000) was performed on specimen of ionic crystals, the loading conditions differ from our case by faster stress growth and application of electric and magnetic fields of major strength. Unfortunately, no acoustic emission measurements were represented in that work. In our experiments the AE responses have been observed by the best way in the experiments with the crossed fields, capacitor discharges and combined effect (G5-54+CD20). In all these cases the low values of longitudinal deformation and low strength properties of the specimens have been observed. In case with combined field influence specimen has failed later than the two others (CR.EMF, CD10). A comparison of shortenings of these three specimens shows that the specimen G5-54+CD20 has the most significant changes of the lateral deformations. In spite of a small sampling (every experiment has been repeated 2 times), the results are valid surely. Actually, the limiting load dispersion was no more than 20% that is without EI influence, the specimens have fractured by the loads from 15 to 18 tons of compressive strength. This fact provokes interest to the group of specimens with high sensitivity to EI. As we see, the combined effects have found a reflection not only in the AE responses and deformations, but significantly decreased the material strength. In this sense the result with the capacitor discharges looks strange, when the discharges' number increase has not affected on the limiting load, although the AE responses have been significant. So, specimens of the same size and from the same material have fractured by different loads and it may depend on the kind of external EI. Another detail is that the most high stress-strain properties have proved to belong to the specimens, which have high flow degree (ε_z). In this case the pronounced dilatancy and high ability to compression resistance is owing to continuous structure transformation at present. But it is necessary to note the fact that concentration of released acoustic energy has fallen on two short periods. From the viewpoint of artificial influence on the geological medium, the listed aspects can play an important role when choosing zones for local removal of an overstress.

All the experiments confirm that starting from loads close to 70% from fracture, when time of forming of second part of the crack is coming (fracture planes), EI becomes efficient and excites several "tides" of AE activization at the level in 15-20 times more than the similar in the P experiment. The example with G5-54+CD on second phase of the main fracture forming is very demonstrative, when several "tides" of gain of the AE intensity initiated by EI stand out very good. At the same time, the effect of the mentioned sources on AE significantly differs in the first case quantity of acts is in 3 times more. The interesting result has been registered with the G5-54, when σ_{max} has increase on 18% by practically the same values of the deformations and AE in comparison to P. So, for example, as in the case of the experiment with CD, maybe a rapid fracture of separate zones with high concentration of stresses following with many shocks of low energy (in our estimations) would be preferable. However, the database of major volume is necessary for the final conclusions.

5. Discussion and the model of mechanism of electric pulses influence

The peculiarities of the AE and deformation responses to effect of electromagnetic pulses, described above, have raised the urgent question about the physical mechanism. The model ought to explain similarity of the effect of acoustic emissive response for such dissimilar materials as granite and rock salt. In the physics of condensed state and in the physical material science the whole variety of instances of electromagnetic fields with structure defects interaction (atom-vacancy defects, charged dislocations, microcracks) is known (Zuev, 1990; Kuksenko et al., 1997; Urusovskaya et al., 2000; Bogomolov et al., 2004; Shpeizman & Zhoga, 2005). By these interactions, and in fact by elementary processes in the certain materials, the electroplastic effect, plastomagnetic effect, cracks stopping and initiation in electric field and etc are explained on the semiqualitative level (Finkel et al., 1975; Finkel et al., 1977; Molotskii, 2000; Alshitz et al, 2008). However such processes are not universal enough and cannot be accepted for the basis of the model of acoustic emissive responses excitement in the materials with various physicochemical and rheological properties by themselves. At the same time the trigger influence of weak vibrations on inelastic deformation has such universality that it is established and well explored at the various scales from centimetric (laboratory) to one-kilometer-wide and more (natural) (Mirzoev,et al., 1991; Sobolev et al., 1993; Belyakov, et al., 1999; Sobolev & Ponomarev, 2003; Kocharyan & Spivak, 2003; Kocharyan et al., 2006). This allows a qualitative model of the possible mechanism of AE activation of loading rock specimens. The model appeals to triggering due to microvibrations produced by the action of electromagnetic pulses (Bogomolov et al, 2004, 2004a). However, when passing from qualitative to quantitative explanation level the following troubles appear. On the one hand, by estimations from the works (Bogomolov et al., 2001, 2004a), the effect of rock's AE activity response is revealed, when the amplitude of pressure oscillation becomes of the order of 10^{-6} from the main stress level. But on the other hand, the well known traditionally considered mechanisms of transformation of electrical energy into mechanical: piezoelectric effect, ponderomotive forces, thermal dilatation when medium heating cannot explain such amplitude of disturbances of stresses σ and deformations ε when acting of pulsed fields with electric field strength $E \sim 1000$ V/m typical for the experiments described above with applying of G5-54 generator as a supply. Actually, in dielectric mediums (specimens of granites, marbles, rock salt) Joule heating is ruled out, and thermal dilatation knowingly cannot play a role. As to a piezoelectric effect, it can provide relative disturbances of σ, ε about 10^{-7} from main compressive loadings only for quartzite specimens with longitudinal piezoelectric modulus $d_E \sim 10^{-14}$ -10^{-12} m/V, where the upper value approaches to the value of d_E modulus for quartz monocrystals (Parhomenko,1965). It follows from the formula of electromechanical coupling $\delta\varepsilon \sim d_E \cdot E \sim 10^{-11}$-$10^{-9}$ with specified typical field strength E. If we take into account the fact that elastic moduli are about 10^{10} N/m^2, then just the value for maximal disturbances $\delta\sigma \sim 10$ Pa, corresponded to10^{-7} from the level of main compressive stress, will be received. For the specimens of granite and marble typical values of d_E are 4 orders lower (Parhomenko, 1965). In the same proportion the $\delta\sigma$ disturbances tied to piezoelectric effect ought to decrease.

For the specimens without piezoelectric properties, the contribution of ponderomotive forces, as evaluated for simplified model of isotropic dielectric, lead to the lesser disturbances of the stress by the same values of $E \sim 1000$ V/m, $\delta\sigma/\sigma_0 \sim 10^{-8}$ from the main compression. This estimate is based on the standard model of the electrodynamics of continua (Landau & Livshitz, 1982) and equivalent to regard of the attraction force of the electrodes fixed on the specimen faces and formed a capacitor. Note that accuracy of received estimations match the parameter of applicability of the known formula for average energy density U of the quasi-harmonic electric field with slowly changing amplitude (Landau & Livshitz,1982):

$$U = \frac{1}{2}\varepsilon_{e0}\frac{\partial}{\partial\omega}(\omega\varepsilon_e)\cdot < E^2 >, \tag{3}$$

where ω - cyclic frequency, ε_e - dielectric permittivity, ε_{0e} =8, 85 10^{-12} F/m, the "e" index is introduced for clear distinction from deformations symbol.

In some works (Shpeizman & Zhoga, 2005; Bogomolov et al.,2006) the attempts have been taken to explain influence of electrical fields on a fracture by their energy contribution with the density (1) in common energy balance in the volume $(2l)^3$ around the crack with a length of 2l. In the work (Shpeizman & Zhoga, 2005) in neglect of the dispersion ε_e the following generalization of Griffith criterion with an electrical polarization is present:

$$\frac{\sigma^2 l}{G} = \frac{2\gamma_S}{\pi} - \left(\frac{2\Lambda}{\pi}\right)\cdot\varepsilon_{e0}\varepsilon_e\cdot < E^2 > l\,, \tag{4}$$

where γ_S - specific surface energy, G - elastic modulus, with <> average by time is marked. In the formula (4) the numerical coefficient Λ is introduced, it takes into account that activation volume, where the electric field significantly changes with crack's concerned "virtual" extension, may slightly differ from adopted in the (Shpeizman & Zhoga, 2005) value $4l \times 4l$. The right part of the (4) just describes the drop of effective Griffith parameter γ_S (energy of new surfaces formation) in the presence of electric field. The further generalization of crack instability condition, allowing to take into account the dispersion of ε_e, involves replacement of the parameter ε_e in (4) by the derivative $\partial/\partial\omega$ $(\omega\ \varepsilon_e)$, according to (3). The value of γ for the rock specimen and ceramics is about J/m², and the parameter $\varepsilon_e<100$. So, the electric field influence on submillimetric cracks becomes essential only when its strength is E~ 100kV/m. It is important that in some heterogeneous mediums, in particular containing bound water in the pore-fractured space, the factor $\partial/\partial\omega$ $(\omega\ \varepsilon_e)$ can reach for hundreds and even thousands during high amplitude electromagnetic disturbances (Chernyak,1987; Svetov, 1995). Meanwhile, the minimal field strength E which is able to influence on stability criteria of the mentioned cracks (l ~ 0,05-1mm) decrease to the value ~10 kV/m. In the experiments with model electromagnetic effects (Bogomolov et al., 2004; Zakupin et al., 2006) such fields have been induced by supplying to the tested specimens pulse voltage of 500-5000 V (from a capacitors battery or inductive discharger). By these effects the activization of crack formation has actually occurred with a delay from hundreds to thousands seconds (Bogomolov et al., 2004a; Zakupin et al., 2006). The delay interpretation may appeal to the fact that the vibrations produced by the electric pulses can influence on microcracks with sizes of the order of 0,01 mm, and this channel has turned out more effective than change of stability condition (4). For the ionic crystals, the electroplastic effect (immediate influence of the field E on dislocation mobility) takes place by the action of more powerful fields E ~ 1 MV/m, (Zuev, 1990). However, by less strength E ~ 1000 V/m, typical for cases of G5-54 generator action, the presented result about AE activity increase can not be explained by the equation (4). It is important to note the availability of compliance in the estimations of field strength threshold, fallen into the range of E ~10-100 kV/m, by which it, on the one hand, is able to shift a bit the critical strength by Griffiths criterion, and, besides, it excites stress oscillations with amplitude of the order of 10^{-7}-10^{-6} from main compression.

To the authors' opinion, the problems of explanation of electric pulses influence with strength E ~ 1kV/m (typical experimental conditions of the effect triggered by the action of G5-54 generator (Bogomolov et al., 2004a; Zakupin et al., 2006)), are related to lack of the quasi-stationary field model. The derivation of the condition (4), proposed in the (Shpeizman & Zhoga, 2005), is based on this model; this condition reflects more rapid change of the medium and average field polarization with a following crack growing by already sustained characteristics ε_e, $\partial\varepsilon_e/\partial\omega$, $\langle E^2 \rangle$. Principle stand of the nonstationarity factor (sharpness of the field E changes) has been realized on the experiments stage yet. As a source of electric pulses the G5-54 generator has been chosen just because it provides steep edges of square pulses, and by this the wave effects may become significant. However, the theoretical description of electromagnetic pulses interaction with growing defects is extremely difficult, like the excitons problem (Knox, 1966). It goes far beyond the scope of this work. Heuristically, to discuss the mechanism of anomalous microvibrations excitation in loading specimens under the effect of electric pulses with steep edges, such "excitation" may be compared to the known in nonlinear optics effect of induced Brillouin scattering. This effect describes light scattering on the time fluctuations of dielectric permittivity, appeared because of the mediums density fluctuations (in particular by rapid deformation), by which the frequency change occurs (Shubert & Vil'gelmi, 1973). Low-frequency analogue of the induced Brillouin scattering effect is a nonlinear interaction of sonic and electromagnetic waves with a resonance on the difference frequency. As a result, the sonic wave with a frequency $\omega_S = (\omega_1 - \omega_2)$ magnifies, where ω_1, ω_2 – frequencies of EM waves, distinguished from spectrum of the pulse by the resonance condition. In the loading specimens when microcrack forming in the surrounding matter volume, undoubtedly, non-uniform deformation disturbances take place. If these disturbances get additional energy and pulse from an external field, and the acoustic Q- factor of the medium is high enough (like of rocks), then oscillations spread and exert a trigger influence on growth of the neighbor crack. Then the process is repeated and, in that way, self-acceleration of the microcracks growth may occur, it reveals in experiments as temporary increase of AE activity. In the range of compressive loadings 0,7-0,9 from fracture in semibrittle materials (rocks) there are zones of plasticity localization and domains (mesovolumes), remained in the bounds of elastic strain (Panin & Grinyaev, 2003; Makarov, 2004). But although within domains microcracks may appear or lengthen, they don't influence on character of the mesovolume deformation in tote, and acoustic Q factor in these areas is higher than in average by over the specimen. In such mesovolumes the most favorable conditions are composed for resonance on the difference frequency (related to nonlinear three-wave interaction).

Let us consider a more detailed interaction of electromagnetic and sonic waves similar to the Brillouin scattering by arrangement of the experiment with Cr. EMF. Electrodes, on which the pulses of the G5-54 generator had been supplied, were fixed on the opposite side faces of tested specimen. At that, the transient processes are described as spreading of two electromagnetic waves with Poynting vectors pointed opposite to each other. In the theoretical model of the induced Brillouin scattering, it corresponds to the case of the "backward scattering" (Shubert & Vil'gelmi, 1973), by which the following conditions must be satisfied for the frequencies ω_S, ω_1, ω_2 and wave vectors $\mathbf{k}_S$, $\mathbf{k}_1$, $\mathbf{k}_2$ of the sonic and electromagnetic waves:

$$\omega_1 = \omega_2 + \omega_S, \quad \vec{k}_1 = \vec{k}_2 + \vec{k}_S \tag{5}$$

Absolute values of the vectors $\mathbf{k}_S$, $\mathbf{k}_1$, $\mathbf{k}_2$ are defined by known expressions:

$$k_S = \omega_S / V_S, \quad k_{1,2} = \omega_{1,2} / C, \tag{6}$$

where V_S - the sonic wave velocity, C - the electromagnetic waves velocity, and for the continuous consolidated medium usually $V_S/C \sim 10^{-5}$. Because the direction of the vector $\mathbf{k}_2$ is opposite to $\mathbf{k}_1$, then it follows from the equations (6), that condition of the sonic wave gain by interaction of electromagnetic waves with frequencies ω_1, ω_2 may be written in the form:

$$\omega_S = \frac{V_S}{C} \cdot (\omega_1 + \omega_2) \approx 2 \cdot \frac{V_S}{C} \cdot \omega_{1,2} \tag{7}$$

For pulses of the G5-54 generator, steep rising edges with width less than 0,1 μs provide excitation of great numbers of the harmonics in the frequency range over or of the order of 10^7 1/s. That is why the condition (7) is compatible with the requirement $\omega_S = \omega_1 - \omega_2$ in the case of close to each other values of ω_1, $\omega_2 \sim 10^7$ 1/s, and then the frequency of sonic waves ω_S falls into the range from hundreds 1/s to 10^3 1/s. A hypothesis about the mechanism of trigger influence of electric pulses on AE through low-frequency vibrations excitement looks like unexpected, because there is a widely held idea that the medium reaction (change in the destruction process) begins at the lowest scales levels, and high frequencies correspond to small sizes. In this connection it is necessary to note that the immediate response of the rock specimens on stimulation by low-frequencies vibrations (with a frequency ~ 200 Hz in the experiments (Bogomolov et al., 2001; Kuksenko et al., 2003), and frequencies 1-2 kHz in the experiments with another microvibrator (Bogomolov et al., 2004a; Mubassarova et al., 2011)) has been demonstrated. Applying a classic (unquantized) representation of the induced Brillouin scattering model (Shubert & Vil'gelmi, 1973), let's estimate amplitude of sonic oscillations, which can be excited by electromagnetic waves taking part in transient processes on the pulses edges. In this representation, in one-dimensional case the original equations for the stress waves $\sigma_{\sim}(t,z)$ and electric field strength can be written in the following form:

$$\frac{\partial^2 \sigma_{\sim}}{\partial t^2} - V_S^2 \frac{\partial^2 \sigma_{\sim}}{\partial z^2} = -\varepsilon_{e0}\, \varepsilon_e \cdot \left(\frac{d\varepsilon_e}{d\sigma} \right) \cdot \frac{G^2}{2\rho} \cdot \frac{\partial^2 E}{\partial z^2} - \Gamma_S \frac{\partial \sigma_{\sim}}{\partial t}, \tag{8}$$

$$\frac{\partial^2 E}{\partial t^2} - C^2 \frac{\partial^2 E}{\partial z^2} = -(\varepsilon_{e0}\, \varepsilon_e)^{-1} \cdot \left(\frac{d\varepsilon_e}{d\sigma} \right) \cdot \frac{\partial^2 (\sigma_{\sim} E)}{\partial z^2}, \tag{9}$$

where ρ - material density, Γ_S - acoustic absorption factor, expressed through Q factor by the ratio $\Gamma_S = \omega_S / 2\pi\, Q$. It is necessary to note that the member with Γ_S, taking into account friction loss of a sonic wave energy, is written in the simplest way, although from the hydrodynamic equations follows a proportionality of this summand $\partial^3\sigma_\sim/\partial z^2\partial t$. According to the (Shubert & Vil'gelmi, 1973), the wave equation (6) models a situation, when, on the one hand, dielectric permittivity disturbances linked to the stress oscillations: $\delta\varepsilon_e \cong (d\varepsilon_e/d\sigma)\sigma_\sim$, and, on the other hand, in a sonic wave density oscillations of (deformations) and stress may be supposed to be proportional. In the usual terms of the experiment in the nonlinear optics these interconnections are undoubted. The right part of the electromagnetic waves spreading equation (9) describes a disturbance of electric polarization. It is believed that a length, where sonic waves subside, is bigger than microcrack dimension (initial disturbance source), but it is much less than the length of electromagnetic waves damping.

Search for solutions of the system (8), (9) in the form of quasi-harmonic waves: sonic wave $\sigma_\sim = \sigma_a \exp(i\, k_S z - i\omega_S t)$ and two electromagnetic waves $E_{1,2} = E_{a1,2} \exp(ik_{1,2}z - i\omega_{1,2}\, t)$, their frequencies and wave numbers satisfy (7), and complex amplitudes are smooth time functions, brings the following resulting expression for σ_a amplitude:

$$\sigma_a = -i\pi\,\varepsilon_{e0}\,\varepsilon_e\left(\frac{d\varepsilon_e}{d\sigma}\right)\cdot G^2\,k_S\cdot\frac{E_{a1}\,E_{a2}^*}{2\rho V_S \Gamma_S}, \quad (10)$$

where with the sign * means the complex conjugation. Actually this is a well known formula of nonlinear optics. Other equations from the system describing the evolutions of harmonics $E_{a1,2}$,(Shubert & Vil'gelmi, 1973), are not necessary for our consideration. The main feature is that the amplitudes of these harmonics are steady in a case of action of enough powerful field. To estimate a maximal value of the amplitude σ_a by the formula (10) the derivative $d\varepsilon_e/d\sigma$, formed its right part, may be changed by the ratio $(\varepsilon_e\text{-}1)/G$ with order accuracy. At the same time presence of the significant density fluctuations in the crack, emitted acoustic waves (AE), neighborhood is automatically taken into account. Conversions in the (10) with applying of the ratio $V_S^2 \cong G/\rho$ and other known formulas bring to the expression:

$$\sigma_a \sim \pi^2 Q\cdot\frac{1}{2}\varepsilon_{e0}\varepsilon_e E_a^2,\quad E_a \sim E_{a1} \sim E_{a2}, \quad (11)$$

which in pictorial form pointed to physical sense of the effect. It has happened that in result of resonance on the difference frequency ω_S the amplitude of stress oscillations may increase in ~ Q times more in comparison to a usual case of quasi-stationary field, when the expression $\delta\sigma \sim \delta U \sim \frac{1}{2}\,\varepsilon_{e0}\,\varepsilon_e\,E_a^2$ by dimensionality and order of the value correspond to the surface density of ponderomotive forces and, simultaneously, to the energy of electric field. Such result of sonic waves gaining may be expected as a natural demonstration of the general physical principle of a resonance, and the harmonic beating with close frequencies ω_1, ω_2 of electric field play the external effect role. For the rocks without large cracks Q factor by atmosphere pressure, Q ~ 100-400, and in compressed station it can reach for several thousand or tens of thousands (Nazarov & Radostin, 2007). If we suppose that in the experiments of G5-54 generator applying the amplitude of high-frequency harmonics has been compared with the pulses amplitude (~ 1000 V/m), then the expression (11) gives the estimate of σ_a ~ 1-10 Pa, fallen just in the range 10^{-7} -10^{-6} of the level of main compressive stress. For the vibrations with amplitudes from this range the mentioned above effect of vibrating stimulation of acoustic emission activity is established. That is why the presented model of the mechanism of microvibrations excitement by the effect of electric pulses to the rock specimens may be considered as a serious candidate for explanation of trigger influence of electric pulses with steep wave fronts on AE, in particular pulses of the G5-54 generator. This model shows a similarity of the effects of acoustic emission responses to the action of weak low-frequency vibrations and electromagnetic pulses, which has been noted in works (Bogomolov et al., 2004, 2004a; Mubassarova et al., 2011). It should be noted that in a real situation, as opposed to one-dimensional task, the σ_a value by resonance interaction of sonic and electromagnetic waves may occur less then estimation (11) because of waste of excited wave divergence and other factors. Proposed model remains a believable hypothesis, its attraction is determined by possibility of removal (or essential decrease) of mentioned above mismatch in the amplitudes of the first-excited microvibrations.

As it has been mentioned above that during microcrack formation or growing (AE events), vibrations and dielectric permittivity disturbances appear, and the last predetermine a possibility of interaction with electromagnetic waves. Thanks to pulse and energy transmission to sonic waves with a resonance (difference) frequency in the medium with high Q factor, the low-frequency vibrations continuing for a time significantly exceeding duration of the high-frequency AE signal may be excited. These vibrations are able to initiate other microcracks growing. But if the undercritical sites sensitive to external influence, are absent (that is naturally by very low background activity, in other words by the loading low level), then a stimulated increase of the events number would not occur. As concerns to the case of high background activity (big loadings), for it the conditions of excited oscillations of dielectric permittivity coherence becomes important. If the following AE event spontaneously appears soon after the first, then besides it the oscillation phase on the frequency ω_C would break because of disturbances from a new crack and the resonance interaction with external waves becomes less effective. Typical time for the fluctuation of a density and dielectric permittivity may be estimated as $\delta t \sim$ 10-50 μs (travel time of sonic wave with rate $V_S \sim$ (2- 4) 10^3 m/s over a specimen of dimension 5-10 cm). Then time of coherence saving including by interaction similar to induced Brillouin scattering will be of the order of $Q \cdot \delta t \sim$10-50 ms (accepted $Q \sim 1000$). By the pulse-repetition frequency of the G5-54 generator 2-10 kHz during this time from 20 to 500 serial pulses may pass, their edges make a contribution in the amplitude of high-frequency harmonics $\omega_{1,2} \sim 10^7$ 1/s. In the case when background activity will exceed 20 1/s, the interval between some events will be shorter 10-50 μs, and real coherence time will shorten. The number of passed fronts will decrease, and in the wake of it, the amplitude $E_{a1,2}$ in the formulas (8-11) will decrease as well. By the corresponding decrease of maximal σ_a from (11), the absence of the effect of acoustic emissive response to electric pulses may be explained. It is necessary to affect by pulses with higher electric field strength or with higher frequency of pulses passing for the effect observing. It is confirmed with several instances of responses observing on gabbro specimens, stood under the loading close to the critical (0.98 from the fracture). In these cases the activity level has amounted from the tens to hundreds of events per second, but effect with maximum possible parameters (the series of 900-volts pulses from the capacitor discharger, $E \sim 5 \cdot 10^4$ V/m) leads to greater activity increase, in 600-800 s after which a relaxation rather than a macrofracture has followed.

The other check of the hypothesis by the independent experimental data involves the aspect of low-frequency vibrations generation before the basic AE response to EM effect. This follows from the model essence. A special experiment has not been conducted, but it is possible to use the results of measurements, when the seismic mini detector (piezoelectric geophone A1605) has been fixed to the granite specimen in the form of parallelepiped for the sake of the hardware adjustment. After the calibration the hardware units have been used for geological acoustic monitoring in the boreholes at the Bishkek test site (Zakupin, 2010), so received by the geophone signals in the range of 10^2 -10^3 Hz may be called as GAE - geoacoustic emission (Gavrilov et al, 2008). The specimen has been tested on the lever-gravitational press by a stepped uniaxial compression with application of EIs (capacitor discharges) on every step. By the previous experiments a character of AE responses of the specimen from this series is known (Bogomolov & Zakupin, 2008). In the measuring sessions by a fixed load either high-frequency channel with AE sensor or low-frequency one with geophone has been turned on, to avoid a cross-effect of primary piezoelectric transducers. For the tested granite specimen by the loads of 85-95% from fracture the AE activity response to EM field pulses (condensing dischargers) has occurred with a delay of 600-1000 s. The change of GAE amplitude level during the session by such pulses effect for 1-5 min is shown on the fig.9. In the session 16 condensing discharges (maximal voltage is 600 V, duration ~ 2ms) with equal intervals in the first 15 minutes have been conducted. Diagrams discontinuity is tied with pause in the registration needed for the data downloading. Increase of the GAE level has been occurred earlier as compared with the typical delay of the response of high-frequency AE activity. It just should be observed, according to discussions about the resonance gain of low-frequency vibrations because of the induced Brillouin scattering. So, the proposed hypothesis (model) of the primary vibrations generation is confirmed by involvement of the new data, it allows understanding a number of aspects of the acoustic emission responses to electromagnetic field action.

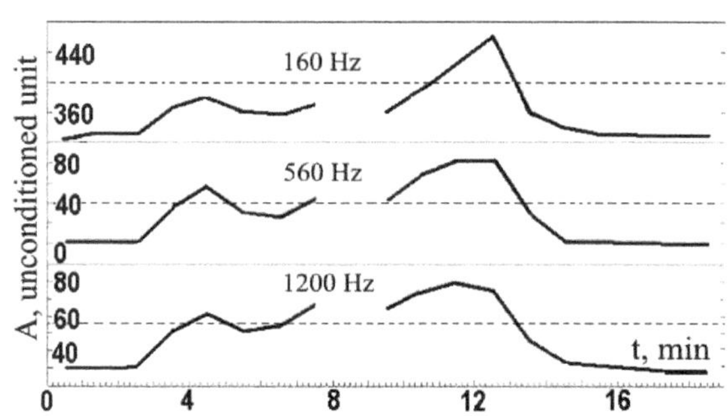

Fig. 9. The time dependence of the low-frequency emission amplitude.

6. Conclusion

The new results of variations of acoustic emission activity of rock specimens (marble) under the effect of electric pulses have reflected the short-term increase of the number of microcracks initiation and stretching acts. The analysis of AE responses to electromagnetic effects has been conducted by a detail research of the macrostrain characteristics of the specimens. As in the previous works, the reaction of rock specimen to EI is a short-term increment of AE activity. From the viewpoint of EI effect on AE and volumetric deformation, the results show the high efficiency of external electromagnetic fields effect on cracking kinetics in loading rocks. The proposed model relates the mechanism of electric pulses influence on the crack formation velocity in loaded rock specimens with initial excitation of low-frequency vibrations in them owing to the nonlinear resonant interaction of electromagnetic waves similar to the effect of induced Brillouin scattering. The model establishes the correlation between the electric pulses amplitude and the parameters of the specimen material (acoustic Q- factor and dielectric permittivity) which is a necessary condition of AE response generation at steady loads. From the viewpoint of analysis and interpreting of laboratory experiments, one can see a convincing reason for geoacoustical surveys in the vicinity of a source of electromagnetic soundings of the Earth's crust.

7. Acknowledgment

The research has partially been supported by: grants of RFBR №10-05-00231a, №11-05-12042; Program 15 of basic research of RAS Presidium (project 4.4); Ministry of education and science RF (state contract 02.740.11.0730)

8. References

Alshitz, V.I.; Darinskaya, E.V.; Koldaeva, M.V. & Pertzhik, E.A. (2008). Magnetoplastic Effect in Nonmagnetic Crystals. // Dislocations in Solids. Ed.by J.P.Hirth. - Amsterdam: Elsevier. 2008. - V. 14. - P. 335-437 ISBN: 978-0-444-53166-7

Belyakov, A.S.; Lavrov, V.S.; Nikolaev, A.V.; Hudzinskii, L.L. (1999) Triggered vibroactions and seismic emission of rocks. *Izvestiya, Physics of the Solid Earth,* Vol.35, No12 (December 1999), pp. 1002-1009, ISSN 1069-3513.

Ben-Zion, Y. & Lyakhovsky, V. (2006). Analysis of aftershocks in a lithospheric model with seismogenic zone governed by damage rheology, *Geophys. J. Int.*, Vol.165, No 1(April 2006), pp.197–210, ISSN 1365-246X

Bogomolov, L.M.; Manzhikov, B.Ts.; Trapeznikov, Yu.A. et.al. (2001). Vibroelasticity, acoustic – plastic effect and acoustic emission of loaded rocks. *Russian Geology and geophysics*, Vol. 42, № 10, (October 2001), pp. 1593-1603, ISSN 1068-7971

Bogomolov, L.M.; Avagimov, A.A.; Sychev, V.N.; Sycheva, N.A. et al. (2005) On manifestations of electric -triggering seismicity at Bishkek site /toward active seismic-electric monitoring/, In: *Active Geophysical Monitoring of the Earth's Lithosphere*, Goldin, S.V, ed., pp. 112–116, Izdatel'stvo SO RAN, ISBN 5-7692-0790-6, Novosibirsk

Bogomolov, L.M.; Il'ichev P.V.; Zakupin, A.S. et.al. (2004). Acoustic emission response of rocks to electric power action as seismic-electric effect manifestation. *Annals of Geophysics,* Vol.47, No 1, (February 2004), pp. 65-72, ISSN 1593-5213

Bogomolov, L.; Zakupin, A.; Tullis, T. et al. (2004a).Acoustic emission measurements to understand transition straining processes and seismicity triggering by power impacts, Proceedings. 8 th Multi-Conference on Systemics, Cybernetics and Informatics, Vol. XII, ISBN 980-6560-13-2, Orlando, USA, July 2004

Bogomolov, L.M.; Adigamov, N.S.; Sychev, V.N. & Zakupin, A.S. (2006). Fenomenological model of multi-scale triggering effects during straining of geomedium with physical field presence, *Preprint HC PAH 1-06*, Research station RS RAS, Bishkek

Bogomolov, L. & Zakupin, A. (2008). Do Electromagnetic Pulses Induce the Relaxation or Activation of Microcracking Rate in Loaded Rocks? *Solid State Phenomena,* Vol.137, pp. 199-208, ISSN 1662-9779

Chelidze, T.; De Rubeis, V.; Matcharasgvili, T. & Tosi, P. (2006). Influence of Strong Electromagnetic Discharges on the Dynamics of Earthquake Time Distribution in the Bishkek Test Area (Central Asia), *Annals of Geophysics*, Vol. 49, No 4–5, (October/December 2006), pp. 961-975, ISSN 1593-5213

Chernyak, G.Ya. (1987). Electromagnetic methods in hydrogeology and engineering geology, Nedra, Moscow.

Finkel', V.M.; Golovin, Yu.I.; Sereda, V.E. et al. (1975). Electrical effects during fracture of LiF crystals in relevance to problem of cracking control, *Physics of the solid state*, Vol. 46, No 5, (May 2004), pp. 770-776, ISSN 1063-7834

Finkel', V.M. (1977). Physical fundamentals for fracture retardation (Fizicheskie osnovy zamedlenija razrushenia), Metallurgia, Moscow.

Gavrilov, V.; Bogomolov, L.; Morozova, Yu. & Storcheus, A. (2008). Variations in geoacoustic emissions in a deep borehole and its correlation with seismicity. *Annals of Geophysics*, Vol.51, No 5/6, (October/December 2008), pp. 737-753, ISSN 1593-5213

Golovin, Yu.I. (2004). Magnetoplastic effects in solid. *Physics of the solid state*, Vol. 46, No 5, (May 2004), pp. 789-824, ISSN 1063-7834

Greshnikov, V.A. & Drobot, Yu.B. (1976). Acoustic emission: application for tests of materials and constructions, Izdatelstvo standartov, Moscow

Gzovskii, M.V. (1975). Grounds of tectonophysics (Oscnvy tektonofiziki), Nauka, Moscow

Heidbach, O.; Tingay, M.; Barth, A.; Reinecker, J.; Kurfeß, D. & Müller, B.(2008). The World Stress Map database release 2008, In: *FZ.WSM.Rel2008*, 2008, Available from: http://dc-app3-14.gfz-potsdam.de/pub/stress_data/stress_data_frame.html

Knox, R.S. (1963). Theory of excitons, Academic Press Inc, ISBN 0-12-607765-7, New York

Kocharyan, G.G. & Spivak, A.A. (2003). The dynamics of rock deformation, PBMC Academkniga, ISNN 5-94628-078-3, Moscow.

Kocharyan, G.G.; Kulyukin, A.A. & Pavlov, D.V. (2006). Specific dynamics of interblock deformation in the Earth's crust, *Russian Geology and geophysics*, Vol. 47, № 5, (May 2006), pp. 667-681, ISSN 1068-7971

Kropotkin, P.N.; Efremov,V.N.; Makeev, V.N. (1987). Stressed state of Earth Crust and Geodynamics, *Geotectonics (Geotektinika)*, No 1, (January 1987), pp. 3-24

Kuksenko, V.S.; Mahmudov, H.F. & Ponomarev, A.V. (1997). Relaxation of electric fields induced by mechanical loading in natural dielectrics. *Physics of the solid state*, Vol. 39, No 7, (July 1997), pp. 1067-1071, ISSN 1063-7834.

Kuksenko, V.S.; Manzhikov, B.Ts.; Tilegenov, K. et al.(2003). Trigger effect of weak vibrations in solids (rocks). *Physics of the solid state*, Vol. 45, No 12, (December 2003), pp. 789-824, ISSN 1063-7834

Lyakhovsky, V.; Ben-Zion, Y. & Agnon, A. (1997). Distributed damage, faulting, and friction, *J. Geophys. Res. - Solid Earth*, Vol. 102, No B12, (December 1997), pp. 27635- 27649, ISSN 0148–0227

Makarov, P.V. (2004). On the hierarchical nature of deformation and fracture of solids and media. *Physical mesomechanics*, Vol.7, No 3-4, (May- August 2004), pp. 21-29, ISSN 1029-9599.

Mirzoev, K.M.; Vinogradov, S.D. & Ruzibaev, Z. (1991). Influence of microseism and vibrations on acoustic emision. *Izvestiya, Physics of the Solid Earth*, Vol.27, No 12, (December 1991), pp. 69-72, ISSN 1069-3513.

Molotskii, M.I. (2000). Theoretical basis for electro- and magnetoplasticity. *Materials Science and Engineering A*, Vol.287, No 2 (August 2000), pp. 248–258, ISSN 0921-5093.

Mubassarova, V.A.; Bogomolov, L.M.; Zakupin, A.S. & Borovsky, B.V. (2011) Peculiarities of AE signals flow of loaded granitic specimens under the influence of weak vibrations. *Vestnik Kyrgyz- Russian slavic university (Vestnik Kyrgyzsko - rossijskogo slavjanskogo universiteta)*, Vol.11, No 4(April 2011), pp. 60-66, ISSN 1694-500X

Nazarov, V.E.& Radostin, A.V. (2007). Nonlinear wave processes in elastic micro-inhomogeneous media (Nelineinye volnovye processy v uprugikh mikroneodnorodnykh sredakh), Institute of Applied Physics of RAS, ISBN 978-5-8048-063-6, Nizhny Novgorod

Pallas-Areny, R. & Webster, J.G. (1991). Sensors and Signal Conditioning. Wiley-Interscience, New-York, ISBN 0471332321

Panin, V.E.; Grinyaev, Yu.V. (2003). Physical mesomechanics: a new paradigm at the interface of solid state physics and solid mechanics. *Physical mesomechanics*, Vol.6, No 4, (August 2003), pp. 7-32, ISSN 1029-9599

Paparo, G.; Gregori, G.P.; Coppa, U.et al.(2002). Acoustic emission (AE) as a diagnostic tool in geophysics, *Annals of Geophysics*, Vol.45, No 2, (February 2002), pp. 401-416, ISSN 1593-5213

Parhomenko, E.I. (1965). Electric properties of rocks, Nauka, Moscow.

Shpeizman, V.V. & Zhoga, L.V. (2005). Kinetics of failure of polycrystalline ferroelectric ceramics in mechanical and electric fields. *Physics of the solid state*, Vol. 47, No 5, (May 2005), pp. 869-875, ISSN 1063-7834.

Shubert, M. & Vil'gelmi, B. (1973). Introduction to nonlinear optic, Part 1, Classical consideration (Transl by Kovner, M.A. from Einfuhrung in die nichtlineare optic, teil 1, BSB B.G. Teubner Verlarsgesellschaft, Leipzig), Mir, Moscow.

Sobolev, G.A.; Spetzler, H.; Koltsov, A. & Chelidze, T. (1993). An Experimental Study of Triggered Stick-slip. *Pure and Applied Geophysics*, Vol.140, No 1 (February 1993), pp. 1-16, ISSN 0033-4553

Sobolev, G.A. & Ponomarev, A.V. (2003). Physics of eartquakes and precursors, Nauka, ISBN 5-02-002832-0, Moscow.

Stavrogin, A.N. & Protosenya, A.G. (1979). Plasticity of rocks, Nedra, Moscow

Stantchitz, S.A. & Tomilin, N.G. (1984). Investigation of temporal parameters of acoustic signals during formation of tensile cracks, In: *Prognoz zemletrjasenij (prediction of earthquakes)*, No 4, Sadovsky, M.A. , ed., Dushanbe- Moscow, pp. 31 46

Svetov, B.S. (1995). "Nonclassical" geoelectrics. *Izvestiya, Physics of the Solid Earth*, Vol.31, No 8 (August 1995), pp. 3-12, ISSN 1069-3513.

Sychev, V.N.; Avagimov, A.A.; Bogomolov, L.M. et al. (2008). On Trigger Effect of Electromagnetic Pulses on Weak Seismicity, In: *Geodynamics and Stressed State of Earth's Interiors (Geodinamika i napryazhennoe sostoyanie nedr zemli*), pp.179-188, Publishing House of Mine Institute SB RAS, Novosibirsk

Tarasov, N.T.; Tarasova, N.V.; Avagimov, A.A.& Zeigarnik, V.A. (1999). The effect of high-energy electromagnetic pulses on seismicity in Central Asia and Kazakhstan, *Journal of Volcanology and Seismology (Vulkanologiya i Seismologiya)*, No 4 -5 (October 1999), pp. 152- 160, ISSN 0203-0306

Tarasov, N.T. & Tarasova, N.V. Spatial-temporal structure of seismicity of the north Tien Shan and its change under effect of high energy electromagnetic pulses. *Annals of Geophysics,* Vol.47, No 1, (February 2004), pp.199-212, ISSN 1593-5213

Urusovskaya, A.A.; Alshitz, V.I.; Bekkauer, N.N. & Smirnov, A.E. (2000). Deformation of NaCl crystals under combined action of magnetic and electric fields. *Physics of the solid state,* Vol. 42, No 2, (February 2000), pp.274-276, ISSN 1063-783.

Utsu, T.; Ogata, Y.; Matsu'ura, R.S. (1995). The centenary of the Omori formula for a decay law of aftershock activity, *Journal of Physics of the Earth,* Vol. 43, No 1 (February, 1995), pp. 1–33, ISSN: 0022-3743

Vinogradov, S.D.(1989). Acoustic method in studies of earthquake physics, Nauka, Moscow

Zakupin, A.S.; Alad'ev, A.V.; Bogomolov, L.M.; et al. (2006). Interrelation between electric polarization and acoustic emission of geomaterials specimens under conditions of uniaxial compression. *Journal of Volcanology and Seismology (Vulkanologiya i Seismologiya),* No 6, (December 2006), pp. 22-33, ISSN 0203-0306

Zakupin, A.S.; Bogomolov, L.M. & Sycheva N.A. (2009). The effect of crossed electric and magnetic fields in loaded rock specimens. *Materials Science and Engineering A,* Vol.521–522, (September 2009), pp. 401–404, ISSN 0921-5093

Zakupin, A.S. (2010) Geoacoustical observations in wells on the territory of Bishkek test site, In: Triggering effects in geosystems, Adushkin A.A. & Kocharyan G.G., pp. 277-285, GEOS, ISBN 978-5-89118-527-2, Moscow

Zuev, L.B. (1990). Physics of electroprasticity of alkali -halogen crystals, Nauka, Novosibirsk

9

Upstream Multiphase Flow Assurance Monitoring Using Acoustic Emission

Salem Al-Lababidi, David Mba and Abdulmajid Addali
Cranfield University,
UK

1. Introduction

Multiphase flow assurance in oil and gas upstream industries covers a wide range of flow conditions. Over the last decade, the investigations and developments of multiphase flow assurance control and metering system for offshore and subsea oil and gas fields have been a major focus for the industry worldwide.

An offshore or subsea production facility consists of several satellite wells. The contents of each well are combined and passed onshore via a common pipeline. Each satellite well produces variable quantities of oil, water and gas during its lifetime. However, if different companies own these wells, their flowrates and compositions in pipelines must be monitored and control before any mixing takes place

Also, the multiphase flow assurance characteristics of each satellite production well such as gas hydrate, scale deposition, sand transportation and management, wax deposition must be monitored and controlled in multiphase flow pipelines.

Within the oil and gas industry, it is generally recognized that the implementation of multiphase flow assurance control and design strategy could lead to great benefits in terms of well testing, reservoir management, production allocation, production monitoring, capital expenditure and operational expenditure.

2. Chapter overview

Recent years have seen the increasing acceptance of deploying acoustic emission and ultrasonic monitoring techniques in oil and gas industries.

This chapter is divided into the followings sections which are described below.

2.1 Review of multiphase flow assurance monitoring techniques

In the first part of this chapter, the definition of multiphase flow, flow regimes and flow map in horizontal pipes are introduced. Then the chapter introduces the reader to the concept of monitoring strategies followed in multiphase flow assurance systems. The chapter also introduces the techniques of multiphase flow measurements currently available including ultrasonic and acoustic emissions.

2.2 Review two-phase gas/liquid slug flow key parameters in pipeline

This part of the chapter describes the slug flow initiation and dissipation process in horizontal pipelines. Empirical correlations for slug flow characteristics including gas void fraction and liquid hold up.

2.3 AE applications for gas/liquid slug flow monitoring in pipeline

This part of the chapter presents the experimental work findings for AE monitoring for gas/liquid slug flow.

3. Multiphase flows

The simultaneous flow of two or more phases in pipe is termed multiphase flow. Multiphase flow systems are of great industrial significance and are commonly found in the chemical, process, nuclear, hydrocarbon and food industries. The subject has received widespread research attention, particularly over the past five decades.

In multiphase flows, the flow behaviour is much more complex than for single-phase flow. The phases tend to separate because of differences in density. Shear stresses at the pipe wall are different for each phase as a result of their different densities and viscosities. The most distinguishing aspect of multiphase flow is the variation in the physical distribution of the phases in the flow conduit, a characteristic known as flow pattern or flow regime.

3.1 Horizontal gas-liquid flow regimes

Gas/Liquid flow regimes in horizontal pipes are summarised in Figure 1, from top to bottom in order of increasing gas flow rate.

Bubble Flow: In bubble flow, small gas bubbles flow along the top of the pipe.

Elongated Bubble Flow: Collisions between the individual bubbles occur more frequently with increasing gas flow rate and they coalesce into elongated "plugs". This is often called plug flow.

Smooth Stratified Flow: The gas plugs coalesce to produce a continuous gas flow along the top of the pipe with a smooth gas-liquid interface typical of stratified flow at relatively low flow rates.

Wavy Stratified Flow: In most situations, the gas-liquid interface is rarely smooth with ripples appear on the liquid surface. The amplitude increases with increased gas flow rate.

Slug Flow: When the amplitude of the waves travelling along the liquid surface becomes sufficiently large enough for them to bridge the top of the pipe, the flow enters the slug flow regime. The gas flows as intermittent slugs and with smaller bubbles entrained in the liquid.

Annular Flow: Occurs when gas flow rate is large enough to support the liquid film around the pipe walls. Liquid is also transported as droplets distributed throughout the continuous gas stream flowing in the centre of the pipe. The liquid film is thicker along the bottom of the pipe because of the effect of gravity.

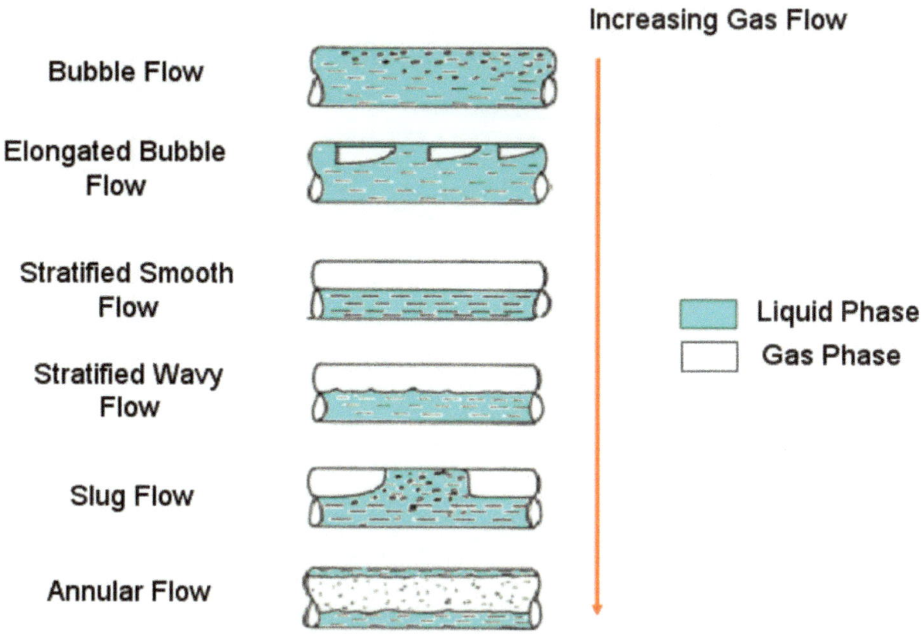

Fig. 1. Flow regimes of gas/ liquid horizontal flow

3.2 Gas-liquid flow regimes map in horizontal flow

The most distinguish aspect of multi-phase flow is the variation of the physical distribution of the phases in the flow conduit, a characteristic known as flow pattern or flow regime. The flow pattern that exists depends on the relative magnitudes of the forces that act on the fluids. These forces such as buoyancy, turbulence, inertia, and surface-tension forces, all vary significantly with flow rates, pipe diameter, inclination angle, and fluid properties of the phases.

Flow pattern is often displayed using a flow pattern map, which is a two-dimensional map depicting flow regime transition boundary. The selection of appropriate coordinates to present clearly and effectively the different flow regimes has been a research topic for a long time. The dimensional coordinates such as superficial velocities are much more generally used in practice as the one by Taitel et al. (1976).as shown in Figure 2.

3.3 Current status in multiphase flow measurements

Multiphase metering methods can be classified into two major groups. The first group is the conditioning methods where the condition of the phases in the pipe is changed (phases partial separations or phases mixing) upstream of the metering point. The second group is the direct method, where the phase parameter measurements are achieved without a pre-conditioning process.

In a partially separation system, as the name suggests, is based on partially separate the flow, usually into predominately liquid and predominately gas streams, before

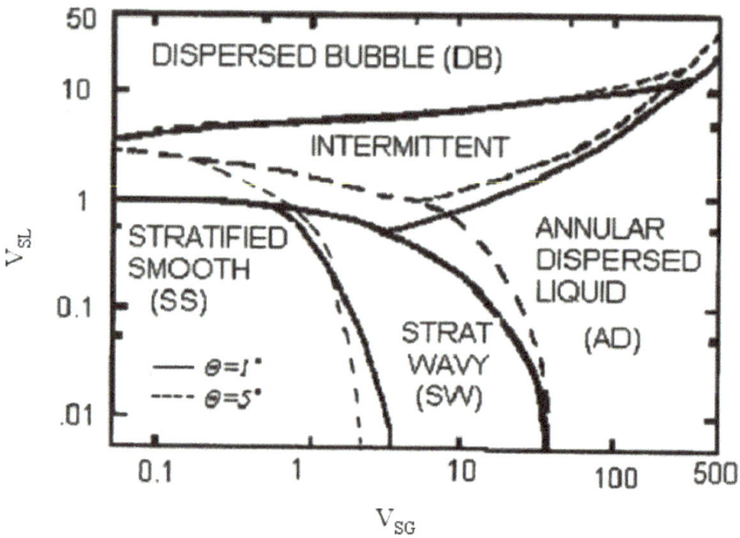

Fig. 2. Horizontal flow regimes map by Taitel et al. (1976)

measurement. Since the separation is only partial, some liquid expected to travel with the gas stream. As a result, each flow stream only needs to be measured over a limited range of phase fractions as shown in Figure 3(a).

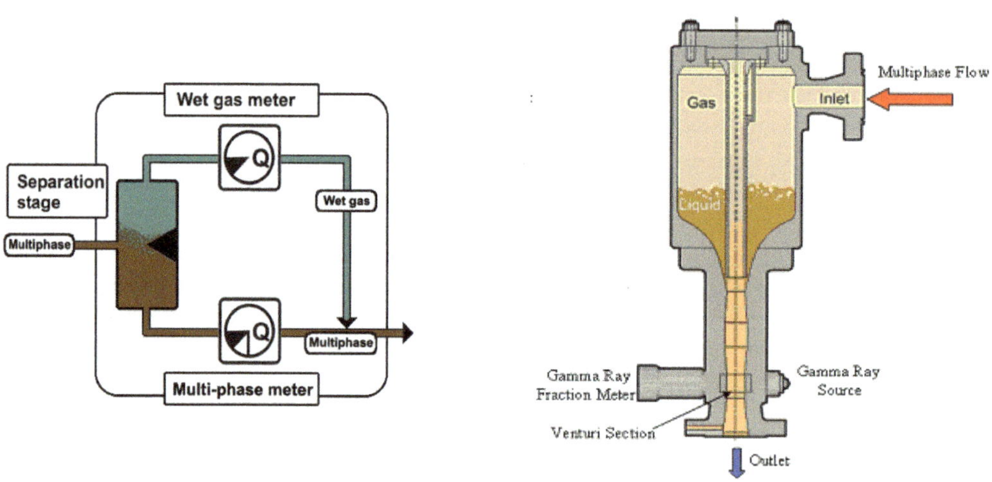

a. Partial Separation System b. Homogenisation System

Fig. 3. Preconditioning system (Thorn *et al.* 1997)

In homogenous based systems, the flow is pre-mixed to try to ensure that all measurements are made on a homogenous flow, hence removing the problem of flow regime dependency and reducing the number of and difficulties of measurement required. Figure 3(b) shows an example of a commercially available three-phase flowmeter which uses this strategy. The Framo multiphase flowmeter uses a tank mixer to homogenize the flow both radially and axially. The homogenized flow then passes through a Venturi meter which is used to measure the velocity of the mixture, and a dual-energy-ray attenuation meter which uses two different energy levels of the Barium 133 isotope to determine the oil, water and gas fractions.

3.4 Measurement strategies – inferential approach

The primary information required by a user of multiphase meters is the mass flow rate of the each phase. An ideal multiphase flow meter would make independent direct measurements of each of these quantities. Unfortunately, direct mass flowmeters for use with two phase flows are rare and do not exist at all for three phase flows.

An alternative to direct mass flow measurement is to use an inferential method. An inferential mass method requires both the instantaneous velocity and cross sectional fraction of each component to be know in order to calculate the individual component mass flowrates and total mixture mass flowrate (Thorn et al. 1997). Figure 4 shows a schematic of such an approach.

Two types of parameters are monitored by a multiphase metering system:

1. Primary parameters:
   - Phase fraction (e.g. void fraction, water cut);
   - Phase velocity (the velocity of each phase as they cannot be assumed to be travelling at the same velocity);
   - Phase density.
2. Secondary parameters:
   - Flow regime (this may be considered as a primary parameter if a flow dependent sensing technique is used);
   - Phase viscosity;
   - Phase salinity;
   - Phase permittivity/ conductivity.

Density information of the oil, water and gas components are readily available from other parts of the production process, e.g. densitometer readings or estimated from PVT diagrams using measured pressure and temperature. Therefore, the problem is to measure the oil, water, and gas velocities and two of the component fractions. The third component fraction (oil fraction) is deduced from the fact that the sum of the three phase fractions is equal to unity.

$$m = a_g \times V_g \times \rho_g \times A_{pipe} + a_w \times V_w \times \rho_w \times A_{pipe} + a_o \times V_o \times \rho_o \times A_{pipe} \tag{1}$$

where

m is the total mass flow rate,
ρ_g, ρ_w *and* ρ_o are the densities of gas, water and oil,
V_g, V_w and V_o are the velocities of gas, water and oil,

A_{pipe} is the pipe cross section area.

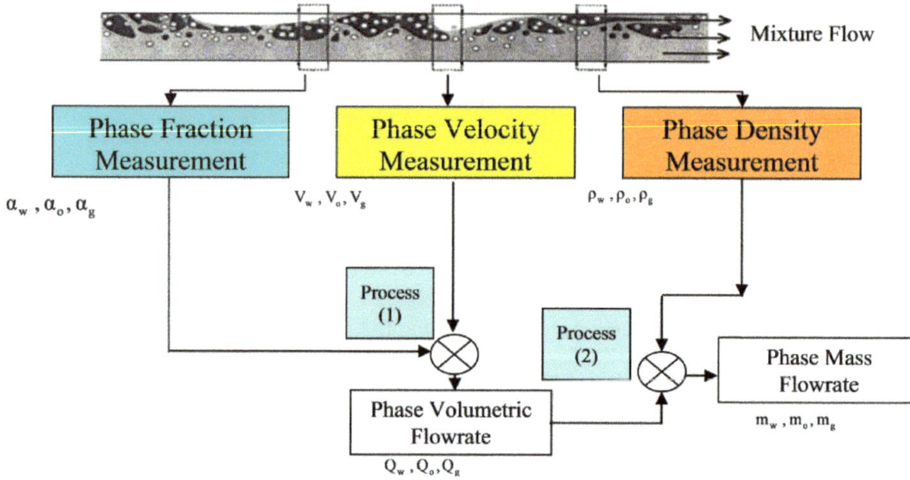

Fig. 4. Inferential multiphase flow measurement strategy (Thorn et al. 1997)

3.5 Direct measurements techniques in multiphase flow

3.5.1 Phase fraction measurement

The two most commonly used methods for measuring gas and liquid fractions in a multiphase flow are based on gamma-ray attenuation and electrical impedance techniques.

3.5.1.1 Gamma ray attenuation (absorption) measurement

The basic principle for the gamma-ray attenuation technique is the fact that the intensity of a collimated beam decreases exponentially as it passes through matter.

There are different gamma ray systems which are used in multiphase flow metering including, single-beam, dual-beam or multiple-beam gamma ray systems. Gamma ray attenuation measurement is applicable to all possible combinations of two-phase and three-phase flows (Corneliussen et al. 2005).

Single-beam gamma ray attenuation (absorption)

The single energy gamma ray attenuation measurement is based on the attenuation of a narrow beam of gamma ray of energy as shown in Figure 5.

Single energy gamma ray attenuation can be used in liquid/liquid system (oil/water) or liquid/gas system. In a pipe, with inner diameter d, the intensity of the beam of radiation after it has passed through the mixture of two-phase flow is given by:

$$I_m(e) = I_v(e) \times exp[-\sum_{i=1}^{2} a_i \times \mu_i(e) \times d] \quad (2)$$

where

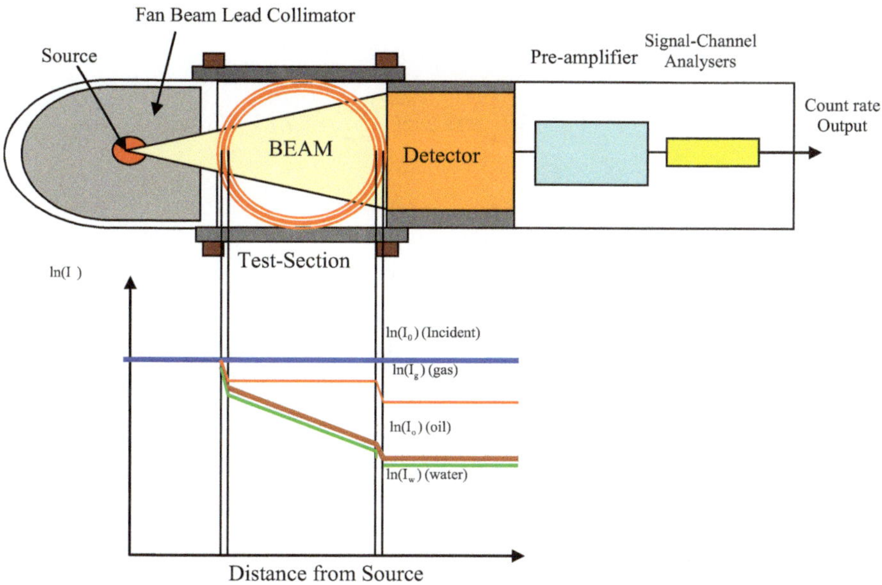

Fig. 5. Single-beam gamma ray densitometer

$I_m(e)$ is the measured countrate,
$I_v(e)$ is the count rate when the pipe is empty,
μ_i is the linear attenuation coefficients for the two phases.

Apart from fractions of the phases a_i, the attenuation coefficients μ_i are also initially unknown. However, the latter can be found by calibration where the meter is filled with individual fluids. In both cases the following two equations for water and oil can be used:

$$I_w = I_v \times exp[-a_w \times \mu_w \times d] \tag{3}$$

$$I_o = I_v \times exp[-a_o \times \mu_o \times d] \tag{4}$$

These two calibration points together with the relation $a_w + a_o = 1$ can be rewritten as an expression for the water fraction in two-phase liquid/liquid mixture (or the water cut):

$$a_w = \frac{ln(I_w) - ln(I_m)}{ln(I_w) - ln(I_o)} \tag{5}$$

Dual-energy gamma ray attenuation (absorption)

In order to determine oil, water and gas fractions, two independent measurements are required using dual or multiple energy technique. This technique has been investigated by a number of researchers (Abouelawafa and Kendall. 1980; Roach and Watt. 1996; Van Santen et al. 1995 and Hewitt et al. 1995).

The basics of dual energy gamma ray absorption measurement are similar to the single energy gamma ray attenuation concept, but now two gamma energies e_1 and e_2 are used. In a pipe, with inner diameter d, containing a mixture of water, oil and gas with fractions a_g, a_w and a_o, the measured count rate $I_m(e)$ is:

$$I_m(e) = I_v(e) \times exp[-\sum_{i=1}^{3} a_i \times \mu_i(e) \times d] \tag{6}$$

where

$I_v(e)$ is the count rate when the pipe is empty,
μ_i is represent the linear attenuation coefficients for the three phases (μ_g, μ_w, μ_o).

For two energy levels, e_1 and e_2, provided the linear attenuation coefficients between water, oil, and gas are sufficiently different, two independent equations are obtained. The third equation is simply the fact that the sum of the three phase fractions in closed conduit should equal to unity as given:

$$a_g + a_w + a_o = 1 \tag{7}$$

A full set of linear equations is given below. R_O, R_W, Rg, and Rm represent the logarithm of the count rates for water, oil, gas and mixture, respectively, at energies e_1 and e_2.

$$\begin{bmatrix} R_w(e_1) & R_o(e_1) & R_g(e_1) \\ R_w(e_2) & R_o(e_2) & R_g(e_2) \\ 1 & 1 & 1 \end{bmatrix} \times \begin{bmatrix} a_w \\ a_o \\ a_g \end{bmatrix} = \begin{bmatrix} R_m(e_1) \\ R_m(e_2) \\ 1 \end{bmatrix} \tag{8}$$

The elements in the matrix are determined in a calibration process by filling the pipe with 100% water, 100% oil, and 100% gas (air) or alternatively by calculations based on the fluid properties. Together with the measured count rates at the two energy levels from a multiphase mixture it is possible to calculate the unknown phase fractions. In Figure 6 shows a typical response triangle with a dual energy source (18 keV and 60 keV) for water, oil and gas mixture.

The corners of the triangle are the water, oil and gas calibrations, and any point inside this triangle represents a particular composition of water, oil and gas (Rafa et al. 1989). The shape of the triangle depends mainly on the energy levels used (thus specific radioactive source), pipe diameter and detector characteristics; however, fluid properties may also influence the triangular shape. If the count levels are too close the triangle will transform into a line and therefore cannot be used for a three-phase composition measurement.

3.5.1.2 Electrical impedance measurements

The main principle of electrical impedance methods for component fraction measurements is that the fluid flowing in the measurement section of the pipe is characterised as an electrical conductor. By measuring the electrical impedance across the pipe diameter (using e.g. contact or non-contact electrodes), properties of the fluid mixture, conductance and capacitance, can be determined. The measured electrical quantity of the mixture then depends on the conductivity and permittivity of the oil, gas and water components, respectively. Figure 7 shows the basic principle of impedance method to measure component fraction.

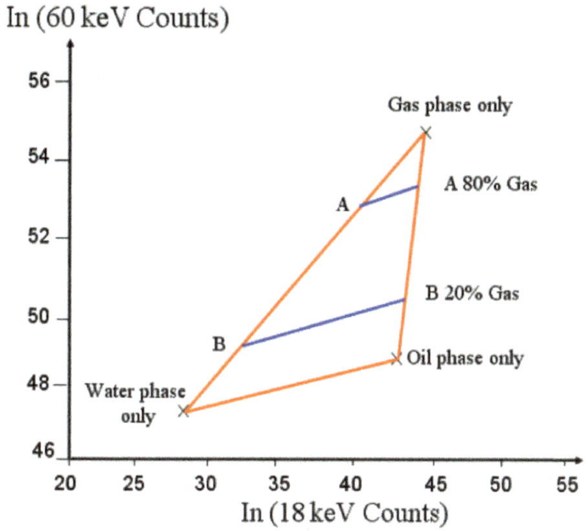

Fig. 6. Dual energy gamma ray response triangle (Rafa *et al.* 1989)

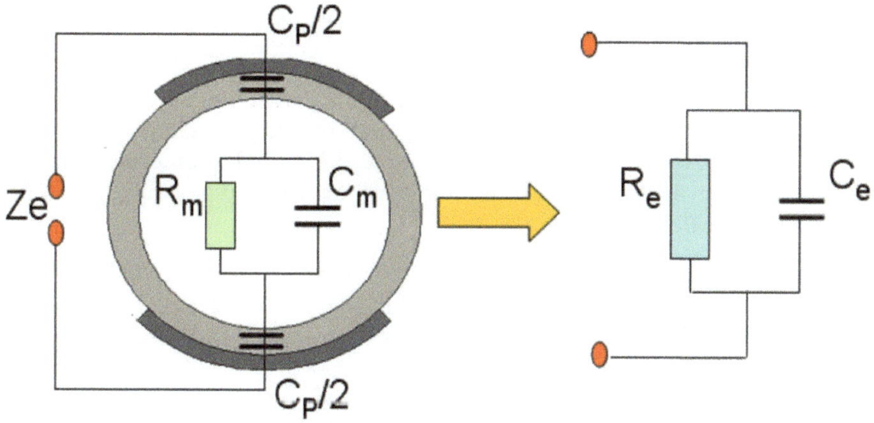

Fig. 7. Impedance for component fraction measurement (Thorn *et al.* 1997)

If the electrical impedance Z_e is measured across two electrodes, between which an oil/water/gas mixture is flowing, the measured resistance R_e and capacitance C_e are given by:

$$R_e = \frac{1+\omega^2 R_m^2 (C_m + C_P)^2}{\omega^2 R_m C_P^2} \tag{9}$$

$$C_e = \frac{[1+\omega^2 R_m^2 C_m (C_m + C_P)^2]\, C_P}{1+\omega^2 R_m (C_m + C_P)^2} \tag{10}$$

The resistance R_m and capacitance C_m of the mixture flowing through the pipe depends on the permittivity and conductivity of the oil, water and gas components, the void fraction and water fraction of the flow, and the flow regime.

The measured resistance and capacitance not only depends on R_m and C_m but also on the excitation frequency ω of the detection electronics and the geometry and materials of the sensor. For a particular sensor geometry (and hence fixed C_P) and flow regime, the measured impedance will be a direct function of the flow component ratio.

In oil continuous mixture R_e is large and can be difficult to measure reliably. For the flows in which water is the continuous phase, a short circuiting effect will occur, caused by the conductive water, if the sensor excitation frequency, fc, is less than:

$$f_c < \frac{\sigma_w}{2\pi\varepsilon_0\varepsilon_w} \tag{11}$$

where

σ_w *and* ε_w are the conductivity and permittivity of the water component respectively.

For process water this would mean a frequency below that of microwave frequencies. Therefore, impedance method is limited to oil or water in continuous phase.

However impedance based methods suffer from two important limitations - they cannot be used over the full component fraction range and are flow regime dependent. This dependency is eliminated by one of two methods: (1) homogenisation of the phase, before the measurement is made and (2) development of electrodes which minimise the dependency upon the flow geometry.

Various methods have been used to reduce the flow regime dependency effects of impedance sensors as shown in Figure 8. In this figure four different designs of non-intrusive sensor are illustrated. These styles have been developed recently, and that to minimise the flow geometry dependency,

Arc electrodes, (Xie et al. 1990), for resistive and capacitive cross-section measurement, Figure 8 (a). In this design the guard electrodes are installed at either side of the electrodes, in order to reduce the sensitivity to axial flow distributions.

Ring electrodes (Andreussi et al. 1988), for resistive measurement and to achieve a uniform electric field structure within the sensing volume, the ring electrodes trade-off a localised cross-section measurement. Figure 8 (b).

Helical electrodes (Abouelwafa and Kendall. 1980), Figure 8 (c). In this model, the electrodes twist round the pipe, to overcome flow geometry dependence.

Rotating field electrodes (Merilo et al. 1977), as shown in Figure 8 (d), this design achieves a similar effect to helical electrodes. Here, three electrode pairs are driven at 120° phase intervals, to produce a rotating field vector in the pipe centre.

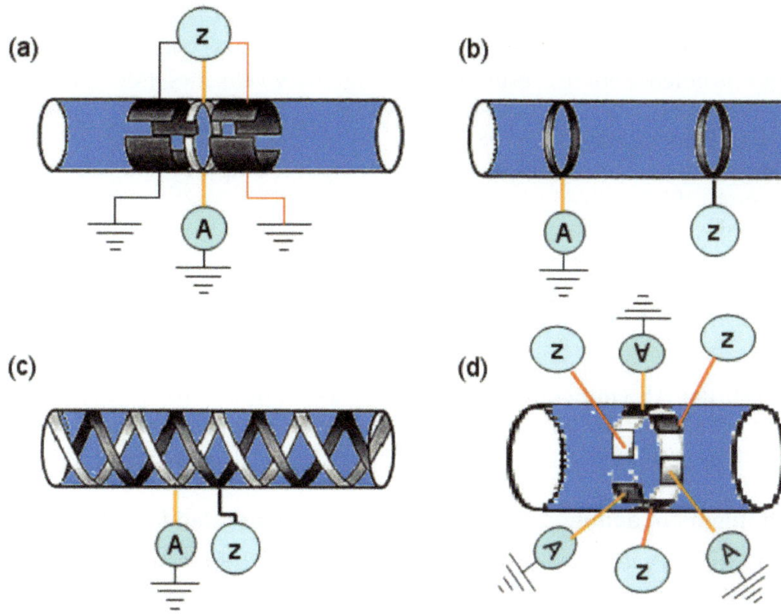

Fig. 8. Non-intrusive impedance measurement

4. AE applications in multiphase flow

4.1 Acoustic emission definition

AE is defined as a physical phenomenon occurring within and/or on the surface of materials (ISO 22096, 2007). The term AE describes the spontaneous elastic energy released by a process in the form of transient elastic waves. Acoustic emissions generated within a material will be manifest as elastic waves on the surface of the material and cover a broad frequency range, typically from 20kHz to 1MHz, which is outside the range of human hearing.

4.2 Sources of acoustic emission from fluid flow processes

AE sources are divided into two main categories; Primary sources are those where the AE originates from mechanisms of deformation and fracture; in metals, sources of AE include crack growth, moving dislocations, slip, twining, grain boundary sliding and fracture, and dislocation of inclusions. Secondary sources of AE include leaks, cavitation, liquefaction and solidification (Miller, 1987).

AE originating in industrial processes are usually generated from one of two main sources: mechanical or hydraulic. Mechanical sources would include; cracks that occurred within a moving part, or friction and rubbing between moving parts of a pump, valves, separators, gears or bearings associated with the process. Hydraulic sources considered to make a contribution to generating AE are listed below:

1. Cavitation - bubble formation, growth and collapse, (Leighton,1994 and Mba et al., 2006),
2. Turbulence noise produced by flow vortices (Derakhshan et al.,1989 and Brennen 1995),

3. Gas, solid and liquid mixture interaction in multiphase flows,(Boyd and Varley 2001)
4. Flow past restrictions and side branches in piping systems (McNulty, 1985),
5. Broadband turbulence energy resulting from high flow velocities, (McNulty, 1985),
6. Intermittent bursts of broadband energy caused by cavitation, flashing, recirculation, and water hammer (Neill et al., 1998),
7. Liquid drops impacting on a liquid surface (rain fall) (Pumphrey and Crum, 1989; Oguz and Prosperetti,1990),
8. Leakage from e.g. pipes (Lee and Lee, 2006),
9. Chemical reactions (Van Ooijen et al., 1978),
10. Foam break up (Lubetkin, 1989), and
11. Oceanic noise due to the breaking of waves (Leighton,1994 ;Vergniolle et al., 1996; Grant 1997)

In the above list, the AE emitted from liquid flow as a direct consequence of gas bubble formation and collapse is an important area for search, even though each source has a different formation mechanism.

4.3 The mechanisms of acoustic emission and sound generated from bubbles

In the literature the earliest known reference relating to the sound emitted from water/air mixtures due to the presence of air bubbles was by Bragg (1921); the murmuring of a brook and the "plunk" of liquid droplets impacting on the water surface were both attributed to entrained air bubbles. The first person to relate the frequency of the sound produced by a bubble formed at a nozzle to the physical parameters involved was Minnaert (1933). Subsequently, a number of investigations were devoted to the sound emitted from two phase flow as a function of bubble size and bubble population, Schiebe (1969), Pandit et al., (1992), Terrill and Melville (2000), Manasseh et al., (2001) and Al-Masry et al., (2005).

Strasberg (1956) suggested that sound emitted by a bubble was the result of oscillatory motion of the bubble wall. He compared volume pulsation and shape oscillation of a bubble. Four orders of oscillation, corresponding to bubble wall oscillation modes are shown in Figure 9. The zeroth mode represents a simple volume pulsation with fixed shape. The first mode corresponded to translational oscillation about a fixed point with fixed volume and shape. The second and third modes of oscillation show changes in shape of increasing complexity with a fixed volume. Strasberg (1956) calculated the natural frequency of the zeroth mode of volume pulsation from Minnaert's (1933) model, see equation 12. The natural frequencies of the other three oscillatory modes were calculated according to Lamb's (1945) model, see equation 13

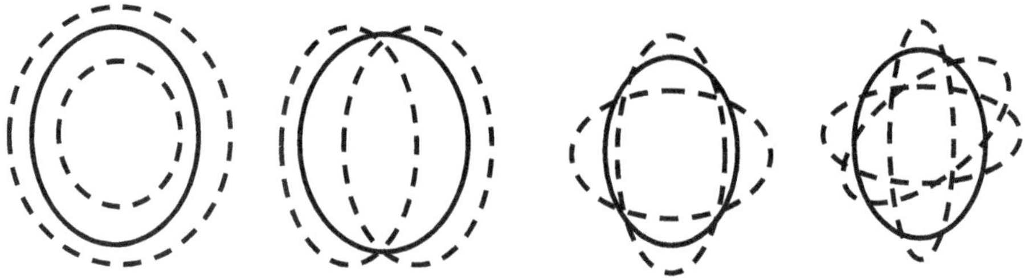

Fig. 9. Four modes of oscillation of a spherical bubble

$$f_0 = \frac{[3\gamma P_0/\rho]^{\frac{1}{2}}}{2\pi R_0} \tag{12}$$

$$f_n = \frac{\left[(n^2-1)(n+2)T / \rho R_0\right]^{\frac{1}{2}}}{2\pi R_o} \tag{13}$$

where:

f_0 is the natural frequency of oscillation for zeroth mode of volume pulsation
f_n is the frequency for the nth mode
P_0 is the static pressure
γ is the ratio of the specific heat of the gas in the bubble
R_0 is the mean radius of the bubble
ρ is the density of the liquid
T is the surface tension

Strasberg (1956) investigated factors influencing the amplitude of the bubble oscillations (the resulting sound pressure) when bubbles were exited by such mechanisms as; (a) bubble formation, (b) bubble coalescence and division and (c) the flow of a free stream of liquid with entrained bubbles past an obstacle and the flow of bubbles through a pipe past a constriction.

5. Slug flow initiation and dissipation

In order to achieve accurate measurement of the volumentric flowrates of slug flows,the hydrodynamic slugs have to be reasonably developed. Thus it is important to understand the slug development and dissipation process. The prediction of the flow conditions at which slug initiation and dissipation occur has received considerable attention in the last two decades (Woods and Hanratty, 1996).

One approach of predicting slug initiation is based on studying and analysing the stability of a stratified flow as investigated by several researchers. Kordyban and Ranov (1970) suggested that the transition from stratified flow to a slug flow might be described through a classical linear stability analysis. Graham et al. (1973) examined the growth of linearly unstable long wavelength disturbances on a flowing liquid. Taitel and Dukler (1976), Mishima and Ishii (1980) and Fan et al. (1993a) considered the evolution of a slug from a finite amplitude wave, with a wavelength in a range that would be stable by the Kelvin Helmholtz mechanism.

Another approach is to examine the stability of slugs travelling over a liquid layer as was investigated by Dukelr and Hubbard (1975) and Ruder et al. (1989). These authors investigated the initiation of a slug using visual observations in two-phase gas/liquid flow in horizontal pipe.

In their slug phenomena description, Dukler and Hubbard (1975) and Taitel and Dukler (1976) presented the development of gas and liquid flow in a pipeline. Near the entrance, the gas tends to flow above a moving stratified liquid layer. However, because of the shear forces created at the pipe wall, the liquid layer tends to decelerate as it moves along the pipe and its height changes gradually towards an equilibrium height which is governed by the pressure force, shear and gravitational forces.

As this occurs, small perturbations on the stratified layer could develop into growing waves. Due to the suction effect caused by an increased gas velocity over these disturbances as shown in Figure 10 (a) and (b), until eventually one of the waves grows to a sufficient size to momentarily bridge the pipe. This process blocks the flow of gas, see Figure 10 (c), and so the upstream pressure builds causing the blockage to be accelerated to the gas velocity.

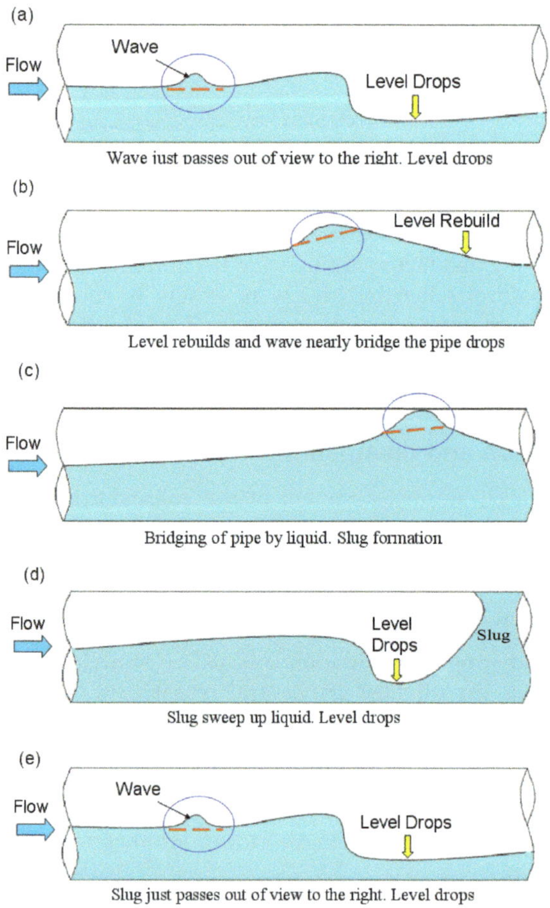

Fig. 10. The process of slug formation by Dukler and Hubbard (1975)

During this stage, the fluid blockage appears to be accelerated uniformly across its cross-section, thereby acting as a scoop, picking up all the slow moving liquid in the film ahead of it (pick-up process) and beginning to grow in volume to become a slug as shown in Figure 10 (d).

Gas may also be entrained in the form of small bubbles, which are deformed by the combined effect of buoyancy forces and the turbulent shear forces created by velocity differences between the slug front and the liquid film. As a result, a dispersion of small bubbles is often produced which may be transported through the body of the liquid slug.

Meanwhile, at the slug tail, liquid and previously entrained gas are released (shedding process) from the slug body. The "shed" liquid decelerates to a velocity determined by the shear stresses at the wall and the interface and becomes a stratified layer as illustrated in Figure 10 (e). The "shed" gas mainly passes into the elongated bubble region above this layer, although a fraction may remain entrained within the liquid film.

As long as the volumetric "pick-up" rate is larger than the "shedding" rate, the slug continues to grow. However, eventually the "pick-up" rate becomes equal to the "shedding" rate and the slug becomes fully developed so that the slug length stabilises. Nydal et al. (1992) experimentally investigated the length of the pipe required to reach quasi-stable flow conditions (slug development distance) and found that it is between 300 and 600 pipe diameters. Once the quasi-stable conditions are reached, the slug length has a mean value between 12 to 15 pipe diameters.

Based on the shedding and pick-up processes, slug flow might be classified into three main states. The pick-up rate is greater than the shedding rate; the slug in this case continues to grow. The "pick-up" rate equals to the "shedding" rate, the slug becomes fully developed so that the slug length stabilises. Finally, when the "pick-up" rate is less than the "shedding" rate, the slug under this condition dissipates.

The slug dissipation process occurs as the gas flowrate and consequently the slug velocity and the degree of aeration of the slug increases. Ultimately the gas forms a continuous phase through the slug body. When this occurs the slug begins bypassing some of the gas. At this point the slug no longer maintains a competent bridge to block the gas flow so the characteristics of the flow changes. This point is the beginning of "blow-though" and the start of the annular flow regime.

The idealised picture of a "stable slug" flow in horizontal pipe is presented in figure 11. The section 'F' represents the front region of the liquid slug body (LSB) and section 'T' represent the tail region of the liquid slug body. In a stable slug flow, the liquid slug body length, L_{LSB} and the elongated bubble (EB) length (L_{EB}) remain essentially constant in the downstream direction.

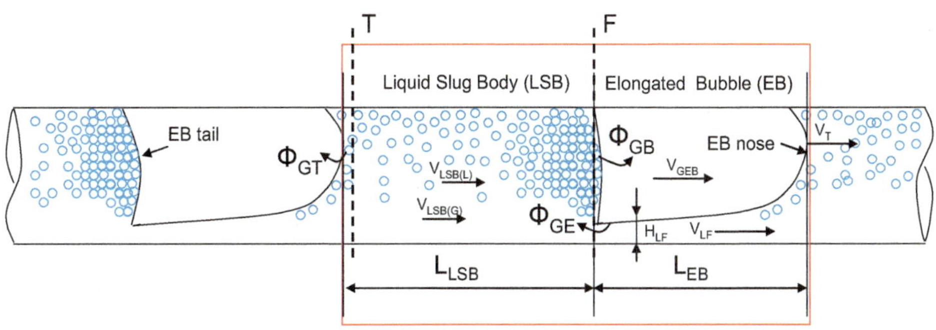

Fig. 11. Schematic description of the EB and LSB in idealised developed slug flow

The gas in the elongated bubble moves at a velocity, V_{GEB}, which is faster than the average mixture velocity, V_{mix} in the liquid slug body. As a result, the liquid is shed from the back of the liquid slug body to form the liquid film layer along the elongated bubble. The liquid in the film at the EB nose may be aerated. Also, bubbles in the liquid slug body coalesce with the elongated bubble interface and are gradually absorbed, which in the fullness of time results in the liquid film becoming un-aerated. The mixture velocity is the sum of the liquid velocity and the gas velocity ($V_{SL} + V_{SG}$).

At the same time, the gas bubbles are fragmented from the tail of the elongated bubble and re-entrained into the front section of liquid slug body 'F' at a rate Φ_{GE}. The fragmentation of the elongated bubble tail and the entrainment of the bubbles into the front section of the liquid slug body are due to the dispersing forces induced by the flow of the liquid film as it plunges into the liquid slug front. However, some of the gas bubbles that are entrained into the front section of the liquid slug body may re-coalesce with the elongated bubble tail at Φ_{GB} rate, resulting in a net gas flow rate of Φ_{GE} - Φ_{GB} out of the elongated bubble tail. The $\Phi_{GT} = \Phi_{GE} - \Phi_{GB}$ is the constant gas flow rate which is entrained to the elongated bubble EB, and eventually is absorbed into the successive elongated bubble. Therefore, the net rate of gas entrainment in the slug body is the result of a balance between bubble injection and rejection.

The gas bubbles entrainment from the elongated bubble tail and their re-coalescence at the successive nose result in an effective elongated bubble translation velocity, V_T, which is higher than the gas velocity in the elongated bubble, V_{GEB}. From the description of the slug flow formation, dispersed gas bubbles can be generated in the liquid slug body region through the formation, coalescence, breakage and collapse of bubbles processes. The entrained gas bubbles in the liquid experience a transient pressure as they move through the hydrodynamic pressure field of the liquid. The transient pressure causes the gas bubbles to oscillate at their natural frequencies; one consequence of which is the generation of sound (Strasberg, 1956).

5.1 Gas void fraction (ε) in slug flow

The liquid holdup in the slug body or gas void fraction (ε) is an important parameter for the design of multiphase pipelines and associated separation equipment. The definition of the liquid holdup is "the flowrate of the liquid phase divided by the total flowrate of both the gas and liquid flowrates". And the definition of the gas void fraction is "the flowrate of the gas phase divided by the total flowrate of both the gas and liquid flowrates". Several gas void fraction correlations were developed, however, the majority of these correlations were obtained based on the slug body liquid holdup. In multiphase flow, the extensively-used correlation developed by Gregory et al. (1978) was obtained from the measurements of liquid holdup, using electrical capacitance probes, in air-water and oil-water flow in horizontal pipes with diameter of 0.0258 m and 0.0512 m. The correlation gives slug body holdup as a function of the mixture velocity V_{mix} only:

$$H_{LSB} = \frac{1}{1 + \left(\frac{V_{mix}}{8.66} \right)^{1.39}} \tag{14}$$

where V_{mix} is the slug mixture velocity in ms^{-1} and is equal to the sum of the superficial gas and liquid velocities.

Malnes (1982) proposed an alternative correlation also based on the same data of Gregory et al. (1978) as:

$$H_{LSB} = 1 - \frac{V_{mix}}{C_C + V_{mix}} \tag{15}$$

Where, the dimensional coefficient C_c is measured as following:

$$C_C = 83\left(\frac{g\sigma}{\rho L}\right)^{0.25} \tag{16}$$

Beggs and Brill, (1973) developed their correlation for the whole spectrum of flow situations using 1" and 1.5" pipe sizes at various angles from the horizontal. The correlation passed on Froude number and the slug body length for the intermittent flow as following;

$$\varepsilon = \frac{0.845E_{LSB}^{0.5351}}{Fr^{0.0173}} \tag{17}$$

where *Fr* is the Froude number, defined as a dimensionless number comparing inertia and gravitational forces, and calculated as

$$Fr = \frac{V_{mix}^2}{gD} \tag{18}$$

Hughmark (1962) proposed correlation for the gas void fraction based on slug mixture velocity and superficial gas velocity, given by:

$$\varepsilon = \frac{V_{SG}}{1.2V_{mix}} \tag{19}$$

Gregory and Scott (1969), developed similar correlation to the one proposed by Hughmark (1962), as following:

$$\varepsilon = \frac{V_{SG}}{1.19V_{mix}} \tag{20}$$

Ferschneider (1983) developed a more complex correlation for slug body holdup H_{LSB} using data obtained from natural gas and a light hydrocarbon oil facility. The facility loop comprised a 0.15 m diameter with 120 m long test section loop and operated at elevated pressure of between 10 and 50 bar. The correlation proposed by Ferschneider (1983) took account for surface tension of the fluids in terms of Bond number (Bo). Bo is a measure of the importance of the surface tension forces compared to gravitational forces.

$$H_{LSB} = \frac{1}{\left\{1+\left[\left(\frac{V_{mix}}{\sqrt{\left(1-\rho_G/\rho_L\right)gD}}\right)\times\left(\frac{Bo^{0.1}}{25}\right)\right]^2\right\}^2} \tag{21}$$

Bo is Bond number and is given by:

$$Bo = \frac{(\rho_L - \rho_G)\times g \times D^2}{\sigma} \tag{22}$$

Nickline (1962) proposed correlation for the gas void fraction based on the measured liquid holdup for air-water flow in 0.05m and 0.09 m horizontal pipe. The correlation is given by the expressions:

$$\varepsilon = \frac{V_{SG}}{(1.2V_{mix} + 0.35\sqrt{gD}} \tag{23}$$

Abdul-Majeed (2000) developed a new correlation to compute the slug void fraction, which only depends on the fluid viscosities and mixture velocity: a data bank set contained several hundreds of slug hold-up points, gathered from seven different resources were used in this correlation, only mixture velocity V_{mix} and liquid and gas viscosities μ_L and μ_G were utilised as following;

$$H_{LSB} = 1.009 - \left(0.006 + 1.3377\frac{\mu_G}{\mu_L}\right)V_{mix} \tag{24}$$

Al-lababidi and Sanderson (2005; 2007) investigated the measurement of the slug body liquid holdup and film liquid holdup using the non-intrusive ultrasonic pulse-echo mode system. The system consists of an ultrasonic pulser–receiver which was used to excite the ultrasonic transducer (2.25 MHz), to receive and amplify the reflected signals and an electronic circuit to measure the time of transmitted and reflected ultrasound wave in liquid phase. The authors concluded that the liquid holdup measurements were validated against the non-intrusive conductivity transmitter and compared with correlations from the literature. However, as the gas void fraction increases, the attenuation and distraction of the transit time ultrasonic signals increases. As a result, the transit time ultrasonic can be considered a good technique for the measurement of the slug body and film liquid holdup at low gas void fraction.

Woldesemayat and Ghajar (2007) performed an intensive literature review of the slug body holdup in multiphase flow in pipeline. However, the majority of the gas void fraction (ε) correlations reviewed in this work and in Woldesemayat and Ghajar (2007) intensive review was experimentally developed based on the direct measurement of the liquid holdup in the slug body rather than measuring the gas void fraction.

Most, if not all, of the available correlations for the average slug void fraction in liquid slugs were found to be unsatisfactory when applied to different geometries from those used in extracting the empirical correlations (Paglianti et al., 1993).This is perhaps not surprising since the above mentioned correlations are derived from fully developed slug flow, and do not account for the transient behaviour of slug growth and collapse (Bonizzi and Issa, 2003).

6. AE applications for multiphase slug flow monitoring in pipelines

6.1 Experimental setup

The two-phase experimental facility used in this work is shown in figure 12. This facility allows measurements over a range of a superficial liquid velocities V_{SL} (0.3 to 2.0 ms^{-1}) and superficial gas velocities V_{SG} (0.2 to 1.4 ms^{-1}). The experimental facility consisted of water and air supply systems, Perspex pipes, steel spacemen test sections, circulating pipes and fittings, Electrode rings for conductivity measurements associated with data acquisition system, and AE acquisition system, The pipeline used in this work is of 2-inch (ID ≈ 50mm) diameter and total length of 22.5m. The majority of the piping system was made from ABS (class E) pipe, and two Perspex sections were installed to allow for visual observations of the flow. AE measurements were taken from a fitted stainless steel pipes of 750mm length and 8mm thickness.

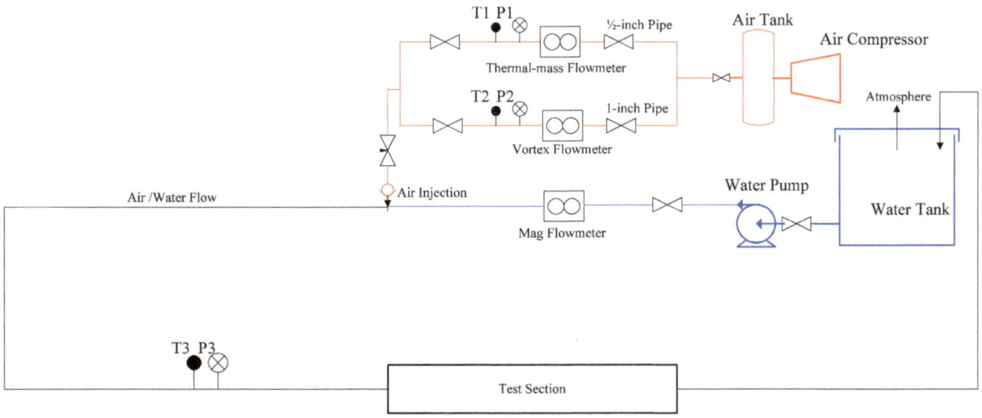

Fig. 12. 2 inch air/water horizontal flow test facility

The flow loop pipeline length was enough to allow the formation of fully developed slugs. Water was pumped to the flow loop using a centrifugal pump with a maximum capacity of 40 m^3/hr and a maximum discharge pressure of 5 barg. The water flow was metered using an electromagnetic flow meter with 0 - 20 m^3/hr range.

A Commercially available AE acquisition system was used for acquiring the data from a Piezo-electric AE sensor, (PICO type) with broadband operating frequency range 150-750 kHz. The AE sensor was placed onto stainless steel pipe (Addali et al. 2007) as shown in the detailed picture ,figure 13. The output AE signals from the sensor were pre-amplified at 60dB and AE absolute energy (Joules) was sampled at 10ms over a time constant of 10msec. The AE energy presented is that received by the AE sensor and it is determined from the integral of the squared voltage signal divided by the reference resistance (10k-ohm in this instance) over the duration of the acquired wave (10msec time constant). It should be noted that AE received at the sensor is significantly less than the energy emitted at the specific source of the AE event; this is attributed to the attenuation of the AE wave across the media (fluid- and structure-borne), and, the dispersive and non-directional nature of AE.

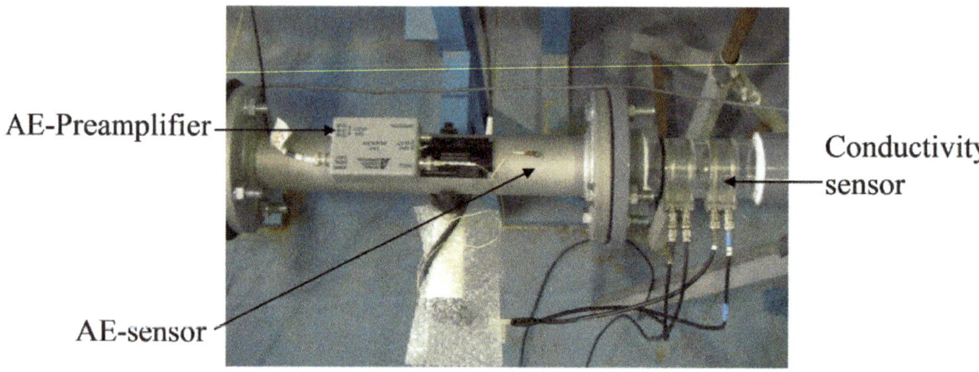

Fig. 13. 2-inch test section (AE sensor and preamplifier) and conductivity sensor

A typical AE time waveform and associated frequency spectrum is presented in figure 14. It is evident that the frequency range of measured AE ranged from 120 kHz to 500 kHz. In this instance the waveform is continuous in appearance though it is fundamentally a mixture of several hundred transient AE bursts. In fact it was noted under slug flow regimes that the entire movement of the slug through the pipe was seen as increasing and decaying AE levels which could be correlated with the slug head and tail respectively, see figure 15.

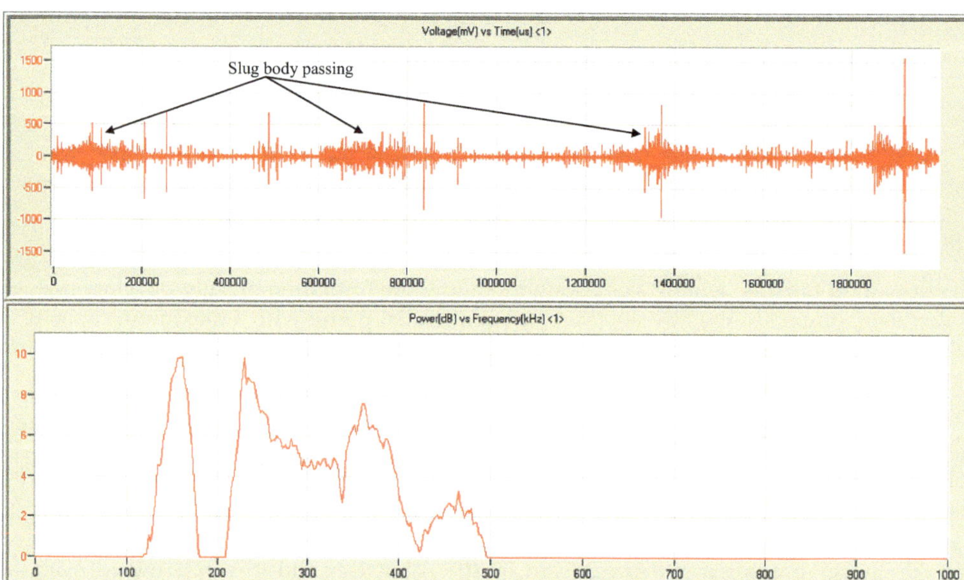

Fig. 14. AE time waveform and associated frequency spectrum for V_{SL} 1.5, V_{SG} 1.0

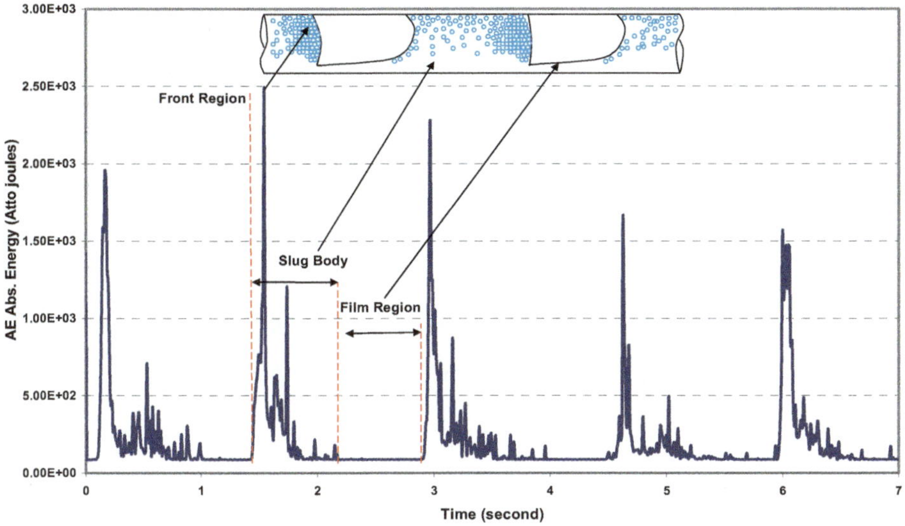

Fig. 15. Typical AE signal from fully developed slug flow

The experiments covered range of superficial water velocities (V_{SL}) from 0.3 ms^{-1} to 2.0 ms^{-1} at an increments of 0.1 ms^{-1}; and superficial air velocities (V_{SG}) from 0.2 to 1.4 ms^{-1} at increments of 0.2 ms^{-1} at each V_{SL}. The V_{SL} and V_{SG} values were achieved by throttling valves downstream of the flow meters, and every test condition was maintained for 120 seconds to develop the flow fully. Data were acquired and analysed for each test condition and comparisons between the conductivity sensor measurement and AE were presented for different gas void fraction of the flow. The test range undertaken is presented in figure 16.

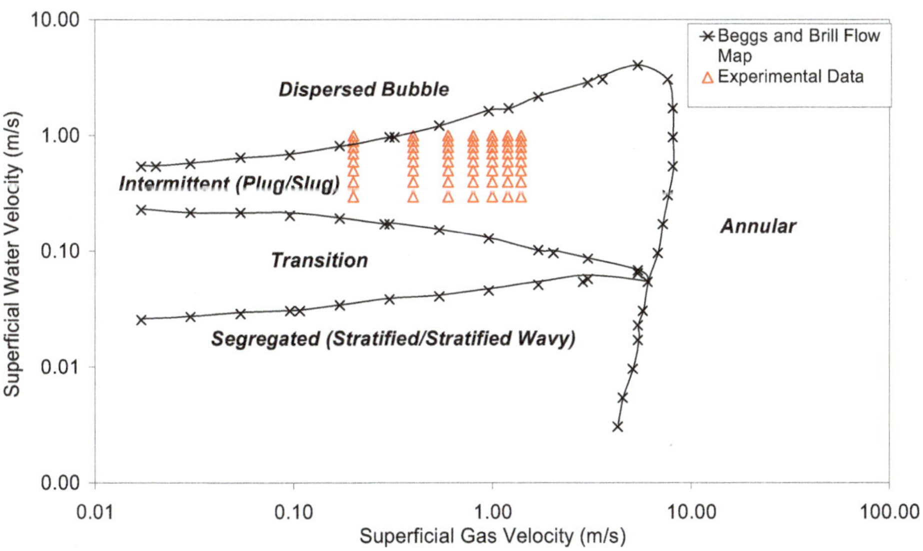

Fig. 16. Test flow regime map

6.2 Result and discussion

6.2.1 Gas bubbles in slug body

In the liquid slug body, a gas bubble in the liquid slug body will be broken down by the turbulent forces exerted on it if its size is larger than a certain value, this value known as the 'Kolmogorov' length scale (Kolmogorov,1949). Based on the extension of Kolmogorov–Hinze theory (Hinze, 1955) for break-up of droplets/bubbles in a turbulent field, several correlations for gas void fraction in the slug body were developed (Barnea & Brauner, 1985; Brauner, 2001; Brauner & Ullmann, 2002). From the Kolmogorov–Hinze theory, the amount of gas bubbles a liquid slug body can hold is dependent on the turbulent intensity of the liquid phase. Therefore, there must be a balance between the total surface free energy of dispersed gas bubbles and the turbulent kinetic energy of the liquid phase.

Adamson,(Adamson, 1990) stated that the surface free energy per unit interfacial area is the work necessary to generate this area, and is equal to the interfacial surface tension between a liquid phase and a gas phase. Assuming gas bubbles are all spherical with a diameter of d_{bubble}, the total surface free energy of the discrete gas bubbles ($E_{surface}$) in the liquid slug body was developed by Brauner and Ulmann (Brauner & Ullmann, 2004) as:

$$E_{surface} = \frac{6\sigma}{d_{bubble}} A_i \ (1 - H_{LSB}) \ L_{LSB} \tag{25}$$

where σ is the interfacial surface tension, A_i is the internal cross-sectional area of the pipe, H_{LSB} is the liquid holdup in the slug body, and L_{LSB} is the length of the liquid slug body. In slug flow, the gas phase is accommodated in the liquid slug body as dispersed spherical bubbles. Brodkey (Brodkey, 1967) noted a critical bubble diameter, d_{crit}, above which the bubbles will be broken up by the turbulent forces and is given as:

$$\frac{d_{crit}}{D} = \left[\frac{0.4\sigma}{(\rho_L - \rho_G)\, g\, D^2 \cos\beta} \right]^{1/2} \tag{26}$$

where D is the pipe diameter, ρ_L and ρ_G are the liquid and gas densities respectively and β is the pipe inclination. Equation (7) was modified by Barnea et al. (Barnea et al., 1982) as:

$$d_{crit} = 2 \left[\frac{0.4\sigma}{(\rho_L - \rho_G)\, g} \right]^{1/2} \tag{27}$$

The turbulent kinetic energy per unit volume of liquid flowing in a pipe is given as (White, 1991).

$$E_{TK} = \frac{1}{2} \rho_L \left(\overline{V_r^2} + \overline{V_\theta^2} + \overline{V_x^2} \right) \tag{28}$$

where ρ_L is liquid density, and , $\overline{V_r^2}$, $\overline{V_\theta^2}$ and $\overline{V_x^2}$ are the radial, tangential and axial velocity fluctuations, respectively . $\overline{V_r^2} = \overline{V_\theta^2} = \overline{V_x^2}$ if the turbulent flow is assumed to be isotropic. Thus, in the liquid slug body, the total turbulent kinetic energy is:

$$E_{TK} = \frac{3}{2} \rho_L \overline{V_r^2} A_i H_{LSB} L_{LSB} \tag{29}$$

Taitel (Taitel & Dukler, 1976) and Chen (Chen at al., 1997), approximated the radial velocity in equation 29 as the friction velocity, whilst Zhang (Zhang at al., 2003) used the pressure gradient including both the shear term and the mixing term to estimate the friction velocity. And as a result, the total turbulent kinetic energy in the liquid slug was estimated by Zhang (Zhang at al., 2003) as:

$$E_{TK} = \frac{3}{2}\left(\frac{f}{2}\rho_{mix}V_{mix}^2 + \frac{D}{4}\frac{\rho_L E_{LF}(V_T - V_{mix})(V_{mix} - V_{LF})}{L_{LSB}}\right)A_i H_{LSB} L_{LSB} \quad (30)$$

D is the pipe diameter, f is the friction factor at the pipe wall for the liquid slug body, ρ_{mix} the mixture density and is given as, ($\rho_{mix} = \rho_L H_{LSB} + \rho_G(1 - H_{LSB})$ where-H_{LSB} is the liquid holdup in the liquid slug body.

For the gas void fraction within the slug body, Zhang (Zhang at al., 2003) assumed that the surface free energy of the discrete gas bubbles, based on the maximum amount of gas the liquid slug can hold, is proportional to the turbulent kinetic energy in the liquid slug body. The calculated values of the surface free energy in air\water slug flow conditions, using equation 25, are plotted against the measured gas void fraction using the conductivity sensor as shown in figure 17. Furthermore, it was noted that the surface free energy was proportional to the amount of the gas bubbles being held in the slug body because of the turbulent intensity force. The significance of this is that a relationship between the measured gas void fraction and the mathematical estimation of surface free energy has been established and this is further explored with AE later in the paper.

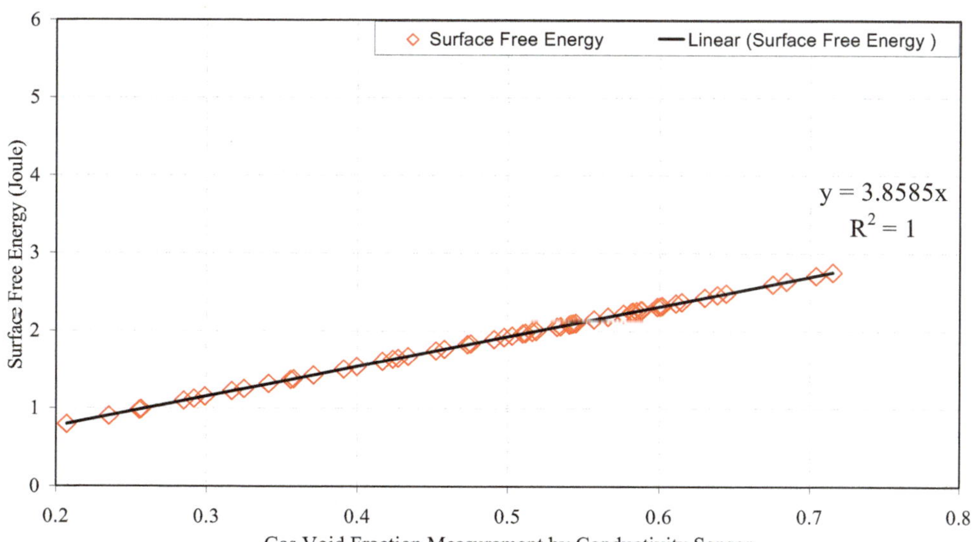

Fig. 17. Surface free energy verse measured gas void fraction in slug body

At a fixed superficial water velocity, for example V_{SL}=0.8 ms^{-1}, increasing the superficial gas velocity resulted in an increase of the measured absolute AE energy, see figure 18. This was observed for all V_{SL} levels investigated. This was not surprising and showed that an increase

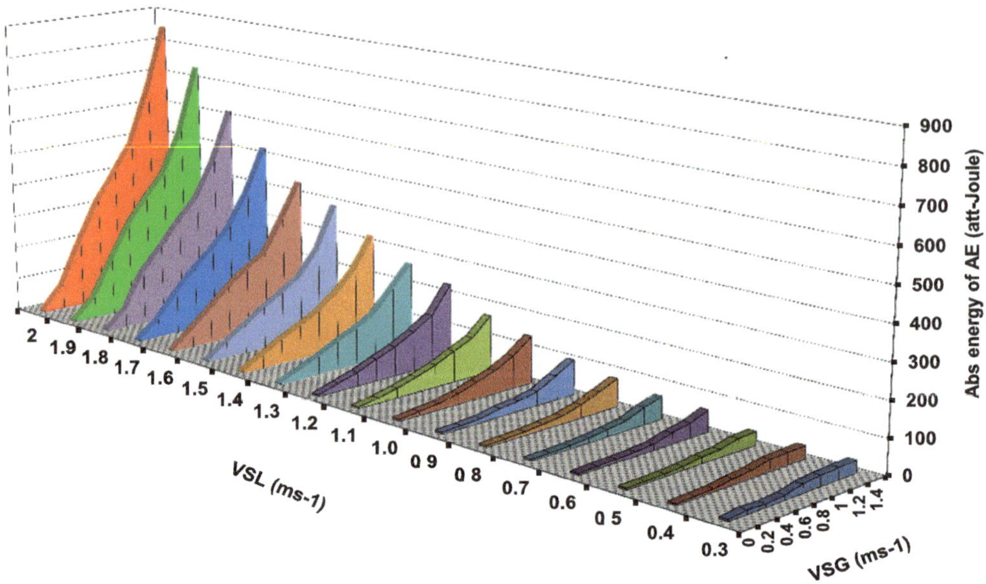

Fig. 18. Contribution of the liquid and air velocities on the increase of the measured AE abs. energy

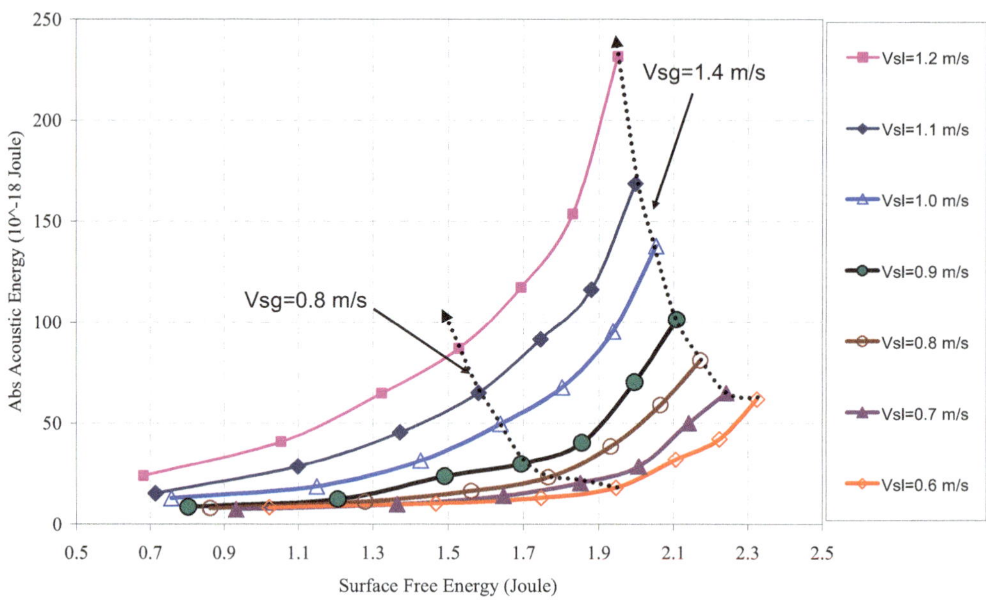

Fig. 19. Contribution of the turbulence kinetic energy on the increase of the absolute acoustic energy

in bubble content, and its associated bubble dynamics, resulted in an increase of AE generated. It was interesting to note that an increase in V_{SL} for a fixed V_{SG} resulted in a relative decrease in surface free energy whilst a simultaneous increase in AE energy was observed; figure 19 describes the relationship between AE energy measured from the AE sensor and the surface free energy calculated from equation 25. This suggests that there are two mechanisms responsible for the generation of AE. The influence of pure liquid flow on generating AE was reinforced experimentally and compared to the electronic noise of the system. The contribution of pure liquid was in average of 0.2 atto-Joules above the electronic noise of the system. This was considered insignificant compared to the influence of gas/liquid flow. This is evident that AE energy levels are mainly contribution of the gas/liquid flow.

It must be noted that turbulence is generated at the front of the slug, where the pick-up process occurs. Turbulence is enhanced by the slowly moving fluid in the film region (defined by the liquid region of the elongated bubble) which is forced to 'plunge' into the relatively faster moving liquid in slug body directly behind, see figure 15. Such turbulence is diffused along the direction of the penetrating film and increases with increased V_{SL}. When the water superficial velocity increases the intensity of the turbulence diffusion increases, and as a result, the associated absolute AE energy increases as illustrated in figure 19. Similarly, an increase in water superficial velocity reduces the gas void fraction for a defined superficial air velocity, which is directly correlated to the free surface energy. Therefore, the authors believe that there are two processes influencing the generation of AE; the free surface energy which is a measure of the bubble content in the liquid and the influence of turbulent diffusion due to high superficial liquid velocities. It can be concluded based on slug body energy analysis, that the absolute AE energy measured in the liquid slug body is dependent on both the surface free energy due to the presence of the gas bubbles and the turbulent kinetic energy.

6.3 Acoustic emission and gas void fraction correlation in slug body

Figure 20 shows the gas void fraction measured by the conductivity sensor is a non-linear function of the absolute AE energy. Also, the data suggests that the AE energy does not change linearly with the gas superficial velocity. The V_{SG} range tested for each V_{SL} presented in figure 20 included 0.2 to 1.2 ms^{-1} at an incremental step of 0.2 ms^{-1}; the lowest AE energy levels were associated with the lowest V_{SG}, increasing with increasing V_{SG}.

Figure 20 provided the basis for establishing a relationship between the ε as a function of the AE energy. A multiple exponential regression resulted in the following relationship:

$$\varepsilon = 0.768\ E_{AE}^{0.003}\ V_{mix}^{-0.69}\ V_{SG}^{0.744} \tag{31}$$

Figure 21 shows the obtained gas void fraction in the liquid slug body from the developed correlation (12) as function of absolute AE energy and slug velocities.

To determine the relative accuracy of the developed correlation (equation 31), three familiar statistical measures of reliability were computed: 1. Percentage error (PE); 2. Average percent error (APE); and 3. Standard deviation (SD). Equations for these measures are:

$$PE = \frac{(Estimated\text{-}Measured)}{Measured} \times 100 \tag{32}$$

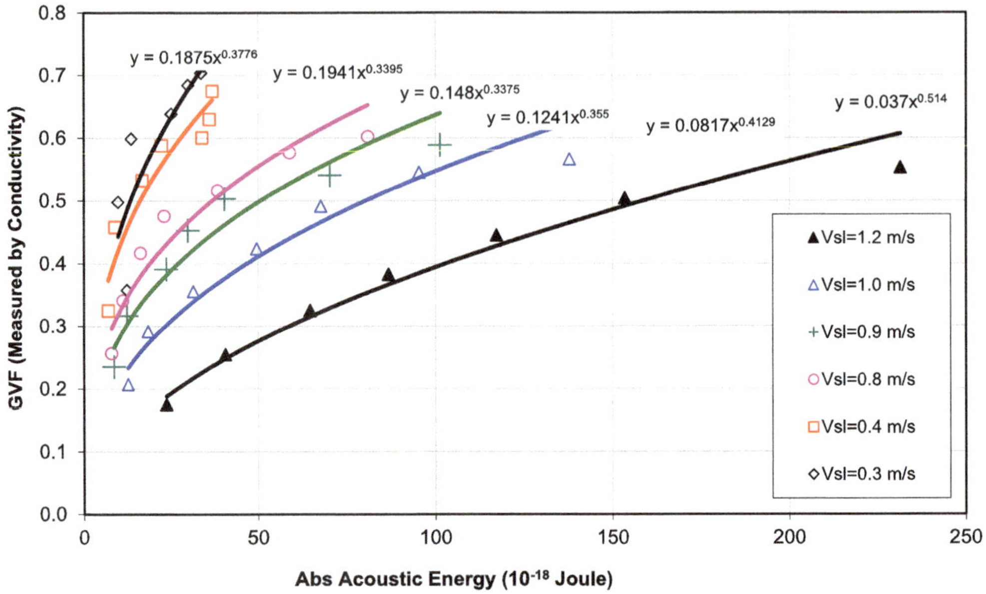

Fig. 20. Absolute AE energy against measured gas void fraction

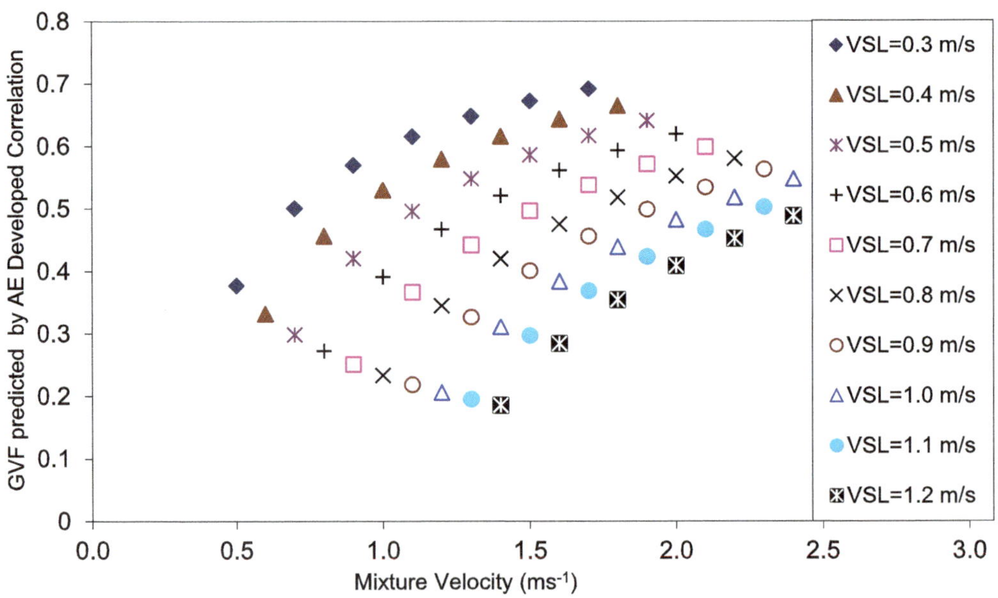

Fig. 21. Gas Void Fraction established with the AE model, equation (15), for a range of flow velocities

$$APE = \frac{\sum_{i=1}^{n} PE}{n} \tag{33}$$

$$SD = \sqrt{\frac{n\sum_{i=1}^{n}(PE_i)^2 - \left[\sum_{i=1}^{n} PE_i\right]^2}{n^2}} \tag{34}$$

The predictions of the proposed correlation, equation 31, and correlations by several researchers (Gregory at al., 1978; Hughmark, 1962; Ferschneider, 1983) are presented in Table 1. The statistical parameters for the proposed correlation are smaller than those listed in Table 1, demonstrating good performance over all correlations.

	APE	SD
This study	-0.49	3.48
Gregory *et al* [31]	12.95	10.80
Hughmark [32]	13.90	10.89
Nicklin *et al* [34]	13.29	14.77

Table 1. Summary of statistical results for correlations

Figure 21 shows the comparison between gas void fraction predicted from the correlations and that obtained from the conductivity sensor. Maximum deviation of the predicted gas void fraction is ±15 % at low measured gas void fraction. Also shown in figure 22 is the large spread of the predicted gas void fraction values at low superficial gas velocity, but this spread of data narrows at higher superficial gas velocities.

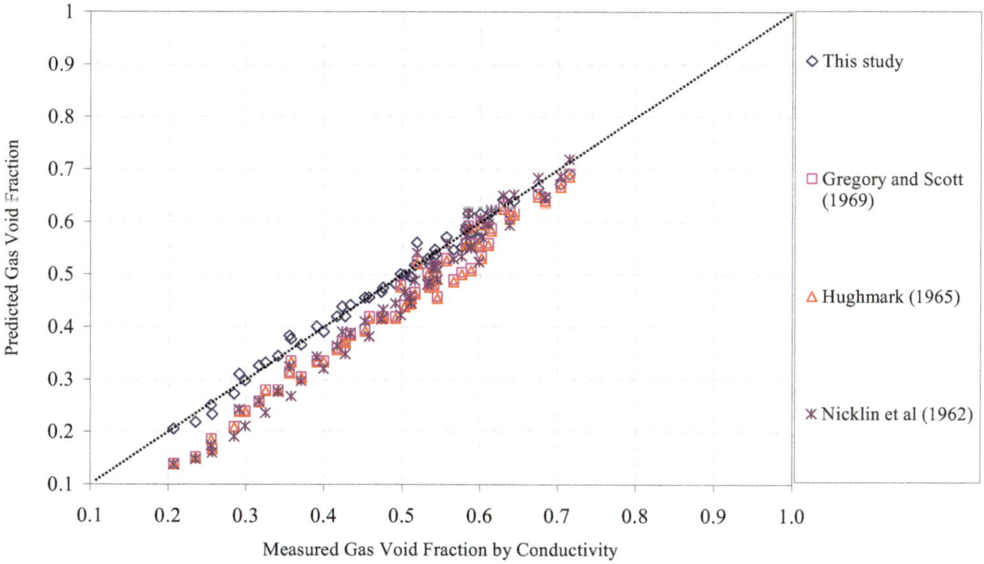

Fig. 22. Comparison of measured and predicted gas void fraction data; proposed correlation

One of the main advantage of the proposed correlation is that the AE signals improves with the higher gas fraction, thus increases the accuracy of the prediction even at higher gas fractions. As a result, the proposed correlation can be successfully applied to two-phase air/liquid systems and which have very high percentage of gas void fraction.

7. Conclusions

This work presents preliminary investigations on the application of AE technology for measuring GVF in two-phase air/water under different flow regimes particularly slug flow conditions. It is evident that the main source of Acoustic Emission is directly correlated with the dynamics of bubbles; its formation, coalescence and destruction. Whilst the movement of bubbles in a fluid is known to be associated with oscillations at frequencies of a few kilohertz, it is thought such movement does not generate energy at the frequencies associated with Acoustic Emissions (AE). As defined earlier, the emissions measured in this investigation cover the frequency range of 100kHz to 1,000KHz. The formation and destruction of bubbles is known to be associated with transient pressure pulses which will excite a broad frequency range which will be detectable in the AE frequency range.

The increase of the superficial gas velocities resulted in an increase in AE due to the increase in bubble content and its associated bubble dynamics. In addition, an increase in the superficial liquid velocity, result in an increase in the intensity of the turbulence diffusion, also generated increased AE energy. It has been demonstrated that two mechanisms are responsible for the generation of AE, the free surface energy which is a measure of the bubble content in the liquid (air velocity) and the influence of turbulent diffusion (turbulent kinetic energy) due to high superficial liquid velocities. Furthermore, statistical analysis of the developed correlation showed the parameters APE and SD to be smaller than the previously reported correlations. Also, a comparison between the predicted and measured gas void fraction values using conductivity sensor agree within ±15 %. Finally, the introduction of a new passive and non-invasive GVF measuring technique utilising AE technology has been proposed for two-phase air-water flows; providing both a quantitative and qualitative values.

7.1 Recommendations for future work

It has been demonstrated that the time domain analysis of AE signal generated from two-phase gas/water flow is a promising tool in monitoring the flow condition. Although the output identifier parameter adopted in this work was the absolute energy levels of the AE signal, the other parameters such as average signal levels (ASL) and root mean square (RMS) of the AE can also be used to monitor the flow conditions. The successfully developed correlation between the gas void fraction (GVF) and AE generated from the flow need to be more generalised by considering several other parameters such as;

- Pipe diameter: Several pipe diameters have to be investigated, and the effect of the pipe diameter needs to be included in the correlation.
- Fluid viscosity: Tests to be carried out by using fluids with different viscosity grades.
- Pipeline inclination: The current research study covered two-phase flow in the horizontal pipelines. Further work should be performed to include the effect of pipeline inclination using inclined and vertical pipelines, which will add significant knowledge to this research.

- Due to the fundamental influence of the bubble content on the generated AE energy, more work has to be done in the frequency domain, in order to investigate the contribution and possibly discriminate the influence of each of the bubble activities such as formation, coalescence and destruction.

8. References

Abouelwafa, M. S. A. and Kendall, E. J. M. (1980). The measurement of component rations in multiphase systems using gamma ray attenuation. *Journal of Physics E: Science Instruments.* 13, 341-345.

Adamson, A.W., 1990. Physical Chemistry of Surfaces, fifth ed. John Willey and Sons Inc.

Addali A., Al-Lababidi S., Mba D. (2007). Application of Acoustic Emission to monitoring two phase flow, *Fourth International Conference on Condition Monitoring, Harrogate, UK*

Al-lababidi, S. and Sanderson, M.L., Closure model for two-phase liquid-gas measurement under slug flow conditions. In: *Proceedings of the 11th International Conference on Flow Measurement, FLOMEKO'03,* Groningen, Netherlands, May 12-14, 2003.

Al-lababidi, S., and Sanderson, M.L. Transit Time Ultrasonic Modelling in Gas/Liquid Intermittent Flow Using Slug Existence Conditions and Void Fraction Analysis". In: *Proceedings of the 12th International Conference on Flow Measurement, FLOMEKO'04,* Guilin, China, Sep12-17 2004.

Andreussi, P. and Bendiksen, K. H. (1989). An Investigation of void fraction in liquid slugs for horizontal and inclined gas-liquid pipe flow. *International Journal of Multiphase Flow,* 15(6), 937-946.

Andreussi, P., Donfrancesco, Di. and Messia, M. (1988). An Impedance Method for the Measurement of Liquid Holdup in Two-Phase Flow. *International Journal of Multiphase Flow,* 14(6), 777–787.

Barnea, D. and Taitel, Y. (1993). A model for slug length distribution in gas-liquid slug flow. *International Journal of Multiphase Flow,* 19(5), 829–838.

Barnea, D., Brauner, N., 1985. Holdup of the liquid slug in two-phase intermittent flow. *International Journal of Multiphase Flow* 11, 43–49.

Beck, M.S., and Plaskowski, A. (1987). *Cross-Correlation Flowmeters: Their Design and Application.* Bristol, UK: Adam Hilger.

Bendiksen, K. H. (1984). An experimental investigation of the motion of long bubbles in inclined tubes. *International Journal of Multiphase Flow,* 10(4), 467-483.

Bendiksen, K. H., Malnes, D. and Nydal, O.J. (1996). On the modelling of slug flow. *Journal of Chemical Engineering Communications,* 141, 71-102.

Bignell, N. and Choi, Y. M. (2001). Volumetric positive-displacement gas flow standard. *Journal of Flow Measurement and Instrumentation,* 12(4), 245–251.

Boyd, J.W.R. and Varley J. (2001), The uses of passive measurement of acoustic emissions from chemical engineering processes, Chemical Engineering Science, Volume 56, Issue 5, pp1749-1767, ISSN 0009-2509.

Brauner, N., 2001. The prediction of dispersed flow boundaries in liquid–liquid and gas–liquid systems. *International Journal of Multiphase Flow* 27, 885–910.

Brauner, N., Ullmann, A., 2002. Prediction of holdup in liquid slugs. In: Heat 2002, Third Int. Conf. on Transport Phenomena in Multiphase Flow, IFFM No. 112, pp. 1–15.

Brauner, N., Ullmann, A., 2004. Modelling of gas entrainment from Taylor bubbles. Part B: A stationary bubble. Int. J.Multiphase Flow Vol. 30, pp. 273-290, 2004

Brennen, C. E., (1995) Cavitation Bubble Dynamics, Oxford University Press, ISBN: 0-19-509409-3.

Brill, J. P., Schmidt, Z., Coberly, W. A., Herring, J. D. and Moore. D.W. (1981). Analysis of two-phase tests in large-diameter flow lines Prudhoe Bay Field. *SPE Journal,* 1981, 363-377.

Brodkey, R.S., 1967. The Phenomena of Fluid Motions. Addisson-Wesley Press.

Cheng, W., Murai, Y., Sasaki, T. and Yamamoto, F. (2005). Bubble velocity measurement with a recursive cross correlation PIV technique. *Journal of Flow Measurement and Instrumentation,* 16 (1), 35-46.

Corneliussen, S., Couput, J., Dahl, E., Dykesteen, E., Frøysa, K. E., Malde, E., Moestue, H., Moksnes, P. O., Scheers, L. and Tunheim, H. (2005). *Handbook of Multiphase Flow.* Norwegian Society for Oil and Gas Measurement (NFOGM) and The Norwegian Society of Chartered Technical and Scientific Professionals (Tekna), Norway, Revision 2, March 2005.

Costigan, G. and Whalley, P. B. (1997). Slug Flow Regime Identification From Dynamic Void Fraction Measurement in Vertical Air-Water Flows. *International Journal of Multiphase Flow,* 23(2), 263-282.

Coull, C. and Sattary, J. (2004). Evaluation of Ultrasonic Technology for Measurement of Multiphase Flow. National Engineering Laboratory, UK, Report No2004/230, November, 2004.

Davies, S. R. (1992). *Studies of Two-Phase Intermittent Flow in Pipelines.* PhD Thesis, Imperial College, London, UK, 1992.

Derakhshan, O., Houghton, J.R., Jones, R.K. and March, P.A., (1989), Cavitation monitoring of hydroturbines with RMS acoustic emission measurement, Symposium on Acoustic Emission: Current Pratice and Future Directions, Mar 20-23 1989 Published by ASTM, Philadelphia, PA, USA, Charlotte, NC, USA, pp. 305.

Dong, F., Xu, Y. B., Xu, L. J., Hua, L. and Qiao, X. T. (2005). Application of dual-plane ERT system and cross-correlation technique to measure gas-liquid in vertical upward pipe. *Journal of Flow Measurement and Instrumentation,* 16(2-3), 191-197.

Dukler, A. E. and Hubbard, M. G. (1975). A model for gas-liquid slug flow in horizontal tubes. *Industrial and Engineering Chemistry Fundamental.* 14 (4), 337-347.

Fan, Z., Lusseyran, F. and Hanratty, T. J. (1993a). Initiation of slugs in horizontal gas-liquid flows. *Journal of American Institute of Chemical Engineering* . 39, 1741-1753.

Fan, Z., Ruder, Z. and Hanratty, T. J. (1993b). Pressure profiles for slugs in horizontal pipelines. *International Journal of Multiphase Flow,* 19(3), 421-437.

Ferschneider, G. (1983). 'Ecoulements Diphasiques Gas-Liquid á Poches et á Bouchon en Conduits. Cited in: Hale, C. P. (2000). *Slug Formation, Growth and Decay in Gas-Liquid Flow, chapter 3, p211.*PhD Thesis, Imperial College, London, UK, 2000.

Ferschneider, G. (1983). 'Ecoulements Diphasiques Gas-Liquid á Poches et á Bouchon en Conduits. Cited in: Hale, C. P. (2000). *Slug Formation, Growth and Decay in Gas-Liquid Flow, chapter 3, p211.*PhD Thesis, Imperial College, London, UK, 2000.

Fossa, M. (1998). Design and Performance of a Conductance Probe for Measuring the Liquid Fraction in Two-Phase Gas-Liquid Flows. *Journal of Flow Measurement and Instrumentation,* 9, 103-109.

Fossa, M., Guglielmini, G., and Marchitto, A. (2003). Intermittent flow Parameters from Void Fraction Analysis. *Journal of Flow Measurement and Instrumentation*, 14, 161–168.

Ghassan, H. and Majeed, A. (1999). Liquid slug holdup in horizontal and slightly inclined two-phase slug flow. *Journal of Petroleum Science and Engineering* 27, 27-32.

Gopal, M. and Jepson, W. P. (1997). Development of digital image analysis techniques for the study of velocity and void profiles in slug flow. *International Journal of Multiphase Flow*, 23(5), 945-965.

Graham, B; Wallis, G. B. and Dobson, J. E. (1973). The onset of slugging in horizontal stratified air-water flow. *International Journal of Multiphase Flow*, 1(1), 173-193.

Gregory, G.A. and Scott, D. S. (1969). Correlation of liquid slug velocity and frequency in horizontal cocurrent gas-liquid slug flow. *Journal of AIChE*, 15, 933-935.

Gregory, G.A., Nicholson, M. and Aziz, K. (1978). Correlation of the liquid volume fraction in the slug for horizontal gas liquid slug flow. *International Journal of Multiphase Flow*, 4(1), 33–39.

Gregory, G.A., Nicholson, M. and Aziz, K. (1978). Correlation of the liquid volume fraction in the slug for horizontal gas liquid slug flow. *International Journal of Multiphase Flow*, 4(1), 33–39.

Gurevich, Y. (2001). Performance evaluation and application of clamp-on ultrasonic cross-correlation flow meter CROSSFLOW. In: *Proceedings of the International Conference of Flow Measurement*, Sao Paolo, Brazil, (date), 2000.

Hale, C. P. (2000). *Slug Formation, Growth and Decay in Gas-Liquid Flow*, PhD Thesis, Imperial College, London, UK, 2000.

Hammer, E. A. and Nordtvedt, J. E. (1991). The Application of a Venturi Meter to Multiphase Flow Meters for Oil Well Production. In: *Proceedings of the 5th Conference on Sensors and their Applications*, London, UK, September 22-25 (Bristol: Adam Hilger).

Hewitt, G. F., Harrison, P. S., Parry, S. J. and Shires, G. L. (1995). Development and Testing of the 'Mixmeter' Multiphase Flowmeter. In: *Proceedings of the 13th North Sea Flow Measurement Workshop*, Lillehammer, Norway, October 24-26 1995.

Hinze, J.O., 1955. Fundamentals of the hydrodynamic mechanism of splitting in dispersion processes. AIChE J. 1, 289.

Hughmark, G. A.,1962 ';Holdup in Gas-Liquid Flows," Chem. Eng.Prog.(April, 1962) 58, 62

Jepson, W. P. and Gopal, M. (1998). Ultrasonic Measuring System and Method of Operation. U.S Patent, Patent No 5,719,329, (February. 17, 1998).

King, M. J. S., Hale, C. P., Roberts, I. F., Fisher, S. A., Lawrence, C. J., Mendes-Tatsis, M. A. and Hewitt, G. F. (1997). Experimental investigations of flowrate Transients in horizontal pipes. In: *Proceedings of the 8th International conference on Multiphase Production*, Cannes, France, June 18-20, 1997.

King, N.W. (1990). Subsea multi-Phase Flow Metering a Chanllenge for the Offshore Industry. In: *Subsea 90 International Conferences*, London, England, December 11-12, 1990.

Kolmogorov, A.N., 1949. On the breaking of drops in turbulent flow. Doklady Akad. Nauk. 66, 825–828.

Kordyban, E. S. (1961). A flow modle for two-phase slug flow in horizontal tubes. *Journal of Basic Engineering, TASME*, 83, 613-618.

Kordyban, E. S. and Ranov, T. (1970). Mechanism of Slug Formation in Horizontal Two-Phase Flow. *Journal of Basic Engineering*. 92, 857-864.

Leighton, T. G. (1994). The acoustic bubble, London: Academic Press.

Letton, W. (2003). Technique for Measurement of Gas and Liquid Flow Velocities, and Liquid Holdup in a Pipe with Stratified Flow. U.S Patent, Patent No 6,550,345 B1, (Aprial 22, 2003).

Lunde, O., Asheim, H. (1989). An Experimental Study of Slug Stability in Horizontal Flow. In: *Proceedings of the* 4th *International conference Multiphase Production*, Nice, France, June 19-21, 1989.

Malnes, D. (1979). Slip relations and momentum equations in two-phase flow. Cited in: Bendiksen, K. H., Malnes, D. and Nydal, O.J. (1996). On the modelling of slug flow. *Journal of Chemical Engineering Communications*, 141, 71-102.

Manfield, P. (2000). *Experimental, computational and analytical studies of slug flow*. PhD Thesis, Imperial College, London, UK, 2000.

Manolis, I. G. (1995). *High Pressure Gas-Liquid Slug Flow*. PhD Thesis, Imperial College, London, UK, 1995.

Mba, D., Rao, Raj B. K. N.,(2006) Development of Acoustic Emission Technology for Condition Monitoring and Diagnosis of Rotating Machines: Bearings, Pumps, Gearboxes, Engines, and Rotating Structures. The Shock and Vibration Digest 2006 38: 3-16.

McNulty, P.J. (1985), PUMP HYDRAULIC NOISE: ITS USES AND CURES, Marine Engineers Review, , pp. 22-23.

Mehdizadeh, P. and Williamson, J. (2004). *Principles of Multiphase Measurements*. Alaska Oil & Gas Conservation Commission, Alaska, USA.

Merilo, M., Dechene, R. L. and Cichowlas, W.M. (1977). Void Fraction Measurement with a Rotating Electric Field Conductance Gauge. *ASME Journal of Heat Transfer* 99, 330-332.

Miller, R. K. (1987),(Ed.) Acoustic emission testing. ASNT Nondestructive Testing Handbook, 1987, Vol. 5 (American Society of Nondestructive Testing), ISBN: 0-931403-02-2.

Mishima, K. and Ishii, M. (1980). Theoretical prediction of onset of horizontal slug flow. *Journal of Fluids Engineering, Transactions of the ASME* . 102, 441-445.

Moore, P. I. (2000). *Modelling of Installation Effects on Transit Time Ultrasonic Flow Meters in Circular Pipes*. PhD Thesis, University of Strathclyde, UK.

Moura, L. F. M. and Marvillet, C. (1997). Measurement of Two-phase Mass Flow Rate and Quality Using Venturi and Void Fraction Meters. In: *Proceedings of the 1997 ASME International Mechanical Engineering Congress and Exposition*, Dallas, TX, USA, November 16-21, 1997, The Fluids Engineering Division, FED, 1997, Vol. 244, 363-368.

Mwambela, A.J. and Johansen, G.A. (2001). Multiphase flow component volume fraction measurement: experimental evaluation of entropic thresholding methods using an electrical capacitance tomography system. *Journal of Measurement Science and Technology*, 12, 1092-1101.

Nadler, M. and Mewes, D. (1995). Effect of the viscosity on the phase distribution in horizontal gas-liquid flow. *International Journal of Multiphase Flow*, 21(2), 253-266.

Nicholson, M. K., Aziz, K. and Gregory, G. A. (1978). Intermittent two-phase phase flow in horizontal pipes: predictive models. *The Canadian Journal of Chemical Engineering*, 56 (6), 653-663.

Nydal, O.J., Pintus, S. and Andreussi, P. (1992). Statistical characterisation of slug flow in horizontal pipes. *International Journal of Multiphase Flow*, 18(3), 439-453.

Olsen, A.B. (1993). Framo Subsea Multiphase Flow Meter System Proc.Sem.Multiphase Meters and their Subsea Application, London, England.

Ong, K.H. (1975). *Hydraulic Flow Measurement Using Ultrasonic Transducers and Correlation Techniques*. PhD thesis. University of Bradford, UK.

Paglianti, A., Andreussi, P.and Nydal, O. J. (1993). The Effect of Fluid Propoerties and Geometry on Void Distribution in Slug Flow. In: *Proceedings of the 6th International Conference on Multiphase Production,Cannes*, France, June 16-18, 1993.

Pan, L. (1996). High Pressure Three-phase (gas-liquid-liquid) flow. PhD Thesis. Imperial College, London.

Rafa, K., Tomoda, T. and Ridley, R. (1989). Flow Loop Field Testing of a Gamma Ray Compositional Meter. In: *Proceedings of the ASME Energy Sources Technology Conference and Exhibition*, paper no.89-Pet-7, Houston, Texas, Jan 22-25, 1989.

Reis, E. dos. and Goldstein Jr. L. (2005). A non-intrusive probe for profile and velocity measurement in horizontal slug flows. *Journal of Flow Measurement and Instrumentation*, 16 (4), 229-239.

Roach, G. J. and Watt, J. S. (1996). Current status development of CSIRO gamma-ray multiphase flow meter. In: *Proceedings of the 14th North Sea Flow Measurement Workshop*, Peebles, Scotland, October 27-30, 1996.

Ruder, Z., Hanratty, P. H. and Hanratty, T. J. (1989). Necessary conditions for the existence of stable slugs. *International Journal of Multiphase Flow*, 15(2), 209-226.

Sanderson, M. L. and Yeung, H. (2002). Guidelines for the use of ultrasonic non-invasive metering techniques. *Journal of Flow Measurement and Instrumentation*, 13: 125–142.

Scott, S. L., Shoham, O. and Brill, J. P. (1986). Prediction of slug length in horizontal large-diameter pipes. In: *Proceedings of the 56th Regional Meeting on the Society of Petroleum Engineers*, Oakland, CA, April 2-4 (SPE 15103).

Singh, G. and Griffith, P. (1970). Determination of the Pressure Drop Optimum Pipe Size for Two-Phase Slug Flow in an Inclined Pipe. Cited in: Hale, C. P. (2000). *Slug Formation, Growth and Decay in Gas-Liquid Flow, chapter 3, p211*.PhD Thesis, Imperial College, London, UK, 2000.

Stanislav, J. F. Kokal, S. and Nicholson, M.K. (1986). Intermittent gas-liquid flow in upward inclined pipes. *International Journal of Multiphase Flow*, 12(3), 33–39.

Steven, R. N. (2002). Wet gas metering with a horizontally mounted Venturi meter. *Journal of Flow Measurement and Instrumentation*, 12(5-6), 361-372.

Stewart, C. (2001). *Instrumentation for the Measurement of Slug Flows*. PhD thesis, University of Strathclyde, UK.

Taitel, Y. and Dukler, A. E. (1976). A model for predicting flow regime transitions in horizontal and near horizontal gas-flow. *Journal of American Institute of Chemical Engineering*. 22(1), 47-55.

Theuveny, B. C., Segeral, G. and Pinguet, B. (2001). Multiphase Flowmeters in Well Testing Applications. In: *Proceedings of the SPE Annual Technical Conference and Exhibition*, New Orleans, Louisiana, USA, September 30-October 3, 2001, SPE Paper 71475.

Thompson, E. J. (1978). Mid-radius ultrasonic flow measurement *FLOMEKO* ed H H Dijstelbergen and E A Spencer, pp 153–61.

Thorn, R., Johansen, G. A., and Hammer, E. A. (1997). Recent developments in three phase flow measurement. *Measurement Science and Technology*, 8(7), 691–701.

Tuss, B., Perry, D. and Shoup, G. (1996). Field tests of the high gas volume fraction multiphase meter. In: *Proceedings of the SPE Annual Technical Conference and Exhibition.* Denver, USA, October 6-9, 1996, SPE Paper 36594.

Van Santen H., Kolar, Z. I. and Scheers, A.M. (1995). Photo Energy Selection for Dual Energy Gamma and /or X-ray Absorption Composition Measurement in Oil-Water-Gas Mixtures. *Journal of Nuclear .Geophysics*, 9(3), 193-202.

Vedapuri, D. and Gopal, M. (2003). Determining Gas and Liquid Flow Rates in A Multi-Phase Flow. U.S Patent, Patent No 6,502,465 B1, (January 7, 2003)

Woods, B. D. and Hanratty, T. J. (1996). Relation of Slug Stability to Shedding Rate. *International Journal of Multiphase Flow*, 22(5), 806-828.

Xie, C. G, Stott, A. L., Plaskowski, A. and Beck, M. S. (1990). Design of capacitance electrodes for concentration measurement of two-phase flow. *Journal of Measurement Science and Technology*, 1, 65-78.

Xu, L.J., Xu, J., Dong, F. and Zhang, T. (2003). On fluctuation of the dynamic differential pressure signal of Venturi meter for wet gas metering. *Journal of Flow Measurement and Instrumentation*, 14(4-5), 211-217.

Yang, W. Q. and Beck, M. S. (1997). An intelligent cross correlation for pipelines flow velocity measurement. *Journal of Flow Measurement and Instrumentation*, 8(2), 77-84.

Zhang, H.J., YUE W.T. and HUANG, Z.Y. (2005). Investigation of Oil-Air Two-Phase Mass Flow rate Measurement using Venturi and Void Fraction Sensor. *Journal of Zhejiang University SCIENCE*,6A(6),601-606.

Zhang, H.-Q., Wang, Q., Sarica, C., and Brill, J. P., 2003, "Unified Model for Gas-Liquid Pipe Flow Via Slug Dynamics—Part 2: Model Validation,. ASME J. Energy Resour. Technol., 125,pp.266-273

10

Damage Assessment in Fibrous Composites Using Acoustic Emission

A.S. Paipetis and D.G. Aggelis
Department of Materials Science and Engineering, University of Ioannina, Greece

1. Introduction

Acoustic Emission is by definition the phenomenon of stress or pressure wave generation during dynamic processes in materials. Failure of engineering materials is often preceded by audible events, as is well known in the case of timber, where the fibrous structure warns of its upcoming failure with the emission of characteristic sounds (Liptai et al. 1972). Early uses of acoustic emission analysis include elastic waves produced by earthquakes. Typical of the acoustic signature of the twinning is what is known as "tin cry", observed by tin smiths. Similar, the martensitic transformation in steel is known to produce clicking sounds. The seminal work by Joseph Kaiser initiated the research in acoustic emission as a result of irreversible energy dissipating processes occurring during the loading of materials and structures (Kaiser 1950). Within a decade, considerable research effort accompanied with extensive improvement in instrumentation led to the specification of the monitoring frequencies well above the audio range, in the area 0f 0.1 to 1 MHz (Dunegan et al. 1964). Today, multi channel acquisition systems corroborated with computing power and advanced signal processing allow for the real time monitorin, location and damage identification in materials and structures (Caprino et al. 2011, Liu et al, 2010).

Acoustic emission (AE) has been used for monitoring of the structural condition and damage characterization for practically all kinds of materials and systems (Grosse and Ohtsu 2008, Mindess 2004, Prosser 2002). It exploits the transient elastic waves emitted during fracture incidences. It is the complete analogue of seismic activity recorded by seismographs, though in shorter range, which extends from a few cm in laboratory specimens to several m in real structures. The emitted waves are recorded by transducers that are fixed on the surface of the component under test. These are usually piezoelectric and transform the pressure felt on their surface to electric signals. The nature of the waves measured is identical to the waves utilized in another non destructive testing (NDT) technique, namely ultrasonic testing (UT). The fundamental difference between AE and UT is the absence of external wave excitation since the emissions originate from the cracking events within the material. Additionally, in AE the material should be stressed in order to produce active cracking events, unlike ultrasonics which do not require mechanical stress on the material. After the wave is transformed to an electric waveform it is normally pre-amplified and digitized in an acquisition board, see Fig. 1.

The presence of cracks in the material alters the load at which irreversible deformation begins. Consequently, the acoustic emission pattern is altered. As a result, the AE technique can be used in laboratory conditions in order to monitor the fracture process of engineering materials and build an understanding on the damage behavior of materials under stress or other degradation factors (Carpinteri et al. 2010, Scholey et al. 2010, Philippidis et al. 1998, Ohtsu and Watanabe 2001, Kawasaki et al. 2010). It is also employed in real structures to monitor the cracking activity under service load (Shiotani et al. 1999, Shiotani et al. 2009, Aggelis et al. 2010a, Kurz et al. 2006), in order to estimate the structural integrity of rock mass related to landslide incidences (Shiotani 2006), of tanks and pressure vessels (Anastasopoulos et al. 2009, Anastasopoulos and Tsimogiannis 2004), while application in composite laminates is extensive (Anastasopoulos and Philippidis 1995, Prosser et al. 1995, Hamstad 1986, Katerelos et al. 2009).

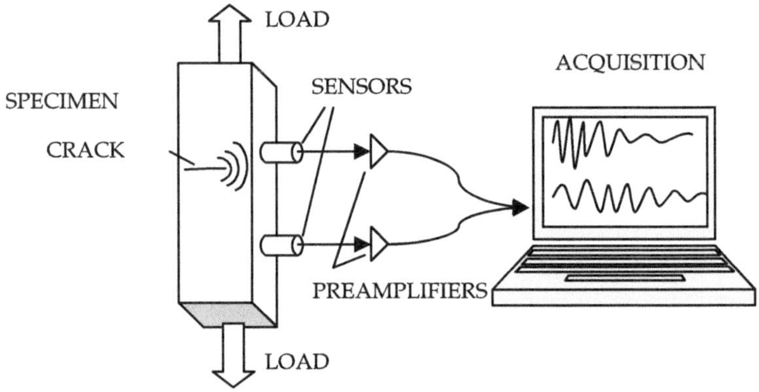

Fig. 1. Schematic representation of AE experiment.

Monitoring of AE behavior supplies information on several levels. One basic function is the characterization of the active crack density. Low AE activity is connected to low density of cracks, since cracking is the source of AE. The number or rate of recorded signals is a parameter of paramount importance and in many cases it is used among other parameters as a criterion for rejection or acceptance of components like pressure vessels and tanks [ASTM E 569, ASME 2005]. However, the total AE activity is not the only information supplied. The shape and content of each recorded waveform is the electric analogue of the elastic wave that impinged on the sensor, excluding the sensor's response. Additionally, this elastic wave depends on the motion of the crack tip during each incidence. Cracking events with high intensity, corresponding to large crack propagation rate, give rise to the emission of waves with high amplitude and energy. Additionally, the relative motion of the sides of the cracks, which defines the cracking mode, influences the shape of the AE waveform. Mode I, or tensile cracking is related to short waveforms with high frequency content, while mode II or cracks resembling shear correspond to longer waveforms and lower frequency content, see Fig. 2 (Ohtsu 2010). This has been shown in a number of studies concerning different kinds of materials like rock (Shiotani 2006), concrete (Ohno and Ohtsu 2010, Soulioti et al. 2009, Aggelis et al. 2011a), metals (Aggelis et al. 2011b) and composite laminates (Aggelis et al. 2010b). It originates from the different elastic wave modes (longitudinal and shear) that are emitted from the mode I excitation (opposing movement of the crack faces) and mode II (sliding movement) respectively (Aggelis et al. 2010c).

In most of technical materials and structures fracture or deterioration process includes both basic cracking modes. Apart from basic cracking modes, it can be argued that every damage mechanism possess a different acoustic signature which is reflected on the pattern of the signal. In this respect, pattern recognition schemes may be employed to correlate distinct "damage entities" to signal attributes (Gutkin et al 2011). The abundance of information in the AE waveforms may allow for "black box" approaches where the observed damage process is correlated to signal attributes (Katerelos et al, 2009, Arumugam et al, 2011).) Alternatively, the knowledge on the effect of specific damage mechanisms on the AE waveform may provide the roadmap for the study of nucleation, evolution and interaction of different mechanisms (Aggelis et al. 2010b).

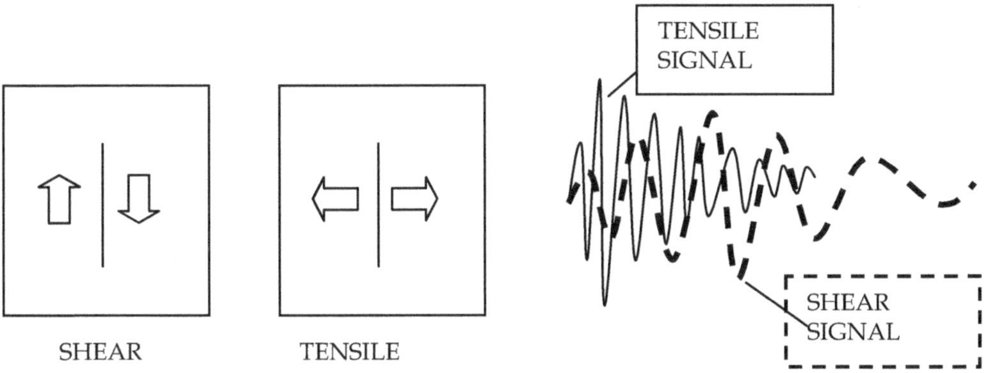

Fig. 2. Cracks modes and schematic representation of waveforms.

A typical AE waveform is depicted in Fig. 3 in detail with its main features. Amplitude, A, is the voltage exhibited by the highest peak of the waveform, which is normally correlated with the intensity of the crack propagation event. Other important parameters are the duration (DUR) and the rise time (RT) of the signal. DUR is defined as the time between the first and the last threshold crossing and is measured usually in μs. RT indicates the delay between the onset of the signal (first threshold crossing) and the time of the maximum peak, again in μs. A parameter that has extensively been used for characterization of the fracture mode mostly depends on the shape of the initial part of the waveform. The ratio of A over RT, which is the "rise angle" or "grade" (Anastassopoulos and Philippidis 1995, Philippidis et al. 1998) of the waveform, is recently utilized in the form of RA value (Ohtsu 2010, Ohtsu and Tomoda 2008, Soulioti et al. 2009, Aggelis et al. 2011a,b). This value is precisely the inverse of grade (rise time over the Amplitude measured in μs/V) (Ohno and Ohtsu 2010). The average frequency, AF is another essential parameter defined by the ratio of threshold crossings over the duration of the signal and is measured in kHz. It is a rough but valuable indicator of the main frequency component and it has been proved sensitive to the fracture mode, with tensile cracking resulting in higher frequency than shear (Ohtsu 2010, Aggelis 2011). Additionally, AE energy (ENE) expresses the area under the rectified waveform envelope. Similarly to the amplitude, it expresses the severity of cracking since generation of a large crack could emit larger amount of energy than a small crack.

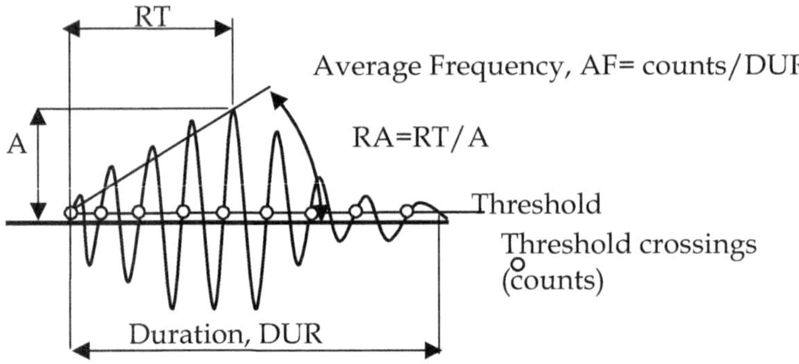

Fig. 3. Typical AE waveform

Fibrous composite materials are the ideal platform for correlating the acoustic signature of service induced damage to the degradation of the material or the structure. This is stemming from the fact that accumulating damage in composites varies in scale, dimension and is correlated to the service life of the material. Typical of the above behavior is damage initiation and propagation in multi axis reinforced composite laminates. In these systems, matrix cracking of the off-axis laminates is considered the primary mode of damage and is the result of complex combinations of thermal and mechanical loading. This damage mode is also referred to as "transverse" or "intra-laminar" cracking, as the cracks are oriented transversely to the loading direction and is the initial mode of damage to appear first in such composite systems. Transverse cracking has been studied extensively by several researchers over the last 25 years (Talreja 1994, Nairn and Hu 1994, Varna et al 1999, Nairn 2000), focussing mainly on cross-ply glass fibre reinforced polymer composite systems.

Failure in composite laminates typically starts by the creation of cracks of the off axis plies. The cracks in the off-axis layers usually propagate along the whole width of the specimens parallel to the fibres direction of the corresponding off-axis laminae. The crack density which is a parameter that characterises these systems and is defined as the number of cracks normalised by the length of the specimen depends on many parameters. Load level, layer orientation with respect to the load direction, temperature change, number of fatigue cycles, laminate stacking sequence, ply thickness and material fracture toughness are the most important parameters that dominate the crack density. Damage development in the form of transverse matrix cracking reduces the stress-bearing ability of individual layers resulting in the deterioration of the laminate's thermo-elastic properties.

In the simplest case of cross ply laminates where reinforcement is at 0 ° and 90 ° to the loading axis, the crack at the 90 ° ply or the transverse crack creates a shear discontinutity at the interface between the 0 ° and the 90 ° ply. The crack may either be deflected at the interface creating a delamination or propagate within the 0 ° ply. This is well known to be a function of the interlaminar shear strength the interface. With increasing load, transverse cracking accumulates and may reach a saturation stage, when the interlaminar shear strength is reach and no further load may be transferred to the neighbouring longitudinal ply. At the same time, longitudinal fiber failure occurs due to the stress concentration at the crack tip which is more pronounced with increasing interlaminar shear strength.

Cumulative failure of the longitudinal fibre is the primary failure mechanism in such systems. The failure process is depicted in Fig. 4.

Due to the succession of the different damage mechanisms, it is realized that real-time characterization of the cracking mode would supply information on the actual fracturing stage, enabling predictions on remaining life of the composite. This is a subject that concentrates strong efforts in the engineering community for decades. The results are not always successful since many factors hinder the reliable classification of the signals. Briefly these include the experimental details like the frequency response of the sensors, the specimen geometry which is not fixed, as well as the effect of propagation which may seriously alter the content of an elastic wave through an inhomogeneous and dispersive medium. It is however shown that in a variety of cases, different damage entities may be readily classified according to their acoustic signature (Katerelos et al 2009, Kostopoulos et al 2007a, Kostopoulos et al 2007b.)

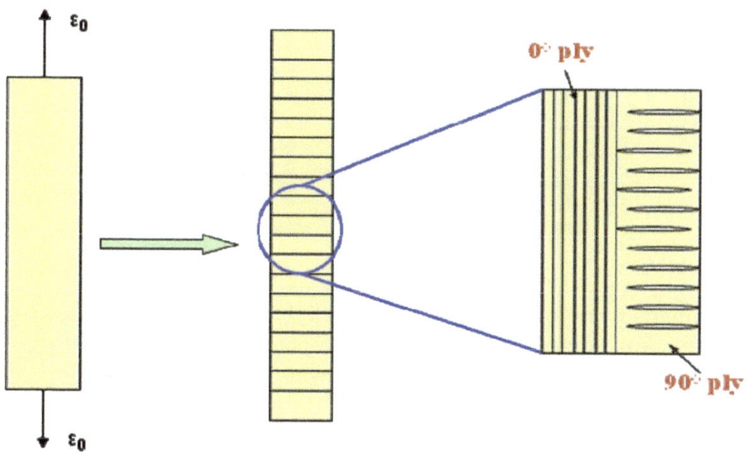

Fig. 4. Failure of cross ply laminates

This chapter describes the basics of an AE monitoring study of cross-ply laminates, emphasizing on the features that prove sensitive to the damage mechanism in small scale laboratory conditions. These composite material systems are excellent for crack monitoring which is performed via the matching of the refractive index of the fibres and the epoxy resin. In this way, they can be made transparent and thus, the optical observation of the crack development is feasible.

2. Experimental methods

The cross-ply laminates were fabricated by hand layup with a sequence $[0°_4/90°_4]_s$, resulting in a number of 16 plies with total specimen thickness of 2 mm. The UD 220 g/m2 (Aero) unidirectional glass fibre fabric was impregnated using the HT2 epoxy resin/hardener matrix system (mixing ratio 100:48) manufactured by R&G Faserverbundwerkstoffe GmbH Composite Technology. A 250x250 mm2 laminate was manufactured and was allowed to cure for 24 h at room temperature. Tensile specimens were subsequently cut according to the ASTM D3039 standard, at a width of 20 mm each. The matching of the refractive index of the glass fibres and the epoxy matrix resulted in the manufacturing of transparent laminates which enabled the optical observation of the failure modes of the composite laminate.

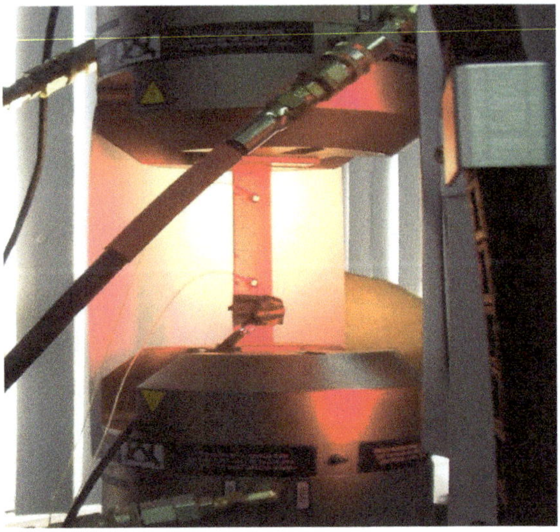

Fig. 5. Experimental set-up

The tensile specimens were loaded in load controlled tension, in a step loading mode. The loading spectrum was a saw-tooth spectrum formed from a sequence of triangular loading/unloading steps. A rate 5 kN/min was employed for both loading and unloading. The maximum load was incremented by 4 kN at each consecutive step. The step loading continued until the tensile failure of the specimen. All tensile tests were performed using an Instron Universal Testing Machine equipped with hydraulic gripping system, under load control, at controlled environmental conditions of 25 °C and 70% relative humidity (ASTM D3039).

For the AE monitoring, two wide band AE sensors (Pico, Physical Acoustics Corp., PAC) were attached on the same side of the specimen), as shown in Fig. 5. Electron wax was applied between the sensors and the specimen to enhance acoustic coupling, while it offered the necessary support to the sensors during the experiment. Due to the specific sensors sensitivity to frequencies from 50 kHz up to approximately 800 kHz, they can capture a wide range of different sources. The distance between the two receivers was 70 mm. The pre-amplifier gain was set to 40 dB. After performing a pilot test, the threshold was also set to 40 dB in order to avoid the possibility of electronic/environmental noise. The signals were recorded in a two-channel monitoring board PCI-2, PAC with a sampling rate of 5 MHz.

3. Results and discussion

The most basic parameter of an AE monitoring is the number of incoming signals, which as aforementioned in many cases of standardized monitoring serves as a criterion for structural integrity characterization. Usually the number of hits increases with load, due to the increase of crack propagation rate as the load increases. This behavior is seen in Fig. 6 concerning the cross-ply laminates subjected to the incremental step loading for the last 5 steps to failure. For each successive loading step, the cumulative AE activity increases considerably. The first cycle produced a few tens of hits, while during the last full cycle (#6) more than 20000 hits were recorded. This increase is reasonable due to the fact that as the load increases, apart from the matrix cracking that becomes extensive, numerous delaminations occur which were not active at the first loading steps. Additionally, the rate of cracking is more rapid and the fracture events happen progressively faster emitting the corresponding signals in rates of hundreds/s. It is mentioned that the number of cumulative AE activity was very repeatable for all specimens and therefore it can act as an indication of the forthcoming failure. However, this is limited to the cases where exactly the same conditions apply, which refer to material, specimen geometry, sensor type and separation distance, as well as specialized user-defined hardware settings concerning acquisition.

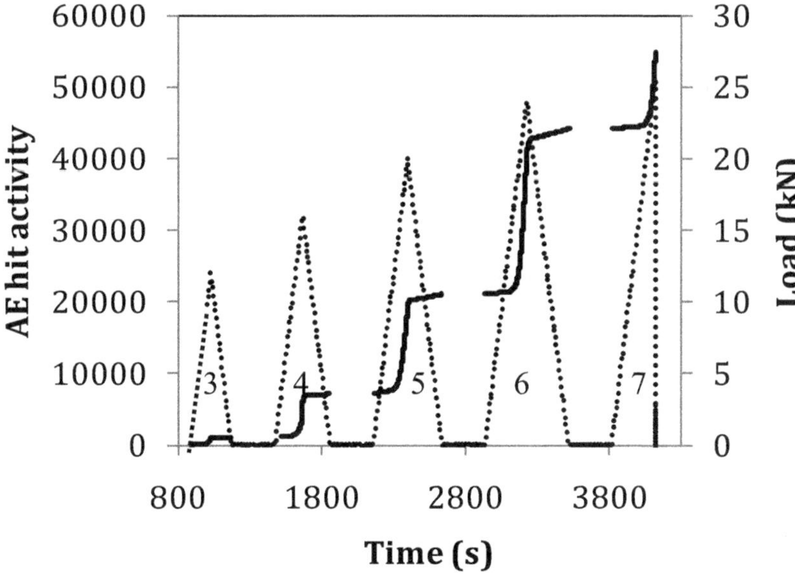

Fig. 6. Typical cumulative AE activity and loading history.

The number of accumulated hits correlates well with the maximum load of each cycle, as can be seen in Fig. 7 following an exponential growth function. The data come from different specimens showing the potential to characterize the sustained load by the number of AE signals under controlled conditions.

The total number of data originates from any different damage mechanism, which may include matrix cracking, delaminations and finally longitudinal fiber breakage. Although it is not straightforward to separate the different populations of data according to their origin, still correlations of this kind are of paramount importance as they enable estimations on the sustained load based on parameters measured in a non invasive way.

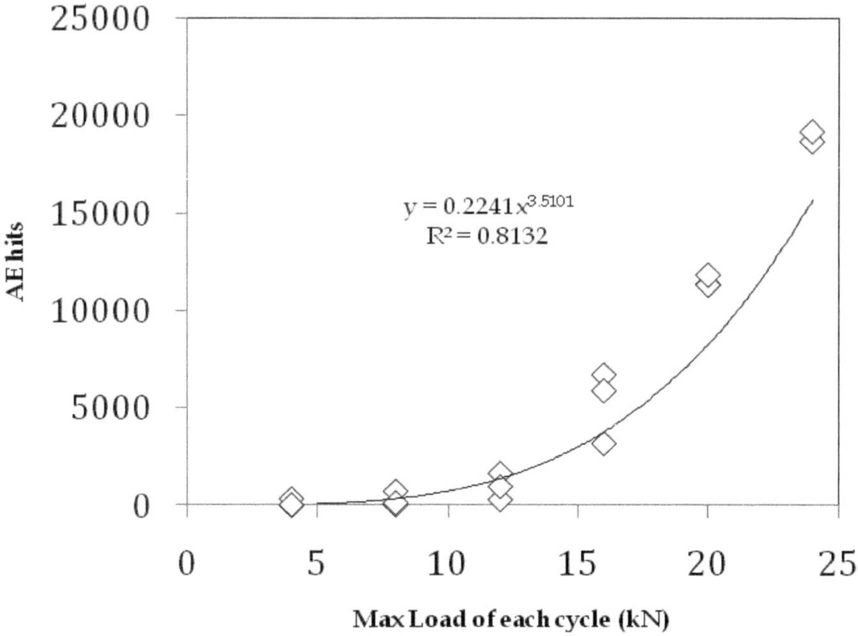

Fig. 7. Correlation between AE hits and maximum sustained load.

Though the general sequence of activation of the fracture mechanisms is known, the degree of overlapping between them cannot be confirmed by the number of the acquired hits. The task of characterization of the signals based on the damage mode can be assisted by the study of the qualitative parameters of the AE waveforms. As aforementioned, the shape and content of the signal strongly depends on the motion of the crack sides and therefore, on the damage mode. Waveform parameters like the RA have been used for characterization of source in different materials, as mentioned above. Actually, the RA value experiences great shifts during the specific loading protocol, as seen in Fig. 8. The RA level increases for each successive step. For the initial low level steps RA is limited below 10 ms/V, while for increasing loading step, the RA continuously rises. At the moment of failure, several signals exhibit RA close to or higher than 200 ms/V. This is characteristic of the gradual shift of the damage mechanism from matrix cracking to delaminations. It is reasonable that the low loads of the initial steps are responsible mostly for matrix cracking while delaminations occur at the highest load levels. The information of Fig. 8 is very informative and indicative of the shift in the damage mode. However, treatment of data in acoustic emission is mainly statistical, since the number of signals can easily reach several thousands or more and therefore, it is more meaningful to discuss on the trends and not on single signals. Therefore, Fig. 9 depicts the RA value as the moving average of the recent 200 hits in order to clear the trend from the cloud of points. Initially, the average RA is contained below 1 ms/V, while for each successive step it exhibits a considerable increase, reaching approximately 10 ms/V just before the moment of ultimate failure. Apart from the increase for each step, it is interesting to note that the RA line increases within each loading step until the load reaches the maximum value. RA starts from a value around 0.5 ms/V and gradually reaches a peak value at the moment of maximum load. Immediately after the load starts to decrease according to the loading protocol, the RA almost instantly drops and remains low until the load starts to increase at the next step. This is characteristic of the shift of damage mechanisms even within each step.

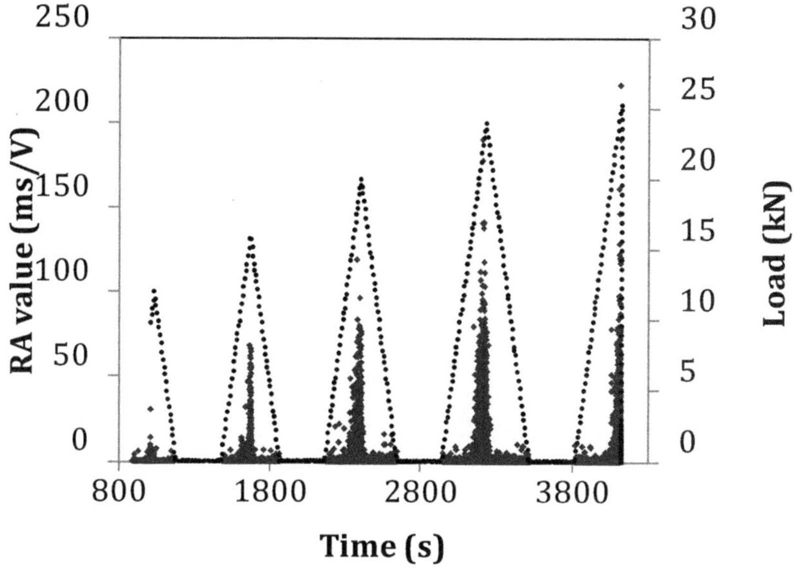

Fig. 8. RA value and Load time history.

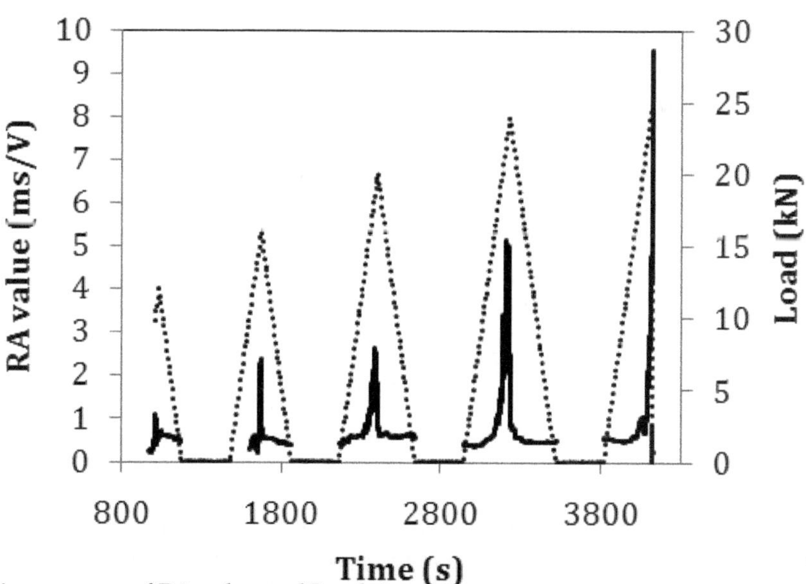

Fig. 9. Moving average of RA value and Load time history.

Averaging the RA of all the AE signals for each successive step results in a very strong correlation as seen in Fig. 10a. As the maximum load is increasing for each step, the RA was shifting to higher values (see Fig. 10a). This increase is considered indicative of the gradually increasing population of delamination events relatively to the matrix cracking ones. Each successive step exhibits higher RA values and at the final step, while the material is approaching final failure, RA exhibits its maximum average value of almost 3 ms/V (see Fig. 10a). As can be seen, the load and the mean RA strongly correlate.

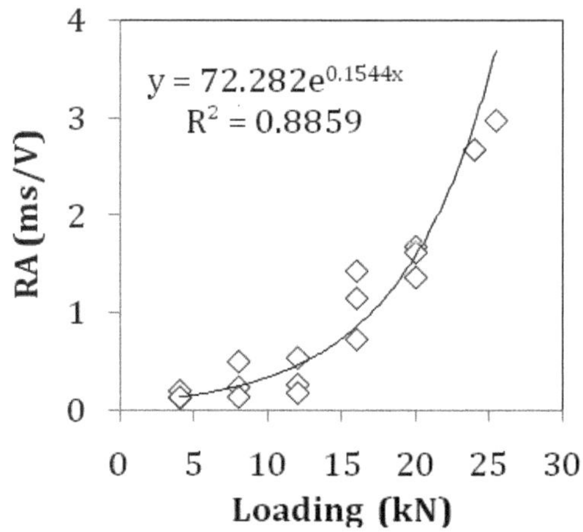

(a)

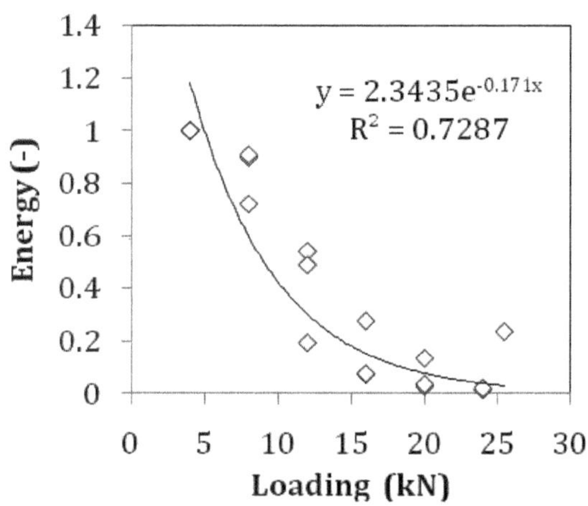

(b)

Fig. 10. (a) RA and (b) average energy vs. maximum load of each cycle

Several other correlations between AE parameters and load can be observed. Indicatively, Fig. 10b shows the development of the average energy of the AE signals according to the maximum load of each cycle. The values are normalized to the energy of the first cycle. It is evident that as loading (and deterioration of the material is accumulated) the average energy shifts to lower values. This may initially seem contradictive to the increasing crack propagation rate. However, it should be considered that the number of AE signals are continuously increasing with load. This implies that the fracture energy is distributed to many more events, reducing their average intensity. Summarizing, the load history and consequently all induced damage mechanisms, leave a distinct signature on the acoustic emission behavior both on the cumulative activity as well as the qualitative characteristics of the acquired waveforms. Monitoring of the AE parameters provides valuable information in respect to the load induced damage of the composite and provide early warning of the forthcoming failure.

It is interesting to note that this kind of classification approach based on RA has been applied in different materials. RA value exhibits an explosive increase when fiber pull-out events result after extensive matrix cracking (Soulioti et al. 2009, Aggelis et al. 2011a,c). The same trend has been seen in concrete containing polymer fibers, as studied by the inverse of RA, namely "grade" of the waveform (Aggelis et al. 2009). Additionally, these parameters have been used for characterization of the development of corrosion cracking (Ohtsu and Tomoda 2008, Kawasaki et al. 2010) as well as fracture of reinforced concrete bars (Ohno and Ohtsu, 2010), while they have been shown effective in rock during landslides (Shiotani, 2006).

In an alternative approach, the correlation of independently observed damage mechanisms may be correlated to the attributes of the recorded acoustic waveforms. Pattern recognition algorithms were utilised to classify and associate the acquired AE signals to the material's damage mechanisms during the monotonic loading of the aforementioned cross ply laminates. As the laminates were transparent, the optical identification and correlation of the transverse cracking to the acoustic activity of the specific damage mechanism enabled the identification of two additional damage mechanisms. These continued to be active until the failure of the specimens. Fig. 11 presents a typical load time curve recorded during the monotonic tensile loading of the composite together with the crack density. This was obtained by taking snapshots during loading at specific intervals. The photographs allowed for the measurement of the cracks as they appeared as lines in the transparent laminates (Katerelos et al 2009).

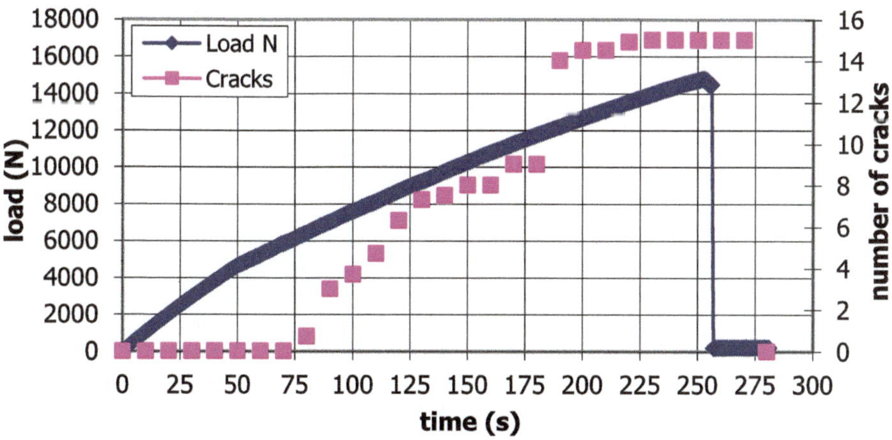

Fig. 11. Load and number of cracks measured in the gauge length for a typical tensile experiment.

Employing pattern recognition algorithms based on the various descriptors from the acquired waveforms, the appropriate clustering that corresponds to the observed damage mechanisms may be attained. Fig. 12 presents the resulting clustering in three classes employing a combination of Maxmin distance and Isodata algorithms (Katerelos 2009). The figure represents a 2D projection of the classes that are completely separated in the 4th dimensional space. The descriptors employed for the specific clustering were amplitude, rise angle, initiation frequency and mechanical load. Cluster validity criteria such as R and τ criteria (Fukunaga 1990, Tou and Gonzales 1974) were applied to confirm the successful separation of the AE population.

Concurrent observation of the optical evolution of damage allowed for the correlation of the separate classes to distinct failure mechanisms. As was observed all acoustic activity initiated shortly before the appearance of transverse cracks. Class 1 was active until fracture. Class 2 became audible when the characteristic knee at the stress curve was recorded at approximately 50 s (Fig. 11) and went silent after the saturation of transverse cracking. Class 3 was audible shortly before the silencing of class 2, and was notably higher in terms of energy content (Katerelos 2009). As is obvious, class 2 may readily be attributed to transverse cracking, class 3 is the primary failure mechanism, or the longitudinal fibre breaking. Class 1 is related to all secondary energy dissipating mechanisms such as interfacial sliding and interfacial bond failure. In this way, the identification of all major mechanisms that precede and accumulate to lead to failure may be monitored and provide information about the structural integrity of the material or the structure.

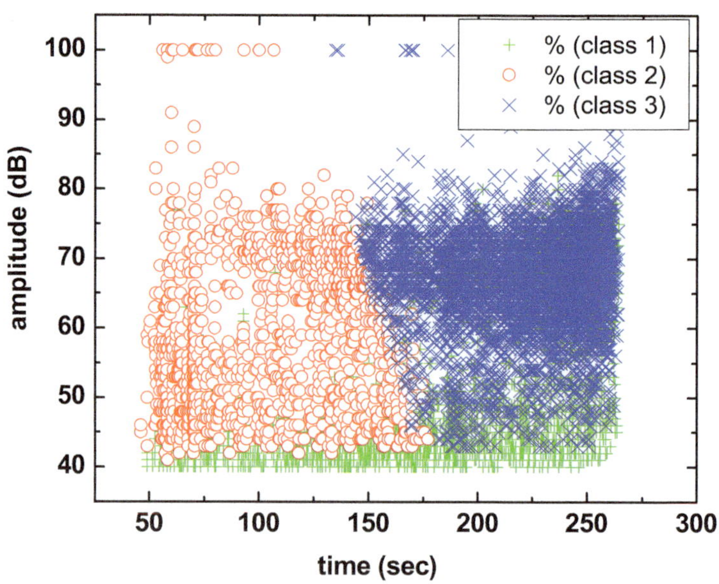

Fig. 12. Separation of resulted classes in 2-D projection.

4. Conclusions

The discussion in the chapter intends to shed light in the fracture process of cross-ply laminates enabling the characterization of the damage stage reliably using simple parameters collected in a non destructive fashion utilizing acoustic emission technique. It is stressed that in laboratory conditions mechanical parameters such as load and displacement are measured simultaneously with AE data. Therefore, any macro-cracks can be easily detected, even by visual observation. The main benefit of AE test and analysis results from the fact that specific AE signatures are closely related to damage mechanisms, which is crucial for real size structures. In an actual structure a crack of similar size will not result in visually detectable deformation. Therefore, in order to detect and identify damage in the early stages AE monitoring would be very effective. Different AE parameters like the energy and RA exhibit strong sensitivity to the fracture mode (tensile micro-cracking and delaminations). Average AE energy near ultimate fracture decreases by more than 90% compared to the initial small scale fracture, while RA increases by at least 20 times. These parameters prove very sensitive to the damage mode and deterioration and therefore, can be used for estimation of the remaining life of the component. The abundance of information in the acoustic waveform is also indicative of the fact that clustering algorithms based on the signal descriptors may be directly attributed to distinct failure mechanisms and yield information about the life stage of the interrogated material.

Studies in the field should continue aiming at the establishment of specific and simple criteria for damage qualification, as well as in the application to real size structures, since propagation through thin plates is dispersive and special care should be given for the interpretation of data collected after long distances.

5. References

Aggelis D. G., 2011. Classification of cracking mode in concrete by acoustic emission parameters, Mechanics Research Communications 38, 153–157

Aggelis D. G., Kordatos E. Z., Matikas T. E., 2011b. Acoustic emission for fatigue damage characterization in metal plates, Mechanics Research Communications 38, 106–110.

Aggelis, D. G., Shiotani T., Terazawa, M. 2010a.Assessment of construction joint effect in full-scale concrete beams by acoustic emission activity. Journal of Engineering Mechanics, 136(7), 906-912.

Aggelis, D. G., Soulioti, D. V., Barkoula, N. M., Paipetis, A. S., Matikas, T. E. 2011c. Influence of fiber chemical coating on the acoustic emission behavior of steel fiber reinforced concrete, Cement and Concrete Composites, 34 (2012) 62-67

Aggelis, D.G., Barkoula, N.M., Matikas, T.E., Paipetis, A.S. 2010b. Acoustic emission monitoring of degradation of cross ply laminates. J Acoust Soc Am., 127(6), EL246-251.

Aggelis, D.G., Matikas, T.E., Shiotani, T. 2010c. Advanced acoustic techniques for health monitoring of concrete structures. In "The Song's handbook of concrete durability", Kim SH, Ann KY eds, pp. 331-378, Middleton Publishing Inc.

Aggelis, D.G., Shiotani, T., Momoki, S., Hirama, A. Acoustic emission and ultrasound for damage characterization of concrete elements. ACI Materials J. 2009, 106(6), 509-514.

Aggelis, D.G., Soulioti, D.V., Sapouridis, N., Barkoula, N.M., Paipetis, A.S., Matikas, T.E. 2011a. Acoustic emission characterization of the fracture process in fibre reinforced concrete, Construction and Building Materials 25 4126–4131.

Anastasopoulos A, Tsimogiannis A. 2004. Evaluation of acoustic emission signals during monitoring of thick-wall vessels operating at elevated temperatures. JAE. 22:59–70.

Anastasopoulos, A., Kourousis, D., Botten, S., Wang, G., 2009. Acoustic emission monitoring for detecting structural defects in vessels and offshore structures, Ships and Offshore Structures, 4(4), 363 - 372.

Anastassopoulos, A., and Philippidis, T. P. 1995. Clustering methodology for evaluation of acoustic emission from composites, J. Acoust. Emiss., 13, 11–22.

Arumugam, J., Kumar, C.S., Santulli, C., Sarasini, F., Stanley, A.J., 2011. A global method for the identification of failure modes in fiberglass using acoustic emission Journal of Testing and Evaluation, 39 (5).

ASME, Section V, Article 12. 2005. Acoustic emission examination of metallic vessels during pressure testing. New York: ASME.

ASTM E 569. 2002. Standard practice for acoustic emission monitoring of structures during controlled stimulation.West Conshohocken, PA: ASTM International.

Caprino, G., Lopresto, V., Leone, C., Papa, I., 2011. Acoustic emission source location in unidirectional carbon-fiber-reinforced plastic plates with virtually trained artificial neural networks, Journal of Applied Polymer Science, 122 (6), 3506-3513.

Carpinteri, A., Cardone, F., Lacidogna, G.,. Energy Emissions from Failure Phenomena:Mechanical, Electromagnetic, Nuclear. Experimental Mechanics 2010, 50, 1235–1243.

Dunegan, H. L., Harris, D. O., and Tatro, C. A., "Acoustic Emission research", Report UCID-4868, Lawrence Radiation Laboratioy, Livermore, California, 1964.

Fukunaga, K.., Introduction to statistical pattern recognition, 2nd English ed. San Antonio, CA, USA: Academic Press; 1990.

Grosse, C. U., Ohtsu, M. 2008. Acoustic Emission Testing, Springer, Heidelberg.

Gutkin, R., Green, C.J., Vangrattanachai, S., Pinho, S.T., Robinson, P., Curtis, P.T. 2011. On acoustic emission for failure investigation in CFRP: Pattern recognition and peak frequency analyses, Mechanical Systems and Signal Processing, 25 (4), 1393-1407.

Hamstad MA. 1986. A review: acoustic emission, a tool for composite materials studies. Exp Mech;26(1):7–13.

Kaiser, J., Ph. D Thesis, 1950.

Katerelos, D. G., Paipetis, A. S., Loutas, T., Sotiriadis, G., Kostopoulos, V., and Orgin, S. L., 2009. "In situ damage monitoring of cross-ply laminates using acoustic emission," Plast. Rubber Compos. 38, 229–234.

Kawasaki, Y., Tomoda, Y., Ohtsu, M. AE monitoring of corrosion process in cyclic wet–dry test. Construction and Building Materials 2010, 24, 2353–2357

Kostopoulos, V., Loutas, T., Dassios, K. 2007a. Fracture behavior and damage mechanisms identification of SiC/glass ceramic composites using AE monitoring, Composites Science and Technology, 67 (7-8), 1740-1746.

Kostopoulos, V., Tsotra, P., Karapappas, P., Tsantzalis, S., Vavouliotis, A., Loutas, T.H., Paipetis, A., Friedrich, K., Tanimoto, T., 2007b. Mode I interlaminar fracture of CNF or/and PZT doped CFRPs via acoustic emission monitoring, Composites Science and Technology, 67 (5), 822-828.

Kurz, J.H., Finck, F., Grosse, C.U., and Reinhardt, H.W. 2006. Stress drop and stress redistribution in concrete quantified over time by the b-value analysis. Structural Health Monitoring, 5, 69-81.

Liptai, R. G., Harris, D. O., and Tatro, C, A., "An Introduction to Acoustic Emission", Acoustic Emission, ASTM STP505, American society for testing and Materials, 1972, pp.3-10.

Liu, Z., Pang, B., Tang, Q., 2010. Acoustic emission source location in laminated plates based on virtual wave front, Yadian Yu Shengguang/Piezoelectrics and Acoustooptics, 32 (3), 493-497.

Mindess S., 2004. Acoustic Emission Methods. In CRC Handbook of Nondestructive Testing of Concrete, Editors Malhotra V.M., Carino N. J., CRC, Boca Raton, FL.

Nairn, J. "Matrix microcracking in composites". In: Kelly, A., Zweben, C., Talreja, R. and Manson, J. – A., editors. Polymer matrix composites. Comprehensive composite materials, vol. 2. Amsterdam: Elsevier; 2000.

Nairn, J. and Hu, S. "Matrix microcracking". In: Pipes R.B., Talreja R, editors. Damage mechanics of composite materials. Composite materials series, vol. 9. Amsterdam: Elsevier; 1994.

Ohno, K., Ohtsu, M. 2010. Crack classification in concrete based on acoustic emission. Construction and Building Materials, 24(12), 2339-2346.

Ohtsu M., Recommendation of RILEM TC 212-ACD 2010. Acoustic emission and related NDE techniques for crack detection and damage evaluation in concrete: Test method for classification of active cracks in concrete structures by acoustic emission, Materials and Structures, 43(9), 1187-1189.

Ohtsu, M, Watanabe H. 2001. Quantitative damage estimation of concrete by acoustic emission. Constr Build Mater, 15, 217–24.

Ohtsu, M., Tomoda, Y. 2008. Phenomenological model of corrosion process in reinforced concrete identified by acoustic emission. ACI Mater J. 105(2), 194–199.

Philippidis, T.P., Nikolaidis, V.N., Anastassopoulos, A.A. 1998. Damage characterization of carbon/carbon laminates using neural network techniques on AE signals. NDT&E Int., 31(5), 329-340.

Prosser WH, Jackson KE, Kellas S, Smith BT, McKeon J, Friedman A. 1995. Advanced, waveform based acoustic emission detection of matrix cracking in composites. Mater Eval;53(9):1052–8.

Prosser, W. H. 2002. Acoustic Emission. In P. J. Shull (ed.) Nondestructive evaluation, theory, techniques and applications, pp.369-444, Taylor and Francis, Boca Raton

Rose, J. L., "Ultrasonic waves in solid media", Camridge University Press, Cambridge, 2004.

Scholey, J. J., Wilcox, P. D., Wisnom, M. R., Friswell, M. I. Quantitative experimental measurements of matrix cracking and delamination using acoustic emission, Composites: Part A 2010, 41, 612–623.

Shiotani, T., 2006. Evaluation of long-term stability for rock slope by means of acoustic emission technique, NDT&E Int, 39(3), 217-228.

Shiotani, T., Aggelis D. G., Makishima O., 2009. Global monitoring of large concrete structures using acoustic emission and ultrasonic techniques. Journal of Bridge Engineering-ASCE, 14(3), 188-192.

Shiotani, T., Shigeishi, M., and Ohtsu, M. 1999. Acoustic emission characteristics of concrete piles. Constr. Build. Mater., 13, 73–85.

Soulioti, D., Barkoula, N.M., Paipetis, A., Matikas, T.E., Shiotani, T., Aggelis D.G. 2009. Acoustic emission behavior of steel fibre reinforced concrete under bending. Construction and Building Materials 23, 3532-3536.

Talreja, R. "Damage characterization by internal variables". In: Pipes, R.B. and Talreja, R., editors. Damage mechanics of composite materials. Composite materials series, vol. 9. Amsterdam: Elsevier; 1994.

Tou, J.T., and Gonzales, R.C., Pattern recognition principles. Reading, MA:Addison-Wesley; 1974.

Varna, J., Joffe, R., Akshantala, N.V. and Talreja, R., 1999. Damage in composite laminates with off-axis plies, Composites Science & Technology,; 59(14), 2139-2147.

11

Experimental Study on Acoustic Emission Characteristics of Concrete Failure Process Under Uniaxial Tension

Sheng-xing Wu, Yan Wang, Ji-kai Zhou and Yao Wang
Hohai University,
People's Republic of China

1. Introduction

Tension is the main mechanical form for concrete structures. The characteristics of concrete failure process under tension is an important part of concrete failure mechanism study [1], because tension of concrete is weak and concrete tends to crack. The post-peak softening response of concrete in direct tension seems primarily related to widening of a single crack, and the post-cracking resistance of concrete may be due to discontinuities in cracking at the submicroscopic level and to bridging of cracked surfaces by aggregates. Acoustic emission, a kind of real-time technique, detects microcrack activities in high sensitivity, thus it is widely used in concrete study. In this paper, the development and evolution of microcracks are studied by analyzing the AE signals of concrete specimens under uniaxial tension.

2. Testing procedure

2.1 Preparation of concrete specimens

Prism specimens in variable cross-section are prepared, and axial tensile force is applied using a steel bar with a diameter of 18mm pre-embedded in the specimen; the steel bar is connected with spherical hinge by thread. Dimension and the shape of concrete specimen are showed in Fig. 1. The materials used are Portland cement of type 32.5, river sand which has a fineness modulus of 2.6, crushed gravel (the maximum grain size is 16 mm). The corresponding content proportion of cement: sand: gravel: water is 1: 1.63: 2.66: 0.55. Concrete mixture is injected in steel molds, and then compacted using a vibrating table. The specimens are demolded the following day and sprayed in water tank for another 7 days, then cured at room temperature until they are tested. The age of specimens is about 150d.

2.2 Loading equipment

The tests are carried out in MTS 322 electro-hydraulic servo material test machine and FlexTest GT multi-channel controlling system. To obtain stable post-peak response of concrete specimens, the concrete specimens' own deformation is used as the control signal of closed-loop controlled test system. Unloading-reloading processes is performed at several points in the softening portion of the stress-strain curve in order to study the generation and development of microcracks, In addition, the stability of response in post-peak situation is

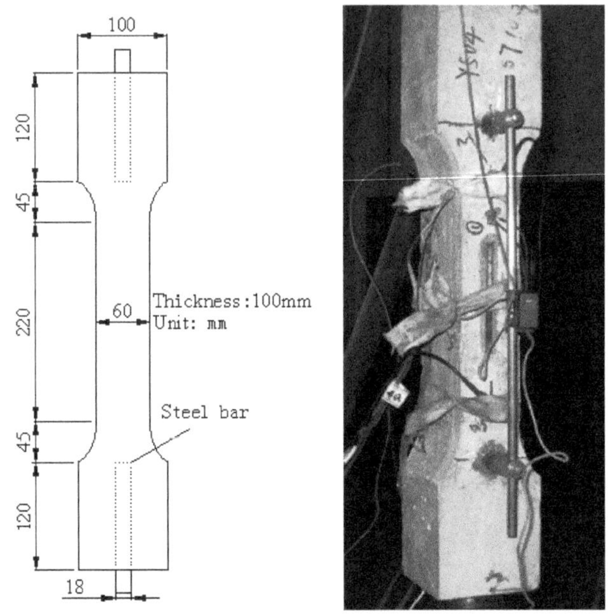

Fig. 1. Dimension and shape of concrete specimens

evaluated. Concrete fracture generally occurs in a particular region due to the development of a large number of micro-cracks. Then, visible cracks appear before the concrete fractures. Irrecoverable deformation occurs in this region, and strain softening phenomenon appears. irrecoverable deformations in this region are different to each other [2].Because the tension failure of concrete is the result of a single main crack growth, gradual failure can be obtained and instability near the peak load can be avoided if the opening of the crack is controlled in a closed-loop manner [3]. However, the final fracture position is unknown before testing for unnotched concrete specimen under direct tension. In order to find the microcrack localized region and to use the deformation of this region as control signal, longer distance for strain measurements are used in this study, see Tab. 1.

Series	Loading type	Transducers	Distance strain measurement
DT01	Monotonous	One broadband transducers	210mm
DT02	Unloading-reloading is performed at 2 points in the softening portion	One broadband and 4 resonant transducers (No. 1-5 in Fig. 2)	210mm
DT03	Unloading-reloading is performed at 3 points in the softening portion	One broadband and 5 resonant transducers (No. 1-6 in Fig. 2)	300mm
DT04	Unloading-reloading is performed at 4 points in the softening portion		275mm

Table 1. Loading types and of the transducer arrangements

2.3 AE data-acquisition system

Acoustic emission monitoring of concrete crack activity is performed using an advanced SAMOS ™ system supplied by PAC Ltd. AEwin ™ software is used to acquire and process the data. Both broadband and resonant transducers (Type: R6α, resonance frequency: 90kHz; and type: PAC-WD, Bandwidth 100kHz~1.0MHz) are attached to specimens using vaseline as coupling agent, then fixed using rubber band. The acoustic emission signals are amplified by a 40dB gain PAC-2/4/6 preamplifier, a Bandpass filter with a range of 10kHz∼2.0MHz is used, the detail for the arrangement of the transducers is shown in Tab. 1.

Resonant transducers are used for 2D AE source location analysis, while the broadband transducer is used to analyze AE parameters. Considering the arrangement of strain measurement, such as strain gauges and extensometer, the arrangement of transducers are finally chosen, as shown in Fig. 2. The thresholds for broadband and resonant transducers are 35dB and 38dB respectively, which are just above the background noise level determined by preliminary tests.

2.4 Loading type

In order to get optimum results from the test, three preliminary tests are carried out, then the final test method is fixed. The results from these tests are not presented in this paper. The results of four concrete specimens are given, the series and loading tapes are shown in Tab. 1. Four strain gauges are attached to the middle of each side of specimens, and the data related to strain is acquired and stored by type DH5922 system and controlled by computer. Preloading of 5kN (about 0.83 MPa) is made in order to check if the deviation of four strain data from each side of the specimens is within 10%, then the extensometer is put on the highest strain side of the specimens.

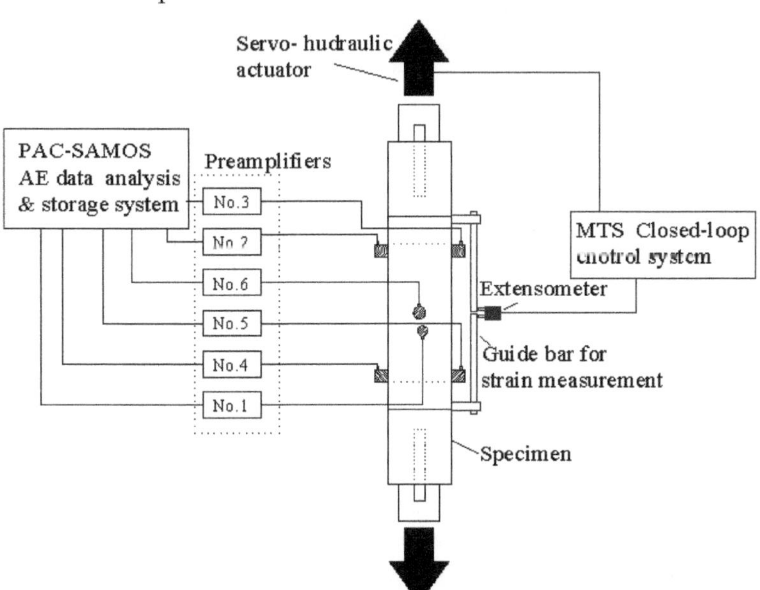

Fig. 2. Schematic diagram of the experimental set-up

3. Acoustic emission analysis

3.1 Acoustic emission parameters

The acoustic emission signal must be parameterized before analysis. The detection and measurement of an AE signal on a channel is called hit[4]. The AE hit parameters concerned in this paper are shown in Fig. 3. In the parameters, amplitude means the maximum AE signal excursion during an AE hit, count means the AE signal excursions over the AE threshold, the duration is defined as the time from the first threshold crossing to the end of the last threshold crossing of the AE signal from the AE threshold. Theoretical research[5, 6] showed that amplitude and duration are the best parameters to characterize the AE source mechanism. these two parameters are used in this paper, and accumulated AE hit number is also used to analyze the AE activity.

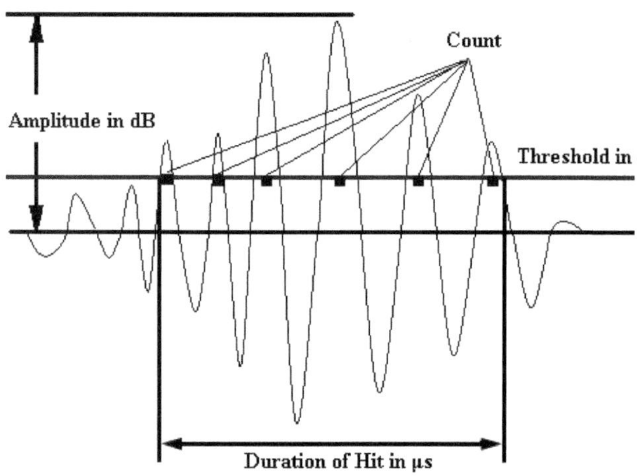

Fig. 3. AE signal parameters

3.2 AE source location

The AE source location can be deduced based on the differences among the arrival time at each transducers, In order to get a better estimate of the first arrival time, the AE signals are smoothed by using a band pass filter. The range of the band-pass filter is from 20 to 400 kHz. For a two-dimensional case, when the wave velocity is known, the calculated and observed value of the difference between the distance from transducer 1 to the AE source and the distance from transducer i to the AE source can be written as $(d1-di)$ and $\Delta t1i \cdot Vp$, respectively. The error between them can be written as:

$$e_{1i} = d_1 - d_i - \Delta t_{1i} V_p \tag{1}$$

Where: $d_1 = \sqrt{(x_e - x_1)^2 + (y_e - y_1)^2}$ is the distance from transducer 1 to the AE source; $d_i = \sqrt{(x_e - x_i)^2 + (y_e - y_i)^2}$ is the distance from transducer i to the AE source; xe, ye are the coordinates of the AE source to be determined; xi, yi (i=1, 2……n) are the coordinates of transducers; Δt1i is the difference of arrival time between transducer 1 and transducer i (i=2, 3, …, n); Vp represents the P-wave velocity.

In an ideal situation, the error expression e1i should be zero. However, in an actual experiment, errors always exist. To minimize the test error, in general, the idea of least square is introduced. The summation of individual errors can be expressed as Eq. (2), then different optimization methods can be employed to calculate the value of x, y and z.

$$e = \sum_{i=2}^{n} (e_{1i})^2 = \sum_{i=2}^{n} (d_1 - d_i - \Delta t_{1i} V_p)^2 \tag{2}$$

A new exhaust based method[7]is used in the paper: first, location zone is divided into mesh, second, the values of objective function of Eq. (2) in each node of the mesh are calculated and compared, then the coordinates are determined when the minimum value is found. The method evaluated the result by comparing the relative value of residual and needn't to set the initial value of iteration, so this method can improve the data utilization ratio.

4. Result and analysis

Tensile strength and the time at which the AE signal for each specimen appears are shown in Tab. 2.

Series	AE detected at	Tensile strength
DT01	1.19MPa (75%)	1.59MPa
DT02	1.67MPa (65%)	2.56MPa
DT03	0.96MPa (33%)	2.92MPa
DT04	0.64MPa (43%)	1.50MPa

Table 2. Tensile strength for concrete specimens

4.1 The description for each loading stages

The strain-stress curve and the accumulated AE hit number are shown in Fig. 4. Three continuous processes are used to describe the different stages of the development of microcracks and mechanical properties[8, 9]:

When the mechanical behaviour of the specimen corresponding to the different stages of cracking is considered, the following [10] is found if the test is performed at an imposed displacement rate (Fig. 4.)

4.1.1 Stage 1(O-A in Fig. 4): Random distribution of microcracks

Before the load reached 1.19MPa (about 72% ultimate stress), the axial stress increased successively with the strain, and is approximately proportional to the displacement, which is the elastic stage of concrete under Uniaxial tension. During this stage, the initial microcracks (due to the shrinkage of the cement paste blocked by the aggregates) are spread, and new microcracks form in the zones of lowest strength (paste-aggregate bond, for example) or in those zones where the local tensile stresses are higher (this is related to the fact that the aggregates and the cement paste have different mechanical characteristics). Because the AE activities are weak, the transducers on the specimen do not detect any AE signals during this stage.

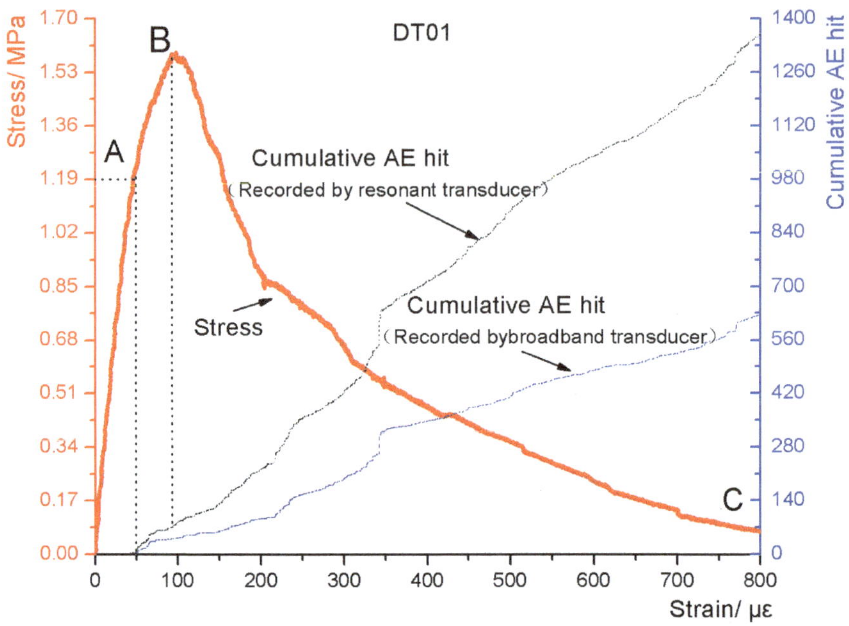

Fig. 4. Tensile stress and accumulated AE hit number versus strain

4.1.2 Stage 2(A-B in Fig. 4): Localization of microcracking

The elastic-plasticity deformation stage is from 1.19MPa to ultimate stress, and the stress-strain curve becomes a little nonlinear, AE signals higher than the threshold are generated in the further development process of microcracks, then recorded by AE system. During this process, the microcracks generated in the stage 1 developed further and macrocracks appeared, and at the same time, the micro-cracks localized phenomenon appeared, the emergence of macrocracks are the beginning of the third stage.

4.1.3 Stage 3 (B-C in Figure 4): Expansion of the macrocracks

When the tensile strength is realized , concrete specimen do not become faulty immediately, but cracking surface gradually been opened (though the naked eye may not be able to observe the cracks), with the increase of deformation the specimen still has some capacity, but the capacity declined rapidly, the stiffness of concrete specimen became smaller gradually and showed "softening" characteristics. At this stage, a stage before the macrocracks at this time continue to spread, performance for the cumulative impact of emissions continued to increase, leading to the loss of specimen bearing capacity.

In addition, as showed in fig. 4, when the strain reached about 350με, the curve of AE hit cumulative number showed suddenly increased for specimen DT01, this phenomenon may be related to the change of internal micro-structure of concrete specimen[10], and may also related to the localization of a single critical microcrack[11]. Sudden change in the stress-strain curves can not be seen clearly, its causes may be part of the micro-cracks in the location and not appear in the extended-distance side of the specimen surface. In contrast with that in the specimen DT03, in the stress-strain curve shortly after the peak, there have also been the same phenomenon, and in the stress-strain curve has appeared on a sudden decline, due to the use of test closed-loop control technology, making micro-cracks in the unstable expansion to a certain degree of inhibition (otherwise for the sudden rupture, leading to the loss of specimens moment carrying capacity).

Thus, through monitoring the change of cumulative curve of AE hit, forecast and describe the development of instable microcrack can be achieved in real-time and high sensitivity.

4.2 AE Characteristics during unloading-reloading process in soften stage

During the test, performing limited times of unloading-reloading processes or changing the deformation rate will not affect envelope shape of the measured stress-strain curve, and the envelope of curve is identify to the curve which obtained from monotonous loading[12]. The unloading-reloading is performed at 3 and 4 points in the softening portion of the stress-strain curve in DT03 and DT04, respectively. It can be seen from Fig. 5, AE hit cumulative curve tend to be a horizon line during the processes of unloading-reloading, showed that there are no obvious AE activities in this processes. The relationship between AE count and axial stress are shown in fig. 6, it is also can be seen clearly that almost not AE activities are found in that processes. The above analysis is described from a qualitative point of view, the AE characteristics in the unloading-reloading processes will be analyzed referred to the stress and strain in the following parts.

4.2.1 Analyzed referring to stress

The Kaiser effect is described as the absence of detectable acoustic emission at a fixed sensitivity level, until previously applied stress levels are exceeded. On the contrary, the presence of acoustic emission, which can be detected at a fixed predetermined sensitivity level at stress levels below those previously applied, which phenomenon is called Felicity effect[4]. During softening stage, it is impossible to load more than previously stress level (tensile strength), therefore, strictly speaking, does not exist Kaiser effect.

The stress value (F_u) corresponding to start further unloading in each loading cycles, and the stress value (F_r) corresponding to the time when AE signals begin to be detectable, are summarized in Tab. 3. It is can be found that, the values of F_r / F_u are less than 1 (Obviously, the ratio of F_r / tensile strength is also less than 1), so the Felicity effect also exists.

Specimen No.	Cycle	F_u(MPa)	F_r (MPa)	F_r/ F_u	ε_u($\mu\varepsilon$)	ε_r($\mu\varepsilon$)	ε_r/ ε_u
DT02	1th	1.152	1.015	0.881	154.234	153.325	0.994
DT03	1th	1.376	1.327	0.965	220.737	224.056	1.015
	2nd	1.266	1.192	0.941	252.109	251.810	0.999
	3rd	1.121	0.928	0.828	296.070	275.287	0.930
DT04	1th	1.184	1.135	0.959	177.858	176.646	0.993
	2nd	1.138	1.068	0.938	190.417	188.982	0.992
	3rd	1.062	1.032	0.972	216.389	218.438	1.009
	4th	0.940	0.860	0.914	254.625	244.775	0.961

Table 3. Acoustic emission Characteristics at the Unloading-reloading process in the softening portion

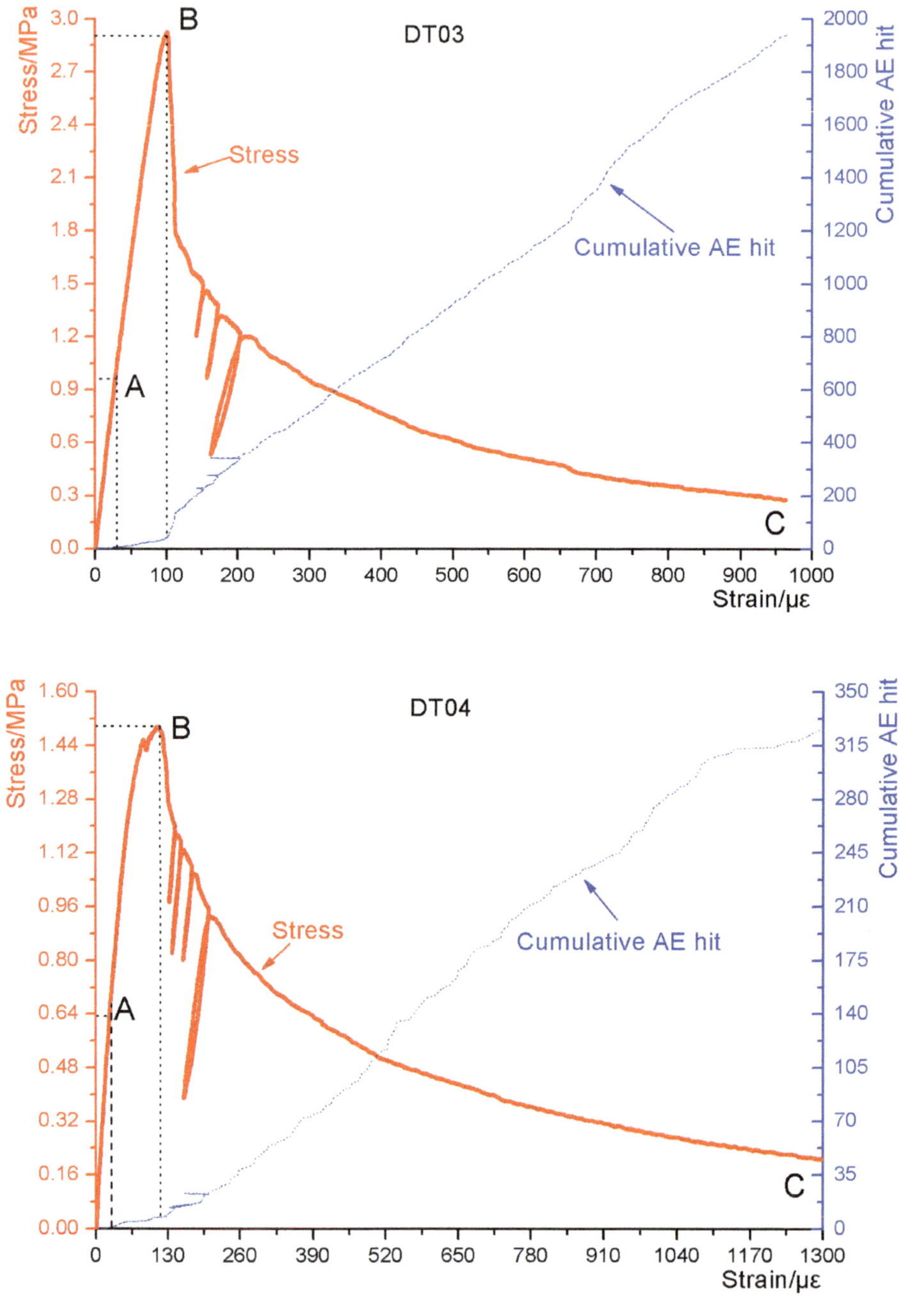

Fig. 5. Tensile stress and accumulated AE hit number versus strain

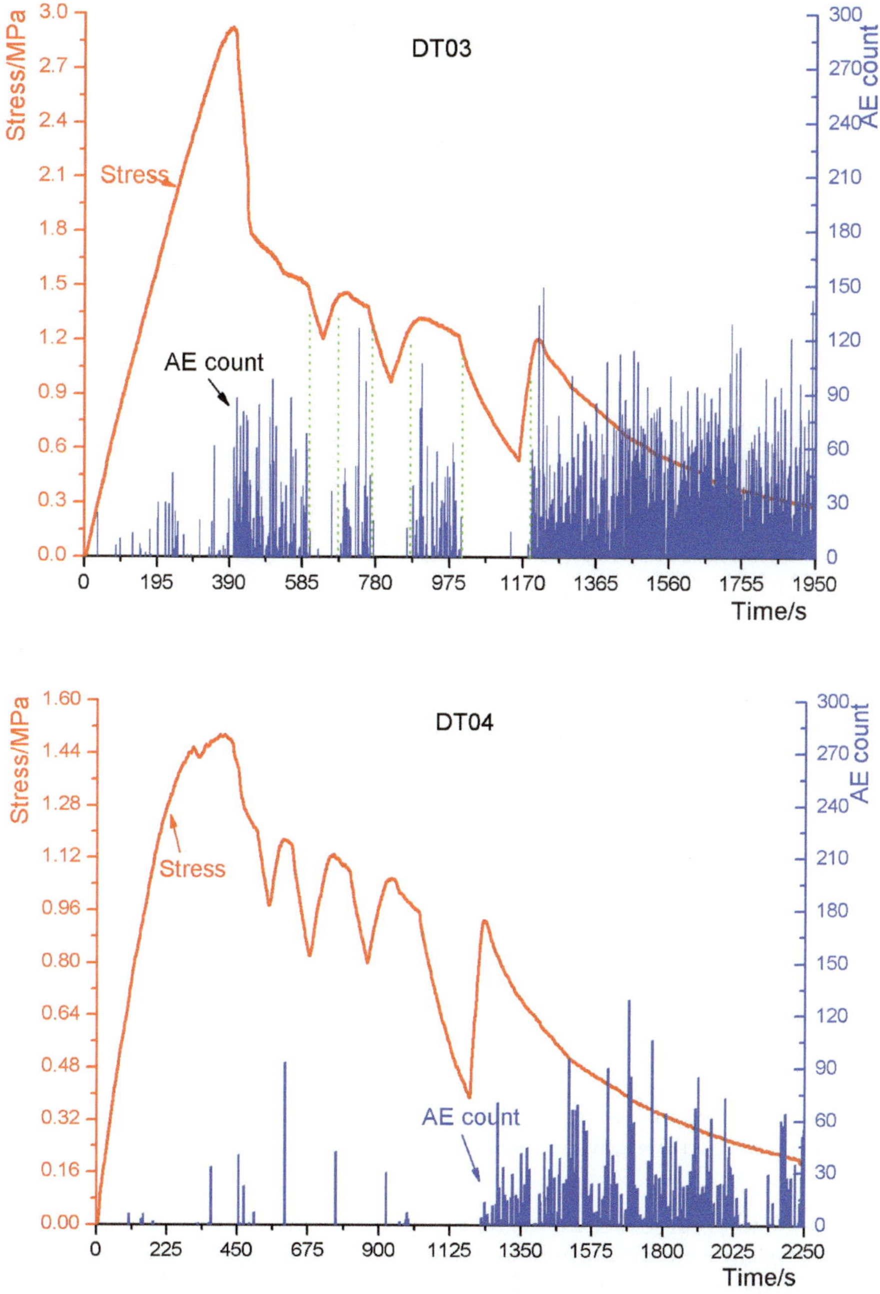

Fig. 6. Tensile stress and AE count number versus time

4.2.2 Analyzed referring to strain (Deformation)

It is can be seen from Figure 7, in the softening phase of the process of loading. The strain increased with the decreased loading (tensile stress), while the strain decreased when further unloading. The strain value (ε_u) corresponding to start further unloading in each loading cycles, and the strain value (ε_r) corresponding to the time when AE signals begin to be detectable, are summarized in Tab. 3. It can be found that, the value of $\varepsilon_r/\varepsilon_u$ during each cycles are around 1, and the value is decline in the last cycle. Hence the acoustic emission can still memory the maximum strain in the loading history, at the same time, the values of $\varepsilon_r/\varepsilon_u$ is larger than the value of F_r / F_u in each load-unloading cycle, therefore, the memory effect of strain of acoustic emission is better than the stress, it is maybe mainly due to the closed-loop servo system used the deformation from the specimens as a loading control signals.

The above properties during the soften stage maybe related to the stress states of concrete specimens: the cracks in the concrete are I-type, and they are mainly vertical to the loading direction. When entering the soften stage, the localized macro-crack has already formed. Then the single macro crack widens in concrete specimens, and there are no mutually dislocation activities. Therefore, there is no obvious AE phenomenon.

4.3 Spatial distribution of AE activities

The mechanical behavior before crack localization, which coincides with the peak stress, can be regarded as intrinsic to the material. That is, as the cracks are small with respect to the volume of the specimen, statistically homogeneous stresses and strains can be defined. Fig. 8 confirms that before the axial stress reaches the limits of stress (before 390 s), the cumulative AE hit curve described in different AE transducers located in different positions of the specimens have no obvious differences.

After localization, since these cracks are no longer assumed to be small with respect to the volume of the specimen, statistically-homogeneous stresses and strains cannot be defined. Therefore, the only relation is between the force and the displacement, which are both global parameters. The behavior is then structural, and this can also be confirmed in Fig. 8. In the soften stage of stress-strain curve, the AE hit number described by the transducer closest to the visible crack (visible crack can be seen in fig. 9) has maximum value. In addition, the locations of AE events calculated by the location method mentioned above and the position of visible crack are shown in Fig. 9. The AE events appear during the post peak stage of soften stage. It is found that most of the AE events are located close to the actual position of the visible crack. Same as the research results from Zongjin Li[13], the obvious microcrack localization phenomenon is observed. This phenomenon can be demonstrated through AE hit cumulative curve in the sensors located in different positions and AE source location technology.

From the stress-strain curves shown in Fig. 10, we can see that the final location of the crack is out of the resistance strain gauge in shorter length, but the crack is located in the range of extensometer. Therefore, the curves expressed by the two measurements are different, and this result can further explain the microcrack localization phenomenon.

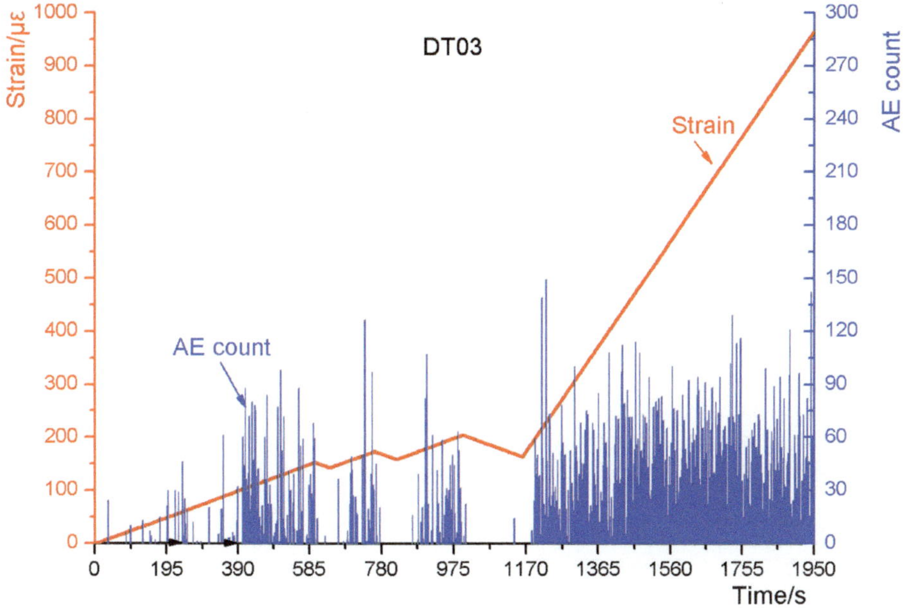

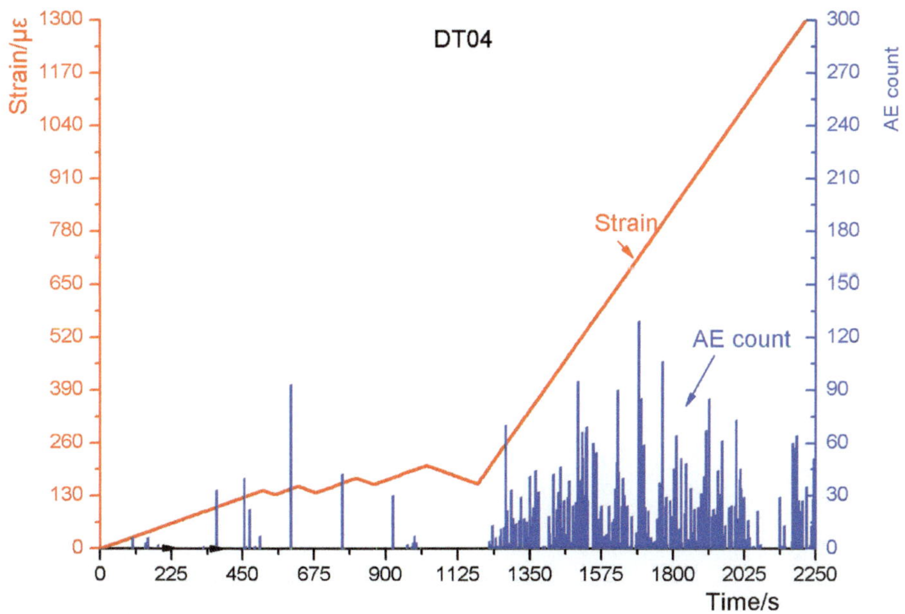

Fig. 7. Tensile strain and AE count number versus time

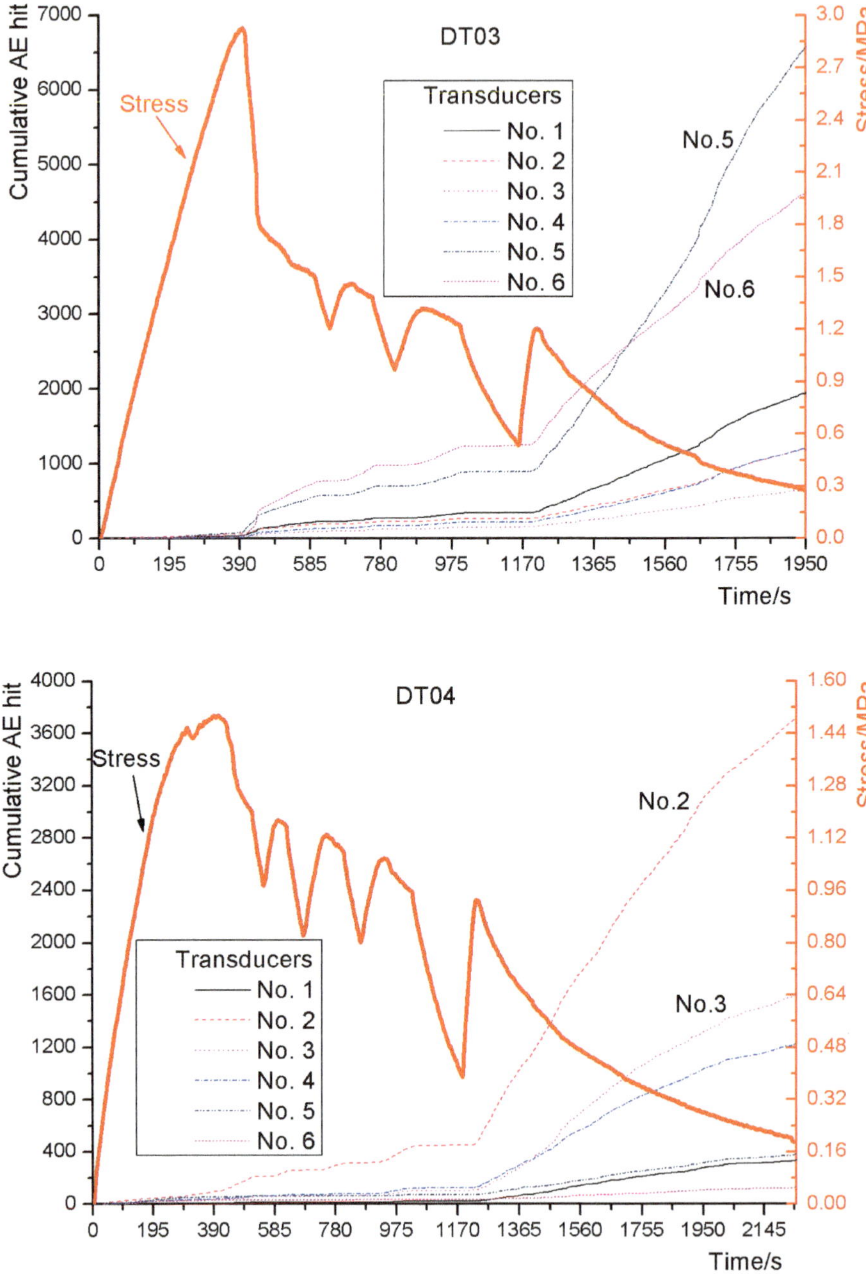

Fig. 8. Tensile stress and accumulated AE hit number versus time for different transducers

To summarize, the material switch to structural properties of concrete under Uniaxial tension can be understood as[9] follows: if the switch is thus performed with the macrocracks after localization, the relation will depend significantly on the positions of the macrocracks in the specimen, which is typical structural behavior.

4.4 AE parameter analysis

Compared to the previous stage, it can be seen from Fig. 11 to Fig. 12 that in the soften stage of concrete specimen under Uniaxial tension, as the crack further opens, AE hits in higher amplitude (greater than 60 dB) and longer duration (more than 600 μ s) are generated. This indicates that the energy released from microcracks tend to increase during the soften stage.

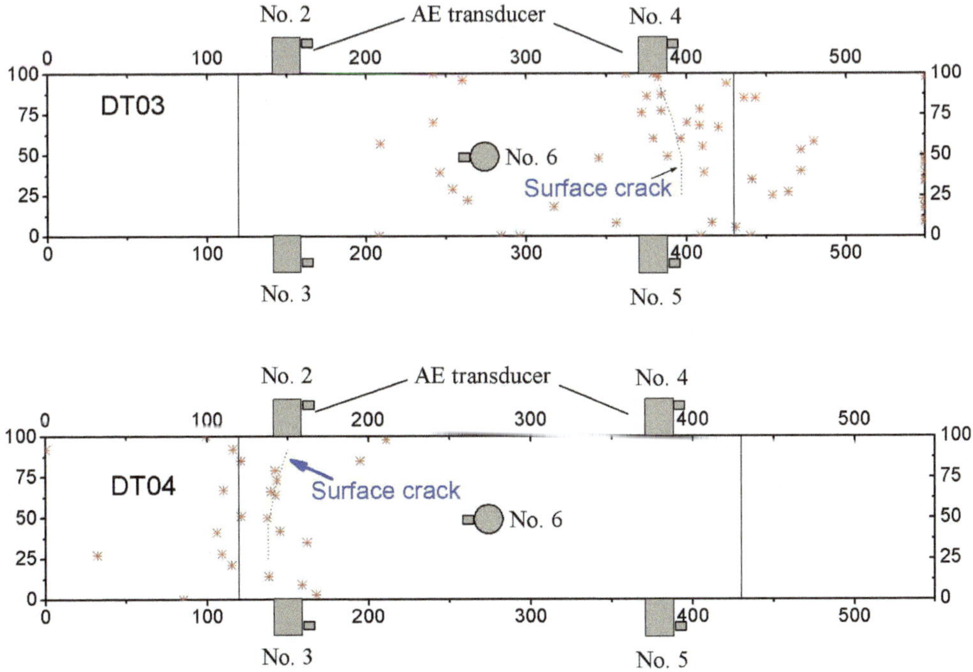

Fig. 9. AE source locations

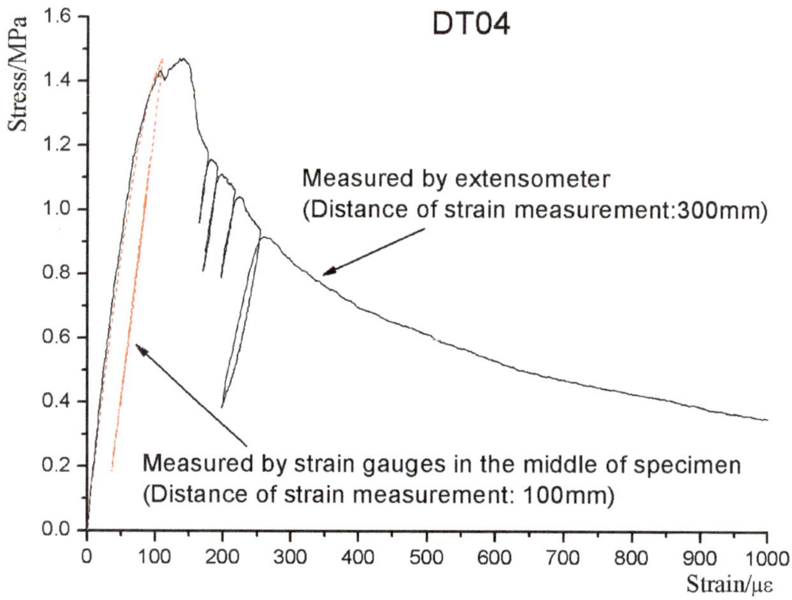

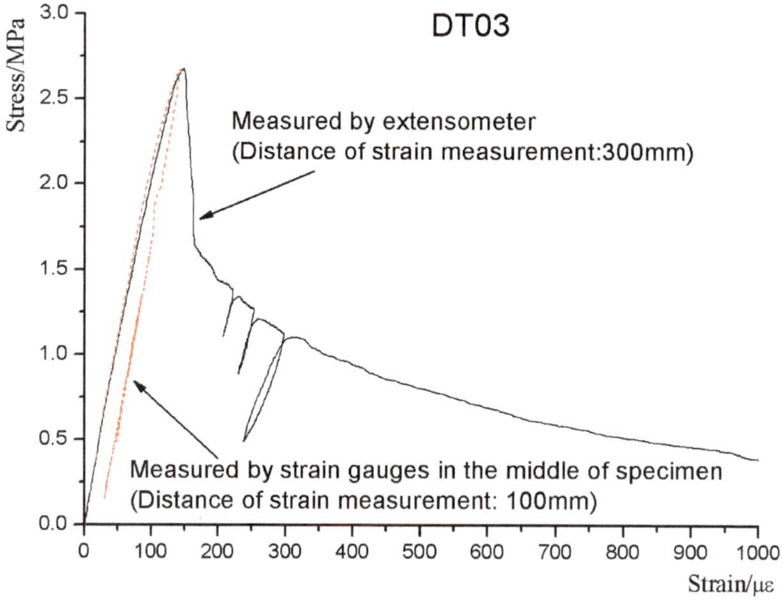

Fig. 10. Stress-strain relationships measured by extensometer and strain gauges

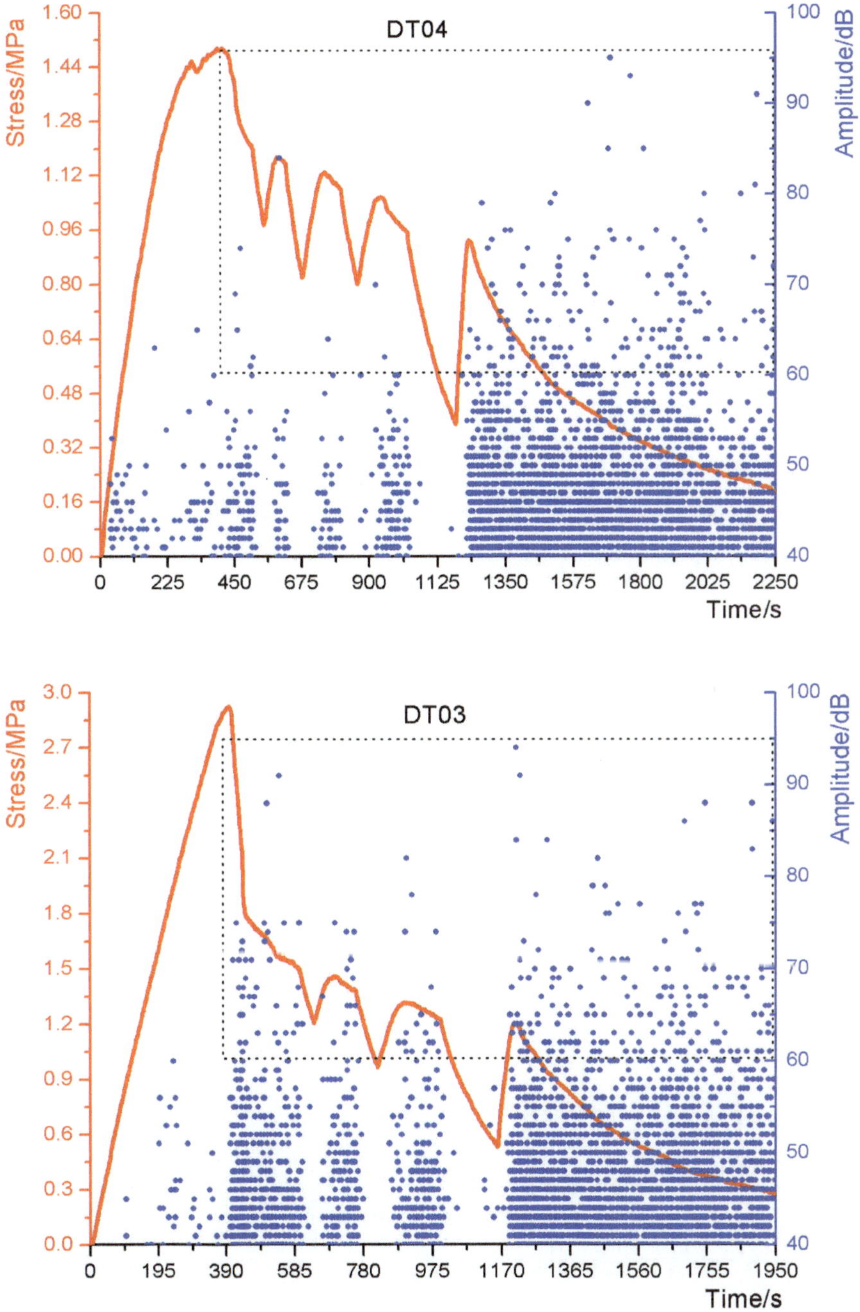

Fig. 11. Tensile stress and amplitude versus time

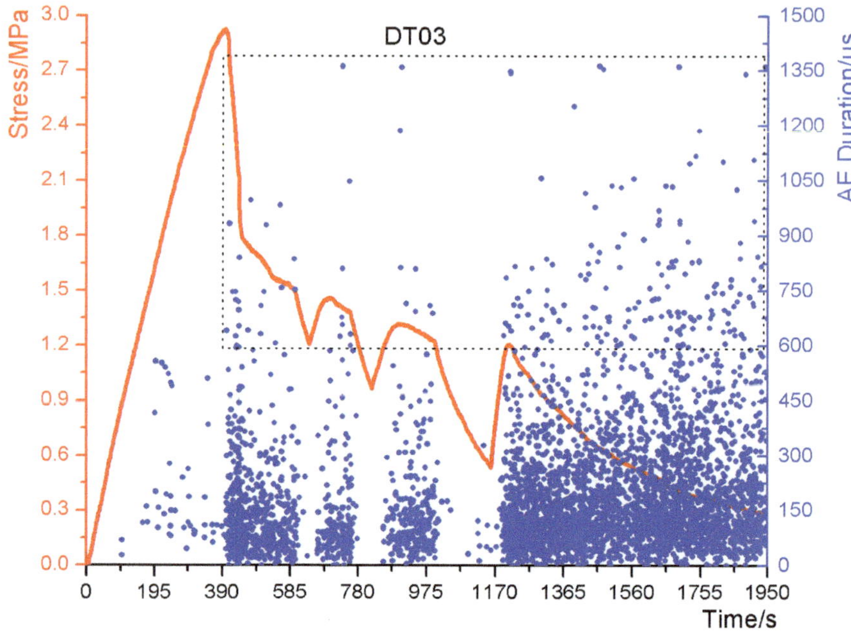

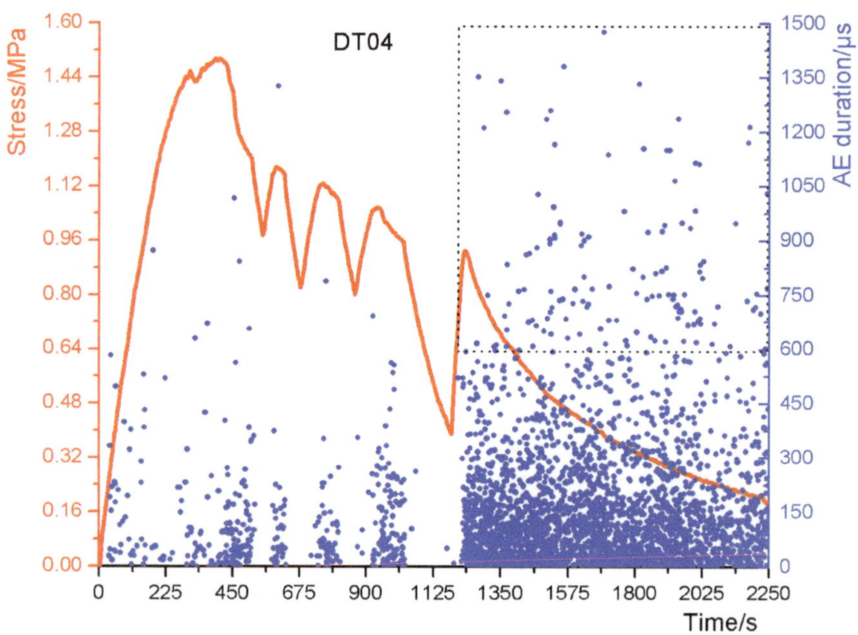

Fig. 12. Tensile stress and duration versus time

5. Conclusions

Through acoustic emission signal analysis of concrete specimens under Uniaxial tension, the AE characteristics of concrete in the whole process of tension are obtained. The conclusions are as follows:

1. Although the width of cracks are too small to be seen by naked eyes, and the specimens are not broken into two parts, the process of developing microcracks can be displayed and demonstrated in real time by AE technology.
2. Good relationship between acoustic emission hit rate and microcracks development in the concrete is found. Therefore, analysis of the timing characteristics of AE activities in concrete is helpful to assess the damage level.
3. During the deterioration stage of soften, unloading-reloading process cannot exceed the tensile strength of the specimen. The Kaiser effect, therefore, does not exist while Felicity effect exists.
4. The maximum strain can still be characterized by AE during the soften process of concrete, and the accuracy is higher than that of the stress memory effect.
5. Before the stage of microcrack localization, AE activities described by the sensors in different position of the specimens are close. No significant differences are observed. During the process of localization of microcracks to the formation of macrocracks, AE activity described by sensors is significantly different, materials-structural characteristics of concrete under Uniaxial tension are further confirmed.
6. AE activities from different sensors and the localization results of acoustic emission source are consistent with the actual distribution of visible crack.
7. The energy released from microcracks activities tend to increase during the soften stage, and AE hits in higher amplitude and longer duration are detected.

6. Acknowledgments

The authors appreciate the support of the National Natural Science Foundation of China (Grant No. 51178162 and Grant No. 51009058).

7. References

[1] Yang Weizhong, Wang Bo. A stochastic damage constitutional model and its application to uniaxial tensile behavior of concrete[J]. Industrial Construction. 2004, 34(10): 50-52, 43(in Chinese).

[2] Qian Jueshi, Wang Zhi, Luo Hui. Determination of strain softening relationship of concrete [J]. Journal of dalian University of technology. 1997, 37(1): S61-S66(in Chinese).

[3] Z. Li, S. M. Kulkarni, S. P. Shah. New Test Method for Obtaining Softening Response of Unnotched Concrete Specimen Under Uniaxial Tension [J]. Experimental Mechanics, 1993, 33(3): 181-188.

[4] ASTM E 1316-99a, Standard terminology for nondestructive examinations [S].

[5] M A Majeed, C R L Murthy. On using peak amplitude and rise time for AE source [J]. Sadhana. 2002, 27(3): 295-307.

[6] K. Malen, L. Bolin. A Theoretical estimate of acoustic-emission stress amplitudes [J]. phys. stat. sol. (b). 1974, 61(2): 637-545.

[7] WANG Yan, WU Sheng-xing, ZHOU Ji-kai, et. al. . 3D Acoustic Emission Source Location Based on Exhaust Method [J]. Nondestructive Testing. 2008, 30( ): ??-??. article in press. (in Chinese).
[8] Rossi, P., Wu, X.. A probabilistic model for material behaviour analysis and appraisement of the concrete structures [J]. Magazine of Concrete Research, 1992, 44 (161): 271-280.
[9] P. Rossi, F. Toutlemonde. Effect of loading rate on the tensile behaviour of concrete: description of the physical mechanisms [J]. Materials and Structures, 1996, 29: 116-118.
[10] Landis E.N. Micro-macro fracture relationships and acoustic emissions in concrete[J]. Construction and Building Materials. 1999, 13(1): 65-72.
[11] Li Z, Shah SP. Microcracking in concrete under uniaxial tension[J].ACI Materials Journal, 1994, 91: 372-381.
[12] Lu Yiyan He Shaoxi. Experimental Research on the Complete Stress-strain Curve with Hoop Tension for SFRC [J]. Journal of wuhan university of hydraulic and electric engineering. 1995, 28 (3): 285-292(in Chinese).
[13] Zongjin Li, Surendra P. Shah. Localization of Microcracking in Concrete under Uniaxial Tension [J]. ACI Materials Journal, 1994, 91(4): 372-381.

12

Analysis of Acoustic Emission on White-Coated Paperboard During a Wedge Cutting Process

Yasushi Fukuzawa and Shigeru Nagasawa
Department of Mechanical Engineering, Nagaoka University of Technology, Nagaoka, Japan

1. Introduction

In our social life, paperboard and its finished products have been become widespread as any recycled materials. The structure of paperboard is composed of multiple plies (normally, lamination structure of 3~8 layers) with the hydrogen-bonded natural fibers which have the vertically beaten shape. The material property of paperboard has an anisotropic characteristic and may be understood as orthotropic-nonlinear elasticity [1].

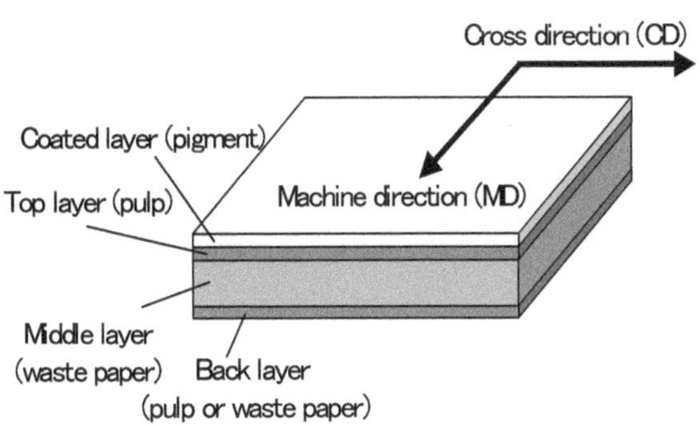

Fig. 1. Paperboard structure.

A typical structure of white-coated paperboard is illustrated in **Fig.1**. A clay pigment layer is coated on the front surface to improve the printing quality. The top layer and back layer tend to be relatively stronger and white, compared with the middle layer. The bonding strength of a ply interface is usually a little weaker than the yielding stress of considered plies.

Die-cutting machine (such as a flat-bed type) consists of a die board, in which the cutting rules and creasing rules are embedded, and a face plate (called as the counter face plate). The flat-bed cutting machine (called as the platen) is widely used in the packaging and printing industry, for processing paperboards, labels, laminated resin sheets, ductile metal film and so on. As other machine types, there are several rotary die cutters and combination (cylinder) type of a half-rotary die with a flat-bed in the packaging industry.

In this book, a simple crank cutting machine as the flat-bed type was considered for the authors' experimental analysis. **Fig.2** shows a schematic of experimental apparatus, which was designed and developed as a small size machine for the purpose of continuous cutting test [2]. A reciprocal motion is used for indentation of a center-bevelled blade (symmetric wedge) into and cutting off the paperboard. The cutting blade is fixed on the upper holder and reciprocally moves with a certain speed. The cutting force is measured by using four load cells mounted underneath the lower base table. **Fig. 3** illustrates a quarter-stroke motion of a cutting blade of the crank machine. At the bottom-dead position, the blade tip is slightly crushed. In this situation, the unevenness of the blade height causes an eccentric tip crushing and it is needed to detect such state of pressure unbalance using a non-destructive inspection method. In this case, the blade is moved in the upward/downward direction resulting from the rotation of the crank mechanism, which has a speed N_C rpm and the length of an eccentric arm e=25mm. The blade tip continues to penetrate the paperboard until it reaches the bottom-dead point (BDP) (Fig. 3 (b,c)) where the paperboard is separated before returning to the position of Fig. 3(d).

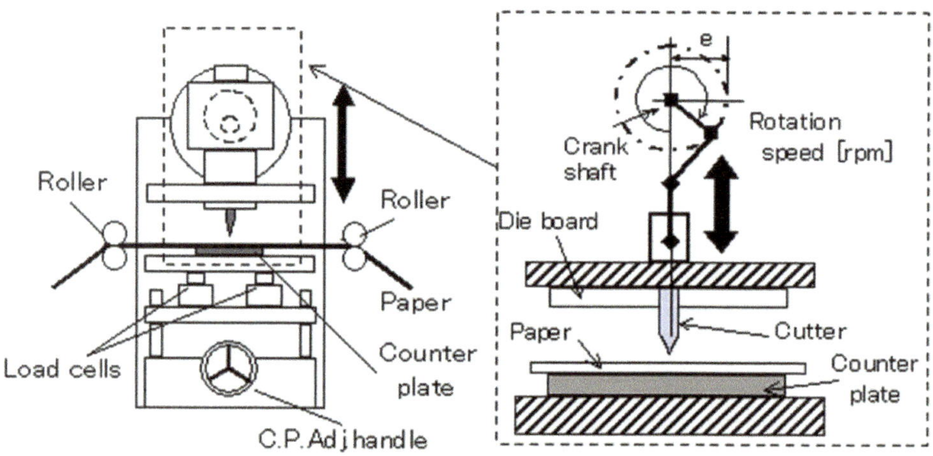

Fig. 2. Schematic outline of die cutting system for paperboard.

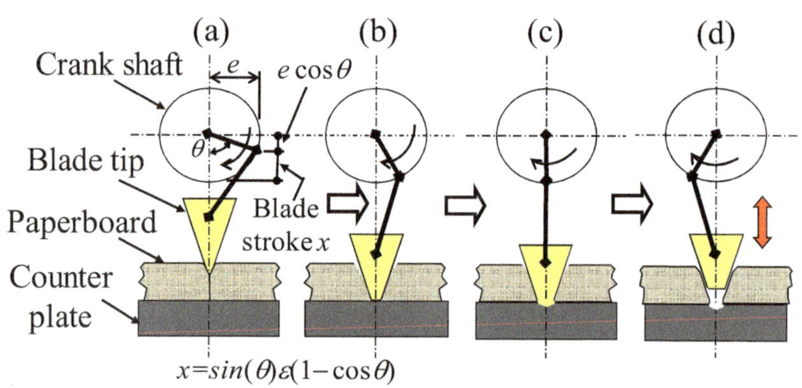

Fig. 3. Mechanic of paperboard cutting on the crank motion.

The crushing of blade tip occurs at the BDP during paperboard separation and this is the position where the severe crushing and eccentric cutting pressure of blade tip are reduced using an expert's adjustment technique (shimming by one's hand working).

From aspects of energy-saving and optimal quality on the material processing, the non-destructive diagnosis with the ageing of cutting tools and the mechanical condition of specified paperboard is important for corresponded companies and our social life.

2. Measurement methods and its object

In the die cutting process of paperboards that used for packaging containers, the following items are important for quality control: (1) adjustment of cutter clearance, shimming of die board; (2) restriction of occurrence of dust chips or string-like chips; (3) prediction of dynamic cutting state with respect to the ageing of cutting tools. In many converting plants of packaging containers, these assignments are empirically solved and maintained by operators' senses. Often it depends on the variation of audible sound from the cutting machine. For increasing productivity and reducing operator's task, any kind of automatic technique for detecting cutting condition is required.

In the author's research, several diagnoses of a tool condition have been investigated. The ageing of a cutting edge nose was acoustically estimated by measuring sound wave during the process of paperboard cutting [3]. In this chapter, a correlation between the blade tip thickness and a solid interior elastic wave that occurs in the cutting tools (the blade and the counter face plate), or the rigid body vibration that occurs in the perimeter where the blade edge collides, is explained by using Acoustic Emission (AE) method.

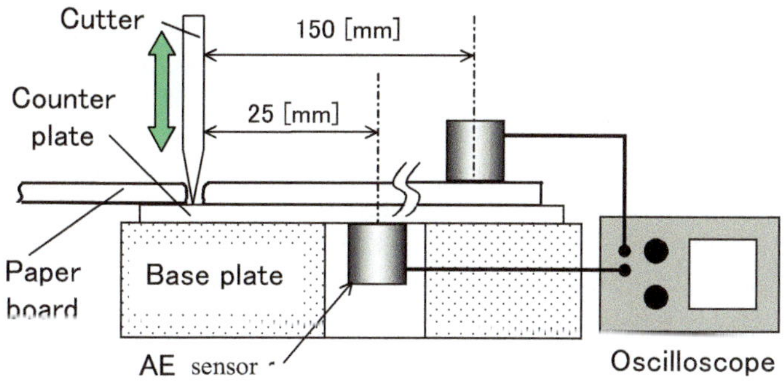

Fig. 4. Detection system of AE signals on the die cutting

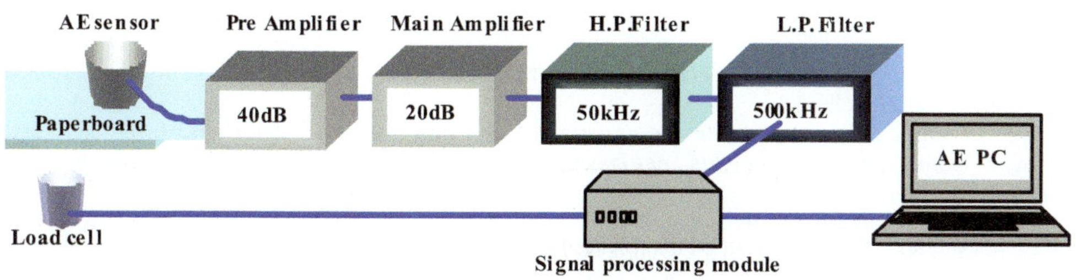

Fig. 5. Measurement and analysis system using AE sensor and load cell

The AE signal was detected at the backside of counter face plate as shown in **Fig. 4**. AE signals measured by an AE transducer were processed as displayed in **Fig. 5**. The amplitude of AE signals were used here. The AE transducers [4] that were a resonant type of 220 kHz (effective frequency range: 1-5 MHz, being products of Physical Acoustics Corporation), were set behind of counter face plate (distance from the cutting edge end, L_a=25 mm) and on the paperboard (distance from the cutting edge end, L_a'=150 mm). The provided voltage signals were recorded as the function parameter of AE. The waveforms of AE signals were detected in a digital oscilloscope and analyzed by using a computer with FFT software.

In order to describe the correlation between the cutting process of paperboard and generation of AE signals, the developed small size die cutting machine shown in Fig. 2 was used for this experiment [2]. The following experimental parameters are examined.

1. Material structure of paperboard: the cutting direction (angle) ϕ =90° (across to MD), 45° and 0° (parallel to MD) to Machine Direction (MD) of paperboard, the thickness direction concerning the coated layer.
2. Machining speed: the rotation speed N_C rpm for the crank arm length of e= 25 mm. The rotation speed was chosen as 5, 10, 20, 30 and 40 rpm for the cutting test.
3. Shape and width of blade tip: a single-straight wedge cutter, the tip thickness t_C μm of which has the apex angle of α.

The specification of paperboard, that of cutting blade, and that of counter face plate were chosen as follows, owing to their popularities:

- Paperboard: White-coated board, of which the nominal basis weight was ρ= 350 g/m^2, the thickness was t= 0.44 mm, the width of web sheet was 100 mm, and the in-plane tensile strength in MD (the machine direction) was σ_B = 33 MPa (the breaking nominal strain of 1.85%, at Room env.: 42%RH, 279K). All the specimens were reserved and the cutting test was carried out at Room env.: the humidity 63%RH, the temperature 303K[1].
- Blade: Thomson's knife (JIS-G3311, SK5) of α=42°, the hardness of tip/core were 680/465 VHN, the initial (virgin) tip thickness of 5μm was modified as two kinds of blunt wedges. One is a naturally crushed tip of t_C= 18 μm, which was dynamically (N_C=80rpm) subjected to 100 times of the maximum line force of f= 27.5 kN/m[2]. Another is artificially made as a trapezoidal shape of t_C=19μm using emery papers. **Fig.6** illustrates two kinds of tip profile. The length of blade was 80 mm, the thickness of core body was 0.71 mm[3]
- Counter face plate: Hardened stainless steel plate (JIS-G4304/5, SUS630) of 1.5mm thickness, which was a rectangular form of 180mm length and 140 mm width. Its surface hardness was 510 VHN.

---

[1] Hopefully, all of reservation of specimens and cutting tests should be considered at the standard room environment, 50%RH, 296K.

[2] After 100 times of punching onto the paperboard, the burrs of crushed tip tend to be sufficiently removed. Namely, round-edge ends are expected to be formed on the crushed tip without any burrs. See the reference [2].

[3] There are several standard knives well known in the world, the tip angle of which is symmetrically α = 30, 42, 53, and 60°. The 42° knives are well used for separation of coated paperboard in Japan. There are the woven-tip type, the side wedge (asymmetric wedge) and the two-line wedge cutter and so on for various purposes.

The cutting test was carried out across to the MD (machine direction) with the direction angle ϕ=0, 45, 90°. The cutting-line force f kN/m and the amplitude of AE signal a mV were measured with respect to the elapsed time t_{ep}.

When the effect of blade tip shape (naturally crushed, artificially mended) was compared with respect to the cutting load response, the cutting direction was only chosen as ϕ= 0° for N_C= 5, 10, 20 rpm[4]. 10 pieces of specimens were prepared and inspected for each cutting condition. The contact condition between the blade tip and the counter face plate was considered as non-lubricant.

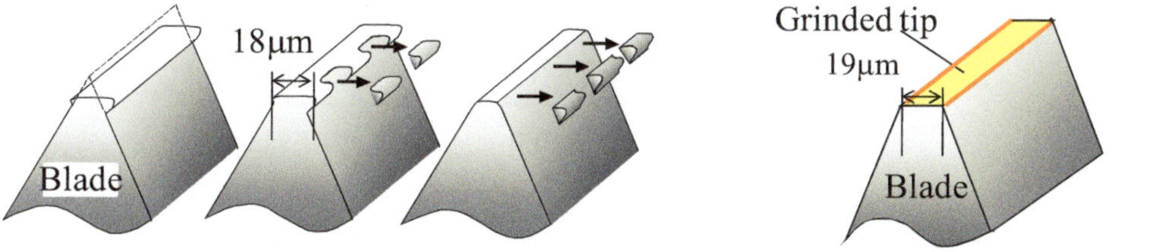

(a) Variance of edge end by continuous cutting (b) Edge end of mended blade

Fig. 6. Preparation of crushed blade tip

Regarding the transient response of cutting force and the AE signal in the cutting duration, the naturally crushed blade was used with the cutting attitude of ϕ=0° when the rotation speed was chosen as N_C = 5, 10, 20, 30, 40 rpm.

3. Outline of cutting characteristics and measurement principle

Before describing the correlation between the AE signal and the cutting line force when the cutting speed is varied for the two kinds of blade tip profile, the relationship between the cutting line force and the modified blade tip thickness should be explained. In a case of the cutting direction ϕ=90°, the basis weight ρ=350 g/m^2, the thickness t= 0.44 mm and the rotation speed N_C= 5 rpm, the authors investigated the cutting line force response [5][6]. The relationship between the cutting line force f and the normalized tip stroke x/t of crankshaft was shown in **Fig.7**.

The origin of the tip stroke x is here defined as x=0 with reference to the bottom dead point (BDP) of the crankshaft. When the blade is in the rising process, the sign of x is defined as positive: x>0, while the blade is in the descent process, the sign of x is defined as negative: x<0. If the blade tip is fairly crushed, the cutting line force response f-x/t has usually two-peaked points f_{C1}, f_{C2} during the cutting process.

For such the experimental relationship between the blade tip profile and the cutting load, there is a technical report written by Grebe and Hofer [7]. The authors have also studied about the effects of blade tip thickness on the cutting line force. The authors prepared artificially a trapezoidal bevel blade, of which the normalized tip thickness was t_C/t, by

[4] Due to a mass effect of lower base table, the dynamic high-speed resonance of implemented load cells was restricted. Therefore, the response of load cells should be mailnly discussed before the second peaked line force.

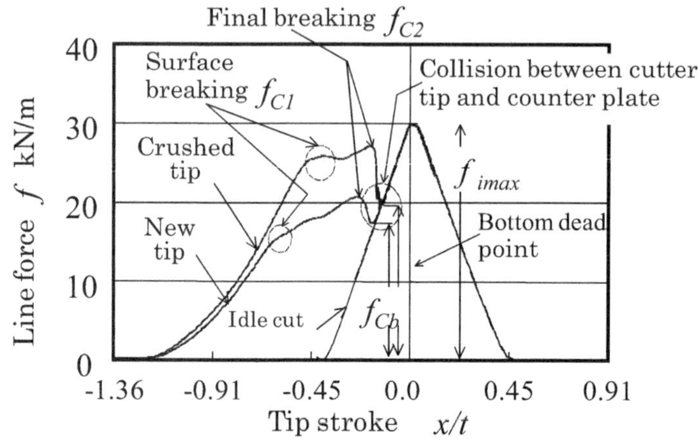

Fig. 7. Relationship between cutting line force and tip stroke (ϕ=90°, N_C=5 rpm)

grinding a new cutting blade. By using the trapezoidal bevel blades, many useful results were obtained [5]. From that work, it is found that trapezoidal bevel blade shows a similar cutting load response to the naturally wrecked blade tip by continuous cutting [6]. When the trapezoidal bevel blade was used for cutting the white-coated paperboard, the relationship between f_{C2} kN/m and t_C/t was approximated with Eq. (1).

The coefficients k_{fc2w} and f_{C20w} were decided from the cutting experiment. For an example, when the trapezoidal bevel blades of α= 42°, t_C= 10~55μm were used for cutting the white-coated paperboard of t= 0.44 mm (ρ=350 g/m^2), under N_C= 5rpm and ϕ=90°, the coefficients were obtained as k_{fC2w}=129 kN/m, f_{C20w}= 19 kN/m.

$$f_{C2} = k_{fC2w}\,(t_C/t) + f_{C20w} \tag{1}$$

Generally, these coefficients appear to be varied with the following conditions: (a) the friction coefficients between blade and paperboard, and/or that of counter face plate and paperboard; (b) the brand and the moisture (water) content of paperboard (mechanical properties); (c) the room humidity and the room temperature; (d) the cutting speed; (e) the shape of blade tip and its surface roughness.

Especially, as the room environment of (c) changes all the friction coefficients and the material properties of paperboard, a person who wants to estimate Eq. (1), must pay attention to maintain the room humidity and temperature for a long duration. Therefore, it is necessary to investigate and decide these coefficients in specified working environment.

At the local maximum (peaked) point f_{C2} in Fig. 6, the necking burst [5] of wedged zone occurs at the lower layers of paperboard, and the cutting line force suddenly decreases. Complete separation of paperboard occurs at the line force f_{Cb} when the blade tip contacts with the

[5] Strictly speaking, since the paperboard is not any continuum body, but a laminated composite material with paper fibers, the necking is generally invisible. However, as a certain kind of slide and pull-out deformation exists among the laminated fibers, we can understand it as an equivalent necking burst.

surface of the counter face plate. Since the blade is driven by the crank shaft, the blade tip moves to a rising process after reaching the BDP line force f_{imax}. Variance of crushed tip thickness t_C fundamentally depends on the magnitude of f_{imax}, while the cutting resistances f_{C1} and f_{C2} depend on t_C. So far, we should understand that f_{imax} indirectly affects the variance of f_{C1} and f_{C2}.

When the cutting line force of the paperboard passes through the peaked point f_{C2}, the blade tip collides on the surface of counter face plate. At this time, the cutting load drop expressed by Eq.(2) occurs in a short time.

$$\Delta f = f_{C2} - f_{Cb} \tag{2}$$

Over here, the dynamic energy, scattered by collision, is equal to the discharged elastic energy expressed by Eq.(3) [8].

$$\Delta E = (f_{C2}^2 - f_{Cb}^{\ 2})/(2k) \tag{3}$$

The stiffness coefficient k is defined as the gradient df/dx of the line force f by the tip stroke x, when the blade tip is elastically pressed against the surface of counter face plate.

Mechanical vibration occurs in the counter face plate by this blade collision, and hence it can be detected by AE transducer. Putting A to the maximum amplitude of detected signal wave, since the square of the A corresponds to the dynamic energy of vibration, the A is related to $\Delta E^{0.5}$ that is a function of the magnitude of f_{C2}. According to the author's experimental results [9][10], the relation of f_{C2} and A and the relation of t_C and A can be expressed as follows:

$$A/A_o = C_w\,(t_C/t) \tag{4}$$

$$f_{C2} = k_{fC2A}\,(A/A_o) + f_{C20A} \tag{5}$$

The coefficients C_w, k_{fC2A} and f_{C20A} are experimentally decided for the specified measuring condition. For an example, when several trapezoidal bevel blades of α=42°, t_C =10~55 μm were used to wedge the white-coated paperboard, mentioned above, the experimental coefficients were k_{fC2A} = 45.8 kN/m, f_{C20A}= 21 kN/m and C_w= 2.3. Over here, the rotation speed was N_C = 5rpm while the blade attitude was ϕ= 90°. A_o is a voltage base signal, derived from a standard evaluation method for AE: "When a protruded pencil lead, of diameter 5mm and length 3mm, is pressed on a target, the generated wave is assumed to have a standard signal strength."

In the experiment [9][10], a specified pencil lead was pressed on the counter face plate, instead of collision of the cutting blade. When the pencil lead broke, we evaluated the maximum value A from the voltage signal measured at the specified AE transducer.

Substituting A/A_o of Eq. (4) to Eq.(5), A/A_o can be erased. The relationship between f_{C2} and t_C/t is derived from this elimination, as shown in the Eq. (6):

$$f_{C2} = k_{fC2A}\,C_w\,(t_C/t) + f_{C20A} \tag{6}$$

Fig.8 shows the approximation of Eqs (1), (4), (6) and the experimental results. The Eqs (1) and (6) are matched overall. You can see how much is the expected accuracy of f_{C2} and t_C/t

from this figure. At least, by using a straight blade of length 80mm under a low speed cutting condition (N_C= 5 rpm), it is possible to estimate the tip thickness t_C/t from the AE signal A/A_o in a fairly good accuracy.

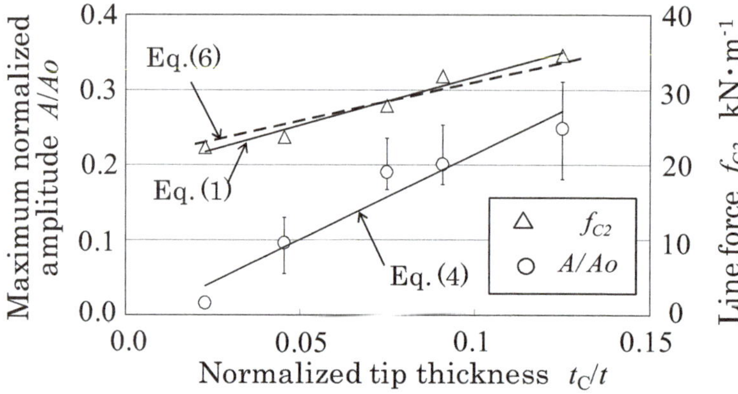

(ρ=350 g/m^2, ϕ=90°, N_C=5 rpm)

Fig. 8. Relationship among AE amplitude, breaking line force and tip thickness

Although we used AE transducer as the sensor device in this experiment, since the piezo-electric element principally detects mechanical vibration, various sensors based on the piezo-electric element can be apply to measure such the solid wave. At least, if AE transducer is successfully applied to the die cutting machine, the contact time of the blade tip with the paperboard and also the contact time of the blade tip with the counter plate can be detected.

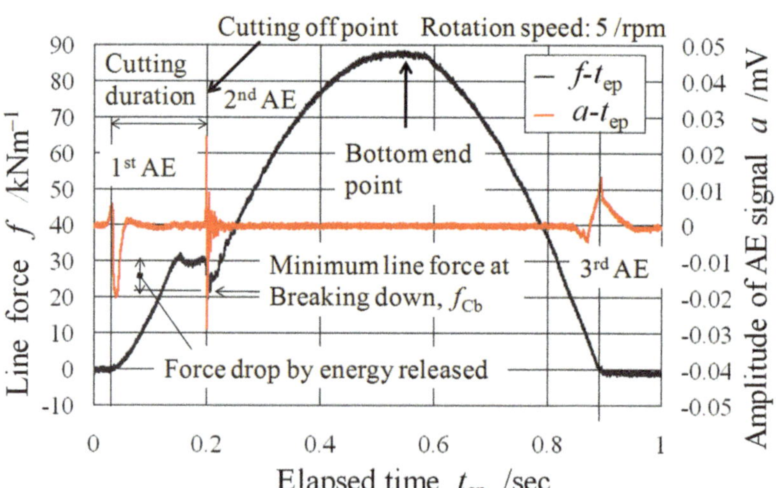

Fig. 9. Cutting line force and AE wave signal with elapsed time

At the setting place of Fig.4, the following AE signals were observed. Typical AE signals detected and its corresponded cutting line force f are shown in **Fig. 9.** In this case, AE signals could be detected at three positions: (1) **1st AE**, the first contact of blade tip on the surface of paperboard stacked on the counter face plate; (2) **2nd AE**, the cutting off point which is coincident to the second peaked load; (3) **3rd AE**, the final detached of blade tip from the counter face plate.

The second AE signal (2) which was detected near the cutting off point, which was caused by the collision between the cutting edge tip and the counter face plate.

Using the three positions (1),(2) and (3), the contact time of blade tip can be estimated [9]. From this detection of contact time of blade tip, the ageing of blade tip height or the time variance of contact with the blade and the counter face plate can be estimated. In other words, AE transducer can be used to diagnose the time variance of the detected signal against the angular position of the crankshaft of the pressing device.

As shown in Fig.8, the tip thickness t_C of crushed blade can be detected by the magnitude of the AE signal, and also the height change of the blade can be detected by observing the time difference of the generated signal. If these two matters are analyzed in the same time, then a highly reliable diagnose of cutting state is possible.

4. Results and discussions

4.1 Sheared profile and 2nd AE signal

Fig.10 shows the normalized amplitude of 2nd AE signal A/A_0 with the cutting speed N_C= 5,10 and 20 rpm for two cases of blade tip shape. Here, the cutting attitude was chosen as ϕ=0°. According to the dynamic load response during cutting process [8], the ratio of dynamic amplitude by static one is theoretically estimated as 1.41. The 2nd AE amplitude at 20 rpm appeared to almost reach such the dynamic state, compared with that at 5 rpm.

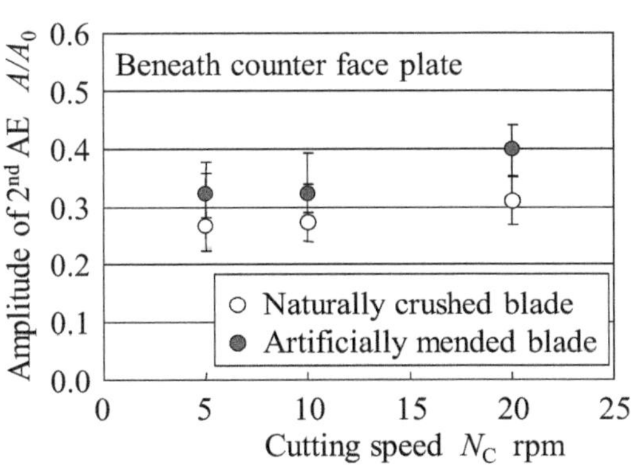

Fig. 10. Amplitude of 2nd AE signals

Fig.11 shows the wear height of upper layer with respect to the cutting speed. **Fig.12** shows representative photographs of sheared section of paperboard, in case of N_C= 5rpm, ϕ=0°. The wear height h_W was measured from CCD microscope photographs. Although there are many unknown phenomena concerning the dynamic high-speed cutting, the sheared profile of paperboard appears to be varied with rotation speed, depending on the blade tip shape.

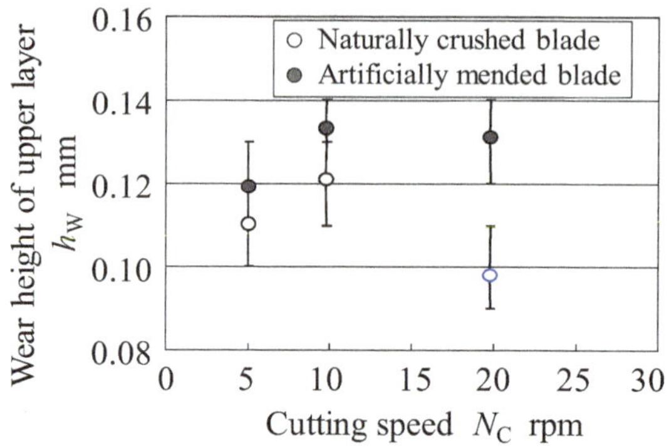

Fig. 11. Wear height of upper layer

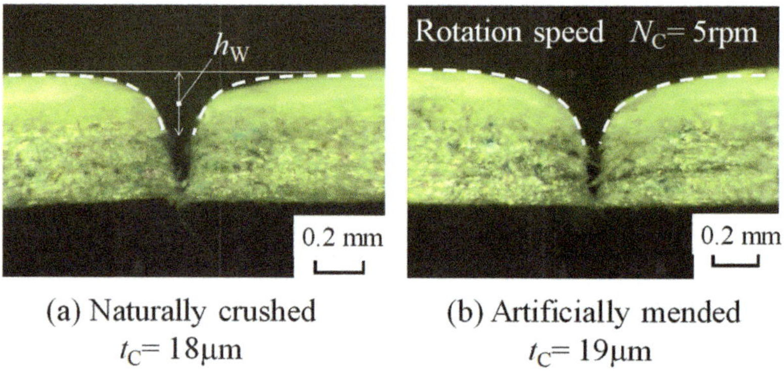

Fig. 12. Side views of sheared section of paperboard (ϕ=0° by CCD microscope)

4.2 Estimation of cutting attitude from 2nd AE signal

From Eqs (1),(5), there is a linear relationship between the amplitude of 2nd AE signal and the peaked cutting line force f_{C2}. Hence, the variance of f_{C2} with ϕ corresponds to that of A/A_0 (or a mV) with ϕ. **Fig. 13** shows the amplitude of second AE signal and the peaked line force f_{C2} in terms of the cutting attitude angle ϕ=0, 45 and 90°. The ratio of 2nd AE signal with ϕ=90° by ϕ=0° was about 2.0, while that of line force f_{C2} with ϕ=90° by ϕ=0° was 1.25. The last value of 1.25 was known from the authors' report [11]. When we consider Eq.(5) with the cutting attitude ϕ, we need to calibrate the coefficients k_{fC2A}, f_{C20A} with ϕ. They are not usually constant with ϕ.

4.3 Spectrum distribution at 2nd AE signal

According to the authors' investigation [12], the crank type, small cutting test machine has several natural frequencies. The upper-half die cutter device has the natural peak points near 0.6, 0.9, 2, and 5 kHz, while the lower-half die cutter device has that near 0.4, 0.8 kHz. When the blade collides to the face plate, some natural peak points are observed near 0.2, 0.5 and 1.3 kHz. **Fig.14** shows the linear spectrum distribution of 2nd AE signal using the naturally crushed blade (t_C=18μm), while **Fig.15** shows that using the artificially mended trapezoidal blade (t_C=19μm).

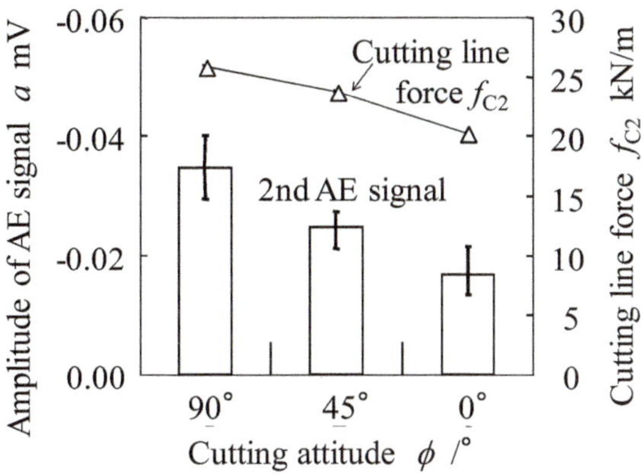

(N_C= 5rpm in case of naturally crushed blade)

Fig. 13. Dependency of cutting line force and 2nd AE on cutting attitude

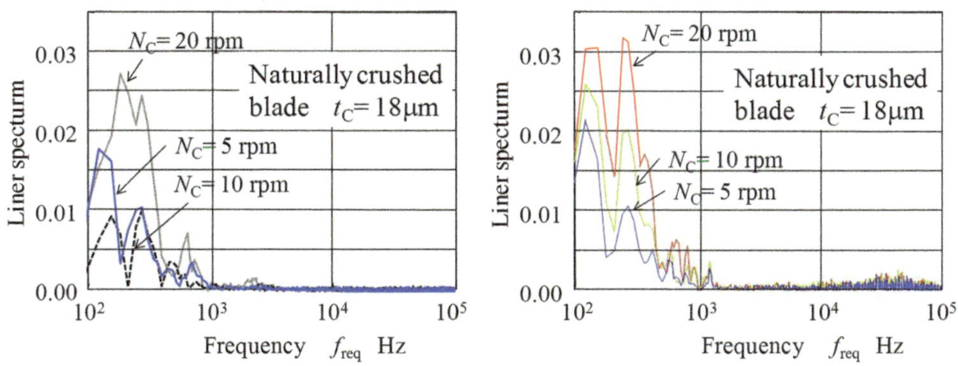

(a) AE signal on paperboard (b) AE signal beneath face plate

Fig. 14. Linear spectrum of 2nd AE signal (naturally crushed blade)

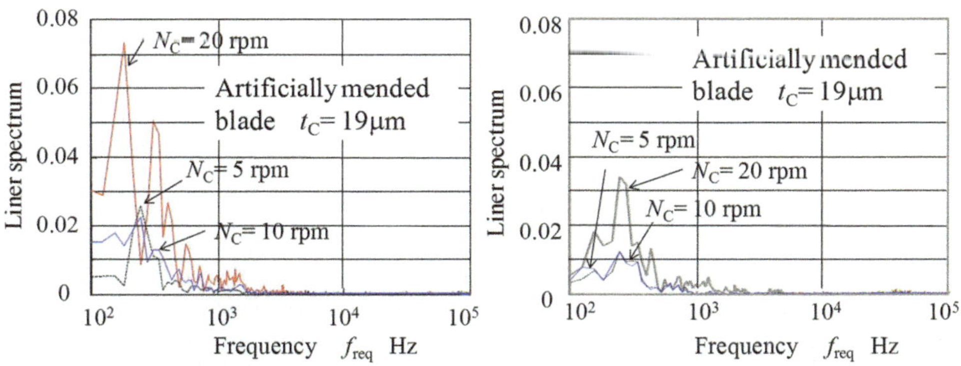

(a) AE signal on paperboard (b) AE signal beneath face plate

Fig. 15. Linear spectrum of 2nd AE signal (artificially mended blade)

It was confirmed that there were the natural peak points of 0.2, 0.5 and 0.8 kHz, which were independent to the rotation speed, in those figures 14, 15. In general, when the rotation speed is increased, the linear spectrum of natural peak points with the 2nd AE signal also tends to be increased due to increasing of the motion energy of the blade crank system [8].

Using the artificially mended blade of 19μm tip thickness in case of N_C= 20 rpm, the spectrum peak points of 0.2, 0.5 kHz were relatively large compared with other cases. This appears to be caused by the increasing of line force f_{C2}. From those results, it is appeared that a smart round-edge shape of blade tip is preferable for performing to restrict the wear height h_W, to reduce the second-peaked cutting line force f_{C2}.

4.4 Transient response of cutting force and AE signal for cutting process

The second AE signal, which corresponded to the second peaked line force f_{C2}, was discussed with respect to the rotation speed N_C in the sections 4.1, 4.3. The transient response of cutting line force was analyzed by the authors when the rotation speed was

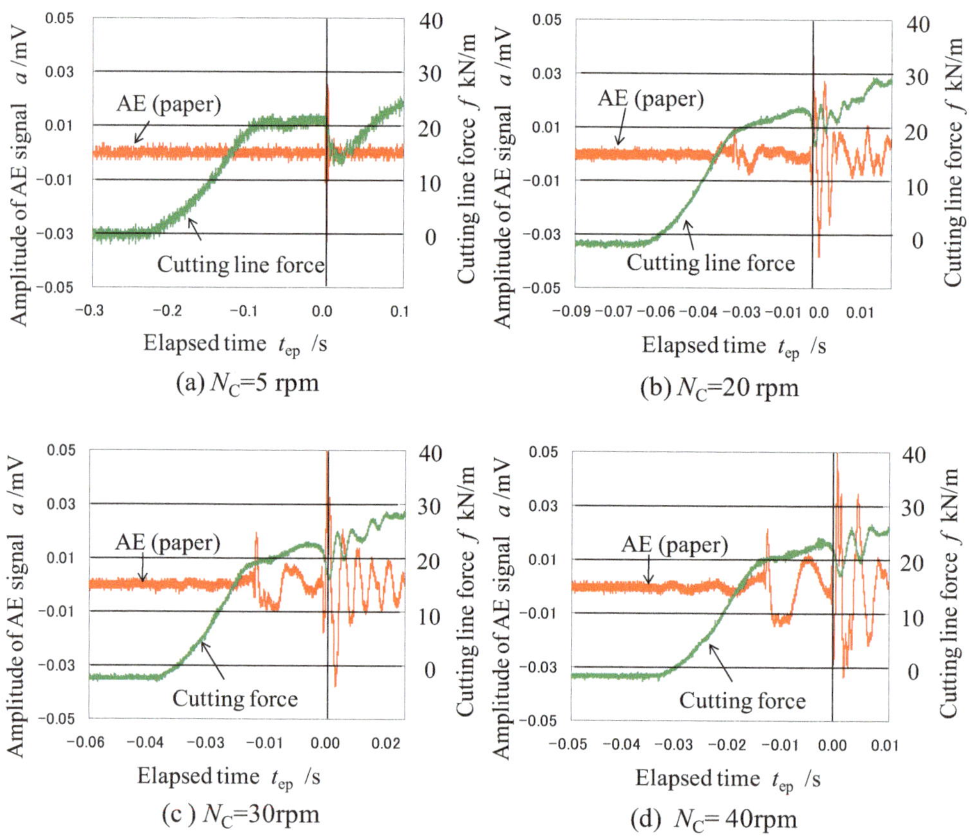

Fig. 16. Transient response of cutting force and AE signal measured on paperboard (ϕ=0°, by naturally crushed blade)

varied from 5 rpm up to 80 rpm[6] [8]. However, the dynamic state of AE signal is not discussed for the cutting duration.

In this section, the cutting line force f kN/m and the amplitude of AE signals a mV, which were measured on the paperboard and beneath the counter face plate, were investigated for the elapsed time t_{ep}. Here, the rotation speed of the crank shaft N_C was varied from 5 rpm up to 40 rpm.

Fig.16 shows the cutting line force f kN/m, which was measured by the load cells with the naturally crushed blade, and the AE signal measured on the paperboard with respect to the elapsed time t_{ep}. Here, the rotation speed N_C was setup to (a) 5rpm, (b) 20rpm, (c) 30rpm and (d) 40rpm, while the cutting attitude was chosen as ϕ=0°. Similarly, **Fig.17** shows the same cutting line force f kN/m and the AE signal measured beneath the counter face plate[7]. From those figures, we can detect several features:

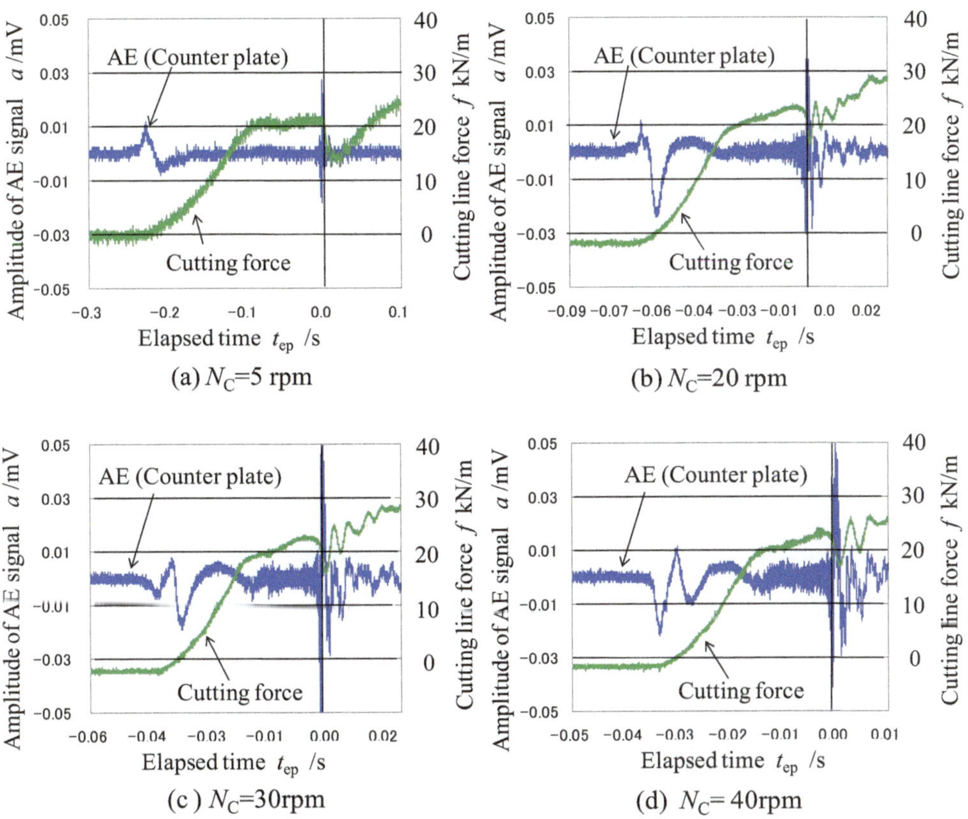

Fig. 17. Transient response of cutting force and AE signal measured beneath face plate (ϕ=0°, by naturally crushed blade)

[6] The dynamic line force was measured by using the strain-gauge method in the past work [8] , due to mass effect of lower base table. The strain-gauge was mounted on the blade.

[7] The AE signals on paperboard and beneath face plate were measured by two sensors at the same time.

1. Comparing N_C=20 rpm with N_C=5 rpm, the AE signal on the paperboard varied (been a little increased and dropped down in a short time) at the first peak point of cutting force f_{C1}. This transient variation was observed for $N_C \geq 20$ rpm. From the physical aspects of sheared process, this transient variation seems to correspond to *the surface breaking* of *the upper layer* (coated layer).
2. When $N_C \geq 30$ rpm, the AE signal on the paperboard was increased in the middle time zone between the timing of f_{C1} and that of f_{C2}, and the AE signal was remarkably decreased at the position of f_{C2}. This seems to mean that *the lower layer* of the paperboard reached *a half-breaking state.*
3. Observing the AE signal beneath the counter face plate, the first AE signal (waveform) which occurs at the initial contact of the blade with the paperboard, varies with the rotation speed. It consists of a positive overshoot and a negative overshoot, and/or several overshoots. This seems to correspond that the face plate is vibrated (bounded) in the vertical direction (out-of-plane) via the compressed paperboard.
4. Observing the AE signal beneath the counter face plate, the stationary level of AE signal (at the early stage before the first AE signal position) is relatively small, while the level of AE signal for the cutting duration (from the position of f_{C1} by that of f_{C2}) is fairly large. This variation seems to be caused by a high-compressive wedging process and a sliding/machining state derived from the wedge friction.

5. Summary

Referring the effect of the blade tip shape on the load response of the paperboard cutting, the principle of identification of the blade tip thickness was surveyed by using the solid acoustics emitted, and the application technique of AE was reviewed in order to diagnose the cutting state. The summary of this work is as follows:

- Measuring the load response of paperboard cutting and the mechanical vibration, caused by collision of the blade tip, the blade tip thickness t_C can be estimated.
- The contact time when the blade tip contacts with the work sheet and the separated time when the blade tip separates from the counter plate can be detected by signal response of AE transducer. By using these timing data, it is possible to diagnose the ageing of the blade tip height and the position adjustment of the die board in the flat bed machine.
- The principle of diagnosing the ageing of irregularity of a cutter height on the flat bed machine was described.

After describing those reviews for the principle of diagnosis, several new problems were discussed for the speed effect of paperboard cutting and the transient response of AE signals measured on the paperboard and/or beneath the counter face plate.

In case of low speed condition: $N_C \leq 20$ rpm, the followings are revealed.

- Magnitude of the second AE signal, which is related to the potential drop from the second-peak line force f_{C2} to the break-down line force f_{Cb}, increases with respect to the rotation speed N_C .
- The wear height of upper layer h_W is sensitively varied with the blade tip shape when N_C is increased. A smart round-edge blade is preferable for reducing the cutting line force f_{C2} and the wear height h_W.

In case of $40 \geq N_C \geq 30$ rpm, the following points are confirmed.

- The first-peak position of f_{C1} and the just-before state (middle zone, cutting duration) with the second-peak position of f_{C2} are detected from the AE signal measured on the paperboard.
- The stationary level of AE signal, measured beneath the face plate, is useful for detecting the cutting duration from the position of f_{C1} to that of f_{C2}.

6. References

[1] Murayama, M., Nagasawa, S., Fukuzawa, Y. and Katayama, I., Orthotropy Effect and Strain Dependency of Paperboard on Load Characteristic of Center Bevel Cutter Indented to Paperboard, *Journal of Materials Processing Technology,* 159(2) (2005), pp. 199-205.

[2] Nagasawa, S., Nagae, S., Fukuzawa, Y., Chaijit, S., Yamashita, Y., Murayama M. and Katayama, I., Effect of surface hardness of counter plate on crushing of blade tip during pushing shear of paperboard, *Journal of Materials Processing Technology,* 192-193 (2007), pp.265-275.

[3] Sadamoto, A., Yamaguchi, T., Nagasawa, S., Fukuzawa, Y., Yamaguchi, D. and Katayama, I., Analysis of sound radiated by paperboard die cutting, *Audio Engineering Society Proceedings of 21st Int. Conference,* St. Petersburg, (2002), pp.136-139.

[4] Gautschi, G., Piezoelectric Sensorics: Force, Strain, Pressure, Acceleration and Acoustic Emission Sensors, *Materials and Amplifiers,* Springer Verlag, (2001).

[5] Nagasawa, S., Fukuzawa, Y., Yamaguchi, T., Murayama, M., Yamaguchi, D. and Katayama, I., Effects of Blade Tip Shape on Thread Dross Occurrence in Paperboard Die Cutting (in Japanese), *SOSEI-TO- KAKOU,* 43(498) (2002), pp.624-628.

[6] Nagae, S., Nagasawa, S., Fukuzawa, Y., Hine, A. and Katayama, I., Effects of Quenching Hardness on Cutting Resistance and Crushing of Blade Tip Indented on Paperboard (in Japanese), *SOSEI-TO-KAKOU,* 45(524) (2004), pp.742-746.

[7] Grebe,W. and Hofer, H., Einige praktische Hinweise zur Schonung der Messer in Bandstahl- werkzeugen (in Germany), *Papierverarbeitung und Druck,* 9 (1973), pp.292-300.

[8] Nagasawa, S., Sudo, A., Fukuzawa, Y., Suzuki, S., Sugita, T. and Katayama, I., Dynamic Characteristics of Blade Cutting Force during Pushing Shear on Paperboard (in Japanese), *SOSEI-TO- KAKOU,* 47(543) (2006), pp.309-313.

[9] Suzuki, S., Fukuzawa, Y., Nagasawa, S., Katayama, I. and Iijima H., Acoustic Emission Characteristics on Variation of Cutter Tip Thickness during Cutter Indentation on Paperboard (in Japanese), *SOSEI-TO-KAKOU* , 46(538) (2005), pp.1061-1065.

[10] Fukuzawa, Y., Nagasawa, S., Suzuki, S., Katayama, I. and Sadamoto, A., Analysis acoustic emission and sound during the paperboard of cutting process, *Journal of Material Processing of Technology* , 192-193 (2007), pp.134-138.

[11] Nagasawa, S., Fukuzawa, Y., Hine, A., Katayama, I., Yoshizawa, A. and Furumi, T., Effects of Edge Clearance and Board Thickness on Shearing Characteristics of Paperboard Die Cutting, *SOSEI-TO-KAKOU,* 41(469) (2000), pp.126-131.

[12] Abdul Hamid, D., Nagasawa, S., Fukuzawa, Y., Komiyama, Y. and Hine, A., AE Analysis on Blade Cutting Pressure Adjustment in Dynamic Cutting of Paperboard, *Journal of Acoustic Emission*, 27 (2009), pp.64-76.

13

EMA Fault Detection Using Fuzzy Inference Tools

Jordi Cusidó i Roura, Miguel Delgado Prieto, Jose Luis Romeral Martínez
MCIA Group, Technical University of Catalonia
Spain

1. Introduction

The challenge in aircraft power system architecture is to move towards the electric power aircraft subsystems including flight control actuation, environmental control system, and utility functions. In this context emerged the concept of "More Electrical Aircraft", which is continuously supported by the new advances in power electronics, electrical motors, digital control and communications.

Traditionally, for primary flight control surfaces only rotary and linear electro-hydraulic actuators (EHA) have been considered, however the later trend is to replace them by electro mechanical actuators (EMA). An EMA has no internal hydraulic fluid, instead using electric motors to directly drive the ram through a mechanical gearbox. Compared to an EHA, the EMA has certain advantages. There is a new issue related to the aforesaid new trend with the aim of improve system reliability. This is the fault detection and diagnosis of the failures that may take place in both, mechanics and electric components. So, it is well assumed that the safety is the main issue for the EMAs development. The motor failures, the damaged bearings, and the eccentricities existing in the drive train, affects on one hand the air gap flux distribution and on the other leads to current and voltage unbalances. However, it is difficult to examine EMA faults by analysing only specific fault harmonics due to fault signal complexity.

To ensure Airplane operation every system needs to be tested and have an associated a Test Program Set (TPS) System. A lot of work has been done in field of TPS and System Design ensuring the testability of a component [1]-[2]. Design for Testability is a major trend in the world of Aerospace and Defence. This paper presents approaches for an one-board fault detection system and TPS fault detection system for EMA by means of Fuzzy Inference Tool.

Presented application

The solution to monitor the condition of electric motor ball bearings with distributed fault proposed in this work is based in problem analysis from a multivariable point of view, what is to say the obtaining of a fault level result from the combined analysis of different signals with the use of fuzzy logic inference techniques. This is proposed to achieve a reinforced diagnose that will be more effective when it comes to detect bearing failure than using just a single signal, especially when it comes to damage severity evaluation.

2. Basic theory

A. Vibration analysis

Vibration analysis is one of the most extended condition monitoring techniques [4]. Despite being a reliable, well studied robust technique, one of its drawbacks is that it requires that the motor under test has a vibration transducer installed, condition that makes its online application expensive. In this work its study is set out as a reference for the other techniques without this kind of restriction.

Vibration frequency components related to specific faults - inner (f_{irf}), outer (f_{orf}) and ball (f_{bsf}) faults - can be calculated using the following expressions [2][4]:

$$f_{irf} = \frac{n}{2} \cdot f_r \cdot \left[1 + \frac{bd}{pd} \cdot \cos\beta\right] \tag{1}$$

$$f_{orf} = \frac{n}{2} \cdot f_r \cdot \left[1 - \frac{bd}{pd} \cdot \cos\beta\right] \tag{2}$$

$$f_{bsf} = \frac{f_r}{2} \cdot \frac{pd}{bd}\left[1 - \left(\frac{bd}{pd} \cdot \cos\beta\right)^2\right] \tag{3}$$

With:

n: number of balls
f_r: rotor speed
bd: ball diameter
pd: bearing pitch diameter
β: contact angle of the ball on the race

B. As some works and standards [5] [6] [7] set out, a RMS vibration value evaluation of the motor also provides a good indicator for motor health, allowing machine overall fault diagnosis - Stator currents

Stator currents analysis (MCSA) represents an interesting alternative method with its own particularities and benefits; the most interesting of them is sparing the access inside the motor making it easy to perform its online fault analysis.

Previous works have shown the existing correlation between vibration and currents RMS values [5]. Although it is a complex function that relates both magnitudes, this work tries to check the RMS currents reliability in order to perform the motor status diagnose.

With this knowledge, it is possible to execute an RMS calculation over the acquired current signals with the aim of performing a more precise and straightforward operation.

C. High frequency common-mode pulses

One of the biggest culprits for bearings failure are common-mode circulating currents, which are generated by switching inverters and expose the motor terminals to high dv/dt. This phenomenon has been widely exposed in [6] and [7].

In this experiment, to limit the acquired signal to only pulses flowing through bearings (the responsible of balls degradation), a motor modification was introduced. All the ball bearing under test were aisled from the motor stator frame but in a point connected to ground through a cable where the pulses were measured. Bearings insulation was achieved by surrounding the piece with a PTFE flat ring with a hole mechanized in it to let the cable pass through.

These currents typically show a frequency range of about 5 MHz with a typical period of 20 μs between bursts.

D. Acoustic emissions

The Acoustic Emission Technique (AET) is a very promising tool that has practical application in several fields and specifically recent important relevance in condition monitoring of machines. [8] [9] [10]

Acoustic emissions (AE) are high frequency elastic waves in the ultrasound range that appear when a material suffers localised plastic deformation. The analysis of this stress waves shows the nature of the original producing source and, therefore, enables the diagnose conducting to the actual element fault type and severity. In the bearings field, AE is a good tool to detect impulsive faults like wear, ball impacts and lubrication problems (like contaminants or degradation).

Acoustic Emission is therefore defined as a radiation of mechanical elastic waves produced by the dynamic local rearrangement of the material internal structure. This phenomenon is associated with cracking, leaking and other physical processes and was described for the first time by Josef Kaiser in 1950. He described the fact that no relevant acoustic emission was detected until the pressure applied over the material under test surpassed the previously highest level applied.

Acoustic Emissions Technique is classified as a passive technique because the object under test generates the sound and the Acoustic Emission sensor captures it. By contrast, Active methods rely on signal injection into the system and analysis of variations of the injected signal due to system interaction.

Then an acoustic emission sensor captures the transient elastic waves produced by cracking or interaction between two surfaces in relative motion and converts their mechanical displacement into an electrical signal. This waves travel through the material in longitudinal, transverse (shear) or surface (Rayleigh) waves, but the majority of sensors are calibrated to receive longitudinal waves.

Wherever the crack is placed, the signal generated travels from the point of fracture to the surface of the material. The transmission pattern will be affected by the type of material crossed and then isotropic material will lead to spherical wave front types of propagation only affected by material surfaces or changes, where the Snell law rules their reflection and reflexion. On Figures 1 and 2 is shown the evolution of acoustic waves inside a Material. On Figure 2 it is shown how appear reflections on waves due to the defect.

The biggest advantage of this method is probably that it is capable of detecting the earliest cracks of the system and their posterior growth, making possible fault detection before any other current method. The main drawback is that it requires additional transducers and a well controlled environment.

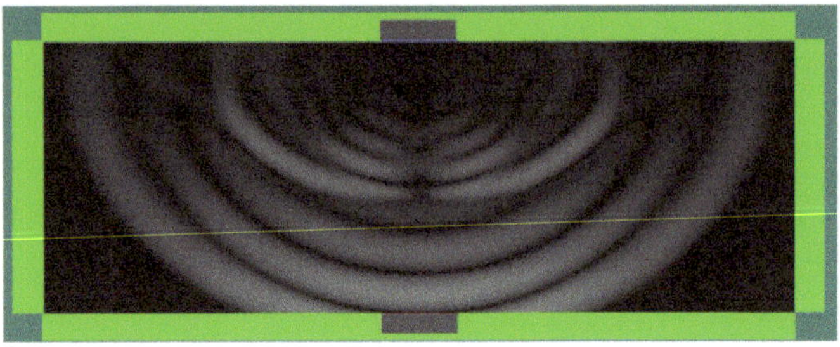

Fig. 1. Acoustic Emission Wave Propagation

Fig. 2. Acoustic Emission Wave Propagation in fractured Material

Bearing measurements and analysis

The hertzian contact stresses between the rolling elements and the races are one of the basic mechanisms that initiate a localized defect. Faults in bearings mainly appear in races and balls. Damages in the bearing races are due to metal fatigue and consequent plastic limit variation. Singular ball defects include cracks, pits, and spalls on the rolling surfaces, at it is shown on Figure 3.

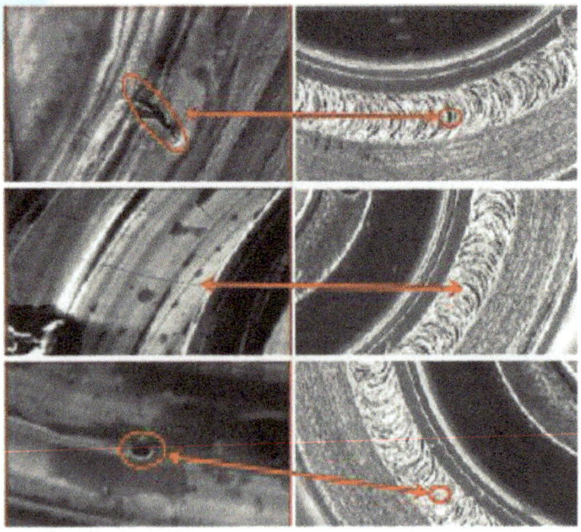

Fig. 3. Bearing Damage Evolution

Bearing race damage is characterized firstly by changes of metal characteristics like elastic limit and later by the appearance of pits and transverse flutes burnt into the bearing race. Bearing race defects lead to changes in high frequency resonance of the metal. Singular defects in rolling element give rise to isolated impacts as the defective surface hits another surface and produces a single detectable AE pulse. Frequency and periodicity of the pulses are related to material characteristics and rotating speed, and are also depending on the type of bearing.

A number of signal processing methods have been used on the time domain to diagnose failures by AE measurements in machinery [13]-[14]. Although these methods are quite simple to apply, it is apparent that they involve a significant expertise in the interpretation of the output [15]. As a conclusion, and despite lubrication and some bearing faults are detectable in the time domain analysis, to benefit most from the high sensitivity of AE to defects filtering and reconstruction from **time - frequency transforms** are proposed in this project, as a way to diagnose the bursts and apply **time - frequency analysis** to perform the feature extraction and characterise the faults. If mean or overall AE parameters obtained from characterisation are considered as fault detection parameters, they can be most suited as a trending parameter where its current value is to be compared to previous ones under similar operational conditions; however more detailed investigations are still required for applying AE for prognosis.

Fig. 4. Main defects on Gears

Gear boxes measurements and analysis

Whilst vibration analysis on gear fault diagnosis is well established, gearboxes are inevitably more complex to monitor using vibration analysis as they contain various shaft support bearings rotating at different speeds and a number of gear teeth interactions which again are operating at different speeds.

Fatigue tests were carried out and they showed that AE detected the first sign of failure when the gear reached 90% of its final life. As the crack progressed, AE amplitude increased. During the final stage of gear tooth fracture, a significantly high amplitude AE burst was detected. On the other hand, the vibration level did not change significantly in the initial stage of crack initiation and propagation until the final stage of failure [16].

As alternative to these vibration monitoring techniques, AE measurements can be made and then do further analyses by using different signal processing techniques, such as cumulative energy count [17], monitoring of rms, standard deviation and duration of AE [18], and kurtosis analysis [19].

Wavelet transforms have been also used for fault diagnosis of gears [20]. By this technique, the time domain AE signals of a rotating machine with normal and defective gears can be processed through wavelet transform to decompose in terms of low-frequency and high-frequency components. The extracted features from the wavelet transform were used as inputs to an artificial - intelligence based diagnostic approach. From these experiments, it is concluded that **AE method offered an advantage over vibration** monitoring techniques, especially for rubbing faults at a low speed. However, difficulty in understanding AE signals, complexity in related signal processing and a lack of industrial development have hampered the manufacture and large scale use of these kind of sensors.

3. Experimental results

3.1 Experimental setup

In order to assess the effectiveness of the methodology proposed in this work, it has been checked by means of experimental data obtained from a motor bench.

The experimental motor bench is based on two identical featured face to face motors, the motor under test and the motor that acts as a load. Between the motors it has been added an screw and a mobile part which is displaced over it. The screw as well as the mobile part has been provided by SKF. The motors are two SPMSMs with 3 pairs of poles, rated torque of 2.3 Nm, 230 Vac, and rated speed of 6000 rpm provided by ABB Group. The motor under test was driven also by an ABB power converter model ACSM1. The drive control for the motor was a vector control, with speed control loop.

The measurement equipment is focused on the acquisition of a stator current, stator common mode currents, vibration and acoustic emissions. The stator currents have been measured by means of a Tektronix current probe model A622. It provides 10–100 mV/A output and can measure ac/dc currents from 50 mA to 100 A-peak over a frequency range from dc to 100 kHz. The stator currents have been acquired by means a PXIe 1062 system from National Instruments sampling at 50 kHz, 100 ms for each measurement.

The vibration measurement has been performed by an ENDEVCO Isotron KS943B.100 triaxial accelerometer with IEPE (Integrated Electronics Piezo Electric) standard output and linear frequency response from 0.5 Hz to 22 kHz with a maximum of 60g. The acceleration data was collected using a specific acquitsition card connected to the PXIe 1062 system from National Instruments sampling at 20kS/s, 10 seconds for each measurement.

3.2 Experimental results

A. Vibrations

With regard to the bearings units under test, there was a healthy one (with very similar vibration levels to other new units tested in previous works) and the other two tested units had different levels of damage due their operation hours, qualitatively evaluated with a shock pulse tester from SPM Instruments.

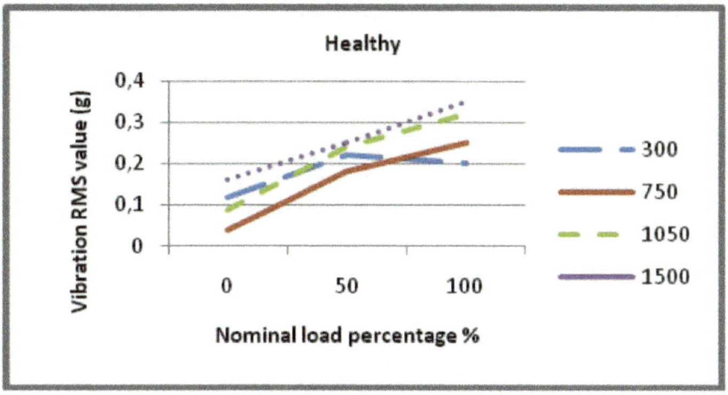

Fig. 5. RMS Vibration for healthy unit, all speeds in rpm and loads.

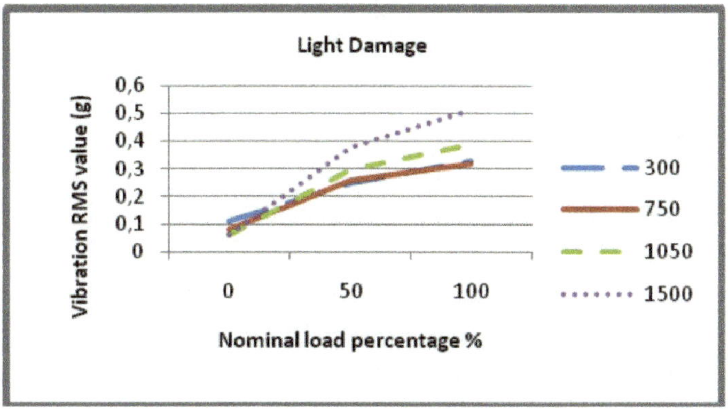

Fig. 6. RMS Vibration for lightly damaged unit, all speeds in rpm and loads.

Fig. 5, 6 and 7 show the evolution of the RMS value of each motor vibration for all speeds and load values tested. The healthy unit shows lower values especially detectable under nominal conditions.

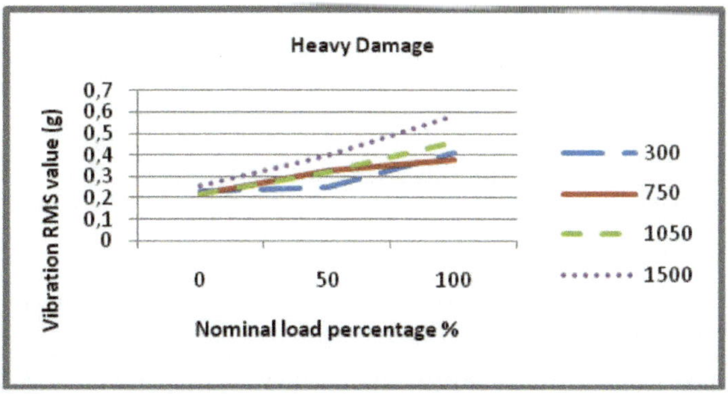

Fig. 7. RMS Vibration for heavily damaged unit, all speeds in rpm and loads.

Clearly, the healthy motor in Fig. 5 shows lower RMS values of vibration in comparison to the other two units. Fig. 7 unit data, which was in the worst operational condition according to the SPM measurements performed, gave also the highest levels of RMS vibration values.

B. Stator currents

To avoid the influence of the main harmonic power value in the RMS measurement, signals have been previously filtered using a band-rejection 5th order Butterworth filter centred in the power supply main harmonic with a bandwidth of 20 Hz between higher and lower cut-off frequencies. Tables 1 and 2 compare the RMS filtered values of the heavily and lightly damaged units with the healthy one.

Heavily Damaged-Healthy (A RMS)				
Load % \ speed	300	750	1050	1500
0	0,004	-0,006	-0,008	-0,007
50	0,036	0,03	0,073	0,044
100	0,018	0,026	0,024	0,024

Table 1. Difference in RMS filtered current value between heavily damaged unit and healthy one used as reference.

Lightly Damaged-Healthy (A RMS)				
Load % \ speed	300	750	1050	1500
0	0,008	0,002	-0,003	-0,003
50	0,002	-0,011	-0,002	-0,005
100	0,02	0,012	0,003	0,014

Table 2. Difference in RMS filtered current value between lightly damaged unit and healthy one used as reference.

A significant difference can be clearly appreciated when the motor is heavily damaged under load condition. Light damage is noticeable under nominal load conditions but its detection does not seem to be easily reliable.

B. High Frequency bearings pulses

Bearings pulses threshold analysis has been executed to validate theories of correlation between bearings state (wear, lubrication, distributed defects, etc.) and pulses discharge over a threshold value.

The results summarized in Figure 8 show that over a defined threshold level healthy bearings undergo a bigger number in comparison to the damaged units. It is noticeable also that this method is able to detect failure at its initial stage if the threshold is correctly placed.

C. Acoustic emission testing

Acoustic Emission acquired data has been statistically classified by means of value binning tools and histogram presentation.

Fifteen sets of data were acquired for each motor and averaged. Fig. 5 shows the results comparing the RMS voltage values acquired for the different units under test.

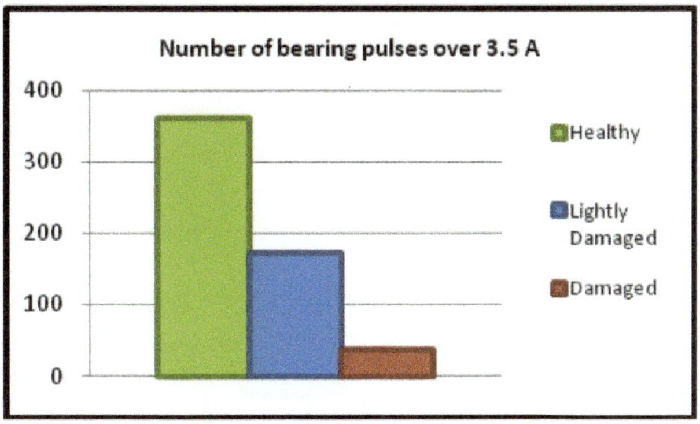

Fig. 8. Number of bearing pulses over threshold value of 3.5 A for all motors under test. Healthy, lightly damaged and heavily damaged.

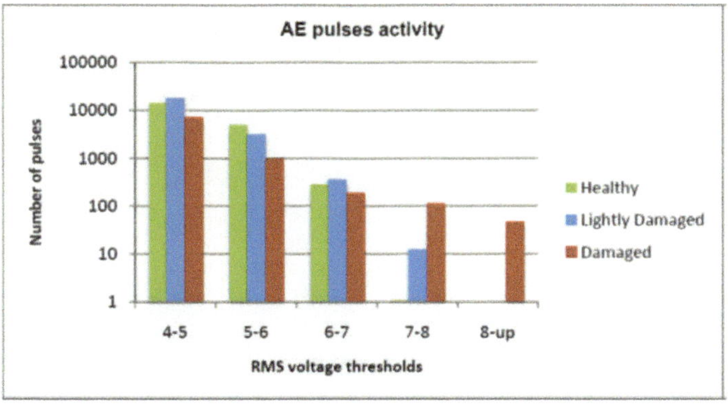

Fig. 9. Acoustic Emission voltage values classification

It is advisable that pulses over 8 V only appeared during the damaged motor testing while under 7 V that unit does not show more activity than the healthy and lightly damaged units.

Then, the fuzzy inference system designed uses as reference the number of pulses that surpass the 7 V value, which is the zone where the distinction of the fault severity of the unit seemed to be more noticeable.

3.3 Fuzzy inference tool

The analysis of the actual bearing status was performed using a fuzzy logic inference implementation [11] [12] which maps given inputs to a single output, the different signals acquired are linked to a damage value scaled from 1 to 3.

> *The membership functions, like Fig. 10, have been obtained through training and validation process, for each signal under analysis using real motor data. MATLAB "Adaptive neuro-fuzzy inference system" tool has been used for this purpose. Fig. 11 shows the obtained relationship between Vibration and Stator Current RMS values against the Failure Level output for a motor speed of 1500 rpm and a load of 0%.*

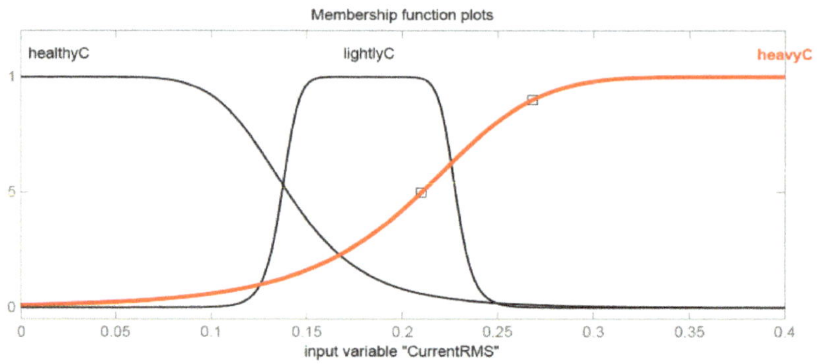

Fig. 10. Membership function plot for Current RMS. (motor speed: 1500 rpm, motor load: 0%)

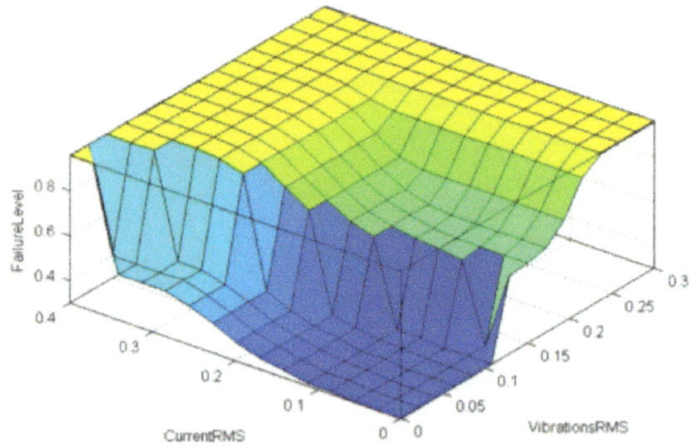

Fig. 11. Plotted surface showing the relationship between the system inputs Vibrations RMS value (g) and Stator Currents RMS value (A) versus the Failure Level output. (Motor speed: 1500 rpm, motor load: 0%)

This process explanation will be properly expanded on the final version of this paper.

To perform the evaluation of the monitoring system designed, fifteen sets of data were collected from the same units and processed. Table 3 summarizes the obtained results.

Unit	Matches	Success %
Healthy	15	100 %
Lightly Damaged	14	93,33%
Heavy Damage	13	86,66%

Table 3. System testing results

All healthy data sets were correctly identified, whilst one of the lightly damaged was recognised as a heavily damaged set and two of the heavily damaged sets were identified as lightly damaged ones. The percentage of success was reasonably high and its improvement is still possible if more data sets are used during the system training stage.

4. Conclusions

This chapter gives an overview of a condition monitoring system that uses a multisensory fuzzy inference approach used to detect faults in EMA Systems. The results show that a multivariable design contributes positively to damage monitoring of EMA, being a more solid solution than just using any of the single signals involved..

The results show that a multivariable design contributes positively to damage monitoring of bearings, being a more solid solution than just using any of the single signals involved..

5. References

[1] IEEE Motor Reliability Working Group, "Report of large motor reliability survey of industrial and commercial installations," IEEE Transactions on Industrial Applications, vol. IA-21, no. 4. 1986.

[2] Howard, "A Review of Rolling Element Bearing Vibration. Detection, Diagnosis and Prognosis", October 1994.

[3] R. Stack, T. G. Habetler and R. G. Harley, "Fault classification and fault signature production for rolling element bearings in electric machines," Industry Applications, IEEE Transactions on, vol. 40, pp. 735-739, 2004.

[4] M. F. Cabanas, M. G. Melero, G. A. Orcajo, J. M. Cano Rodríguez, Juan Solares Sariego; "Tecnicas para el Mantenimiento y diagnóstico de Máquinas Eléctricas Rotativas", Marcombo, Oviedo, Spain, 1996.

[5] C. M. Riley, B. K. Lin, T. G. Habetler and G. B. Kliman, "Stator current harmonics and their causal vibrations: a preliminary investigation of sensorless vibration monitoring applications," Industry Applications, IEEE Transactions on, vol. 35, pp. 94-99, 1999.

[6] A. Binder, A. Muetze, "Systematical Approach to Bearing Current Evaluation in Speed Drive Systems," ICEM2002.

[7] D. Busse, J. Erdman, R. Kerkman, D. Schlegel, and G. Skibinski, "Characteristics of shaft voltage and bearing currents," Industry Applications Magazine, IEEE, vol. 3, pp. 21-32, 1997.

[8] D. Mba and Raj B. K. N. Rao, "Development of Acoustic Emission Technology forCondition Monitoring and Diagnosis of Rotating Machines: Bearings, Pumps, Gearboxes, Engines, and Rotating Structures," *The Shock and Vibration Digest*, vol. 38, pp. 3, January 2006. 2006.

[9] N. Tandon, A. Choudhury, A review of the vibration and acoustic measurement methods for detection of defects in rolling element bearings, *Tribology International* 32 (8) (1999) 469–480.

[10] A. M. Al-Ghamd and D. Mba, "A comparative experimental study on the use of acoustic emission and vibration analysis for bearing defect identification and estimation of defect size," *Mechanical Systems and Signal Processing*, vol. 20, pp. 1537-1571, 10. 2006.

[11] X. Lou, K.A. Kenneth and A. Loparo, "Bearing fault diagnosis based on wavelet transform and fuzzy inference", Mechanical Systems and Signal Processing, Volume 18, Issue 5, p. 1077-1095.

[12] S. Ballal, Z.J. Khan, H. M. Suryawanshi, R.L. Sonolikar, "Adaptive Neural Fuzzy Inference System for the Detection of Inter-Turn Insulation and Bearing Wear Faults in Induction Motor", IEEE Transactions on Industrial Electronics, 2007, Vol.54 Issue 1.

[13] Miettinen, J., Pataniitty, P., "Acoustic Emission in Monitoring Extremely Slowly Rotating Rolling Bearing". Proceedings of COMADEN 1999, 12th International Congress and Exhibition on Condition Monitoring and Diagnostic Engineering Management, July 6-9, 1999, Sunderland, U.K., pp. 289 - 297

[14] Morhain, A. and Mba, D., 2003, "Bearing Defect Diagnosis and Acoustic Emission," Journal of Engineering Tribology, I Mech E, Vol. 217, No. 4, Part J, 257–272

[15] Holroyd, T. J.;"Condition monitoring of very slowly rotating machinery using AE techniques", COMADEN 2001, Manchester, 4th of September, 2001, Manchester, 8p

[16] Singh, A., Houser, D. R., and Vijayakar, S., 1999, "Detecting Gear Tooth Breakage Using Acoustic Emission: A Feasibility and Sensor Placement Study," Journal of Mechanical Design, Vol. 121, 587–593.

[17] Miyachika, K., Zheng, Y., Tsubokura, K., Oda, S., Kanayama, Y., Koide, T., Namba, C., and Hayashi, T., 2002, "Acoustic Emission of Bending Fatigue Process of Supercarburized Spur Gear Teeth," Progress in Acoustic Emission XI, The Japanese Society for NDI, pp. 304–310

[18] Siores, E. and Negro, A. A., 1997, "Condition Monitoring of a Gear Box sing Acoustic Emission Testing," Material Evaluation, Vol. 55, No. 2, 183–187

[19] Badi, M. N. M., Engin, S. N., and Schonfeld, D., 1996, "Fault Classification of a Model Drive Line Using Time Domain Data," in Proceedings of COMADEM 1996, Sheffield, UK, pp. 43–50

[20] Al-Balushi, K. R. and Samanta, B., 2000, "Gear Fault Diagnostics Using Wavelets and Artificial Neural Networks," in Proceedings of COMADEM 2000, Houston, TX, pp. 1001–1010

14

Denoise and Recognition of Friction AE Signal

Deng Aidong and Jiang Zhang
National Engineering Research Center of Turbo-Generator Vibration,
Southeast University,
China

1. Introduction

The friction between rotor and stator of a rotating machine is the common fault, as well as a huge problem to be studied and solved. Acoustic emission (AE) technology provides a new solution to friction detection. However, due to that AE signals cause waveform distortion during transmission in complex structure of a rotating machine, and they are interfered by strong noises of operating environment, AE technology is technically challenged in respect of friction detection. It will be an interesting job to discuss new denoise technologies and research steady identifiction models.

With respect to the problem that friction AE signals generated by rotating machines are easily interfered by strong noises, this Chapter presents a generalized morphological filter (GMF) based on the fundamental morphological transformation and the combination of those transformation of mathematical morphology. GMF adopts gradient method for iterative computation of weight coefficients of generalized morphological open-closing (GMOC) filter and generalized morphological close-opening (GMCO) filter, so as to find out the weight coefficient value to give the best denoise effect. Then the best weight coefficient value will be used for linear combination of generalized morphological filters to achieve the optimal denoise effect.

With respect to identification study, this Chapter presents an AE identification method that combines cepstral coefficient and fractal dimension together as a Gaussian mixture model (GMM) of characteristic parameters. Such identification model will model the probability density functions contained in eigenvectors of different modal waves in friction AE signals, and cluster these eigenvectors. Each clustering will be considered as a multi-dimensional Gaussian distribution function. Take the mean, covariance matrix and probability of each clustering as the training format of every modal wave, and then put the eigenvectors of friction AE signals to be detected into each format during identification. Weigh the likelihood probability of Gaussian model with maximum ratio combination method to obtain a total likelihood probability. Once the value of total likelihood probability is larger than the given threshold, we can be sure that there is friction AE.

The object to be studied in this Chapter is a rotor test bed equipped with friction device, which can simulate different degrees of friction faults. Use broad band AE sensor and high-speed data acquisition card to acquire complete waveforms of friction AE signals, and then superpose noises on friction AE signals acquired for further study.

2. Adaptive generalized morphological filter based on the gradient method (AGMF-G)

Mathematical morphology [1][2][3] applies set theory to image analysis, which processes an image as a whole package of operands without considering the details, or defining each and every image point of such operands. Set theory can provide a relative mathematical system describing image space combination. Therefore, mathematical morphology can be used to analyze geometric characteristics and structure forms of image objects, making it an extremely important tool for image analysis and processing. Currently mathematical morphology has been extensively applied to fields including image processing, pattern identification, computer vision, and among others.

2.1 Fundamental operations of mathematical morphology

Binary image refers to that the grey scale of an image only consists of 0 and 1. Traditional image processing takes 1 as the grey scale of image object, and 0 as the grey scale of background, and notes X as the set composed of those points in image which grey scale is 1. Mathematical morphology image processing mainly analyzes set X. Morphology uses a method of subjective "probe" interacting with objects to analyze X. "Probe" is a set either which is called "structuring element" by morphology and determined by analyst based on analytical purpose. Mathematical morphology has defined two basic transformation types referring to "dilation" and "erosion" respectively.

1. Set A and B as subsets of N-dimensional Euclidean space, and note $A \oplus B$ as A is dilated by B, therefore

$$A \oplus B = \{x \in E^N | x = a + b, a \in A, b \in B\} \tag{1}$$

2. Note $A \Theta B$ as A is eroded by B, therefore

$$A \Theta B = \{x \in E^N | x - b \in A, a \in A, b \in B\} \tag{2}$$

Based on dilation and erosion these two fundamental operations, we can structure a morphological operation cluster, namely all operations composed of multiple operations and set operations (union set, intersection set, supplementary set, etc.) of these two fundamental operations. Two most important multiple operations are morphological opening operation and morphological closing operation. Generally speaking, dilation and erosion are unrecoverable operations that either erosion first or dilation first is unable to recover the objects, but generate a new morphological transformation.

3. Note $A \circ B$ as opening operation between A and B, therefore

$$A \circ B = (A \Theta B) \oplus B \tag{3}$$

4. Note $A \bullet B$ as closing operation between A and B, therefore

$$A \bullet B = (A \oplus B) \Theta B \tag{4}$$

$A \circ B$ consists of union sets of subsets in A that are congruent to structural element B, or we can say that for each x in $A \circ B$, there is a translational B_y of structural element in A making $x \in B_y$, namely A has a geometric structure no less than B in close proximity to x.

2.2 Grey scale morphology

Grey scale images are the most circumstances in practice. Contrary to binary images, a grey scale image is the one which contains more than two grey scale values of pixels. Normally a real-valued function in consecutive or digital space is used to describe the grey scale images. Serra [3] adopts the concept of function umbra to apply binary morphology to grey scale images, so that a set representation of grey scale images is established.

One-dimensional signals can be represented with a set. One d-dimensional function $f(x)$ can be represented with a $d+1$-dimensional set to define its function umbra:

$$U(f) = \{(x,a) : a \le f(x)\} \tag{5}$$

That is to say, in a d-dimensional space, umbra is the set less than all the points expressed in function $f(x)$. With the concept of function umbra, we may apply binary morphology to grey scale morphology for signal processing. Normally, when $a = -\infty$, function $f(x)$ is able to be restructured by its umbra set, namely:

$$f(x) = \max\{a : (x,a) \in U(f)\}, \qquad \forall x \tag{6}$$

Let $f(x)$ as one-dimensional input signal of definition domain $D_f \subseteq E$, and $g(x)$ as structural element of definition domain $D_g \subseteq E$. Dilate or erode the umbra of function $f(x)$ with the umbra of function $g(x)$ to generate the umbra of a new function, which can be directly represented with $U(f \oplus g) = U(f) \oplus U(g)$ and $U(f \ominus g) = U(f) \oplus U(g)$. The formulas for functional operations of dilation and erosion are:

$$(f \oplus g)(x) = \max_y\{f(x-y) + g(y)\} \tag{7}$$

$$(f \ominus g)(x) = \min_y\{f(x+y) - g(y)\} \tag{8}$$

In the formulas above there are $x \in D_f$ and $y \in D_g$. There is a more specific definition of Equations (7) and (8) that set $f(x)$ as a one-dimensional original signal which definition domain is $\{0,1,\ldots,N-1\}$, set $g(x)$ as a structure element which definition domain is $\{0,1,\ldots,M-1\}$, and set the origin at 0, so that dilation and erosion can be defined as follows:

$$(f \oplus g)(n) = \max_{\substack{m=0,1,\ldots,M-1 \\ n=0,1,\ldots,N+M-2}} \{f(n-m) + g(m)\} \tag{9}$$

$$(f \ominus g)(n) = \max_{\substack{m=0,1,\ldots,M-1 \\ n=0,1,\ldots,N-M}} \{f(n+m) - g(m)\} \tag{10}$$

Similar to binary morphological transformation, two important morphological operators of grey scale can be obtained through the combination of dilation and erosion of grey scale: opening operation and closing operation, which are defined as follows:

$$(f \circ g)(x) = ((f \Theta g) \oplus g)(x) \tag{11}$$

$$(f \bullet g)(x) = ((f \oplus g) \Theta g)(x) \tag{12}$$

2.3 Morphological filtering

The fundamental concept of morphological filtering is based on the characteristics of geometric structures of signals that it matches or partially amends signals with predefined structural elements to extract signals and suppress noises. Morphological filter is made of fundamental morphological transformation combination.

Opening operation is able to remove isolated structures like tittles, burrs and small bridges, as well as suppress peak noises (positive impulse) in signals; while closing operation is able to level up small grooves, holes and small cracks, as well as suppress valley noises (negative impulse) in signals. In order to simultaneously suppress positive and negative impulse noises in signals, we may construct morphological open-closing (OC) filter and morphological close-opening (CO) filter by using structural elements with the same size and through open-closing operation with different order cascades [4][5]. The definitions are as follows:

Set $f(x)$ as input discrete signal and g as structural element, therefore, morphological open-closing filter and morphological close-opening filter shall be defined as follows:

$$Oc(f) = (f \circ g \bullet g)(x) \tag{13}$$

$$Co(f) = (f \bullet g \circ g)(x) \tag{14}$$

Due to anti-expandability of morphological opening and expandability of morphological closing, for a morphological open-closing filter, the opening operation conducted first not only removes positive impulse noises, but also enhances negative impulse noises. If we conduct closing operation with the same structural element, we may not actually remove all negative impulse noises; similarly, a morphological close-opening filter using the same structural element may not actually remove all positive impulse noises. Considering the selection of width of structural element is the major factor to affect denoise, in order to filter impulse interference from data, the width of structural element shall be larger than the width of maximum impulse of data. Therefore, we have to improve these two types of filters that we select structural elements with different sizes based on the definitions of morphological opening operation and morphological closing operation, so that the next structural function will be wider than the previous one. Thus, we may structure generalized morphological open-closing (GMOC) filters and generalized morphological close-opening (GMCO) filters [6] which are defined as follows:

Set $f(x)$ as input discrete signal, g_1 and g_2 as structural elements, and $g_1 \subseteq g_2$, then generalized morphological open-closing filters and generalized morphological close-opening filters shall be defined respectively as follows:

$$GMOC(f) = (f \circ g_1 \bullet g_2)(x) \tag{15}$$

$$GMCO(f) = (f \bullet g_1 \circ g_2)(x) \tag{16}$$

However, due to that generalized morphological filtering method is still composed of opening operation and closing operation, the two types of generalized morphological filters aforesaid still have statistics offset problems causing output amplitude of open-closing filter relatively low, while output amplitude of close-opening filter relatively high. In most circumstances, it's impossible to gain the best filtering effect by using it alone [7]. Considering the amplitude of friction AE signals of rotor varies a lot, and there are multiple noise components in data, we may use a weighted array of these two generalized morphological filters to effectively erase noises. When weight coefficient is 0.5, the weighted array of two filters is:

$$GMF(f) = 0.5 \times GMOC(f) + 0.5 \times GMCO(f) \tag{17}$$

The filter defined by Equation (17) is called generalized morphological filter (GMF).

2.4 Adaptive generalized morphological filter

In a generalized morphological filter, the weight coefficient of both generalized open-closing filter and generalized close-opening filter is 0.5, and such weight coefficient stays the same during filtering process. The fixed weight coefficient makes filtering results hard to adaptively achieve the best. In order to optimize filtering results, we may use gradient method to conduct iterative computation of weight coefficient, so as to obtain the optimal weight coefficient value.

Set input data as $x(n) = s(n) + u(n)$ and $n = 1,2,\ldots,N$. $s(n)$ represents an ideal signal without any noise, while $u(n)$ represents all types of noise interference. The output of generalized open-closing filters and generalized close-opening filters are:

$$y_1(n) = Goc(x(n)) = (x \circ g_1 \bullet g_2)(n) \tag{18}$$

$$y_2(n) = Gco(x(n)) = (x \bullet g_1 \circ g_2)(n) \tag{19}$$

g_1 and g_2 represent structural elements, therefore, filter output is:

$$y(n) = a_1(n)y_1(n) + u_2(n)y_2(n) = \sum_{i=1}^{2} a_i(n)y_i(n) \tag{20}$$

Where, $a_i(i = 1,2)$ represents weight coefficient.

Mean square deviation of output signal is:

$$E[e^2(n)] = E[|s(n) - y(n)|^2] = E[\left|s(n) - \sum_{i=1}^{2} a_i(n)y_i(n)\right|^2] \tag{21}$$

With gradient method and by gradually amending weight coefficient $a_i(i = 1,2)$, filter output $y(n)$ can be closest to the ideal signal $s(n)$ with respect to mean square deviation. To simplify the computation, we may replace $E[e^2(n)]$ with the square of a single error sample $e^2(n)$. Then the gradient of $e^2(n)$ to weight coefficient $a_i(i = 1,2)$ shall be:

$$\nabla e^2(n) = \left[\frac{\partial[e^2(n)]}{\partial a_1(n)} \quad \frac{\partial[e^2(n)]}{\partial a_2(n)} \right]^T = [grad_1 \ grad_2]^T \tag{22}$$

Iterative computation of adaptive generalized morphological filtering based on gradient method is as follows:

Step 1. Conduct generalized open-closing operation and generalized close-opening operation for input signal $x(n)$ to obtain $y_1(n)$ and $y_2(n)$. Therefore, filter output at the k th iteration is

$$y^k(n) = a_1^k(n)y_1(n) + a_2^k(n)y_2(n) = \sum_{i=1}^{2} a_i^k(n)y_i(n) \tag{23}$$

Where $a_i^k(i=1,2)$ represents weight coefficient at the k th iteration.

Step 2. Single error sample is:

$$[e(n)]^k = y^{k-1}(n) - y^k(n) \tag{24}$$

Replace $s(n)$ with $y^{k-1}(n)$, then the gradient shall be:

$$[\nabla e^2(n)]^k = \left[grad_1^k \ grad_2^k \right]^T \tag{25}$$

Step 3. Determine coefficient β_i^k. There are multiple methods to calculate β_i^k, including the famous HS method presented by Hestenes and Steifel [8]:

$$\beta_i^k = \frac{(\Delta grad_i^{k-1})^T grad_i^k}{(\Delta grad_i^{k-1})^T p_i^k} \tag{26}$$

FR method presented by Fletcher and Reeves [9]:

$$\beta_i^k = \frac{(grad_i^k)^T grad_i^k}{(grad_i^{k-1})^T grad_i^{k-1}} \tag{27}$$

And PRP method presented by Polak, Ribiere and Polyak individually [10][11]:

$$\beta_i^k = \frac{(\Delta grad_i^{k-1})^T grad_i^k}{(grad_i^{k-1})^T grad_i^{k-1}} \tag{28}$$

Where $\Delta grad_i^{k-1} = grad_i^k - grad_i^{k-1}$. In Equation (26), p_i^k represents the i th component of direction vector at the k th iteration, $i = 1,2$.

Here we use FR method to calculate coefficient β_i^k.

Step 4. Calculate each component of direction vector $p^k = [p_1^k \ p_2^k]^T$:

$$p_i^k = -grad_i^k + \beta_i^k p_i^{k-1} \tag{29}$$

Then weight coefficient shall be:

$$a_i^{k+1} = a_i^k + \mu p_i^k \tag{30}$$

Where μ represents step parameter. The selection of μ will significantly affect filtering results that a relatively big μ will cause convergent oscillation to unstabilize the system, while a relatively small μ will compromise convergent speed of system for weight coefficient can not be effectively adjusted. There is an optimal μ. Fig. 1 shows the schematic diagram of such algorithm.

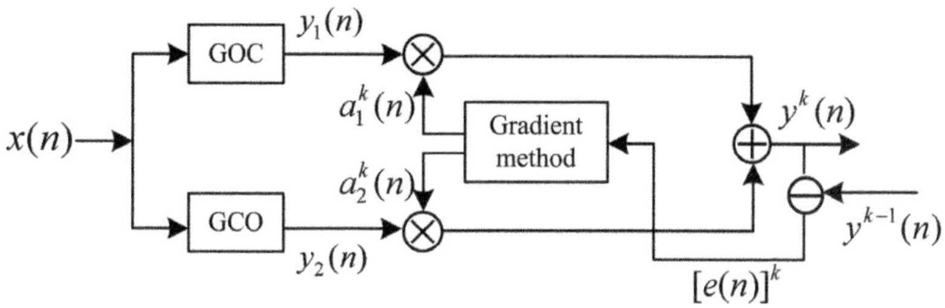

Fig. 1. Schematic Block Diagram of Adaptive Generalized Morphological Filtering Algorithm Based on Gradient Method

In Equations (23) to (30), the superscript k represents the times of iteration. $k = 0,1,2\ldots$ and initial weight coefficient is $a_i^0 = 0.5$. Let coefficient $\beta_i^0 = 0$, then $p_i^0 = -grad_i^0$ $(i = 1,2)$. Let $[e(n)]^0 = x(n) - y^0(n)$ for the first iteration. Adaptive amendment of weight coefficient can be achieved through Equations (23) to (30).

2.5 Applications of AGMF-G to denoise of friction AE signals

2.5.1 Experimental conditions

On the rotor friction test bed, the screw on friction support is adjusted to contact the rotor, so as to generate a friction source. We use UT-1000 sensor for the experiment which frequency response ranges from 60kHz to 1000kHz, preamplifier gain is 40dB, and resolution is 18 bit A/D. Rotating speed of rotor on test bed is set at 1500r/min, and sampling frequency is 1MHz. Fig. 2 shows a rotor friction test bed and Fig. 3 shows a friction AE waveform sampled on the rotor test bed.

2.5.2 Selection of structural elements

The results of morphological filtering are greatly related to the structural elements applied. The selection of such structural elements depends on the forms of signals to be processed, which structures shall be as similar to the form characteristics of signals to be analyzed as possible [12]. The selection of structural elements mainly refers to the form, width and height of a structural element. Common structural elements are linear, triangular, circular and polygonal. According to the waveform characteristics of friction AE signals, here we select triangular structural elements to discuss. The width of structural elements g_1 and g_2 in generalized open-closing filter and generalized close-opening filter respectively are 3 and 6.

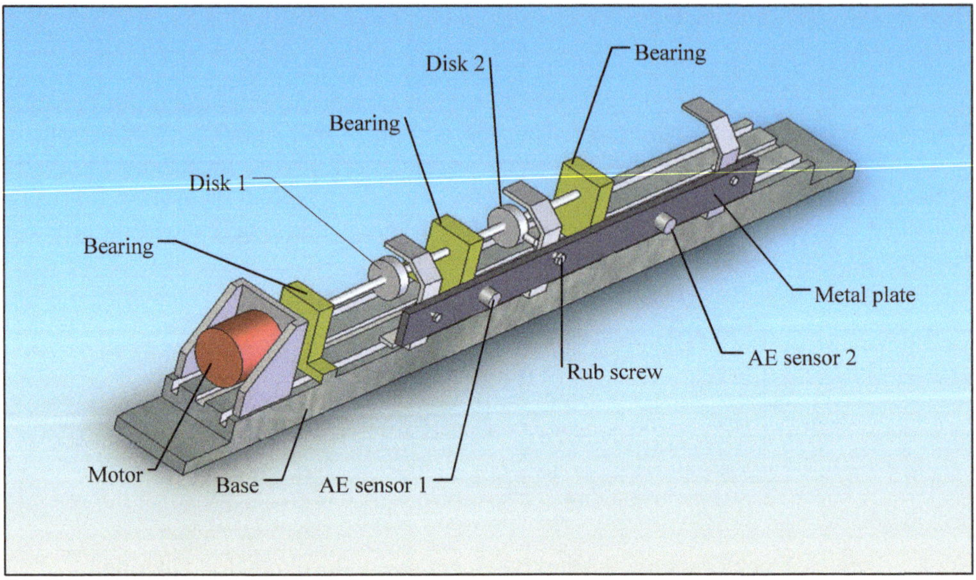

Fig. 2. Rotor Friction Test Bed

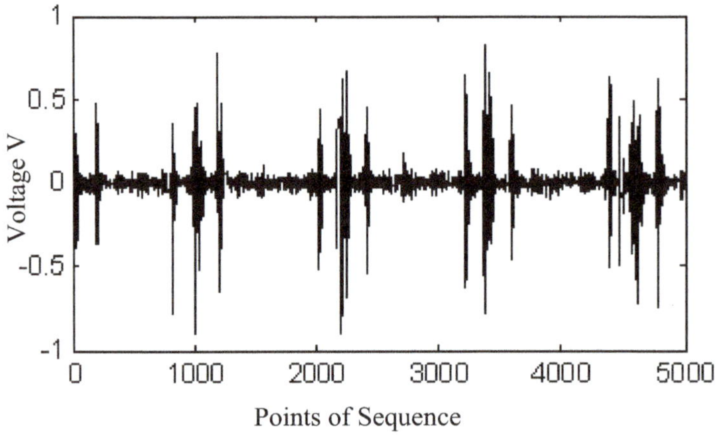

Fig. 3. Friction AE Oscillogram

Experimental simulation indicates that the height As of structural element and step parameter μ significantly affect filtering results. The analysis on how height As of structural element and step parameter μ affect denoise results is as follows.

2.5.3 Experimental analysis

The denoise effect of adaptive generalized morphological filter based on gradient method depends on the values of both height As of structural element and step parameter μ of a structural element.. Set $A\max$ as the amplitude of AE signals. Figs. 4, 5 and 6 show the relation between ratio $As / A\max$ and signal-to-noise ratio (SNR) of denoise results when superpose random white noises which SNRs are 0dB, 5dB and 10dB respectively, as well as different step parameters μ .

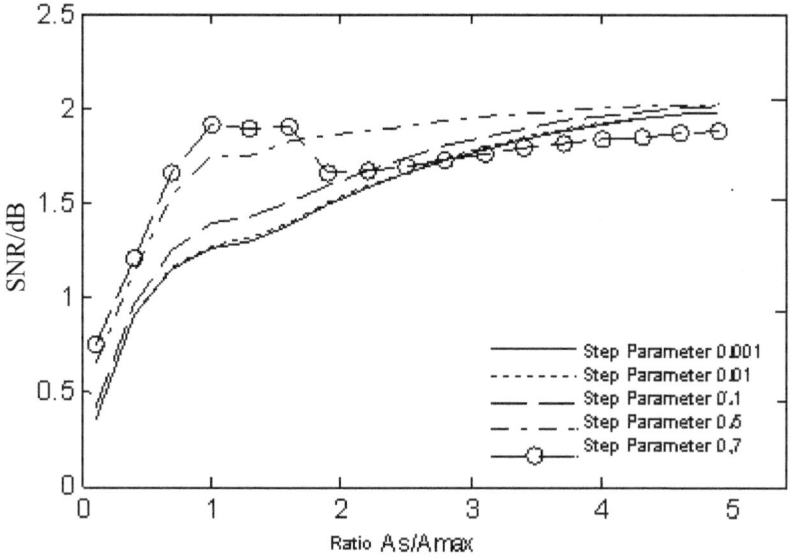

(a) Step Parameter Lower than 0.7

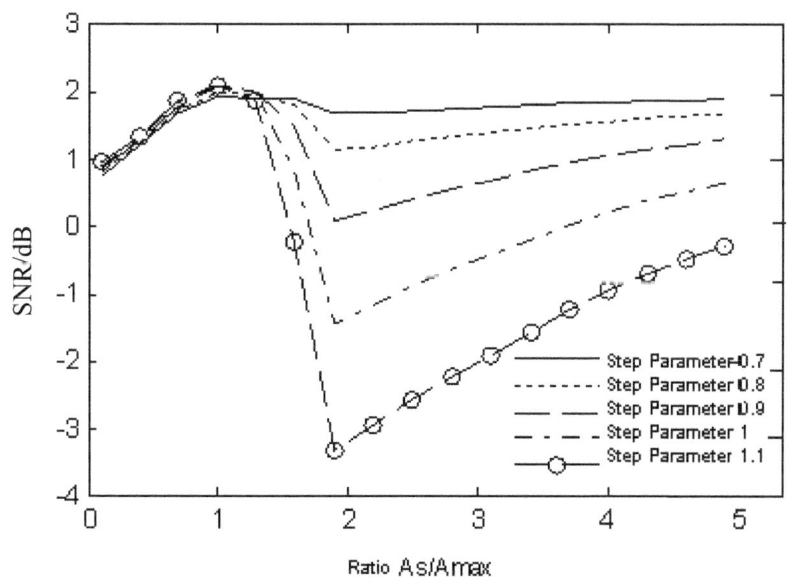

(b) Step Parameter Higher than 0.7

Fig. 4. How Ratio of Step Parameter to As/Amax Affects Denoise SNR when SNR is 0dB

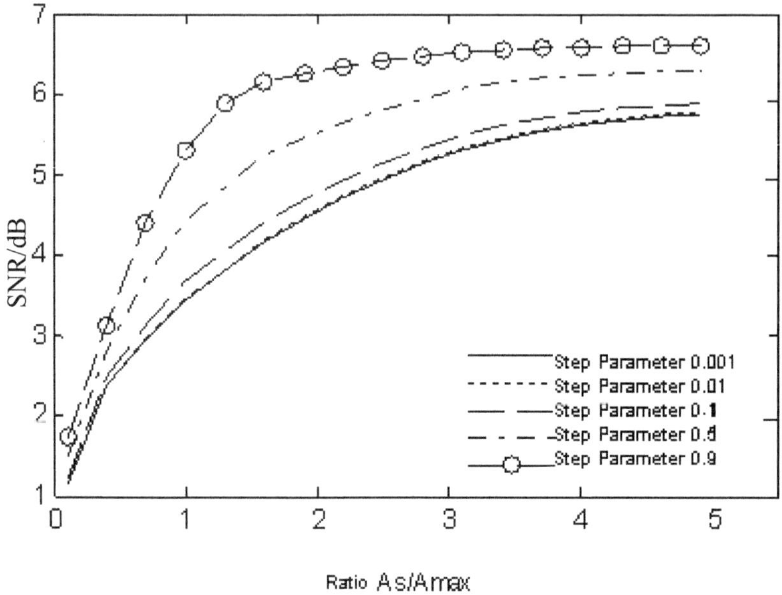

(a) Step Parameter Lower than 0.9

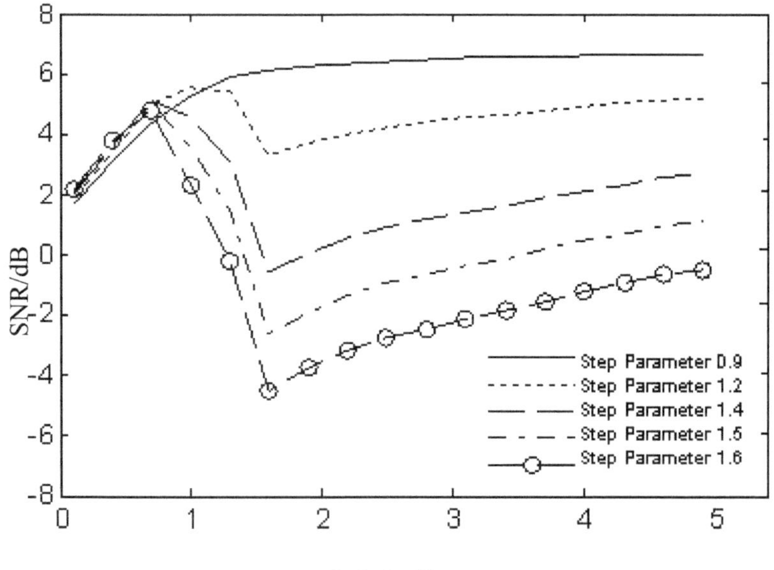

(b) Step Parameter Higher than 0.9

Fig. 5. How Ratio of Step Parameter to As/Amax Affects Denoise SNR when SNR is 5dB

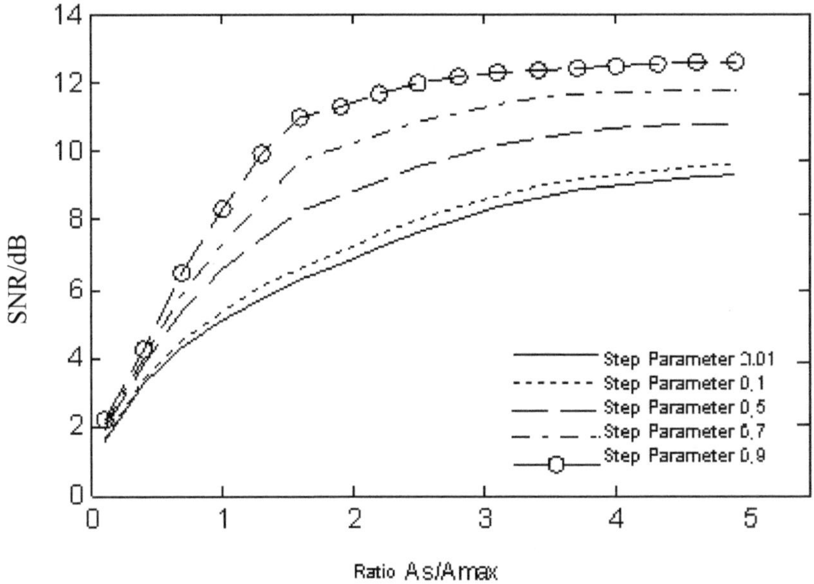

(a) Step Parameter Lower than 0.9

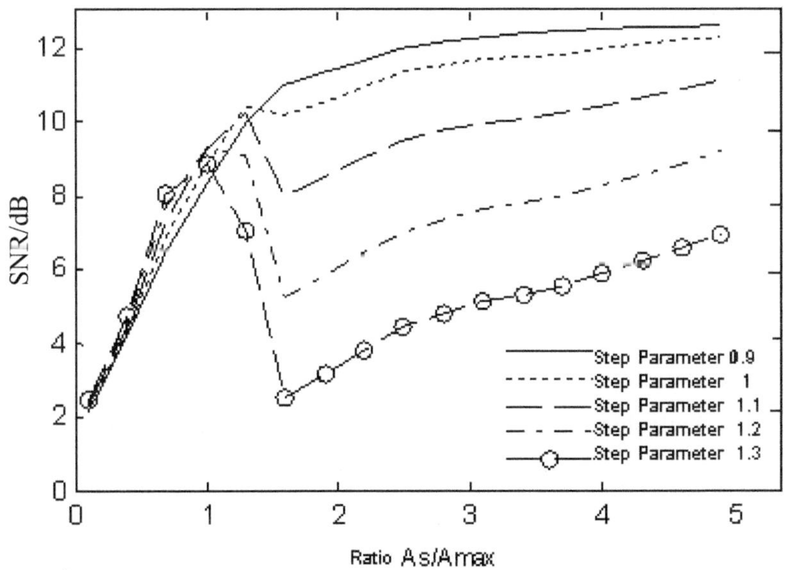

(b) Step Parameter Higher than 0.9

Fig. 6. How Ratio of Step Parameter to As/Amax Affects Denoise SNR when SNR is 10dB

Figs. 4-6 indicate that step parameter μ and ratio $As / A\max$ jointly affect denoise results. There is an optimal step parameter μ that either μ is higher or lower than such optimal vale, denoise results are worse than those when μ is at its optimal value. The above figure shows that when SNRs are 0dB, 5dB and 10dB respectively, the optimal values of step parameter μ are 0.7, 0.9 and 0.9 respectively.

The value of ratio $As / A\max$ also significantly affects denoise results. When step parameter μ is lower than the optimal one, denoise results get better as ratio $As / A\max$ increases, see Figs. 4(a), 5(a) and 6(a). However, no matter which value of ratio $As / A\max$ is, denoise results are the best when μ is at its optimal value. When step parameter μ is higher than the optimal one, denoise SNR drops first and then increases as ratio $As / A\max$ increases, meaning that denoise results vary, see Figs. 4(b), 5(b) and 6(b). However, no matter which value of ratio $As / A\max$ is, denoise results are inevitably worse than those when μ is at its optimal value.

Therefore, denoise results are ultimately determined by whether step parameter μ is at its optimal value. When step parameter μ is at its optimal value, denoise results get better as ratio $As / A\max$ increases, see the curves when step parameters μ are 0.7, 0.9 and 0.9 respectivley in Figs. 4, 5 and 6. As ratio $As / A\max$ keeps increasing, the amplitude of denoise SNR tends to 0, which makes few denoise differences. If step parameter μ is at its optimal value, the minimum ratio $As / A\max$ that makes the amplitude of denoise SNR tends to 0 is the optimal $As / A\max$. Figs. 4-6 show that when noise SNRs are 0dB, 5dB and 10dB, optimal ratios $As / A\max$ are 1, 3 and 4 respectively.

Table 1 indicates the comparison of performances among four denoise algorithms including GMOC, GMCO, GMF and AGMF-G, from which we find that denoise results obtained with AGMF-G is better than those with other three denoise algorithms.

Input Noise/dB	SNR/dB			
	GMOC	GMCO	GMF	AGMF-G
0	0.5002	0.2943	1.2596	1.9201
5	3.3396	2.7512	5.2581	6.5238
10	7.3927	5.9341	9.4739	12.4047

Table 1. Performance Comparison of Friction AE Signal Denoise Algorithms of Morphological Filtering at Different SNRs

3. Identification of friction AE based on GMM

3.1 Modal Acoustic Emission (MAE) and Extraction of Characteristic Parameters

MAE is a type of AE signal processing technology based on guided wave theory [13][14]. MAE thinks that the elastic waves generated by AE sources in materials being detected under loading effect are guided wave signals with diversified frequencies and modes, waves with different modes have different transmission speed and frequencies in transmission media, and some modal waves have frequency dispersion effect [15]. By acquiring AE signals with high-resolution broad band sensor, we may find out the composition mode in AE wave, so that we can connect these modal waves with AE source mechanism to provide prior knowledge for AE signals judging fault-source mechanism (namely fault type identification). Because different modal waves have different transmission speeds, positioning accuracy will be significantly improved if we use the same modal wave as basis during positioning process. Besides, a large number of non-AE sources or noises do not contain the characteristics of modal waves in AE signals, therefore, it'll be easy to distinguish AE signals from noises.

In this section, we analyze modal characteristics of AE waves during collision and friction excitation, as well as transmission characteristics of AE waves in two different routes of rotor bed. We propose use a mixed parameter composed of logarithmic cepstral parameter and fractal dimension as characteristic parameters of AE signal identification, so as to establish an AE signal identification system of GMM.

According to MAE theory, AE signal $\xi(t)$ received by the sensor can be

$$\xi(t) = \sum_{i=0}^{N-1} \alpha_i \xi_i(t - \tau_i) \tag{31}$$

In the above equation, AE signal $\xi(t)$ is superposed by N modal waves $\xi_0(t-\tau_0), \xi_1(t-\tau_1), ..., \xi_{N-1}(t-\tau_{N-1})$, and τ represents the time delay from modal wave ξ_i to sensor. Set a_i as 0 or 1. If it is 0, then it means that the sensor doesn't receive such modal wave. $\xi_i(t)$ represents a narrow band random process, namely

$$\xi_i(t) = \xi_{ci}(t)\cos\omega_i t - \xi_{si}(t)\sin\omega_i t \tag{32}$$

$\xi_{ci}(t)$ and $\xi_{si}(t)$ represent in-phase component of quadrature component of narrow band random process $\xi_i(t)$, and ω_i represents its central angle frequency. Equations (31) and (32) show that in a frequency domain, an AE source signal can be considered as the sum of multiple separable narrow band random processes. All modal waves have different speeds during transmission process. Due to frequency dispersion and attenuation, as well as these modal waves start to separate as transmission distance increases, some modal waves will disappear. Thus, we may construct a group of filters with central frequency of modal wave as the center, and band width no more than that of a narrow band random process. For modal waves vary slowly in a frequency domain, energy output of filters also vary slowly. While noises and interference vary randomly, we may distinguish AE signals from noises with the energy output by signals coming through this group of filters.

When a friction fault occurs in rotor system, AE waveforms only show burst characteristics for simple-point friction, while common AE waveforms generated by partial friction show continuous characteristics. Due to waveform distortion and attenuation during transmission, it's impossible for time domain parameter to effectively describe signals, so that intrinsic characteristics of signals can only be identified through time-frequency conversion.

3.1.1 Extraction of cepstral coefficient

AE source signals reach the sensor through transmission in material media, signals received by the sensor are not only affected by media channels, but also present multi-path effects. Therefore, AE signals received by the sensor are affected by multiplicative interference of the channel that characteristics of AE signals and multiplicative interference must be separated with signal processing. Traditional denoise methods can only process additive interference, while homomorphic filtering especially cepstral technology can obtain excellent results.

Cepstrum is able to interpret relevant characteristics and compensate the distortion of convolutional channels, and it's also able to separate and extract original signals as well as transmit system characteristics, which shows an outstanding robustness in a noise environment. Besides, based on first-order statistical quantity, cepstrum is able to suppress noises [16]. Here we use cepstral coefficient as characteristic quantity of identification model.

The process of extracting cepstral coefficient is shown in Fig. 7.

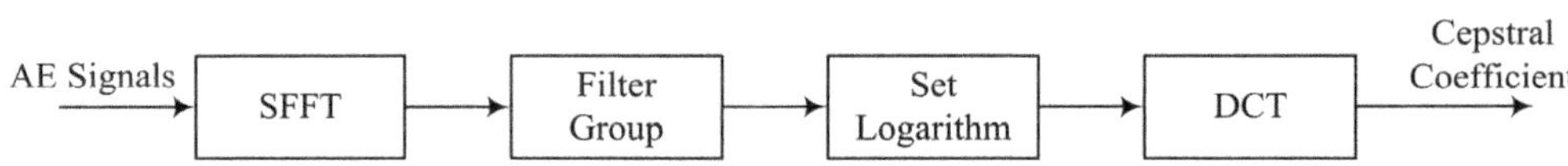

Fig. 7. Extraction of Cepstral Coefficient

Step 1. Conduct short-time Fourier transform (STFT) to AE source signal $\xi(t)$ to obtain its frequency spectrum.

$$X(k,\omega_k) = \sum_{m=-\infty}^{+\infty} \xi[m]w[k-m]\exp(-j\omega_k m) \tag{33}$$

Step 2. Conduct bandpass filtering to energy spectrum in frequency domain with a group of triangular filters. This can be considered as weighing amplitude spectrum with frequency responses of a group of filters. Weighted energy spectrum $E_{\text{mel}}(k)$ refers to Equation (34). Central frequencies of this group of bandpass filters are in a logarithmic scale order, and two bottom point frequencies of filter triangles are equal to the central frequencies of two adjacent filters. Generally, set the number of filters N as 24.

$$E_{\text{mel}}(k) = \frac{1}{A_l}\sum_{k=L_l}^{U_l} |V_l(\omega_k)X(k,\omega_k)|^2$$

$$k = 1,2,....,N \tag{34}$$

$$A_l = \sum_{k=L_l}^{U_l} |V_l(\omega_k)|^2 \tag{35}$$

In Equation (35), $V_l(\omega)$ represents the frequency response of the lth logarithmic scale filter, and L_l and U_l represent the minimum frequency and the maximum frequency of each filter in a non-zero value-taking interval. A_l is introduced mainly for normalized processing of filters based on band width.

The test shows that the main frequency of rotor friction AE signals ranges from 5kHz to 500kHz. Frequency points contribute differently to AE judgment. HF signals have bigger attenuation during transmission, so the proportion of HF components shall be reduced. Introduce a logarithmic function so that the central frequency of a filter can be:

$$\frac{\ln[1+a(f_i - f_{\min})/(f_{\max} - f_{\min})]}{\ln(1+a)} = \frac{i}{24},\ i = 1,2,....,N \tag{36}$$

In Equation (36), a represents coefficient ($a > 0$), f_i ($i = 1,2,...,N$) represents central frequencies of N filters, and $f_{\min}$ represents lower frequency limit. Due to that low frequency range (below 100kHz) is interfered by other industrial noises, therefore, here we take 100kHz. $f_{\max}$ represents upper frequency limit which is 300kHz here.

With Equation (36), we can obtain

$$f_i = \left\{ \frac{\left[\exp\left(\frac{i}{24}\ln(1+a)\right) - 1\right] \times 200}{a} + 100 \right\} \text{kHz} \tag{37}$$

Set N=24 in practical. The bigger a is, the lower LF filter band width is, and the higher HF filter band width is. Set a =10 for identification system analysis later on.

Step 3. Set a logarithm for the output of filter group, and then conduct discrete cosine transform (DCT) of 2N points to obtain an modificatory cepstral parameter.

$$C_n = \sum_{k=1}^{N} \log E_{\text{mel}}(k) \cos[\pi(k-0.5)n/N],\ n = 1,2,....,L \tag{38}$$

In Equation (38), L represents the number of cepstral coefficients, which normally is set to be 12-16, and here we take L=12. For 0-order cepstral coefficient represents frequency spectrum energy, we don't use 0-order cepstral coefficient for measurement and definition of spectral distortion.

3.1.2 Fractal dimension extraction of AE waveforms

Friction AE signals of rotor system possess fractal characteristics. Fractal dimension not only can be considered as a characteristic parameter of friction AE signal identification, but also can be considered as an index to determine the intensity of friction. DENG Aidong [17] presented a fractal dimension algorithm based on wave length. He used a box which dimension was δ to cover the whole AE wave curve. Set $N_\delta(F)$ as the smallest number of curves covered by the box which side length is δ, and set $l_i(\delta)$ as the curve length in the i th box, so that AE wave length can be

$$L(\delta) = \sum_{i=1}^{N_\delta(F)} l_i(\delta) \tag{39}$$

The curve length $l_i(\delta)$ inside each box is equivalent to: $l_i(\delta) = k_1\delta$ where k_1 represents coefficient that can be set between 1 and 1.5 in practical. Set δ_0 as k times of the minimum sampling interval (namely the box which dimension is δ): $\delta_0 = k\delta$ where k is an integer. The algorithm is as follows

$$D = A - \frac{\ln L(\delta)}{\ln \delta_0} - \frac{B\delta_0}{L(\delta)\ln L(\delta)} \tag{40}$$

In Equation (40), D is called the wave length fractal dimension (WLFD), where $A = 1 + \ln k_1 / \ln \delta_0$, $B = r(\delta / \delta_0 - 1)$, and parameters A, B, k are determined with fractal Brownian curve.

In practical, we may conduct framing processing to noise-bearing friction AE signals that have been sampled and quantified. Because the number of sample points in each frame is relatively low, it causes fractal dimension of frame varying greatly and high variance. In order to improve the accuracy of fractal dimension, we may conduct medium filtering to fractal dimension of frame first to filter HF components, so that the variance of fractal dimension can be reduced. For noise-bearing data in each frame, we may equally divide them into m (m is an odd number) data segments which will be in $D_{i-v}, \ldots, D_{i-1}, D_i, D_{i+1}, \ldots, D_{i+v}$ order. Set the data which serial number is in the middle as output of medium filtering

$$Y_i = Med\{D_{i-v}, \ldots, D_{i-1}, D_i, D_{i+1}, \ldots, D_{i+v}\}, \quad v = \frac{m-1}{2} \tag{41}$$

3.2 GMM-based identification system

GMM approaches any distribution through linear combination of several Gaussian probability density functions that it can well describe the spatial distribution and its characteristics of training data in parameter space. In a friction AE identification system, GMM models probability density functions contained in eigenvector of different modal waves, and clusters through these eigenvector. Each clustering can be considered as a multi-dimensional Gaussian distribution function. Calculate the mean, covariance matrix and probability of each clustering and set as the training format of each modal wave, and then put the characteristic sequences of AE signals to be detected into the format of each modal wave to obtain the maximum posterior probability, namely the modal wave identified by correspondence。Here we only have to know if AE signals exist rather than understand the details of modal waves at receiving end. Therefore, combine the output likelihood ratios of all modal wave models to obtain a total likelihood ration, and then judge based on such total likelihood ratio. See Fig. 8 for identification system model.

A model is composed of training stage and identification stage. Although AE signals are non-stationary, they can be considered stationary in a short time window that can be analyzed with stationary process. Therefore, we may divide AE signals into several frames with short time intervals, and then we use overlapping segmentation method to achieve smooth transition between frames. Here we divide the whole AE signal sequence to be trained or identified into several short time intervals that each with 512 points and 50% overlapping.

Based short time intervals, we extract a 12-dimensional logarithmic cepstral parameter (remove No. 0-dimensional cepstral parameter) and fractal dimension of wave length according to Equations (38) and (41) to form a 13-dimensional AE signal characteristic parameter.

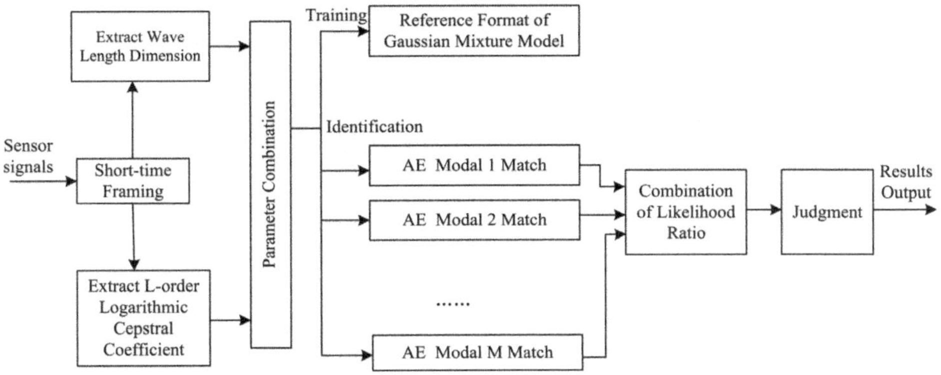

Fig. 8. GMM-based AE Identification Model

In a rotor system, AE waves may show up in multiple patterns during transmission including flexural waves, spreading waves and torsional waves, so that each type of wave in the model can be trained and identified with a corresponding Gaussian model.

GMM is the weighted sum of M modal densities. The likelihood ratio corresponding to vector $\vec{\mathbf{x}}$ extracted from AE signals can be expressed as M Gaussian components

$$p(\vec{\mathbf{x}} \mid \lambda) = \sum_{i=1}^{M} a_i b_i(\vec{\mathbf{x}}) \tag{42}$$

Here $\vec{\mathbf{x}}$ represents a D-dimensional random vector; $b_i(\vec{x})$ $(i = 1, \cdots, M)$ represents density function of each modal; and a_i $(i = 1, \cdots, M)$ represents the weight of i th Gaussian component. Each modal density represents a Gaussian function of D-dimensional variable with respect to mean vector $\vec{\boldsymbol{\mu}}_{\mathrm{i}}$ and covariance matrix $\boldsymbol{\Sigma}_{\mathrm{i}}$

$$b_i(\vec{\mathbf{x}}) = \frac{1}{(2\pi)^{D/2} \mid \boldsymbol{\Sigma}_{\mathrm{i}} \mid^{1/2}} \exp\{\frac{1}{2}(\vec{\mathbf{x}} - \vec{\boldsymbol{\mu}}_{\mathrm{i}})' \boldsymbol{\Sigma}_{\mathrm{i}}^{-1} (\vec{\mathbf{x}} - \vec{\boldsymbol{\mu}}_{\mathrm{i}})\} \tag{43}$$

Where mixture weight meets the following condition

$$\sum_{i=1}^{M} a_i = 1$$

Complete parameters of AE identification model of Gaussian mixture density are obtained through mean vectors, covariance matrixes and parameterized mixture weights of all modal densities. Together, these parameters can be expressed as

$$\lambda_i = \{a_i, \vec{\boldsymbol{\mu}}_{\mathrm{i}}, \boldsymbol{\Sigma}_{\mathrm{i}}\},\ i = 1, \cdots, M \tag{44}$$

Each modal wave can be expressed as one GMM and its model parameter λ_i. For the sequence of T test vectors $\mathbf{X} = (\vec{\mathbf{x}}_1, \vec{\mathbf{x}}_2, \cdots, \vec{\mathbf{x}}_{\mathrm{T}})$, its GMM probability is

$$P(\mathbf{X} \mid \lambda) = \prod_{i=1}^{T} p(\vec{\mathbf{x}}_{\mathrm{t}} \mid \lambda) \tag{45}$$

AE signals may go through modal transform and multi-modal coexistence during a certain period of time. The optimization can not be achieved with Equation (45). Therefore, the following shall be conducted

$$p(\vec{\mathbf{x}}_t,\lambda)=\sum_{i=1}^{M}a_i p(\vec{\mathbf{x}}_t\mid\lambda_i) \tag{46}$$

Equation (46) adopts the combination of maximum ratios. a_i represents the weight of branch channel, related to the probability $p(\vec{\mathbf{x}}_t\mid\lambda_i)$ of branch channel. Branch channel with bigger probability has larger weight coefficient, vice versa. To simplify the analysis, we may set $a_i=p(\vec{\mathbf{x}}_t\mid\lambda_i)$, thus Equation (46) may be

$$p(\vec{\mathbf{x}}_t,\lambda)=\sum_{i=1}^{M}p(\vec{\mathbf{x}}_t\mid\lambda_i)^2 \tag{47}$$

Put likelihood probability $p(\vec{\mathbf{x}}_t\mid\lambda)$ of each frame obtained with Equation (46) into Equation (45), we may get a total likelihood probability $P(\mathbf{X}\mid\lambda)$. It can be considered that AE signals exist if $P(\mathbf{X}\mid\lambda)$ is larger than the threshold.

3.3 Test analysis

3.3.1 Test analysis on GMM input

See Fig. 2 for test device. Set sampling set as 2MHz, and set the number of sampling data points of single trigger as 32768.

When friction occurs in the rotor, such friction actually contains tangential friction and normal collision force. AE signals we observe are waveforms stimulated by the combination of collision and friction between rotor and stator. Experimental research shows that AE modal waves stimulated by collision and friction respectively are different that collision mainly stimulates flexural waves while friction mainly stimulates spreading waves[18]. And it's hard to distinguish these two types of waves in friction AE signals acquired in practice. For the sake of simplification, we may generally classify friction AE signals based on their waveforms during transmission and their fractal dimension curves in certain structures and distances, so as to determine GMM model input type.

Fig. 9 and Fig. 10 indicate friction AE waveforms and fractal dimension curves of waveforms at distance end and near end respectively, from which we find that fractal dimension can effectively distinguish AE signals and noises. When AE signals are transmitted far enough, some frequency components are separated, which means that although there are waveform peaks, waveforms stretch a lot and are relatively flat. With fractal curves, we may divide them into two modal types as source input of identification model, see Fig. 9.

If friction AE source is at the near end of sensor, the in respect of waveforms, we find that almost all modal components gather together, and the sharp composite wave formed contains plenty of frequency components, see Fig. 10. Therefore, the whole wave packet shall be input as a model source of GMM.

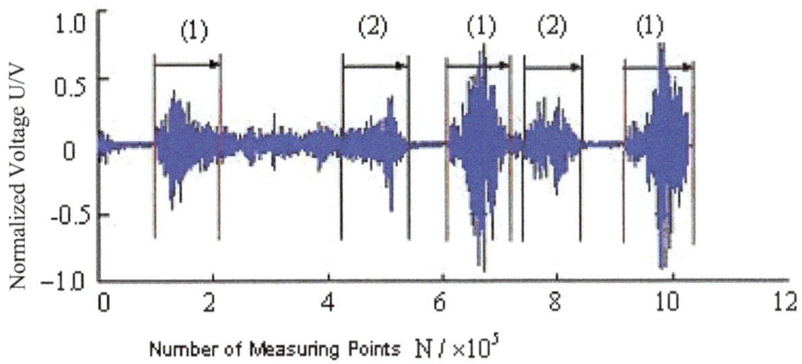

(a) AE Signal Waveform at Distance End and Division of Two Modals

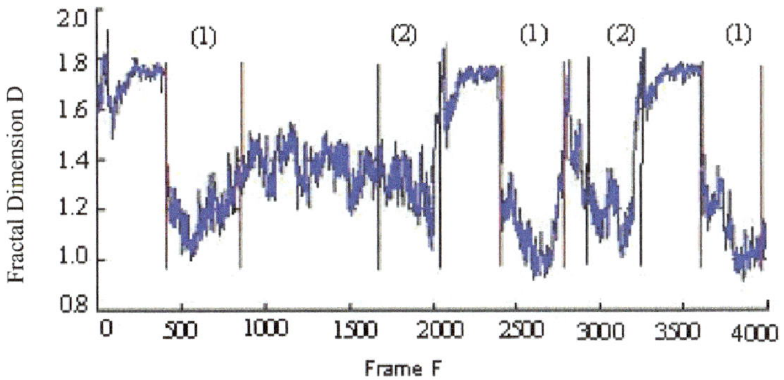

(b) Logarithmic Wave Length Dimension Curve of AE Signal Waveform at Distance End and Division of Two Modals

Fig. 9. Modal Characteristics of AE Waveforms at Distance End

Based on the analysis above, friction AE waveforms can be classified into three models. Take the situation of no AE into consideration, we may set the number of models as $M = 4$.

3.3.2 Performance analysis on AE identification model under noise environment

Acquire near end AE data 3cm away from friction source and distance end AE data 40cm away from friction source for 10 seconds on rotor friction test bed for training (rotating speed = 1450r/min). Then acquire AE test data for 10 seconds when rotating speeds are 500r/min and 1800r/min respectively. Superpose Gaussian white noises and workshop noises with different SNRs on all test data, and then identify them with the above mentioned model. For results, see Fig. 11. It shows that identification rate at near end is higher than that at distance end, identification performance under Gaussian white noise environment is better than that under workshop noise environment, and identification rates at different rotating speeds are almost the same.

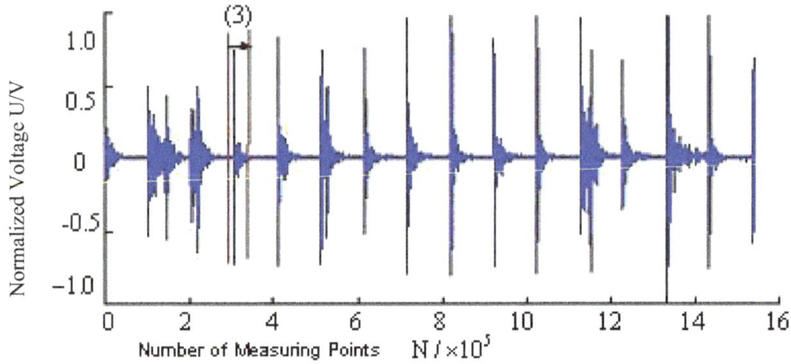

(a) AE Signal Waveform at Near End

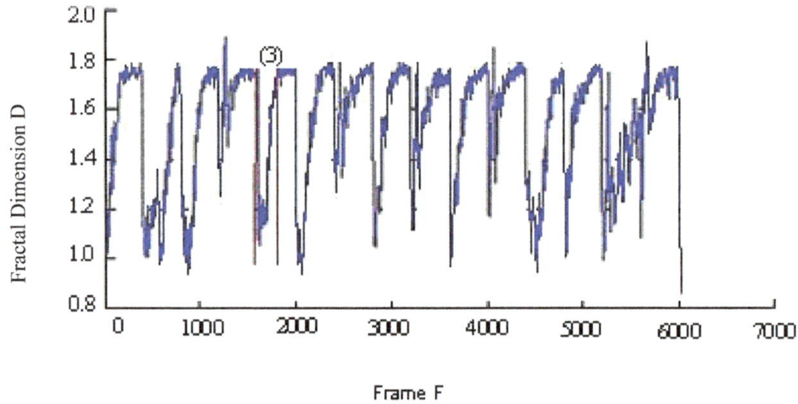

(b) Wave Length Fractal Dimension Curve of AE Signal at Near End

Fig. 10. Modal Characteristics of AE Waveforms at Near End

Compare the identification system of mixed characteristic parameters combining 12-order and fractal dimension to the identification system of mixed characteristic parameters only using 12-order cepstral coefficients. Table 2 shows that the identification performance of former one is improved at different SNRs, especially at a low SNR it's improved significantly. This means that mixed characteristic parameters not only effectively improves identification performance, but also are able to suppress Gaussian white noises to a certain degree.

Characteristic Parameters	Identification Rate r/(%)				
	5dB	10dB	20dB	30dB	∞
Ceptral Coefficient	36.7	62.9	81.1	86.2	96.8
Cepstral Coefficient + Fractal Dimension	46.2	74.3	86.9	95.4	98.1

Table 2. Comparison of Identification Rates under Gaussian White Noise Environment Based on Different Characteristic Parameters and at Different SNRs

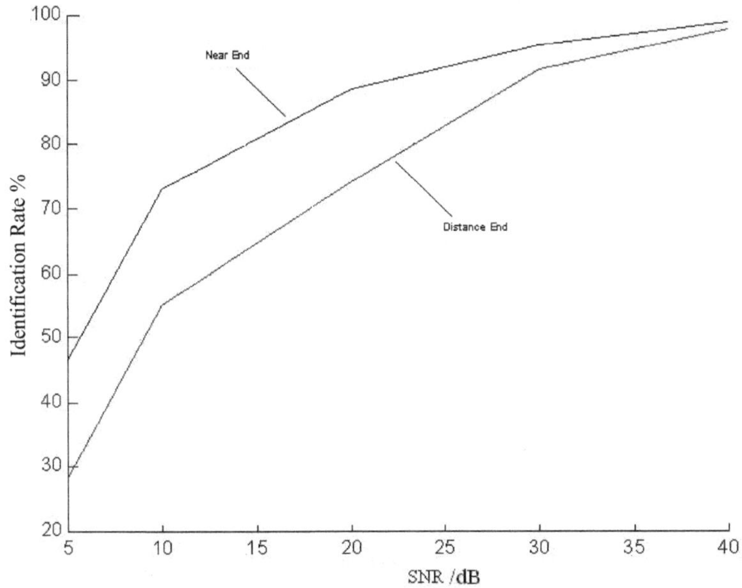

(a) Friction Identification Rate at Different SNRs and Under Gaussian White Noise Environment (500r/min)

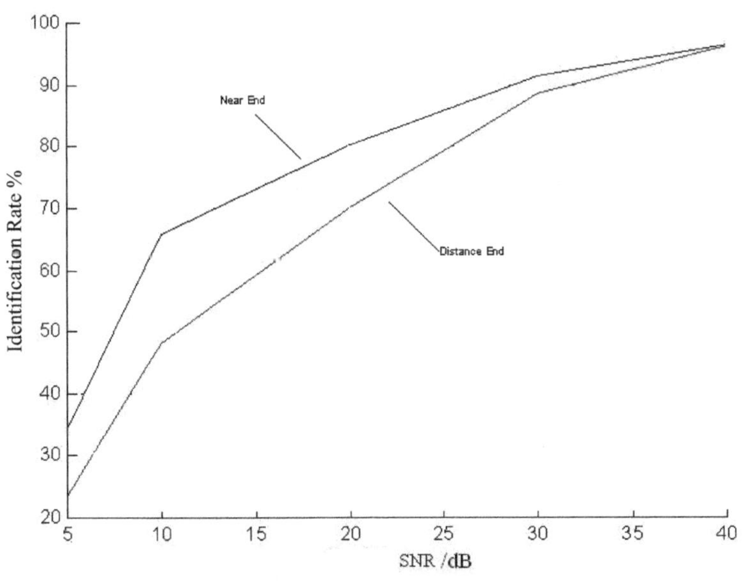

(b) Friction Identification Rate at Different SNRs and Under Gaussian White Noise Environment (1800r/min)

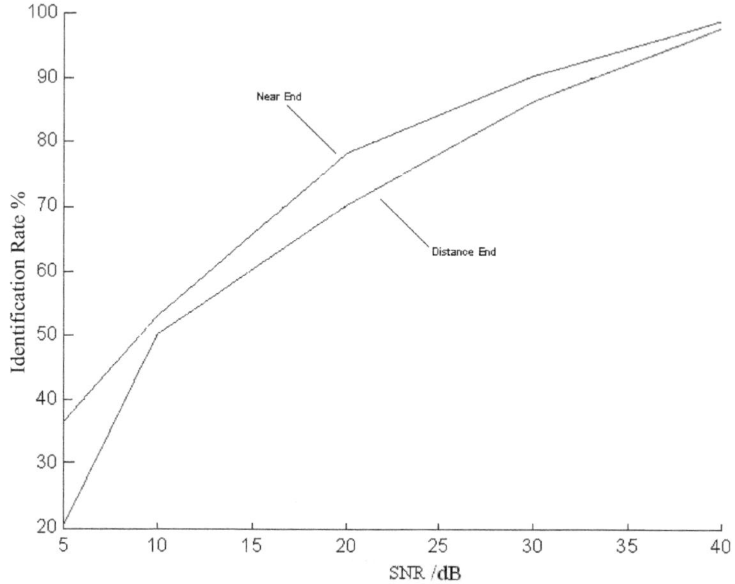

(c) Friction Identification Rate at Different SNRs and Under Workshop Noise Environment (500r/min)

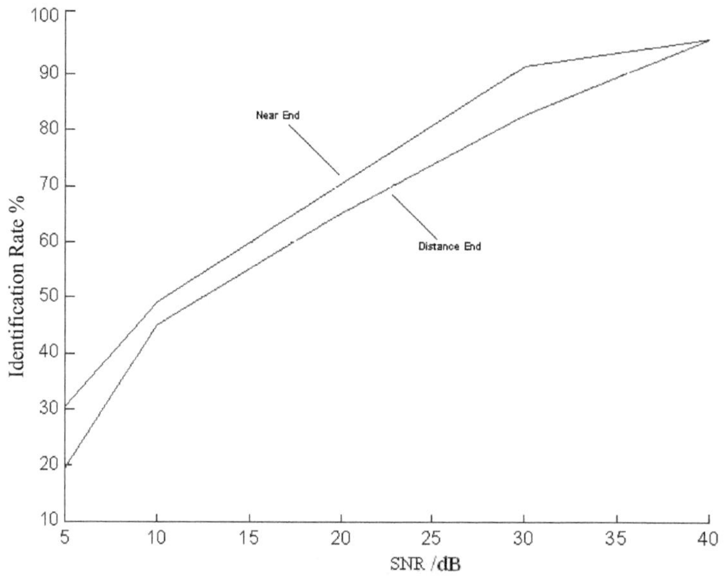

(d) Friction Identification Rate at Different SNRs and Under Workshop Noise Environment (1800r/min)

Fig. 11. AE Identification Rates under Different Noise Environments

4. Conclusion

This Chapter presents an adaptive generalized morphological filtering denoise algorithm based on gradient method, and has applied it to denoise of friction AE signals. Through experimental analysis, we may select appropriate step parameter μ and ratio $As / A\max$. Denoise effects generated by adaptive generalized morphological filtering algorithm based on gradient method are better than those generated by other traditional morphological filtering algorithms, which is a new solution to denoise of AE signals. In addition, the selection of optimal step parameter μ still has normal statistics characteristics that it needs to be determined through multiple computations and experiments. Therefore, this filtering algorithm still can be improved.

We have classified the modals of AE waveforms based on how friction AE signals transmit in a rotor system structure. Then we combine cepstral coefficient and fractal dimension of each modal wave as mixed characteristic parameters to represent friction AE signals, and then construct a friction AE identification system of GMM. Test results show that this model is able to obtain a relatively high identification rate under noise environment, and is able to suppress Gaussian white noises to a certain degree, which can be used as an approach to identify friction.

5. Acknowledgement

This work was supported by the National Natural Science Foundation of China (51075068).

6. References

[1] J. Serra, L. Vincent. An Overview of Morphological Filtering. Circuits, Systems, Signal Process. 1992, 11(1): 47-108.
[2] G. Matheron. Random Sets and Integral Geometry. New York: Wiley, 1975.
[3] J. Serra. Image Analysis and Mathematical Morphology. New York: Academic, 1982.
[4] Maragos P, Schafer R W. Morphological Filters-Part I: Their Set Theoretic Analysis and Relations to Linear Shift Invariant Filters. IEEE Trans. on Acoustics, Speech and Signal Processing, 1987, 35(8), 1153~1169.
[5] Maragos P, Schafer R W. Morphological Filters-Part II: Their Relations to Median, Order-Statistics, and Stack Filters, IEEE Trans. on Acoustics, Speech and Signal Processing, 1987, 35(8), 1170~1184.
[6] Zhao Chunhul, Sun Shenghe. An Adaptive Weighted Combination Filter Based on Morphological Opening and Closing Operators. Acta Electronica Sinica, 997, 25(6): 107-111.
[7] Hu Aijun, Tang Guiji, An Liansuo. De-noising Technique for Vibration Signals of Rotating Machinery Based on Mathematical Morphology Filter. Chinese Journal of Mechanical Engineering, 2006, 42(4): 127-130.
[8] Hestenes M R, Stieefl E L. Methods of Conjugate Gradients for Solving Linear Systems. Journal of Research of the National Bureau of Standards, 1952, 5(2):409—432.
[9] Fletcher R, Reeves C. Function Minimization by Conjugate Gradients. Computer Journal, 1964, 7(1): 149-154.

[10] Polak B, Ribiere G. Note Surla Convergence des Methodes de Directions Conjuguees. Rev Francaise Imofrmmat Recherche Opertionelle, 1969, 16(1): 35-43.
[11] Polyak B T. The Conjugate Gradient Method in Extreme Problems. USSR Computational Mathematics and Mathematical Physics, 1969, 9(1): 94-112.
[12] Zhao Wenbin, Zhou Xiaojun, Lin Yong. Application Research of Generalized Morphological Filter in Vibration Signal Processing. Transactions of the CSAE, 2008, 24(6):203-205.
[13] Gorman M R. Plate Wave Acoustic Emission. Journal of Acoustic Society of America (JASA), 1991, 90(1): 358-364.
[14] Prosser W H, Gorman M R. Extensional and Flexural Waves in a Thin-Walled Graphite/Epoxy Tube. Journal of Compos Mater 1992, 26(20): 2016-2027.
[15] Dunegan H L. Modal Analysis of Acoustic Emission Signals. Journal of Acoustic Emission. 1998, 15(1): 1-4.
[16] Kotnik B, Vlaj D, Kacic Z. Robust MFCC Feature Extraction Algorithm Using Efficient Additive and Convolutional Denoise Procedures//ICSLP'02 Proceedings, Denver, USA, 2002: 455-448.
[17] DENG Aidong, BAO Yongqiang, GAO Wei. Research on Fractal Dimension Algorithm of Rub-Impact Acoustic Emission Signal in Rotating Machinery. Chinese Journal of Scientific Instrument, 2008, 29(6): 1285-1289.
[18] Deng Aidong, Tong Hang, Jiang Zhang, Gao Wei. Characteristics of Rub-Impact AE Signals in Rotating Machinery Based on Modal Analysis. Journal of Southeast University (Natural Science Edition), 2010, 40(6):1232-1237.

15

Delayed Fracturing: Acoustic Emission Analysis of Nucleation and Transformation of an Ensemble of Mesoscopic Cracks in Deforming Heterogeneous Materials

Albert Leksovskii
Ioffe Physical – Technical Institute, Russian Academy of Sciences, St. Petersburg, Russia

1. Introduction

Acoustic emission is accompanied by many physical processes and phenomena, including discrete structural changes in solids during deformation. Information about these processes is very important for determination of the resource strength and durability of elements of machines and constructions in order to monitor and prevent technogenic catastrophes. Submitted experimental results demonstrate the unique capabilities of acoustic emission method in the analysis of nucleation and transformation of an ensemble of micro- and mesocracks. This allows us to formulate adequate, physically based models of the process of material damage in the course of its operation.

A lot of conventional engineering materials, such as polymers, metals, composites, as well as coarsely heterogeneous materials (rocks, concrete etc.), response on the applied deformation through the nucleation of microscopic and mesoscopic cracks. Such cracks could appear in a short time after the load application, and so cracked material might work a long time. Under condition of long-term operation under loading, these cracks in many cases allow one to use a concept of damaging that is deterioration of physical and mechanical properties. There is very limited information on the nucleation and transformation of the crack ensemble up to now. How far and under which conditions do cracks of micro- and mesoscopic size determine altogether the performance of material and the critical state of loaded body? Our study is to elucidate the real properties of both individual mesocracks and the ensemble of such cracks.

The cracks of micro- and meso-scale play, far and away, the role of indicators of local critical state of structure. These defects, which, undoubtedly, accumulate in a solid under the action of mechanical stress, are regarded as damages. A lot of microscopic studies of metals evidence (Botvina, 2008, among many others) that cracks of meso-scale are presented in multitude in vicinity of stress concentrators, in localities of heterogeneous development of inelastic deformation. And the ensembles of such cracks, what is their role? The ensemble/cluster of cracks is an entity that gives rise to the probability of the direct elastic interaction between defects in heterogeneous materials that exhibit microplasticity. Is this an unavoidable formation that specifies a necessary stage of the main crack nucleation? Does one might regard such the ensemble as an idiosyncratic embryo of the main crack? Or else any other physical mechanism underlies the organization of the critical state of deformable solid?

The diffraction techniques (Regel', Slutsker & Tomashevskii, 1974, Kraus at al., 1993, Schors at al., 2006,) give precious few information on the nucleation and development of the microcrack ensemble with the exception of the fact that these ensembles are formed quite rapidly (say, in creep mode), and then their further evolution becomes, really, exhausted. As a rule, it is postulated that even the cracks of such a scale can be regarded a priori as conventional stress concentrators (Arias et al., 2003, Bouchbinder at al., 2004, Nasseri at al., 2006 among others); to our best knowledge, the latter postulate was not confirmed experimentally.

Moreover, this approach was not justified in SEM in situ experiments (Leksowskij & Baskin, 2011). Contrariwise, the ensemble of a limited number of microcracks appearing in a short time interval at a certain stage of the microcrack growth inhibits the steady crack propagation.

As regards conventional engineering materials, there are no any appropriate instrumental opportunities to monitor in real time not only the interactions between micro- and meso-cracks but also the evolution of the crack ensemble as a whole.

However, one could expect that the application of amplitude-time statistical analysis of acoustic emission (AE) signals in a mode of linear location might substantially extend our conceptualization of multiple-site microdamaging in deforming heterogeneous materials with uniform structure. A great difference in physical and mechanical properties between the host material and reinforcing fibers ensures preferred breaking of high-modulus elements under conditions of common deforming, thus providing the unambiguous identification of the AE source from measurements of signal amplitudes. In the case of composites, one can establish the structure, vary the properties of elements, control the nucleation of the collection of cracks of the given size, investigate some statistical aspects of nucleation and development of the ensemble of such cracks; using a linear location method, even the nucleation and evolution of a particular, limited group of cracks could be traced both in time and in the given sample cross-section. In addition, one could consider not only the behavior of a model sample but also a response of the real engineering material with the same structural elements.

2. Samples and equipment

The samples for a SEM in situ study were made of hot solidified epoxy resin and shaped to a double trowel with the working part of 10×6×(0.7-1) mm, and they had artificial stress concentrator (incision). In order to eliminate the surface charge, the samples were preliminary saturated by iodine. Finally, they were dark-cherry-coloured, quite brittle; their ultimate strength was about 25-35 MPa. The samples' geometry and the thermal treatment provided the stressed state close to that of plain strain type.

The composites based on either carbon or boron fibers of 7-8 μm and 100 μm in diameter, respectively, were used for an AE study. The selected composites differed in the scale of structural elements and, correspondingly, in the size of defects generated under stretching.

Let us consider consistently how the AE technique allows one to obtain direct information on the nucleation of defects of the scale of structural elements, on ensemble of such defects, on cooperative effects in behavior of defects under conditions of active stretching of special model samples as well as samples of real composite materials.

The main equipment set for carrying out the AE study was developed at the Kurchatov Atomic Energy Institute. The analogue module provided the processing speed not less than 10^3 pulse/s, so, the time interval τ between resolved pulses in train was not shorter than 1 ms. The dynamic amplitude range was not less than 60 dB; the operating frequency band was 0.1-1.1 MHz with nonuniformity of amplitude frequency response ± 3 dB; the noise level as reduced to preamplifier input was not more than 3 μV. The output of digital block included the following data: signal amplitude A (μV), number of oscillations (i. e. number of excesses of discrimination threshold), signals/events duration θ (ms), intensity $\dot{N}$ (number of AE signals/events per time unit, N/s), signal energy ($W = V^2\theta$), signal power $\acute{W}$ ($V^2 \times \theta$/s), energy per pulse (W/pulse); time difference between signals detected by antenna transducers (for determination of coordinates of signal source). In the case of 3D presentation, the knowledge of the latter parameter enhances substantially the information content and reduces a field of hypothetical speculations.

The samples were model materials composed of polymer and metal matrices reinforced by uniaxially oriented carbon or boron fibers. In the case of epoxy-based CFRP (binders ED-20, EDT-1, and EFNB; carbon strip ELUR-P 0.1), the samples were shaped to plates of 160×(7-10)×(0.3-1) mm. The volume content of fibers with diameter of ~7-8 μm was $V_f \approx 60\%$.

The boron-aluminium samples D16T-B were prepared by diffused pressure welding. The size of the test section was 2.1 × 4.3 × 22 mm; the volume content of fibers with diameter of 100 μm was 1, 2.5, and 18 %. There were approximately 30 fibers in the cross-section of the sample D16 –B (2.5%). Samples of the composite based on the aluminium alloy AMG61 were prepared by plasma-sputtering technique; boron fiber content was 43 %. The deformation rates were 0.017, 0.118, or 0.525 mm/min at 20, 200, or 300^0 C. Two piezoceramic transducers with resonance frequency of 400±50 kHz were fixed on the test section at the distance of 80-100 mm one from another.

3. SEM in situ experiment

Let us consider briefly some previously unknown features related with the nucleation and evolution of individual microcracks .

Certain aspects of the nucleation and development of micro- and mesocracks are available from in situ studies of model samples made of conventionally "structure-free" polymer with the help of the SEM technique.

The experiments have shown the following: i) microcracks nucleate in an explosive fashion, and then their initial velocity of propagation decreases in approximately two orders of magnitude (fig. 1a); and ii) the crack nucleation is the event with the inevitable relaxation of the finite-amount, elastically accumulated energy (fig. 1b).

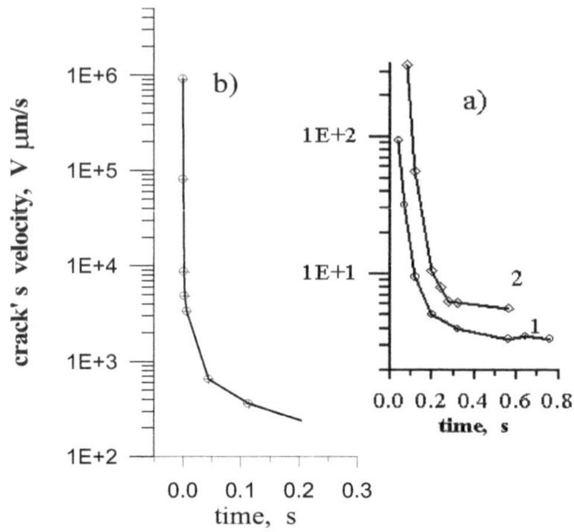

Fig. 1. Instant rate variation of newly nucleated "natural" microcrack *a)* at the initial (*1*, l_0 = 0.5 μm) and final (*2*, l_0= 0.9 μm) stages of crack formation, and *b)* in the case of crack initiating in the bulk polymer due to a rupture of high-modulus *d* = 100 μm boron fiber.

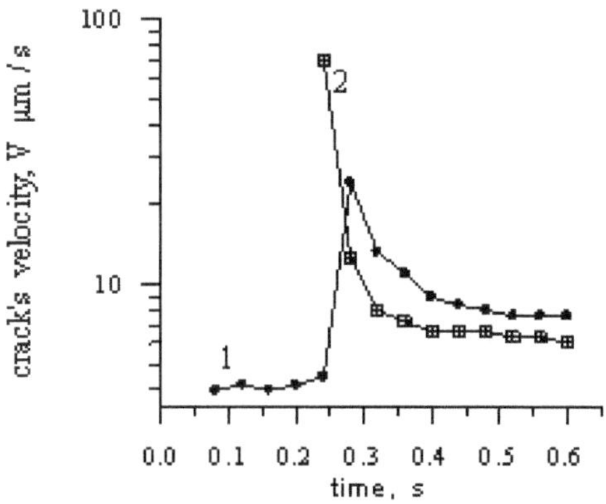

Fig. 2. Variation of the propagation velocity of an "old" crack (1) of 5 μ in length caused by the nucleation of a new crack (2) of 1.5 μ in length in its neighborhood.

The experiment has shown (see fig. 2) that the strain energy that releases due to the generation of microcracks in the course of the relaxation-evoked redistribution exerts the short-term (but considerable) change of the velocity of neighbouring "old" microcrack. It was found also that the final size of a (micro-) mesocrack depends on the dissipative properties of material (see fig. 3).

From the data presented in fig. 4, one can conclude that the dissipative properties of material influence the crack propagation rate and on the rate of local stress redistribution. The distribution maximum is not at all in the time interval of elastic interactions (1–3 μs) but falls in the range of 10^2 μs or even 10^3 μs in dependence of the dissipative properties of the matrix. Thus, one can state that the dissipative properties of the matrix could play the role of a regulator of stress redistribution (Tishkin, Gubanova, Leksowski & Yudin, 1994)

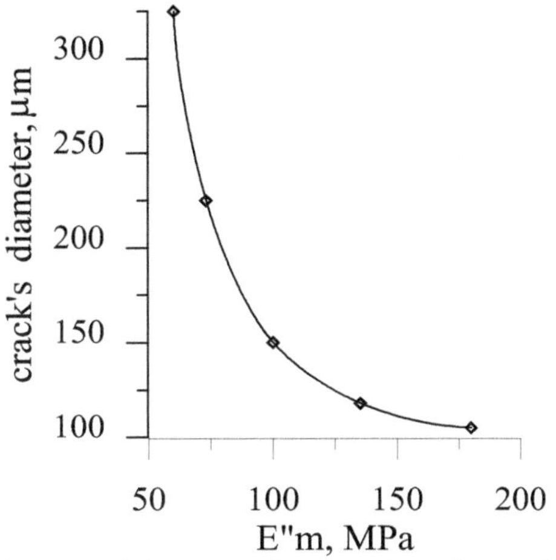

Fig. 3. Dependence of the critical diameter d_{cr} of crack in the model composite on the maximum modulus of mechanical losses E''_m of the Rolivsan binder used as a matrix.

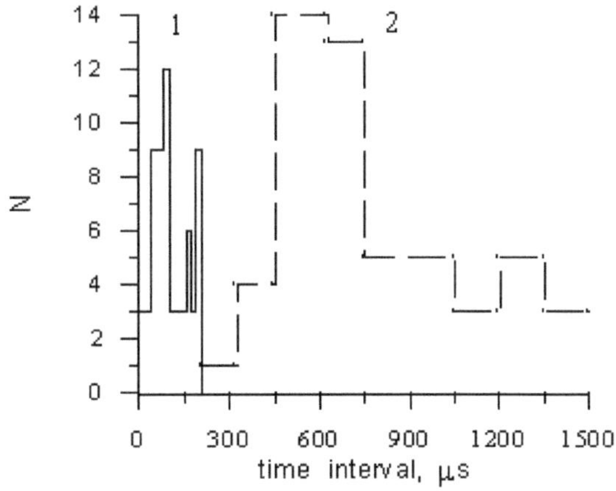

Fig. 4. Histograms of the numeric distribution of time intervals between acoustic emission signals for the carbon plastics based on (1) PEI-N and (2) PEI matrix.

The data shown in fig. 4 demonstrate that the breakage of neighboring fibers in real carbon fiber reinforced plastics, CFRP, ($V_f \sim 60\%$ carbon fibers) may occur in the course of the relaxation redistribution of local stresses [4]. In this case, the time interval between neighboring breaks ranges from tens of microseconds to hundreds of milliseconds. These times are much longer than the time of the redistribution of the elastic stress (for the given CFRP structure, the temporal range of elastic interaction is 1-3 μs) but they are sufficiently short to neglect any increase of the external stress in these experiments. In other words, the dissipative properties of materials affect the rate of the stress relaxation redistribution, thus delaying a break of neighboring fiber after the primary fiber rupture. This evidences directly the prevailing role of the relaxation mechanism of local stress redistribution in the competition of elastic mechanism in microdamaging.

The in situ SEM data and model experiment confirmed the fact of the presence of both the dynamic and relaxation stages in development of newly-born (micro-) mesocracks. It is very important to note, that the explosion-like formation of a microcrack is related to a discrete change in the local strain energy with the inevitable subsequent transient relaxation process.

4. Ensemble of cracks of the scale of structural element

4.1 Cracks of 10 μm

Figure 5 shows a fragment of the load-extension diagram for the special model ED-20-based sample with a single fiber, and results of AE signals detection under stretching with the rate of 0.017 mm/min. As the fiber reaches its ultimate strain (~1.1 %), its fragmentation begins. Number of detected AE pulses and coordinates of failures calculated from the location data coincide quite well with the same data on number of failures and coordinates found in direct microscopic measurements. 50-60 failures were detected along the test section length of 40 mm. The statistical processing of lengths of broken fibers in a series of identical samples gives the estimated average value equal to ~ 280 μm.

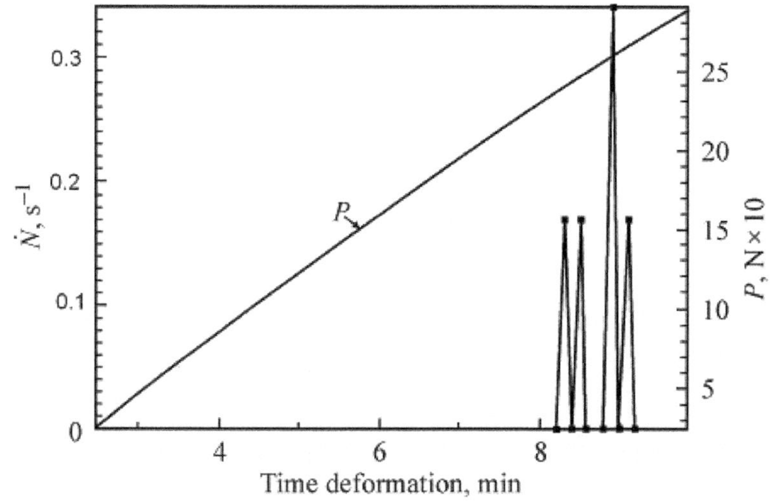

Fig. 5. Variation of AE intensity $\dot{N}$ from loaded sample containing a single fiber.

The amplitude distribution of AE signals is depicted in fig. 6. The value of the AE signal amplitude from a single failure (as reduced to input of preamplifier) was 90-130 μV. It should be noted that the observed dispersion of signal amplitudes could be due to a number of causes, such as variation of sample strength along its length, signal decaying, and inaccuracy of fiber positioning in the sample relatively the axis of loading. The latter inaccuracy results in emerging of a shear component and, correspondingly, reduces the value of ultimate stress. The average amplitude value, $\bar{A}$, which is equal to ~ 100 μV, could be regarded as first approximation when estimating the number of simultaneously (here within 1 ms) broken fibers.

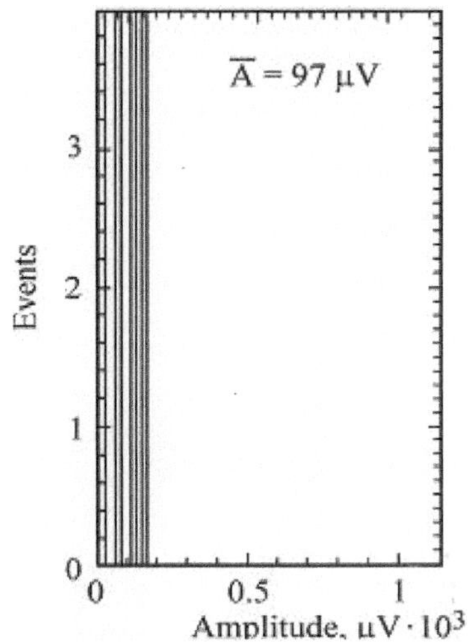

Fig. 6. Amplitude distribution of AE signals.

Now we shall consider the behavior of uniaxially reinforced sample that contained not one but a few thousands of carbon fibers, the volume content of which, V_f, was about 60 %. A fragment of the microscopic structure of pressed plastics is depicted in fig. 7.

We shall consider how the accumulation of fiber failures, that is crack-like damages, occurs in the sample stretched along the direction of reinforcing. What happens in heterogeneous system when in spite of a single crack comparable in size with structural elements, a certain multitude of cracks appears?

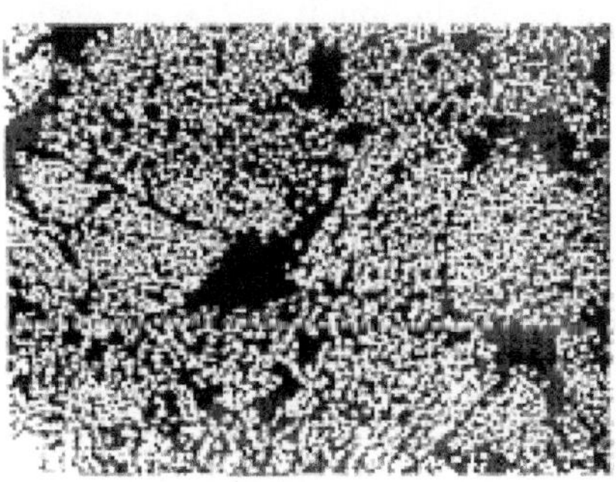

Fig. 7. A fragment of the cross-sectional sample CFRP

The location of AE signals (see Fig. 8) demonstrates that the process of fiber failures under constant rate deformation flow neither gradually nor avalanche-like but consists of alternate stages of acceleration and deceleration. A break of the sample occurred in the region with coordinate (5-13) mm. One can see that this locality did not exhibit any particularities during all the deformation cycle. There were neither multiple failures of fibers nor any other signs of "preliminary preparation" of the macroscopic failure. The fiber breakage process in adjacent zones is well pronounced at particular stages of deforming. The intensive process of fiber breakage was observed at coordinates (21-13) mm in the time interval 1 to 5 minutes.

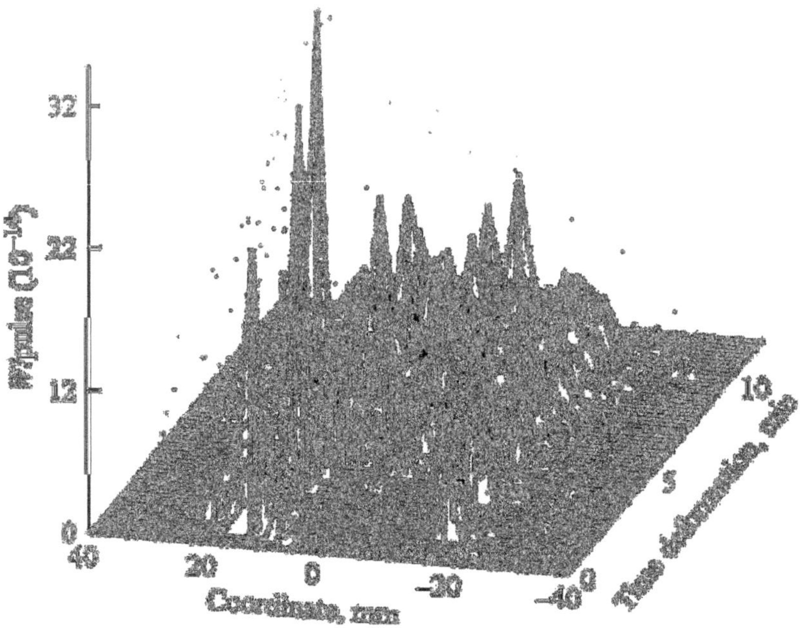

Fig. 8. Variation of average energy in pulse (AE event) over cross-section of CFRP with fiber content of 60 % as deformation increases. The deformation rate 0.22 mm/min.

Nevertheless, there was not a further process localization in this particular volume. The appearance of a group of cracks could reduce the effective elastic module of this part of the sample with inducing the load redistribution. It will be seen below that the locality and time of macroscopic damage are not due to previous damaging of any domain but caused by the dynamics of the energy dissipation processes, which leads to the exhausting of the energy capacity of system.

Let us consider this process in details with the help of the statistical analysis of trains of AE signals from different sites.

On should expect the existence of the interrelation between AE events because the AE source is related, predominantly, with the crack nucleation. According to the data of in situ SEM experiment (Leksowskij & Baskin, 2011), a microscopic crack cannot nucleate without leaving a trace in its nearest environment; such an event is accompanied by either a short-term acceleration of neighboring crack or a nucleation of the new one.

The using of equipment which provides the precise determination of pulse arrival allows one to perform the AE time analysis by constructing the function of distribution of time intervals between pulses. To obtain correct data from the statistical analysis, one should provide sufficiently representative sampling (70-100 signals); in addition, the process itself should be essentially stationary. For example, the intensity variation in the train of AE pulses, as a rule, should not exceed 20 %. In the case of complete independence of AE signals, the process is regarded as poissonian with the following analytical representation $\rho(\Delta\tau) = m \times \exp(-m\Delta\tau)$, where m is the event frequency, $m = \dot{N}$ (AE intensity); $\Delta\tau$ is the time interval between events; $\rho(\Delta\tau)$ is the probability density.

Now we shall consider the distribution of time intervals between AE pulses in certain sample cross-sections and in different instances of deformation in order to reveal some particular features that would evidence the presence of correlation and interrelation between failures of reinforcing fibers. Figure 9a shows an example of Poisson distribution of time intervals between AE pulses from a deforming CFRP detected at the time interval (480-510) s in zone (5-13) mm taken from the whole diagram W/pulse (l, t) (fig. 8).

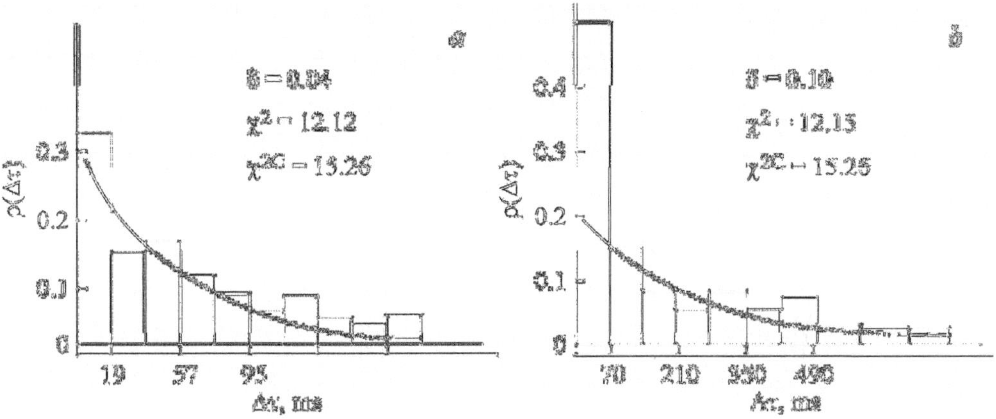

Fig. 9. Distribution of time intervals between pulses in zone of future failure at different stages of deforming.

One can see that in this case the distribution of time intervals between pulses does not differ from the theoretical (poissonian) exponential distribution. This means that the fiber failures process in this locality is independent, poissonian process. However one could expect that such a situation takes place not always and not in any place. Figure 9b shows the distribution of time intervals between pulses in the same part of the whole diagram (5-13) mm but at the later stage of deforming (590-616) s. This distribution differs from the theoretical (poissonian) one. The main deviation is observed in the time interval ~ 70 ms. In such a case, one may suggest the poissonian process of crack accumulation (Tishkin and Leksowski, 1988) with a certain local ordering though the events in complete flow remain independent. The parameter $\delta = | 1 - \Delta\tau / \sigma_{\Delta\tau} |$ may serve as a quantitative measure of such ordering (deviation from poissonian-like process) (Braginski, Vinogradov and Leksowskii , 1986). Here $\overline{\Delta\tau}$ is the average time interval between pulses in the given sampling, and $\sigma_{\Delta\tau}$ is the mean-square deviation. It was revealed that such analysis allows one to obtain more detailed information on the evolving fracture process. Figure 10 shows a variation of the above defined parameter of ordering δ that characterizes the correlation between AE events in deforming CFRP in zone of future failure as well as in two neighboring zones (AE diagram in fig. 8).

One can see in fig. 10 that in the last ~ 30% of the total time of deforming of CFRP sample, the process of fiber failure / crack nucleation begins to pass from the stage of independent, non-connected events to the stage of correlated interactions. This process is more pronounced in localities situated in the vicinity of the zone of future macroscopic failure. Nevertheless, the value of the parameter δ is stable in the future-failure-zone (5-13 mm), and

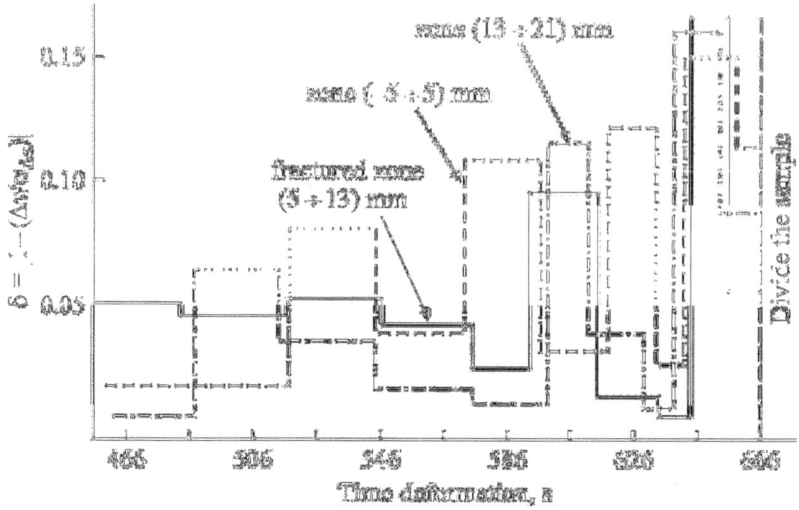

Fig. 10. Variation of the degree of ordering (δ) in the train of AE events from three neighboring local volumes in one-directional CFRP at the deformation rate equal to 0.22 mm/min.

a steady trend of its multi-fold increase is observed. So, there is a growth of ordering and correlation in the train of AE events: each event affects virtually the future ones. This trend is not monotonic but exhibits the pulsating character with periods of decay and growth of tens seconds' length. A certain phase opposition was observed in variations of the degree of ordering in AE trains from local zones in the vicinity of the position of macroscopic damage. The increase of the degree of correlation of AE events (fiber failures /crack nucleation) in a particular zone was accompanied with the deceleration of the fracture process in an adjacent zone characterized by a relative stagnancy. At the same time, it is worthy to note that even a simple conservation of a certain level of correlation in the train of AE signals during a few tens seconds under conditions of the external load increase evidences the effective discharge of the elastic energy concentration in a local volume. It is the reduction of the level of local stresses resulted from intensive structural reconfigurations that suppresses the driving force of local fracture process with exciting it in neighboring sites. Every increase of the value of the parameter δ means the increase of either the total "length" or the number of correlated fiber failures. This is a kind of start of the avalanche-like process that, as one could expect, must result in macroscopic failure. However, every time it turns out to be a false start. One should pay attention to some fine details of the formation of the pre-starting, critical state of deforming sample. In the time interval 630 to 645 s, one can see an additional decrease of the parameter δ in the zone of future failure; simultaneously, the degree of correlation rises sharply in adjacent zones. One more feature: there was a significant drop of the correlation in all three zones during last ten seconds of the deformation cycle. This means that the stress redistribution over neighboring zones takes place in this time interval.

One can see in fig. 11 that the length of trains of correlated crack formation (fiber failures) in the zone of macroscopic failure (5-13) mm at the final stage of deforming (510-660) s is equal to 50-60 μm (2-3 structural elements). There were strong detected signals neither in zone of

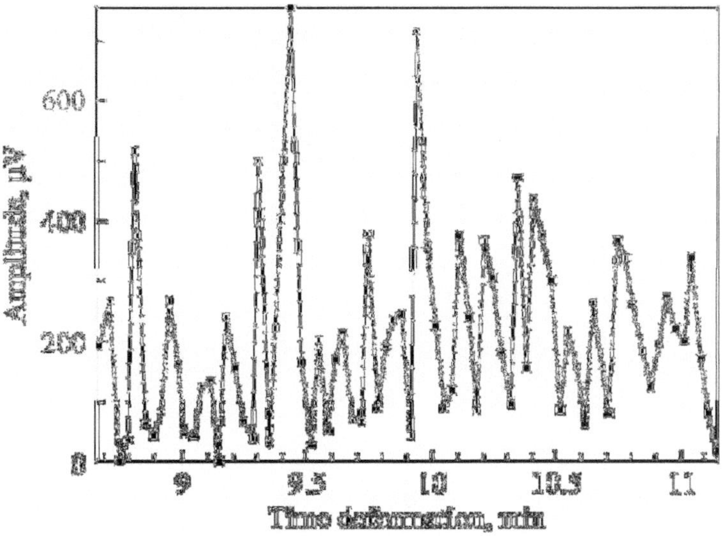

Fig. 11. Variation of AE amplitudes in zone of macroscopic failure during last 150 seconds of deforming of unidirectional carbonplastic.

macroscopic failure, nor in adjacent or any other zones in the last instance (1 ms). The data acquisition system was capable to detect amplitudes that exceed the maximum signals shown in fig. 11 in two orders of magnitude. The actual start of macroscopic failure is, really, out of the sensitivity of this equipment.

So, the statistical analysis of the train of AE signals from deforming composite with small-scale structural elements (7-9 μm) performed in the mode of linear location demonstrated the following:

- the increase of the degree of correlation of AE events (fiber failures / cracking) in a particular sample domain is accompanied with the deceleration of this process in neighboring zones;
- this evidences the effective discharge of concentration of elastic energy in a local volume under conditions of growing external load;
- it is the decrease of the level of local stress resulted from the intensive process of microstructural reconfigurations that suppresses the driving force of this process for a certain time thus inducing the similar process in adjacent domains;
- the existence of ensemble of correlated cracks is neither necessary nor sufficient condition for the formation of main crack.

4.2 Cracks of 100 μm

We shall consider this problem by the example of the composite made of aluminum alloy reinforced by high-modulus, high-strength fibers with the strength distribution shown in fig. 12. Fiber diameter was equal to 100 μm; at large volume content of fibers ($V_f > 30\%$) this composite is very sensitive to neighboring failures. In the case of small volume content, the multiple crashing of fibers takes place, what, as is known, produces in matrix a limited

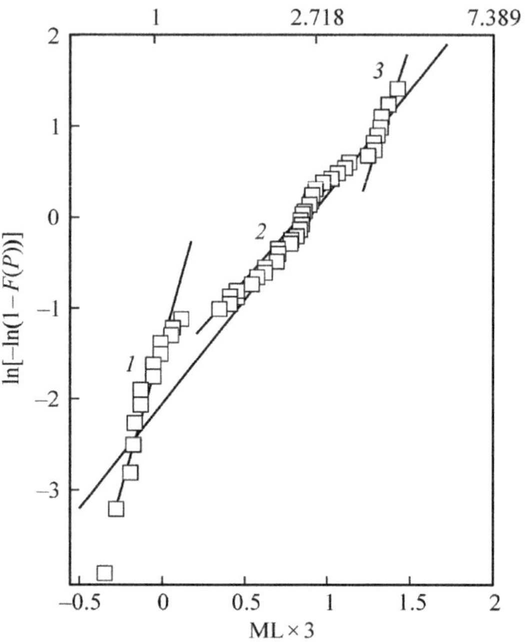

Fig. 12. Weibull distribution of boron fiber strengths

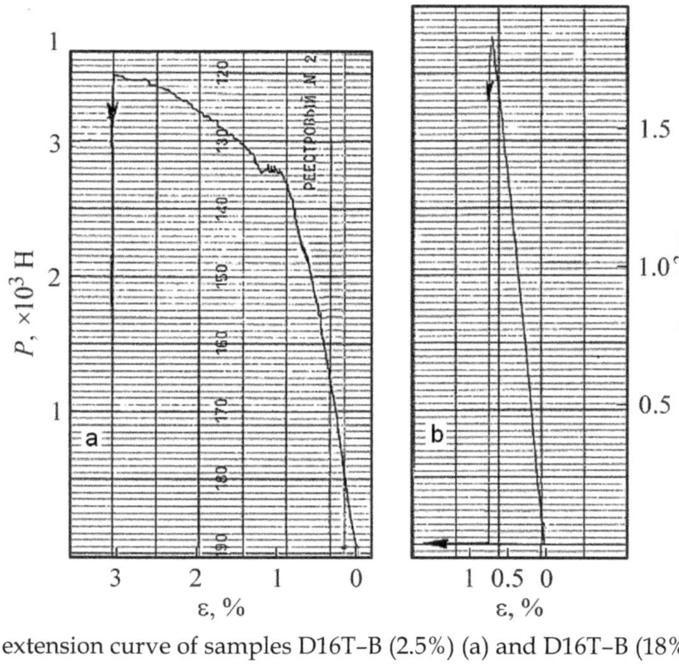

Fig. 13. Load extension curve of samples D16T–B (2.5%) (a) and D16T–B (18%) (b); deformation rate was 0.017 mm/min.

number of cracks at the initial stage of stretching but does not affect the further course of the deformation curve (fig. 13). The dimensions of defects of continuity in the low-modulus host, which appear due to failures of high-modulus fibers, are, unambiguously, close to that of structural elements. Their behavior might be monitored reliably by the method of AE signal location, which allows one to trace the course of events in real time scale. A great part of experiments was carried out with samples containing small amount of reinforcing fibers (V_f = 1 – 2.5%); in this case, the main load was carried by matrix.

The AE signals related to fiber failure are well discriminated from signals from other sources by their amplitude (fig.14), therefore, it will be implied below (unless otherwise specified) that the necessary selection was performed, and the analyzed AE is caused exclusively by fiber failures.

According to the data of metallographic analysis (fig. 15), multiple fiber failures occur in the interval 4000 to 6000 s; then the mayor load is carried by the matrix. A total of fiber failures ranges from 600 to 800.

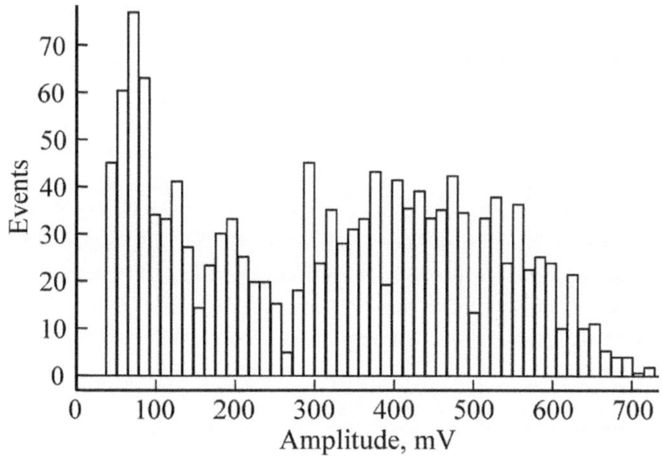

Fig. 14. AE signal amplitude distribution in deforming composite D16T – B(2.5%)

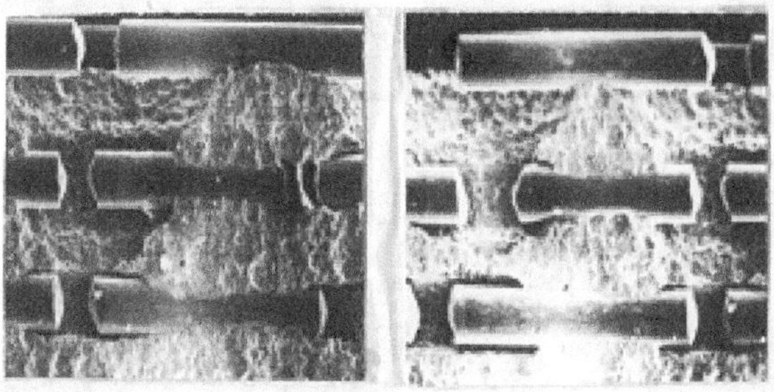

Fig. 15. Fragments of etched sample surface after extension.

Figure 16a shows a plot of the AE activity $\dot{N}$ (number of signals per unit of time) over all time of deformation at deformation rate 0.017 mm/min. The final distribution of failures along the sample length, as revealed from the location data (fig. 17), is sufficiently uniform, therefore, the critical fiber length, 1_{cr}, might be about 1 mm in agreement with the metallography (fig. 15). The performed analysis showed that independently of the used deformation rate, the region of multiple failures is related with the variation of deformation of ~ 2.5%.

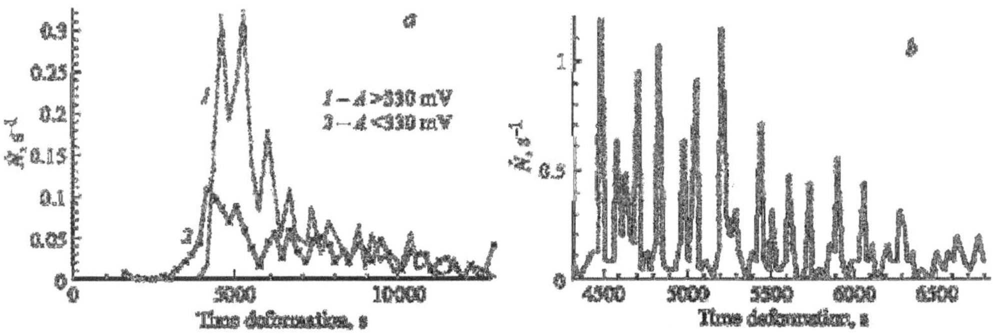

Fig. 16. AE activity ($\dot{N}$, s^{-1}) at deformation rate 0.017 mm/min; (a) shows total time output, (b) shows output in the interval of multiple failures.

The yield of the AE activity in the region of multiple crashing is depicted in fig. 16b in the extended time scale. One can see that in all cases the "fine structure" of the peak of activity manifests itself in splitting of the peak into 12-15 narrow features. The number of signals in these features varies anywhere from 25 to100 % of the total number of fibers in the sample cross-section. The location data confirm the localization of failures in limits of the peak $\dot{N}$ in a single section of 2-3 mm. As recalculated into the value of deformation, the distance between the peaks does not depend also on the deformation rate and is equal to 0.1-0.2 % all the time.

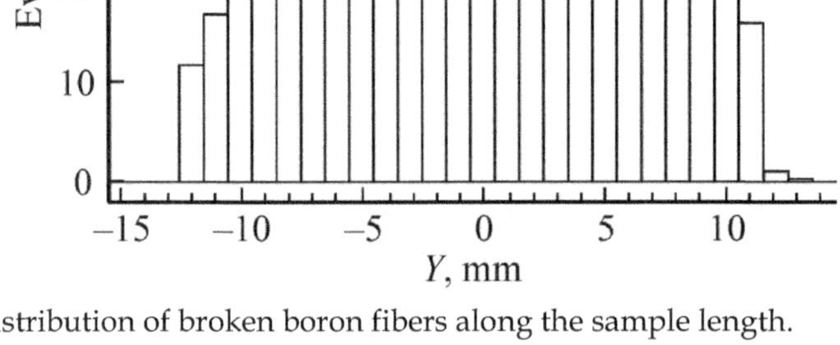

Fig. 17. Final distribution of broken boron fibers along the sample length.

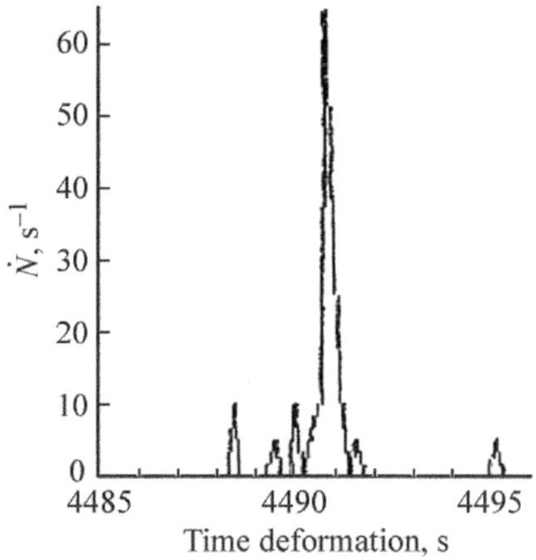

Fig. 18. AE activity in the time interval 4485 to 4495 seconds (see also fig. 16b)

The widths of the peaks themselves (see fig. 18) change approximately 8 times when the deformation rate varies 40 times. In our opinion, this gives one a reason to regard the peak width as invariant not only in the scale of deformation but also in the scale of time. This means that the multiple fiber failure in a single cross-section can be regarded as the process occurring at constant stress and constant strain, that is as the correlated process (Leksowskij at al., 2008). At the same time, it is clear that this process cannot be determined only by the elastic stress redistribution since in this case it would be over in a few microseconds and manifest itself as a very high amplitude signal. Under the given specific conditions, the used equipment would detect the AE signal train consisting of signals separated by intervals shorter than 1 ms as a single signal. As far as in our case the intervals between signals were much longer, one should conclude that the process of correlated fiber failures in the sample cross-section is determined by the relaxation properties of matrix.

Now we shall consider the process of failure of boron fibers / crack formation in this model composite sample in the 3D representation (see fig. 19).

One can see that the process of correlated fiber failures starts in one of the cross-sections in accordance with the Weibull distribution of fiber strengths. However, this process under the given conditions of experiment (deformation rate, temperature) does not cover all fibers in the given cross-section and, all the more so, does not lead to sample failure in this particular place but induces the development of the similar process in the neighboring cross-section. The coordinates of the active cross-section change in one way mode successively and continually with the accuracy limited by 1_{cr}, therefore, the process of fiber failures returns to the initially most active cross-sections. It seems that such a behavior is caused by the fact that the accumulation of failures (mesodefects of continuity) leads to the decrease of the local effective elasticity modulus.

We have seen (fig. 1-2) that the crack nucleation represents an explosive-like event with the inevitable relaxation of the finite-amount, elastically accumulated energy, and that the dissipative properties of material affect the crack size and the rate of local stress redistribution. In the case of suppressed or insufficient dissipative properties (Leksowskij &

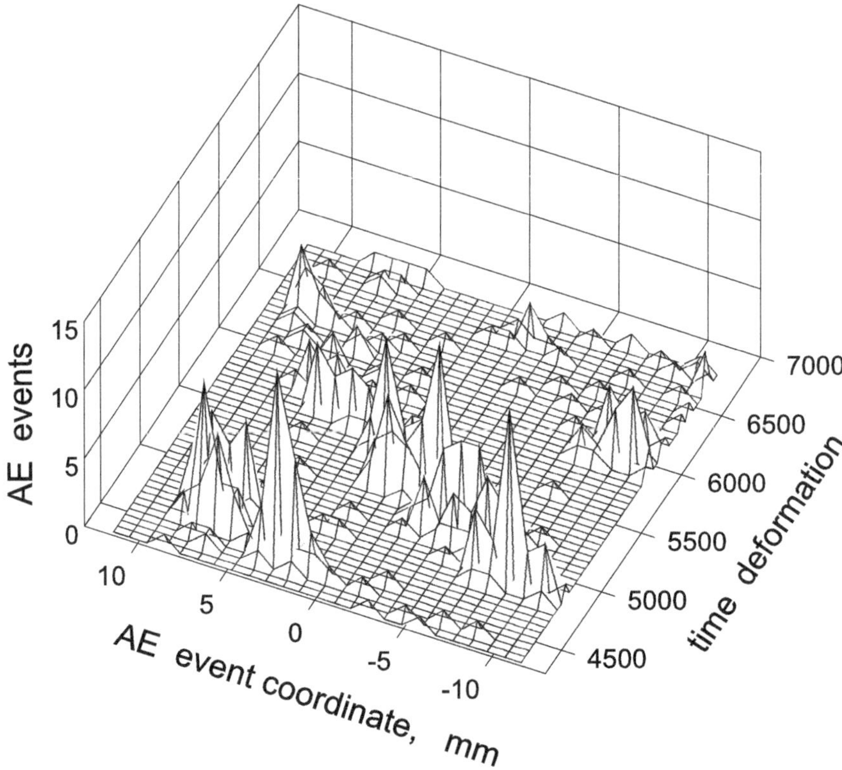

Fig. 19. 3D AE diagram of the fiber failure distribution in the course of sample deforming.

Baskin, 2011), the elastic, linear interaction between defects takes place, and the macroscopic fracture occurs. Therefore, one can control the cracking either through suppressing, minimizing the relaxation component of the local stress redistribution or by gaining the dissipative properties. We shall demonstrate below that in the general case one should consider the dissipative properties of both the nearest environment and the solid as a whole.

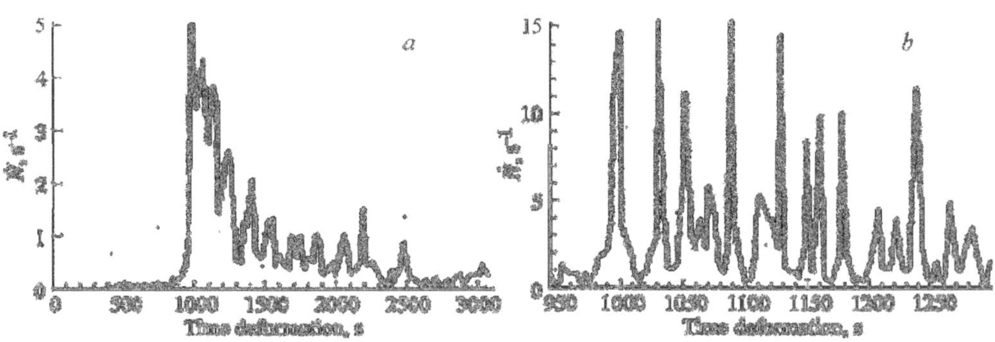

Fig. 20. AE activity $\dot{N}$ over all time of deformation of sample D16T – B(1 %) at deformation rate 0.118 mm/min (a), and the same in the time interval of multiple crashing (950 to 1300 s) (b).

When comparing the data presented in fig. 16 and fig. 20, one can see that the increase of the deformation rate a few times (from 0.017 to 0.118 mm/min) resulted in the noticeable increase of crack formation. Consequently, the restriction of the realization of the dissipative properties of matrix leads to the increase of correlated fiber failures per unit of time of deforming.

Figure 21 shows the 3D AE diagram for this sample. In this case, the crashing process started in a certain cross-section and then progressed simultaneously and successively in two different directions.

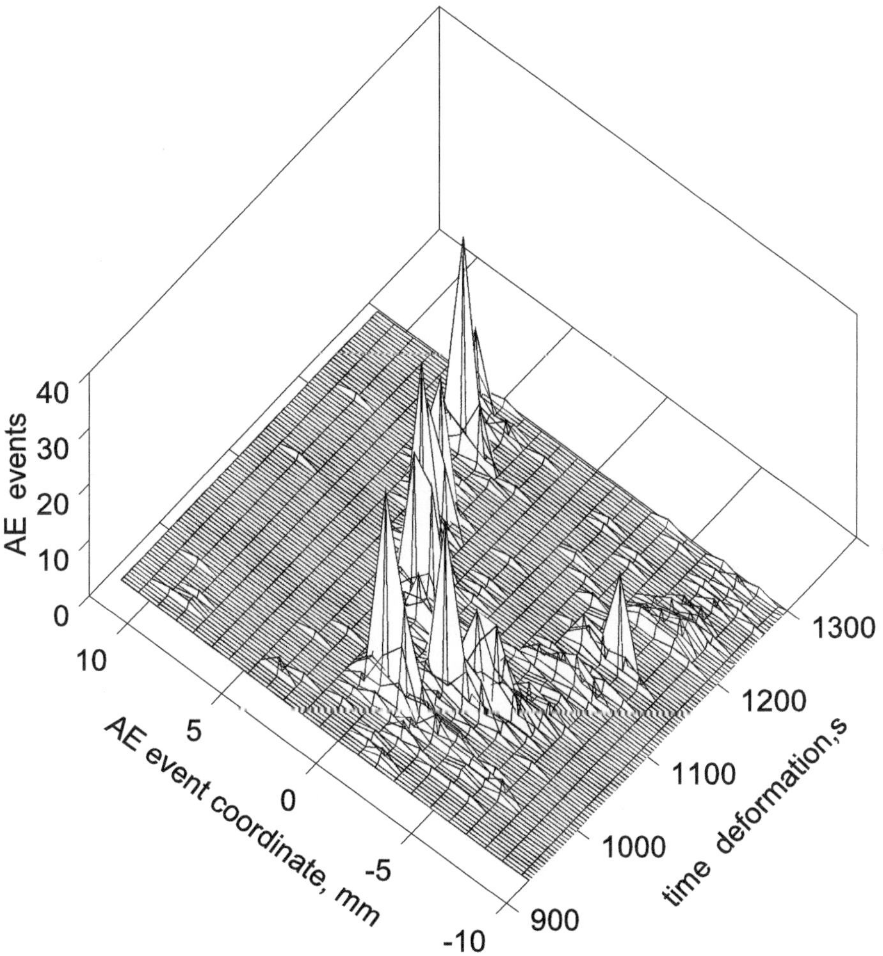

Fig. 21. 3D AE diagram of the fiber failure distribution in the course of deforming of sample D16T – B(1 %) at deformation rate 0.118 mm/min

As the deformation rate increased to 0.545 mm/min (see. fig.22), the process of fiber failures /cracking advanced still faster.

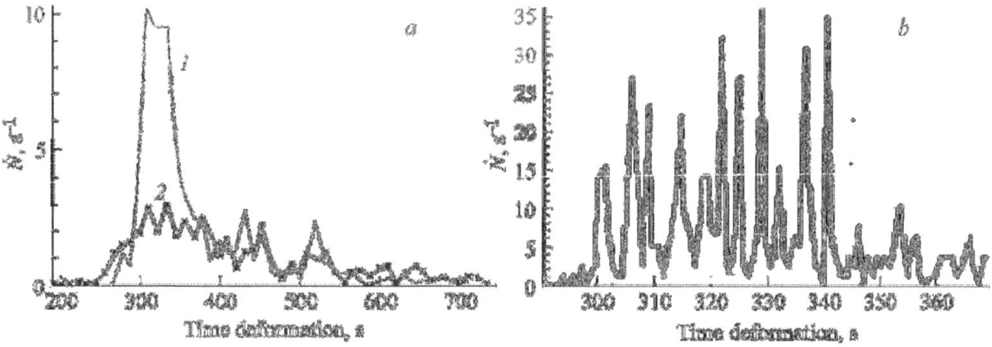

Fig. 22. AE activity $\dot{N}$ over all time of deformation of sample D16T – B(2.5%) at deformation rate 0.545 mm/min (a), and the same in the time interval of multiple failures (300 to 370 s) (b). *1* - amplitude AE signals > 270 mV, *2* - amplitude AE signals < 270 mV .

One can see in fig. 23 that at elevated testing temperature (300° C) the process of correlated fiber failure fairly degenerates. Whilst at 20° C and at the same deformation rate of 0.118 mm/min one could see 9-10 bursts of correlated fiber failures, at 300° C, the number of such clusters reduced to 2-3, and the number of fiber failures in each cluster decreased markedly. The 3D AE diagram for this sample as deformed at 300° C is depicted in fig. 24, in which one can see clearly the mentioned particularities. Under conditions providing a more efficient realization of the dissipative properties of matrix in relation to the redistribution of local stresses, the probability of correlated failures is ultimately minimized, and discrete, isolated damages are prevailing in the sample.

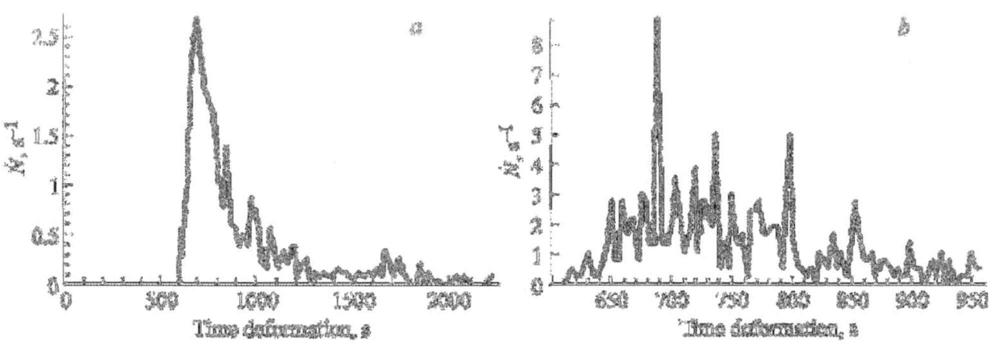

Fig. 23. AE activity $\dot{N}$ over all time of deformation of sample D16T – B(2.5%) (300° C) at deformation rate 0.118 mm/min (a), and the same in the time interval of multiple failures (650 to 950 s) (b).

In our opinion, the data given in figs 16-24 evidence unambiguously the dependence of possibility and "depth" of the development of the process of correlated cracking (fiber failures) on realization of the dissipative properties of material in the course of local stress redistribution. It is emerged that under particular conditions of experiment one can determine the dependence of the time delay of failure of fiber situated in the vicinity of fractured fiber on the amplitude of AE signal, which specifies the energy of elastic deformation released from broken fiber. Figure 25 shows such the dependence for the data depicted previously in fig. 22.

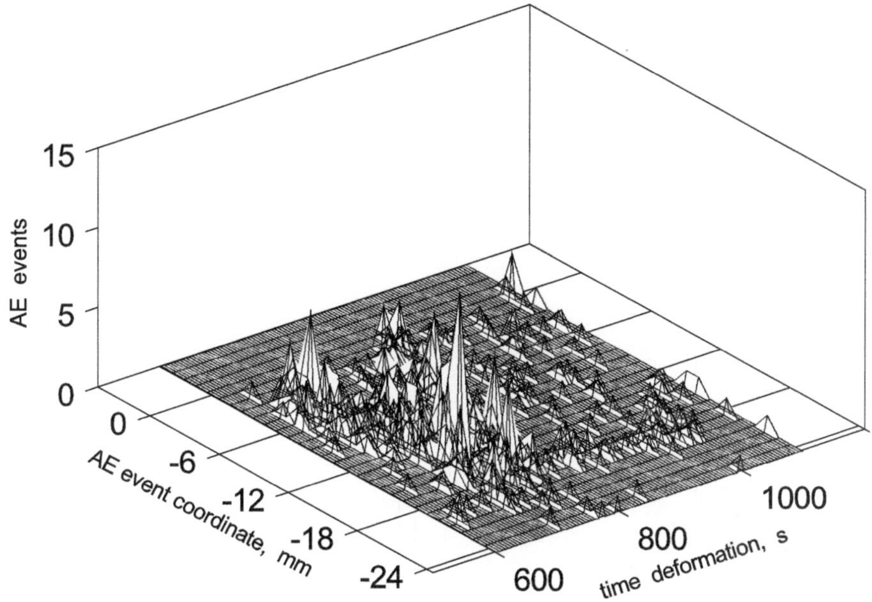

Fig. 24. 3D AE diagram of the fiber failure distribution in the course deforming of sample D16T - B(2,5 %) at deformation rate 0.118 mm/min and at elevated testing temperature 300⁰ C.

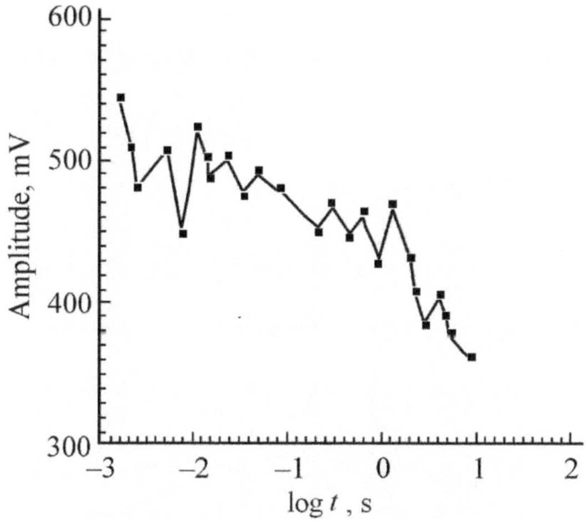

Fig. 25. Time delay of failure of fiber situated in vicinity of fractured fiber versus amplitude of AE signal.

This dependence (under the given conditions of the experiment) illustrates, in fact, the long-range influence of newly-forming defect on a similar neighboring structural element. It should be noted that this dependence is essentially a development of the results of experiments in situ (see fig. 2) for "structure-free" polymer with the help of the SEM technique as well as figures 3 and 4.

Surely, this result should be taken into account when constructing the physically grounded model of deformable heterogeneous solid.

When considering the general problem of emergence and evolution of the ensemble / cluster consisting of cracks of size close to that of structural elements, one should not overpass a very important aspect concerning the dissipative capability of solid as a whole. It was shown above that multiple starts of correlated crack nascence (each "attempt" could produce from a few unities to two tens of such cracks) are interlocked by a possibility to redirect this process to neighboring domains, which has conserved by the time their dissipative and deformation properties. In the simplest composites with a large volume content of high-strength and high-modulus fibers, the dissipative capabilities are, generally, sufficiently low.

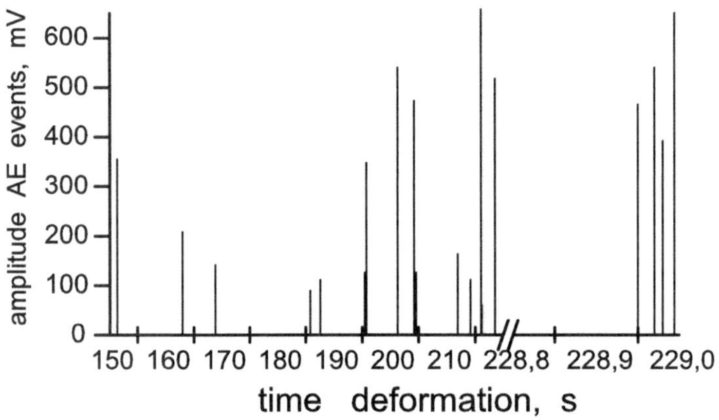

Fig. 26. Acoustic emission from the composite AMG61-B(43%) deformed at 0.118 mm/min, at 20°C.

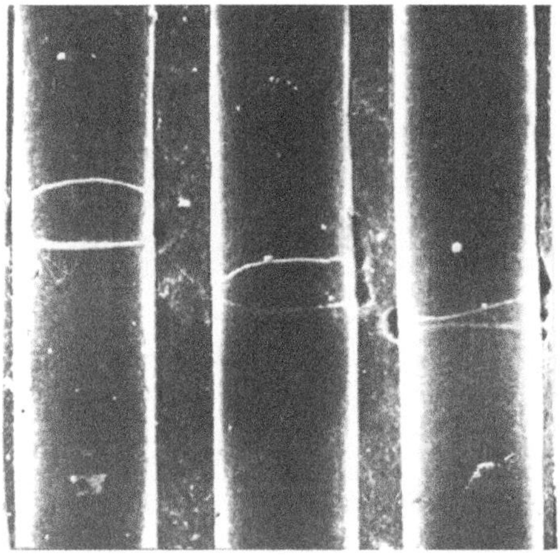

Fig. 27. Example of correlated failure of boron fibers in the sample with high volume content of fibers.

Figure 26 exhibits all the train of AE signals from the composite AMG61-B(43%) deformed at the rate 0.118 mm/min at 20°C. Total 21 signals were detected; 10 of them originated from fiber failures (in accordance with the amplitude distribution depicted in fig. 14). Four final signals at 228.953 s, 228.967 s, 228.980 s and 228.994 s represent a sequence of correlated failures. The characteristic time of the stress increase at the fiber neighboring with the broken one is Δt = 12 - 15 ms.

Figure 27 shows an example of the correlated breakage of boron fibers of diameter of 100 μm.

5. Conclusion

To finalize, let us resume the results. The AE method in a mode of linear location was applied for studying the process of nucleation of cracks of size close to that of structural elements in composites deformed at constant rate. Two model unidirectional composites were used. The high-modulus composite was composed of carbon fibers (60 % by volume) embedded in polymer host, while the low-modulus one consisted of 1 to 2.5% of boron fibers embedded in the matrix based on aluminum alloy. Under conditions of combined deforming, a great difference in elastic modulus between the host and the reinforcing fibers provides the foremost failure of the high-modulus material and, correspondingly, the unequivocal identification of the sources of AE signals.

In the case of high-modulus composition, the multiple cracking at the initial stage of deforming runs as a poissonian-like process of independent, unconnected events, while at the last third of deformation cycle it runs as the process of correlated interaction. It is widely accepted that it is the last process leads to the formation of the main crack. Our experiment demonstrated another trend. As starting in a random domain, this process decelerates soon but, inevitably, a similar process becomes induced in a neighboring domain. A few attempts to organize the start of the main crack might occur.

When deforming a low-modulus composition, such as Al-B (2.5%), this trend is still more pronounced. Owing to the high energy release from failures of boron fibers, more significant long-range interactions take place, and the process of fiber failures manifests itself as the correlated process from the beginning of deforming.

However, multiple starts of the process of correlated crack nucleation (each "attempt" could produce from a few cracks to two tens of cracks) become obstructed by a possibility of redirecting of this process over adjacent domains, which, by that time, conserved still their dissipative, deformational capacities. The number of "false starts" as well as the length of a train of correlated failures might be governed by changing the conditions of running of dissipative processes. In the case of the composition containing a great amount of boron fibers (AMG61-B(43%)), such possibilities are, in fact, lacking. Therefore, the correlated failure of 3-4 fibers with time discontinuity of 12-15 milliseconds causes the macroscopic failure.

The performed experiments allows one to suggest that the higher the excess of energy absorption capability over the elastic strain energy released when a defect continuity nucleates, the greater number of meso/micro cracks might appear, and the interaction between them would be determined by the relaxation properties. Thereby, the stage of multiple crack formation is not critical à priori. The ensemble/ cluster of cracks is neither unavoidable nor necessary stage in the formation of the main crack. During the basic time of deforming, the discrete structural changes occur under conditions of sufficient dissipative properties/ energy absorption capability. When this capacity becomes suppressed down to the critical level, a few explosion-like nucleation of mesocracks occurs as a result of emerging quasi-elastic interaction followed by the transition to the overcritical stage.

6. Acknowledgments

Author would like to thank A. Abdumanonov, G.N. Gubanova, A.P. Tishkin and V.E. Yudin for their assistance in performing model, acoustic emission measurements and useful discussion.

7. References

Arias, M. L., Frontini, P. M. & Williams, R.J.J. (2003) Analysis of the damage zone the crack tip for two rubber- modified epoxy matrices exhibiting different toughenability. Polymer. Vol. 44, p.p.1537-1546

Botvina, L. R.. (2008) *The destruction: kinetics,* mechanisms and *general laws,* Nauka, Moscow,2008 [in Russian]

Bouchbinder, E., David Kessler, D. & Procaccia, I. (2004) Crack-microcrack interactions in dynamical fracture. Phys. Rev. E 70, 046107

Braginski, A. P., Vinogradov, A. Yu. and Lekswskii (1986) *Acoustic amplitude*-frequency analysis of *the kinetics of deformation of metal glass.* Soviet Technical Physics Letters, Vol. 12, No. 9, p.p. 459-460

Kraus, R. Payer, A. & Wilke W. (1993) Acoustic emission analysis and small-angle X-ray scattering from microcracks during deformation of ETFE composites, J. of Materials Science, 28, p.p. 4047-4052

Leksowskij, A. M. and Baskin, B.L (2011) Some Aspects of Nucleation and Extension of Microskopic and Mesoscopic Cracks and Quasi-Brittle Fracture of Homogeneous Materials. Physics of the Solid State, Vol. 53, No, 6, p.p. 1223-1233

Leksowskij, A. M., Baskin, B.L, Tishkin, A.P. and Abdumanonov, A. A. (2008) Nonlinear Correlated Interaction of Mesodefects and Transition to Macrofracturing, Solid State Phenomena, Vol. 137, No.9, p.p. 9-14.

Nasseri, M., Mohanty, B. & Young, R. (2006) Fracture toughness measurements and acoustic emission activity in brittle rocks, Pure and Applied Geophysics, vol. 163, pp. 917-945,.

Regel', V. R., Slutsker, A. I., Tomashevskii, E. E. (1974) Kinetic Nature of Solids, Nauka, Moscow [in Russian]

Schors, J., Harbch., K.-W., Hentschel, M. and Lange., A. (2006) Non-Destructive Micro Crack Detection in Modern Materials, ECNDT - We.2.2.2

Tishkin,A. P., Gubanova, G. N., Leksowski, A. M. and Yudin, V. E. (1994) Acoustic emission data on delayed damage processes in the vicinity of defects fiber-reinforced plastics. J. of Materials Science, Vol. 29, p.p. 632-639

Tishkin, A. P. and Leksowski, A. M. (1988) *Correlation of the flow of acoustic emission events.* Soviet Technical Physics Letters, Vol. 14, No. 8, p.p. 636-638.

16

Monitoring of Grinding Burn by Acoustic Emission

Paulo Roberto de Aguiar, Eduardo Carlos Bianchi and Rubens Chinali Canarim
University Estadual Paulista Julio de Mesquita Filho (UNESP) – Bauru campus, Brazil

1. Introduction

In the metal and mechanical industry, grinding is usually the last step of the manufacturing chain, when it comes to precision components. This process is used in the fabrication of parts of a wide variety of materials, which require low surface roughness, strict dimensional and shape tolerances, while maintaining maximum tool life and the shortest operation time, along with minimum costs.

Damage of the workpiece during grinding is very costly, since it has already a high aggregated value from the previous manufacturing processes. The most common types of damage in grinding are burning, cracking, and undesirable residual stresses. In the case of metals, the most common cause of damage is excessive heat generation on the ground surface.

The high temperatures generated in the grinding zone may cause several types of damage to the workpiece, e.g., burns (in the case of metals), excessive hardening of the surface layer, with possible rehardening and increased brittleness, undesirable residual tensile stresses, reduced fatigue strength, and cracking (Malkin, 1989).

The combination of acoustic emission signals and cutting power have been used successfully to determine indicative parameters of burning (Kwak & Ha, 2004). These signals, properly manipulated, allow for the implementation of a burn control system in real time, optimizing thus the grinding process (Dotto et al., 2006). This would be highly beneficial for companies that strictly depend on this process, since the requisite of quality and international competitiveness increases continually with globalization (Brinksmeier et al., 2006).

On the other hand, the growing interest in the use of artificial intelligence for the solution of engineering problems is visible from the considerable number of articles published in the last decade. These problems are normally difficult to solve analytically or through mathematical modelling, and usually require human intelligence (D.T. Pham & P.T.N. Pham, 1999).

Thus, the present chapter aims to present results of statistical tools and fuzzy modelling to detect burn in grinding, by digitally processing the acoustic emission signals generated during the process.

2. Monitoring and process control

The implementation of intelligent processes in industries utilizing computer numerically controlled machining is increasing rapidly. However, these systems are not enough reliable to operate without human interference so far. It is common to observe operators of CNC machines correct the process parameters or identify the end of the tool life (Aguiar et al., 1999).

When it comes to grinding, there are three main goals related to process monitoring: detection of problems during machining; provision of information necessary to optimize the process; and the contribution to the development of a database needed to determine the control parameters (Inasaki, 1999). For example, in external plunging grinding, many parameters are involved and need to be determined, which are related to the choice of the grinding wheel and coolant. The wheel and work speeds, and feed rate are one of these aforementioned parameters. Among these, the feed rate is the most critical in influencing the results.

The choice of the grinding cycle, the satisfactory end of the operation and spark-out time are other important parameters to be considered, in order to obtain the desired surface quality. The information obtained during monitoring may be used, thus, to minimize the grinding cycle time and increase the overall process quality (Inasaki, 1999).

The use of acoustic emission (AE) to monitor and control the grinding process is a relatively recent technology (Bennett, 1994), besides being more sensitive to the grinding condition variations, when compared with the force and power measurements (Webster et al., 1994), standing as a promising technique to the process monitoring.

The relatively easiness of digitally processing the root mean square (RMS) of the acoustic emission signal has led to approaches in which this type of statistic is employed. However, the inherent average operation involved in determining the RMS makes it to a certain extent insensitive to impulsive events, such as cracks and burn of the workpieces, although this kind of parameter carries a lot of useful information.

2.1 Burn control in grinding

Grinding burn occurs during the cutting process when the amount of energy generated in the contact zone produces a sufficiently high increase in temperature in order to cause a localized phase change in the workpiece material. This occurrence can be observed visually from the discoloration of the part surface (Nathan et al., 1999; Kwak & Song, 2001). Also, burning is expected when a critical temperature (approx. 720°C for steels) is exceeded in the grinding zone (Nathan et al., 1999).

The root mean square (RMS) value of the acoustic emission signal has been the main parameter studied in previous researches on grinding over a carefully selected frequency band. This signal has provided a reasonable parameter, because it is rich in sound waves carrying a lot of useful information (Spadotto et al., 2008).

Thermal models for grinding can be taken as an example of the theory of a moving heat source with various boundary conditions for estimating the distribution of temperature inside the grinding zone. This distribution is used to predict the generation of residual stresses, the onset of burning of the workpiece surface, or other characteristics related to surface integrity. However, these models require parameters that are often not readily known in the production environment. Moreover, several properties of many materials and of the cutting fluid are not known exactly in the predominating grinding conditions. In addition, the production operation does not require a high precision "absolute" model, because such precision is not reproduced in industrial practice (Ali & Zhang, 2004). What is needed, is a "relative" model that can guide the user about what should be done and how to do it, because there will always be a certain degree of trial and error on the shop floor, and this relative model will be a good starting point (Shaw, 1996).

2.2 Statistical parameters for burn detection

Acoustic emission and grinding power signals provide a variety of information about the grinding process. However, more rigorous analyses can be obtained by the signal processing with the help of statistical parameters. Using mathematical manipulation software, these signals can be processed to obtain information such as the RMS value, standard deviation, autocorrelation, FFT, etc. (Wang et al., 2001; Tönshoff et al., 2000).

Acoustic emission (AE) can be defined as the elastic stress waves generated by the rapid release of deformation energy, inside a material subjected to an external stimulus. These stress waves produce surface displacements that can be detected by a piezoelectric sensor, which transforms the displacements into electrical signals (Tönshoff et al., 2000). Their frequency range varies from 50 kHz to 2MHz, which is above the range of many noises originating from sources outside the grinding process. Thus, it is a sensitive and adequate method to monitor the grinding process (Tönshoff et al., 2000; Liu et al., 2006).

The parameter that has been studied predominantly in previous researches using acoustic emission has been the root mean square (RMS) of the AE signal (AERMS), filtered over a carefully selected frequency band. This signal has been a reasonable study parameter, since the grinding process is very rich in sound waves, thus containing a lot of available acoustic information (Aguiar et al., 1999).

The RMS value of AE can be expressed by equation (1) (Kim et al., 2001).

$$AE_{RMS} = \sqrt{\frac{1}{\Delta T}\int_{0}^{\Delta t} AE^2(t).dt} \tag{1}$$

where Δt is the integration time constant.

The mean-value deviance (MVD) statistic was used successfully in burn detection (Wang et al., 2001), and is defined by equation (2).

$$T_{mvd}(X) = \frac{1}{M}\sum_{k=0}^{M-1} \log\left[\frac{\overline{X}}{X_k}\right] \tag{2}$$

where $\overline{X}$ is the mean value of $\{X_k\}$; $2M$ is the total number of FFT bins, and X_k is the kth magnitude-squared FFT bin.

3. In-process grinding monitoring by acoustic emission

In the following sections, results are presented for the evaluation of the surface integrity of AISI 1045 steel parts during grinding, using properly processed acoustic emission signals. Furthermore, results from several digital signal processing tools are shown, where those with amplitude independent are stressed and then the power of the AE signal does not affect the signal characteristic. This is explained due to the fact that the power of the AE signal may undergo variations during the grinding process, which have nothing to do with the part condition than its geometry. The result analyses from the digital signal processing as well as a discussion of the investigation are presented.

3.1 Materials and methods

The experimental tests were carried out upon a surface grinding machine (Sulmecânica RUAP H 1055-E - Brazil) where raw acoustic emission signals were gathered for fifteen (15) different runs, at 2.5 million of samples per second rate. The AISI 1045 steel has been used for the tests. The major parameters were kept constant during the runs. However, the depth of cut was varied from low to severe material removing. All workpieces were evaluated after machining, and the burn marks were then identified. The setup for these tests is shown in Figure 1.

The grinding parameters include: wheel peripheral speed: 27.94 m/s; tool: aluminium oxide grinding wheel (38A80-PVS) manufactured by Norton Abrasives, with dimensions of: 296.50 mm external diameter, 40.21 mm width; workpiece speed: 0.044 m/s; workpiece dimensions: 98.58 x 80.12 x 8.74 mm; and coolant type: oil emulsion (4% concentration on water).

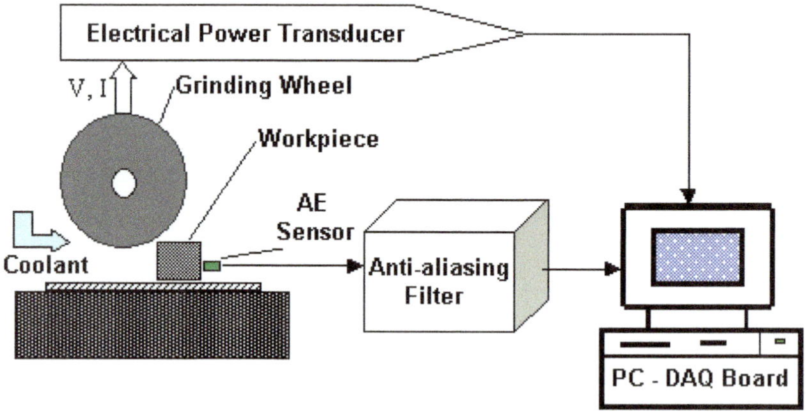

Fig. 1. Experimental setup

Data was gathered from a fixed acoustic emission sensor (Sensis PAC U80D-87), which was fixed on the part holder. The data acquisition board (National Instruments, PCI-6011) was set up to work at 2.5 million of samples per second, with 12 bits precision per sample. Table 1 shows details of carried out tests.

Test	Depth of cut [μm]		Cutting Profile	Comments
1	10			No burn
2	30			Slight burn
3	20			Severe burn
4	90	10	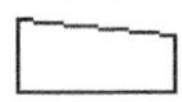	Severe burn
5	20	2,5	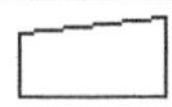	Severe burn
6	40	5	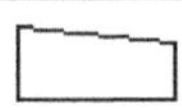	Severe burn
7	15		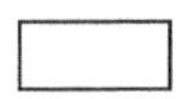	Burn at middle

Table 1. Tests with AISI 1045 Steel

3.2 Signal processing

From acoustic emission data available on the binary files, several routines developed in MATLAB for digitally processing the signals were utilized, where many statistical correlations such as kurtosis, skewness, autocorrelation, RMS, among others were employed and are described as following.

3.2.1 Root mean square (RMS) of acoustic emission signal

For a given time t, the RMS value of a raw acoustic emission signal can be expressed by:

$$AE_{rms} = \sqrt{\frac{1}{T}\int_{t-T}^{t} AE_{raw,t}^{2}.dt} = \sqrt{\frac{1}{N}\sum_{i=1}^{N} AE_{raw,i}^{2}} \quad (3)$$

where T is the integration time and N is the discrete number of AE data within the interval T. In this work, T was considered equal to 1 ms (Webster et al., 1996).

3.2.2 Constant false alarm rate (CFAR)

Constant false alarm rate (CFAR) is a statistic tool employed in detection of events, which is described by (Nuttal, 1997):

$$T_{pl}(X) = \sum_{k=0}^{M-1} X_k^{\upsilon} \quad (4)$$

Where X_k corresponds to the kth FFT of $X(t)$, υ is a variable exponent and $2M$ corresponds to the data base vector size to get the FFT calculated. Although υ between 2 and 3 provides a good performance for a wide frequency band of the studied signal, this statistic needs pre-normalized data. Thus, due to the acoustic emission signal variations during the process, the constant false alarm rate (CFAR) is utilized (Nuttal, 1996). This statistic is based on the supposition of flatness of the acoustic emission signal. An alternative version of this tool was employed due to system distortions, which is expressed by the following equation (Wang, 1999).

$$T_{bcpl}(X) = \frac{\sum_{k=n1}^{n2} X_k^{\upsilon}}{\left(\sum_{n1}^{n2} X_k\right)^{\upsilon}} \tag{5}$$

Values for M=1280, and a frequency range between 300 and 700 kHz were considered.

3.2.3 Kurtosis and skewness statistics

The measurement, if the distribution tail is longer than other, is made by skewness. In case of kurtosis, the tail size is concerned. Both statistics are utilized in this work aiming to find an indicator to the acoustic emission variations. Thus, abrupt changes in the AE signal such as those in which burn occurs may result in peaks for these statistics. Equations 6 and 7 show the way of calculating kurtosis and skewness of an x signal, respectively.

$$K = \sum \frac{(x-\mu)^4}{N\sigma^4} - 3 \tag{6}$$

$$K = \sum \frac{(x-\mu)^3}{N\sigma^3} \tag{7}$$

where μ is the mean of x, N the number of samples in the range considered and σ the standard deviation.

3.2.4 Mean value dispersion (MVD) statistic

The form of MVD statistic is also employed, but in a more convenient form (Wang et al., 1999), as shown in equation 8.

$$MVD = \sum_{k=n1}^{n2} \log \left(\frac{\frac{1}{n_2 - n_1 + 1} \sum_{l=n1}^{n2} X_l}{X_k} \right) \tag{8}$$

where X has the same meaning as to CFAR statistic, as well as n_1 and n_2.

3.2.5 Ratio of power (ROP)

It is instinctive to think about the different behaviours expected for a satisfactory or non satisfactory part by observing the frequency spectrum of the acoustic emission signal. Hence, for each block of acoustic emission data, ROP is given by equation 9.

$$ROP = \sum_{k=n_1}^{n_2} \frac{|X_k|^2}{\sum_{k=0}^{N-1} |X_k|^2} \tag{9}$$

where N set to 1024; n_1 and n_2 in the range of 300 to 700 kHz were chosen.

3.2.6 Autocorrelation

The time correlation of a function ϕ_{xy} is defined in equation 10 (Oppenheim et al., 1997).

$$\phi_{xy}(t) = \int_{-\infty}^{+\infty} x(t+\tau)y(\tau)d\tau \tag{10}$$

where ϕ_{xx} is commonly referred to as autocorrelation.

3.3 Results and discussion

The figures for each workpiece tested were obtained from the digital signal processing of acoustic emission signals, in which the statistics previously described were employed. The results from tests 1, 5 and 7 for ABNT 1045 steel are presented as shown in Figures 2 to 4, respectively. In all figures below, the horizontal axis corresponds to time (in seconds), and the vertical axis to Volts (multiplied by a constant).

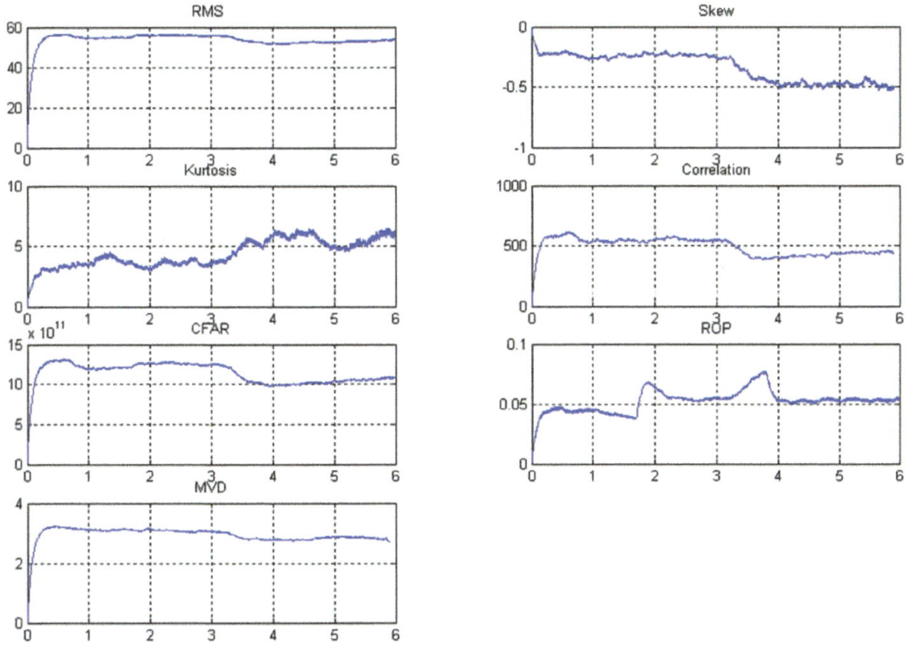

Fig. 2. Results for test 1 – AISI 1045 steel with no burn

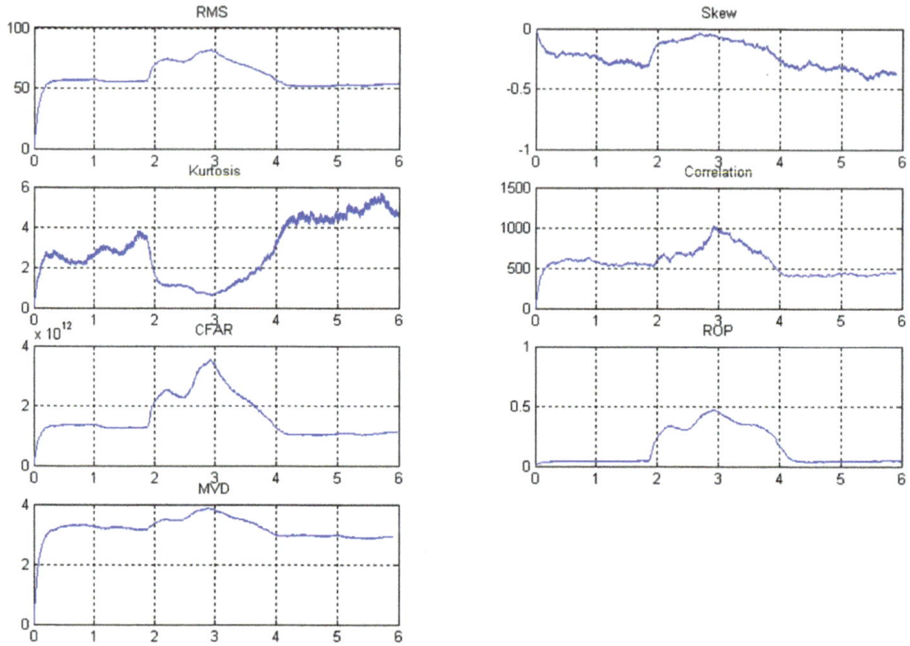

Fig. 3. Results for test 5 – AISI 1045 steel with severe burn in practically the whole workpiece

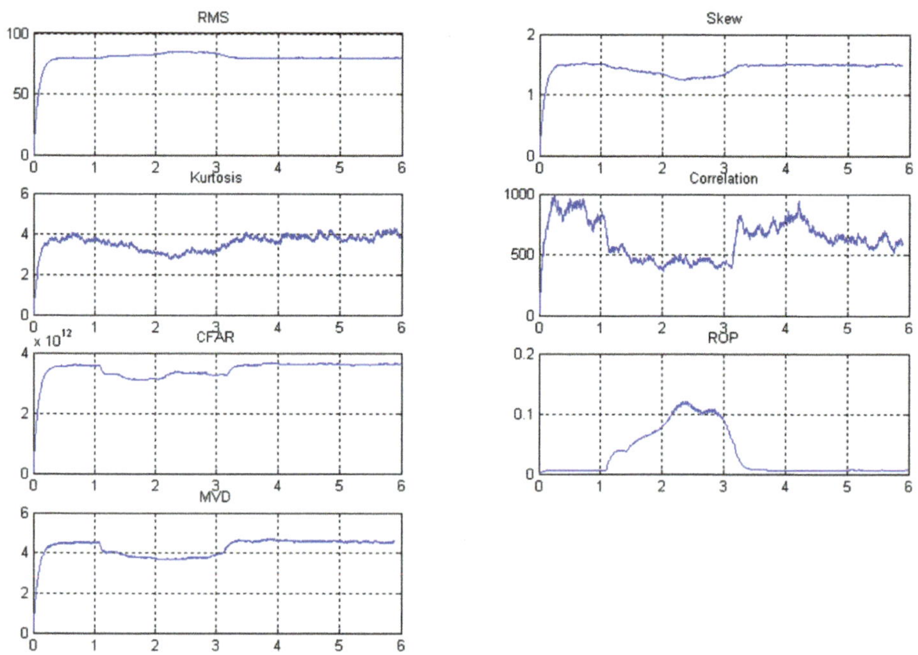

Fig. 4. Results for test 7 – AISI 1045 steel with burn in the midst of the part;

From the results for the ABNT 1045 steel, it can be observed that RMS providade a pretty steady level for the non-burning workpiece, during all over the grinding pass, while the signal had good variation when severe burn occurred, such as in Figure 3. Skewness and kurtosis presented variation when burn took place, but positive and negative amplitudes were observed, which are not useful for an indicator parameter to burn.

Surprisingly, the ROP turned out to be a good indicative to burn, since its behavior has shown quite sensitive to the studied phenomenon. Besides, its level is low to those non-burning parts and high to the burning ones. Additionally, it has well characterized the beginning of contact between the wheel and part as well as the end of the grinding pass.

The MVD tool presented a behavior similar to the RMS statistic, but not as satisfactory as RMS, due to the low level obtained for test 7. The autocorrelation statistic was very sensitive to burn, for the most tests performed, but for a few it has shown itself useless, by virtue of the decrease observed when burn occurred.

Similarly to the autocorrelation, CFAR has behaved quite satisfactory to burn detection, for most of the tests, not providing signal decreases at all, except for test 7, where a decrease was observed during the grinding pass. This behaviour, however, did not compromise the utility of CFAR tool, for the level of test 7 has kept higher than the non-burning test.

4. Acoustic emission and fuzzy logic to predict grinding burns

The use of fuzzy logic, which reflects the nature of qualitative and inexact reasoning of humans, enables specialist systems to be more flexible. With fuzzy logic, the precise value of a variable is substituted by a linguistic description represented by a fuzzy set, and inferences are made based on this representation (D.T. Pham & P.T.N. Pham).

Fuzzy logic has numerous applications in engineering, where the command of knowledge is usually imprecise. Interesting results have been achieved in the area of machining processes and control, although other sectors have also benefited from this tool. Several engineering applications can be cited, such as welding arc height control (Bigand et al., 1994); control of robotic hands with multiple fingers (Bas & Erkmen, 1995); prediction of surface roughness of ground components (Ali & Zhang, 1999); control of grinding burn (Ali & Zhang, 2004), among others. However, the development of an intelligent system for burn detection, prediction and classification still poses a challenge for researchers.

Based on these practical aspects, a fuzzy model was proposed with 37 absolute rules and eight relative rules for predicting burn of ground workpiece surfaces. This model is designed for practical application, i.e., an operator can infer from the model, engineers can use it in process planning, and the model can be part of an intelligent adaptive control, without the need for additional information (Ali & Zhang, 1999).

A fuzzy clustering method was proposed for evaluating the degree of grinding burn damage using burn colour spots on the workpiece side surface. This method can replace the complicated wet grinding thermometry method. The results can be used to evaluate the performance of cutting fluids for restricting grinding burn damage (Ge et al., 2002).

Others presented a new method of grinding burn identification using a wavelet packet transform to extract features from acoustic emission signals and employing fuzzy pattern recognition to optimize features and identify the grinding status. Experimental results showed that the accuracy of grinding burn recognition was satisfactory (Liu et al., 2005).

Other works employing fuzzy logic in the area of manufacturing can be found, including estimation of residual stress induced by grinding (Ali & Zhang, 1997), prediction of surface roughness of ground components (Ali & Zhang, 1999), classification of the condition of the grinding wheel's cutting ability (Lezanski, 2001), etc. However, only few researches have been published about grinding burn using fuzzy logic. Therefore, this work explores not only the application of fuzzy logic but also the fusion sensors and grinding burn parameters not used to date.

Therefore, the next sections represent the work which aimed to investigate burning in the grinding process based on a fuzzy model. The inputs of the models are obtained from the digital processing of the raw acoustic emission and cutting power signals. The parameters to be obtained and used in this work include the mean-value deviance (MVD), which proved efficient in grinding burn detection (Wang et al., 2001), grinding power, and root mean square (RMS) of the acoustic emission signal (Dotto et al., 2006).

4.1 Materials and methods

The grinding tests were carried out with a surface grinding machine (Sulmecânica RUAP-H1055-E, Brazil) equipped with an aluminium oxide grinding wheel (Norton ART-FE-38A80PVH). A fixed acoustic emission sensor (Sensis DM-42) was placed near the workpiece, and the electric power consumed by the wheel three-phase induction motor was measured with an electrical power transducer.

The workpieces for the grinding tests consisted of SAE 1020 laminated steel bars with dimensions of 150mm length, 10mm width and 60mm height. Grinding was performed along the length of the workpiece.

The power transducer consisted of a Hall sensor, to measure both the electric current and the voltage in the electric motor terminals. Both signals were processed in the power transducer module by an integrated circuit, which delivered a voltage signal proportional to the electrical power consumed by the motor. The acoustic emission and power signals were then sent to the data acquisition board (National Instruments, PCI-6011) installed in a personal computer. The signals were captured by LabVIEW software and stored in binary files for further processing and analysis. The acoustic emission sensor had a broad-band sensitivity of 1.0 MHz. Its amplifier also filtered the signal outside the range of 50 kHz to 1.0 MHz. Figure 5 shows a schematic diagram of the grinding machine and instrumentation used.

The tests were carried out in 12 different grinding conditions, and the degrees of burn (no burn, slight burn, medium burn, and severe burn) were evaluated visually on each workpiece surface. Dressing parameters, lubrication and peripheral wheel speed were controlled to ensure the same grinding condition in the three repetitions of each test. The workpiece speed was set at 0.033 m/s and the wheel speed at 30 m/s. The latter was kept constant by adjusting the frequency of the induction motor on the frequency inverter, since the diameter of the grinding wheel decreased as the tests progressed.

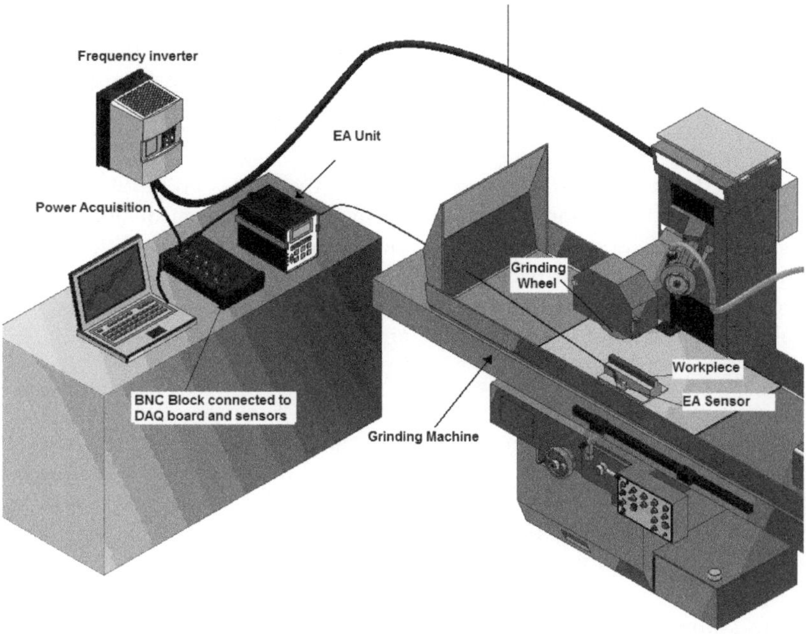

Fig. 5. Test bench layout

The overlapping ratio (U_d), which is the relationship between the effective cutting width and the axial dressing feed rate, was set to 1, and the dressing conditions were kept constant in all the tests. An emulsion oil (4% concentration on water) was used. Each run consisted of a single grinding pass of the grinding wheel along the workpiece length, at a given grinding condition. The acoustic emission and grinding power signals were measured in real time at a rate of 2.0 million samples per second, and stored in binary data files. It is important to mention that the raw acoustic emission signal was acquired instead of the root mean square, generally used.

The digital signals were processed after the 12 tests were carried out, and the data files stored. The processing of acoustic emission signals generated the aforementioned statistics, i.e., RMS and MVD.

4.2 Construction of input vectors

The percentage of burn and the degree of burn of each workpiece were determined using a software previously developed (Dotto et al., 2006). This software analyzes the surface condition of a workpiece, based on a photograph of the machined workpiece. Thus, aided by this software, reliable input data were extracted to represent the levels: no burn, slight burn, medium burn, and severe burn.

The signals were processed using MATLAB. RMS and MVD statistics were generated based on the raw acoustic emission signals and the grinding power signals, which served to build the input vectors for the fuzzy models.

This procedure was carried out for all the tested workpieces. However, it should be noted that among the inputs considered in this work (RMS, MVD and power), some were better than others, in order to define differences in the degrees of burn. Therefore, the use of fuzzy logic is attractive for this application, since it is based on the levels of imprecision generated by these inputs.

The files contained a set of burn data associated with the processed inputs. For example, the RMS signal of acoustic emission is represented by the level no burn, slight burn, medium burn and severe burn. The same situation applies to the MVD and grinding power statistics. The data vectors were standardized, i.e., all the vectors contained the same number of collected points. This allows comparisons to be made always among the same points of the input variables without distortions occurring among points.

Subsequently, the data that had been separated by level of burn were combined in a single file of a given input. To exemplify, the file of the RMS signal of AE, which previously consisted of separate files of no burn, slight burn, medium burn and severe burn, now had a single vector containing the data of the aforementioned files. This new vector is in the order of burn of the workpiece, i.e., the vector contains the no burn, slight burn, medium burn and severe burn data, respectively, in this position of the vector. The vectors for the MVD statistic and grinding power were constructed in the same way.

4.3 Fuzzy modeling

One of the advantages of using fuzzy logic is the possibility of transforming natural language into a set of numbers, allowing for computational manipulation. Linguistic variables can be defined as "variables whose values are words or sentences in natural or artificial language" (Zadeh, 1965). Linguistic variables assume values called linguistic values, e.g., the values of no burn, slight burn, medium burn and severe burn are relative to the burn variable of the workpiece. Figure 6 shows a general model of a fuzzy inference system (FIS).

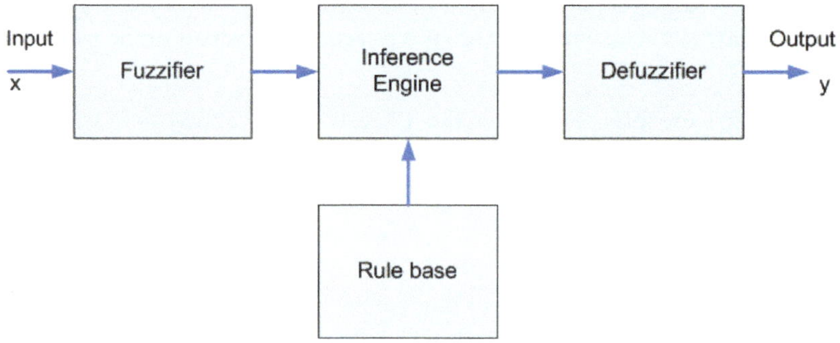

Fig. 6. Block diagram of a fuzzy inference system

It can be observed from the figure that the FIS includes four components: the fuzzifier, inference engine, rule base, and defuzzifier. The rule base contains linguistic rules that are provided by experts. It is also possible to extract rules from numeric data. Once the rules have been established, the FIS can be viewed as a system that maps an input vector to an output vector. The fuzzifier maps input numbers into corresponding fuzzy memberships.

This is required in order to activate rules that are in terms of linguistic variables. The fuzzifier takes input values and determines the degree to which they belong to each of the fuzzy sets via membership functions.

The inference engine defines mapping from input fuzzy sets into output fuzzy sets. It determines the degree to which the antecedent is satisfied for each rule. If the antecedent of a given rule has more than one clause, fuzzy operators are applied to obtain one number that represents the result of the antecedent for that rule. It is possible that one or more rules may fire at the same time. Outputs for all rules are then aggregated. During aggregation, fuzzy sets that represent the output of each rule are combined into a single fuzzy set.

The defuzzifier maps output fuzzy sets into a crisp number. Given a fuzzy set that encompasses a range of output values, the defuzzifier returns one number, thereby moving from a fuzzy set to a crisp number. Several methods for defuzzification are used in practice, including the centroid, maximum, mean of maxima, height, and modified height defuzzifier. The most popular defuzzification method is the centroid, which calculates and returns the center of gravity of the aggregated fuzzy set.

4.4 Input vectors

The aforementioned vectors of RMS, MVD and grinding power represent the numerical inputs of the fuzzy system. These inputs were transformed into fuzzy inputs. Figure 7 shows the RMS vector of acoustic emission (expressed in k*Volts), containing all the levels of burn and no burn observed in the tests. The k constant is a conversion constant that depends on the number of bits of the data acquisition board.

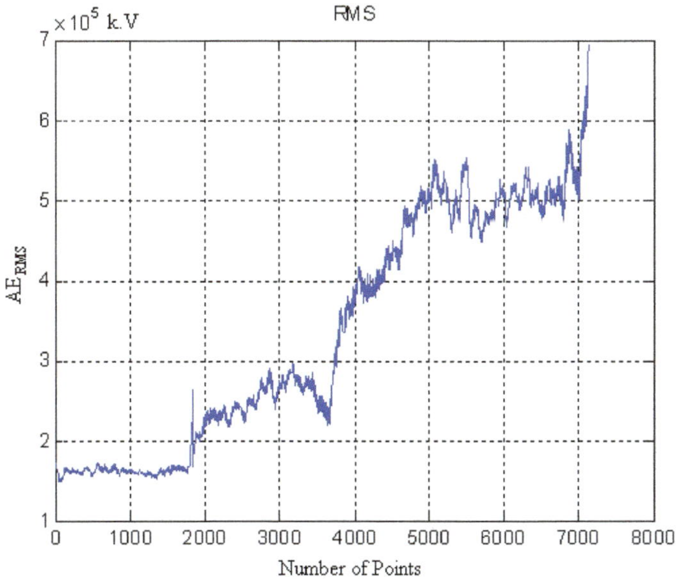

Fig. 7. Acoustic emission (RMS)

A clustering process was applied to each numerical input, which consisted of determining 4 subclusters (Ge et al., 2002). Clustering serves to group input vector data into 4 major groups that represent the burn levels of the workpiece. Each group thus determined contains a cluster center which best represents the group. For the statistics or signals employed here, the clusters were very close to the mean values of each group. Thus, the point that best represents a burn group is the mean value of the range of points of a burn level.

After this representative point was attributed and determined, the data vectors were normalized with values of 0 to 1. The numerical data vectors were then transformed in fuzzy system vectors, and the value of 1 was attributed to the cluster center. This value diminishes as the statistic values move away from this cluster point. With this process, the further away from the cluster center, the lower the relevance of the information and the lower the value attributed to the normalization of the data.

Figure 8 shows the fuzzified acoustic emission RMS input into the system by means of membership functions. The horizontal axis refers to the value of the amplitude of the statistic. In other words, the higher the amplitude of the statistic the greater the scope of the burn level in the fuzzified set.

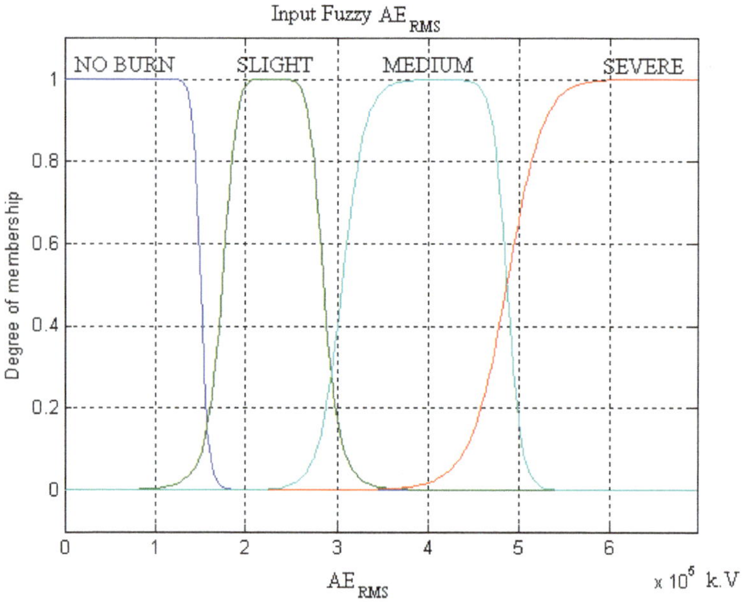

Fig. 8. Fuzzified input of the RMS acoustic emission signal

These membership functions help in converting numeric variables into linguistic terms.

4.5 Output vector

A single output vector was then created with the fuzzified input vectors to the fuzzy system. This output vector was created according to the various burn levels the workpiece would undergo. The fuzzy output has a totally encompassing set of burn levels of a surface. The output set contains the following levels: no burn, super slight, very slight, slight, slight +, almost medium, medium, medium +, almost severe, severe, severe +, and total damage.

This scope is due to the fact that the model is not restricted only to 3 or 4 burn levels. By extending these levels, the result was decentralized to a broader and strongly detailed vision of the problem. The fuzzified output vector represents the fuzzy output set.

4.6 Inferences and rule base of fuzzy models

Based on the inputs and the fuzzified output, an inference was established between both (Hu & Tzeng, 2003). First, 3 logic system models were created. The first model had two inputs: the RMS of AE and the grinding power (GP). The other two models had 3 inputs: the RMS of AE, the GP, and the MVD statistic. What differentiates these two models is the inference of the rules used.

The rule sets extracted for all the models were based on the experience of specialists in the subject, and are therefore based on a typically fuzzy system. The first model is a simple control of two inputs. The rule base is shown in Table 2. Each input analyzed in all the models had the same weight relative to the system, i.e., all the inputs affect the system equally.

GP\AE RMS	NO BURN	SLIGHT	MEDIUM	SEVERE
NO BURN	NO BURN	VERY SLIGHT	MEDIUM	SEVERE
SLIGHT	VERY SLIGHT	SLIGHT	MEDIUM	SEVERE
MEDIUM	MEDIUM	MEDIUM	VERY SEVERE	TOTAL DAMAGE
SEVERE	SEVERE	SEVERE	TOTAL DAMAGE	TOTAL DAMAGE

Table 2. Rule base for model 1. Combination of GP and the RMS value

The second model is based on an inference system of three inputs, combined in pairs. This system is widely valid, since, if one of the inputs has an incorrect value, it will not affect the output to any appreciable degree. The rules for this model are shown in Table 3.

The third and last model developed here is a fuzzy control with three fuzzy inputs, namely: the RMS value of the acoustic emission signal, the grinding power (GP), and the MVD statistic of the signal. What differentiates this model from the second one is the part of rule inference. The rule set was developed from a triple combination of inputs, thus providing a total of 64 well distributed rules. Table 4 shows the rule set for model 3.

4.7 Defuzzification of the proposed systems

Three models of fuzzy systems were presented. Fuzziness helps us to evaluate the rules, but the final output of a fuzzy system has to be a crisp number. The input for the defuzzification process is the aggregate output fuzzy set and the output is a single number.

GP\AE RMS	NO BURN	SLIGHT	MEDIUM	SEVERE
NO BURN	NO BURN	VERY SLIGHT	MEDIUM	SEVERE
SLIGHT	VERY SLIGHT	SLIGHT	MEDIUM	SEVERE
MEDIUM	MEDIUM	MEDIUM	VERY SEVERE	TOTAL DAMAGE
SEVERE	SEVERE	SEVERE	TOTAL DAMAGE	TOTAL DAMAGE

AE RMS \MVD	NO BURN	SLIGHT	MEDIUM	SEVERE
NO BURN	NO BURN	SLIGHT	MEDIUM	VERY SEVERE
SLIGHT	SLIGHT	SLIGHT	VERY SEVERE	SEVERE
MEDIUM	MEDIUM	VERY SEVERE	VERY SEVERE	TOTAL DAMAGE
SEVERE	SEVERE	SEVERE	TOTAL DAMAGE	TOTAL DAMAGE

GP\MVD	NO BURN	SLIGHT	MEDIUM	SEVERE
NO BURN	NO BURN	SLIGHT	MEDIUM	MEDIUM
SLIGHT	SLIGHT	SLIGHT	MEDIUM	VERY SEVERE
MEDIUM	MEDIUM	MEDIUM	SEVERE	TOTAL DAMAGE
SEVERE	SEVERE	VERY SEVERE	TOTAL DAMAGE	TOTAL DAMAGE

Table 3. Rule set of model 2, paired combination of the fuzzy inputs

The center of gravity (COG) or centroid defuzzification method was used in each model. It finds the point where a vertical line would cut off the aggregate set into two equal masses. A reasonable estimate can be obtained by calculating it over a sample of points and expressed by equation 11.

$$COG = y_{crisp} = \frac{\sum_{i=1}^{n} \mu_A(y_i) y_i}{\sum_{i=1}^{n} \mu_A(y_i)} \tag{11}$$

where $\mu_A(y_i)$ is the membership function and y is the variable to be defuzzified.

It is the most widely used technique because, when it is used, the defuzzified values tend to move smoothly around the output fuzzy region.

4.8 Results and discussion

The proposed models were developed using the "Fuzzy Logic" toolbox of MATLAB. Initially, a model was built with only two fuzzy inputs. Its rule base was edited and placed in the software toolbox.

RMS	GP	MVD	OUTPUT
NO BURN	NO BURN	NO BURN	NO BURN
NO BURN	NO BURN	SLIGHT	SUPER SLIGHT
NO BURN	NO BURN	MEDIUM	VERY SLIGHT
NO BURN	NO BURN	SEVERE	SLIGHT
NO BURN	SLIGHT	NO BURN	SUPER SLIGHT
NO BURN	SLIGHT	SLIGHT	SLIGHT
NO BURN	SLIGHT	MEDIUM	SLIGHT +
NO BURN	SLIGHT	SEVERE	ALMOST MEDIUM
NO BURN	MEDIUM	NO BURN	SLIGHT +
NO BURN	MEDIUM	SLIGHT	ALMOST MEDIUM
NO BURN	MEDIUM	MEDIUM	MEDIUM
NO BURN	MEDIUM	SEVERE	MEDIUM +
NO BURN	SEVERE	NO BURN	MEDIUM
NO BURN	SEVERE	SLIGHT	MEDIUM +
NO BURN	SEVERE	MEDIUM	ALMOST SEVERE
NO BURN	SEVERE	SEVERE	SEVERE
SLIGHT	NO BURN	NO BURN	SUPER SLIGHT
SLIGHT	NO BURN	SLIGHT	VERY SLIGHT
SLIGHT	NO BURN	MEDIUM	SLIGHT
SLIGHT	NO BURN	SEVERE	SLIGHT +
SLIGHT	SLIGHT	NO BURN	VERY SLIGHT
SLIGHT	SLIGHT	SLIGHT	SLIGHT +
SLIGHT	SLIGHT	MEDIUM	SLIGHT +
SLIGHT	SLIGHT	SEVERE	ALMOST MEDIUM
SLIGHT	MEDIUM	NO BURN	ALMOST MEDIUM
SLIGHT	MEDIUM	SLIGHT	MEDIUM
SLIGHT	MEDIUM	MEDIUM	MEDIUM
SLIGHT	MEDIUM	SEVERE	MEDIUM +
SLIGHT	SEVERE	NO BURN	ALMOST MEDIUM
SLIGHT	SEVERE	SLIGHT	MEDIUM
SLIGHT	SEVERE	MEDIUM	MEDIUM +
SLIGHT	SEVERE	SEVERE	SEVERE
MEDIUM	NO BURN	NO BURN	SLIGHT
MEDIUM	NO BURN	SLIGHT	SLIGHT +
MEDIUM	NO BURN	MEDIUM	ALMOST MEDIUM
MEDIUM	NO BURN	SEVERE	MEDIUM +
MEDIUM	SLIGHT	NO BURN	SLIGHT +
MEDIUM	SLIGHT	SLIGHT	ALMOST MEDIUM
MEDIUM	SLIGHT	MEDIUM	MEDIUM
MEDIUM	SLIGHT	SEVERE	SLIGHT +
MEDIUM	MEDIUM	NO BURN	ALMOST MEDIUM
MEDIUM	MEDIUM	SLIGHT	MEDIUM
MEDIUM	MEDIUM	MEDIUM	MEDIUM +
MEDIUM	MEDIUM	SEVERE	ALMOST MEDIUM
MEDIUM	SEVERE	NO BURN	MEDIUM +
MEDIUM	SEVERE	SLIGHT	ALMOST SEVERE
MEDIUM	SEVERE	MEDIUM	SEVERE
MEDIUM	SEVERE	SEVERE	SEVERE +
SEVERE	NO BURN	NO BURN	ALMOST MEDIUM
SEVERE	NO BURN	SLIGHT	MEDIUM
SEVERE	NO BURN	MEDIUM	MEDIUM +
SEVERE	NO BURN	SEVERE	ALMOST SEVERE
SEVERE	SLIGHT	NO BURN	MEDIUM
SEVERE	SLIGHT	SLIGHT	MEDIUM +
SEVERE	SLIGHT	MEDIUM	ALMOST SEVERE
SEVERE	SLIGHT	SEVERE	SEVERE
SEVERE	MEDIUM	NO BURN	MEDIUM +
SEVERE	MEDIUM	SLIGHT	ALMOST SEVERE
SEVERE	MEDIUM	MEDIUM	SEVERAL +
SEVERE	MEDIUM	SEVERE	TOTAL DAMAGE
SEVERE	SEVERE	NO BURN	SEVERE
SEVERE	SEVERE	SLIGHT	SEVERE +
SEVERE	SEVERE	MEDIUM	TOTAL DAMAGE
SEVERE	SEVERE	SEVERE	TOTAL DAMAGE

Table 4. Third fuzzy model. Triple combination of inputs

4.8.1 Model 1: Two-input control

Based on the rules, a 3D surface was generated, which allowed to visualize the effects of the rules on the system inputs and output. Figure 9 shows this surface for the first model. This surface indicates how regular the newly developed rule base is. If it is not completely consistent, disproportions and coarse irregularities are visible on the generated surface.

Continuity, transition and symmetry are characteristics well expressed in Figure 9. As can be seen, as the values of the RMS signal and grinding power increase, a prediction of high burn level on the surface is generated.

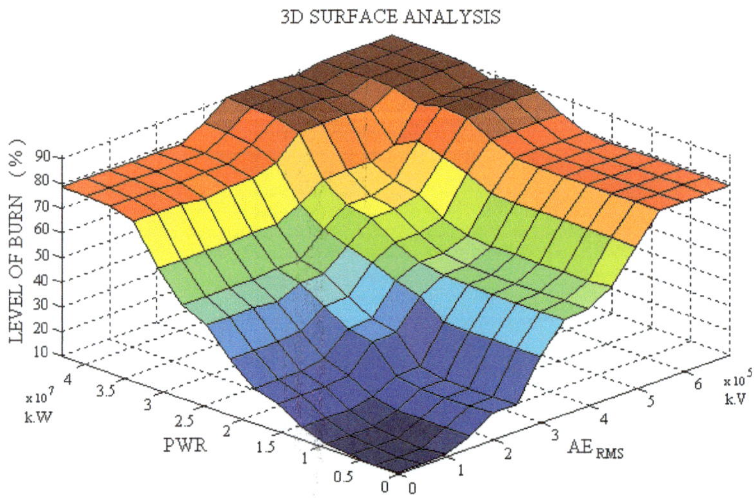

Fig. 9. 3D analysis of the surface generated by the rule system of model 1

Once the consistency of the system inference domain has been verified, it is possible to predict the burn level of the workpiece surface, based on values of the applied RMS signal and on the system electric power. The simulation was carried out in MATLAB, and is based on a low RMS signal and GP. The analysis resulted in a no burn prediction with a maximum value of 9.61% of surface burn. This simulation can be made with varied values, presenting different levels of prediction.

The advantage of model 1 is its simplicity. With only two fuzzified inputs having the same weight, model 1 produces a reliable and stable prediction in the system. However, if one of the variables presents an incorrect value originating from innumerable real factors such as reading errors, etc., it will affect the prediction of the control.

4.8.2 Model 2: Three-input control combined in pairs

Model 2 was created to minimize the aforementioned error, i.e., an input variable containing an incorrect value of the system. This second model, on the other hand, has three fuzzified inputs. The idea of this model is that its rule base be combined in pairs, generating a total of 48 rules (Table 2). With the two main inputs of RMS and GP, the MVD statistic added to this model served to aid the system by increasing its robustness.

If one of the input variables were to present an incorrect value, the combinations of the other two would ensure a logical result for the control. To exemplify, if the RMS is very high, the GP is very low and the MVD statistic is also very high, the result will indicate a prediction of high burn. The simulation will assume that the GP value, which is outside the standard, is incorrect. Thus, the incorrect value will exert little influence in the analysis.

The three-dimensional analysis of the surface generated by the rules system of this model is shown in Figure 10. This analysis indicates that the rule base continues to be logically valid. As can be seen from the generated surface, the inputs of the RMS value of the acoustic emission signal and the grinding power (GP) exert a stronger influence on the system than the MVD statistic that was introduced.

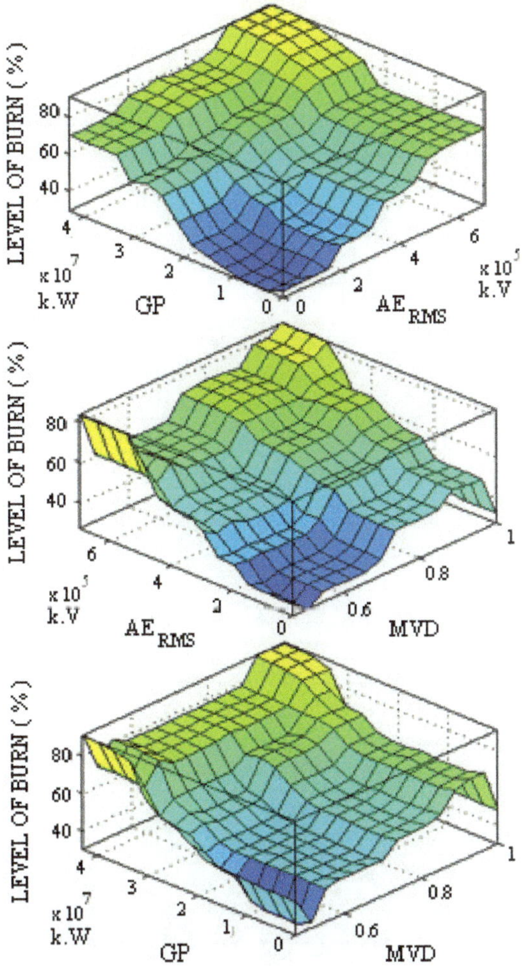

Fig. 10. 3D analysis of the surface generated by the rules system of model 2

The simulation of model 2 is performed following the same standard as model 1. A very high RMS value, a very low GP value, and a high MVD value result in a prediction of severe burn of the workpiece, with a value of 70.3% burn.

4.8.3 Model 3: Three-input control combined in triplets

The third model is based on the same three inputs described earlier, RMS, GP and MVD. The difference between this model and model 2 is the rule system. Its rules are based on the combination of three variables, i.e., all the inputs affect the resulting level of prediction. With a total of 64 rules described in Table 3, the third model possibly offers the best contribution of input signals in relation to the output. The advantage of this system in relation to that of model 1 is its more comprehensive complexity, which enhances its capacity to predict the burn level.

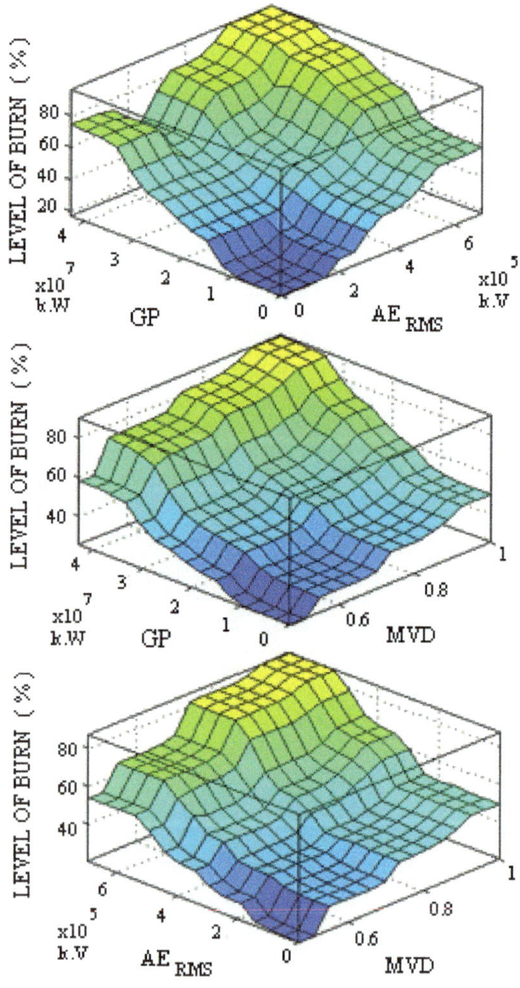

Fig. 11. 3D surfaces generated by the rules of model 3

Figure 11 depicts the 3D surfaces generated by the rules. As can be seen, the rules bases are valid due to their symmetry and flexibility. The main change in these surfaces compared to those of model 2 in Figure 11 is with respect to the RMS vs. MVD and GP vs. MVD analysis.

Previously, there was a high value in the prediction analysis with high RMS and low MVD. Now there is a value that provides equality, i.e., low RMS and low MVD generate a low value in the prediction analysis.

This model is simulated in the same way as the previous processes. A high value of RMS, GP and MVD generate an output with a prediction of 97% of burn of the workpiece, resulting in total damage of the machined material.

4.9 Conclusions

As can be seen, acoustic emission signals can be widely applied in monitoring grinding burn. The undeniable significance of this phenomenon stimulated the researchers to study different methods predict and even to quantify burn.

From digital processing of the raw acoustic emission signal, for the AISI 1045 steel, the results showed that several statistics have worked quite well to burn detection, as is the case of RMS, CFAR, ROP e MVD. Nevertheless, skewness and kurtosis presented an interesting behaviour regarding the signal waveform and the variation along the grinding pass, though they were not effective to detect burn. These features may be better explored in further investigation. The autocorrelation, on its way, has shown itself ineffective to detect burn.

In terms of fuzzy modeling, the results of the proposed models were substantially validated. The imprecise information about statistical values of acoustic emission and electric power signals was transformed into reliable burn prediction values. Models 1, 2 and 3 differ from each other, enabling them to adapt to different circumstances.

The models may be attractive to the practicing engineer who would like to get fast answers for on-line intelligent control and optimization. In its current state, the models are limited to SAE 1020 steel and oxide grinding wheel, but they can easily be extended to other types of workpiece and tool materials.

5. Acknowledgment

The authors are indebted to Univ. Estadual Paulista (UNESP - Bauru campus), and to FAPESP (Brazil) and CNPq (Brazil) for their financial support of the researches.

6. References

Aguiar, P. R.; Willet, P. & Webster, J. (1999). Acoustic emission applied to detect workpiece burn during grinding, In: *Acoustic emission: Standards and Technology Update, ASTM STP 1353,* S. J. Vahaviolos, (Ed.), 107-124, American Society for Testing and Materials, ISBN: 0-8031-2498-8, West Conshohocken, Pennsylvania, USA

Ali, Y. M. & Zhang, L. C. (1997). Estimation of residual stresses induced by grinding using a fuzzy logic approach, *Journal of Materials Processing Technology*, Vol.63, No.1-3, pp. 875-880, ISSN 0924-0136

Ali, Y. M. & Zhang, L. C. (1999). Surface Roughness Prediction of Ground Components using a Fuzzy Logic Approach, *Journal of Materials Processing Technology*, Vol.89-90, pp. 561-568, ISSN 0924-0136

Ali, Y. M. & Zhang, L. C. (2004). A fuzzy Model for Predicting Burns in Surface Grinding of Steel, *International Journal of Machine Tools & Manufacture*, Vol.44, No.5, pp. 563-571, ISSN 0890-6955

Bas, K. & Erkmen, A. M. (1995). Fuzzy preshape and reshape control of Anthrobot-III 5-fingered robot hand, *Proceedings of the International Conference on Recent Advances in Mechatronics, ICRAM'95*, pp. 673-677, ISBN 975-518-063-x, Istanbul, Turkey, August 14-16, 1995

Bennett, R. T (1994). *Acoustic emission in grinding*, Master Thesis, University of Connecticut, 1994

Bigand, A.; Goureau, P. & Kalemkarian, J. (1994). Fuzzy control of a welding process, *Proceedings of the 1994 IMACS International Symposium on Signal Processing, Robotics and Neural Networks, SPRANN'94*, pp. 379-342, ISBN 2950290833, Villeneuve d'Ascq, France, April 25-27, 1994

Brinksmeier, E.; Aurich, J. C. ; Govekar, E. ; Heinzel, C.; Hoffmeister, H. W. ; Klocke, F. ; Peters, J. ; Rentch, R. ; Stepenson, D. J. ; Uhlmann, E. ; Weinert, K. & Wittmann, M. (2006). Advances in modeling and simulation of grinding process. *CIRP Annals - Manufacturing Technology*, Vol.55, No.2 , pp. 667-696, ISSN 0007-8506

Dotto, F. R. L. ; Aguiar, P. R. ; Bianchi, E. C. ; Serni, P. J. A. & Thomazella, R. (2006). Automatic system for thermal damage detection in manufacturing process with internet monitoring. *Journal of Brazilian Society of Mechanical Science & Engineering*, Vol.28, No.2, pp. 153-160, ISSN 1678-5878

Ge, P. Q.; Liu, W. P.; Liu, Z. C.; Lu, C. H. & Li, J. F. (2002). Fuzzy clustering analysis of the grinding burn damage level of a workpiece surface layer, *Journal of Materials Processing Technology*, Vol.129, No.1-3, pp.373-376, ISSN 0924-0136

Hu, Y. C. & Tzeng, G. H. (2003). Elicitation of classification rules by fuzzy data mining, *Engineering Applications of Artificial Intelligence*, Vol.16, No.7-8, pp. 709-716, ISSN 0952-1976

Inasaki, I. (1999). Sensor fusion for monitoring and controlling grinding processes, *The International Journal of Advanced Manufacturing Technology*, Vol.15, No.10, pp. 730-736, ISSN 0268-3768

Kim, H. Y.; Kim, S. R.; Ahn, J. H. & Kim, S. H. (2001). Process monitoring of centerless grinding using acoustic emission, *Journal of Materials Processing Technology*, Vol.111, No.1-3, pp. 273-278, ISSN 0924-0136

Kwak, J. S. & Ha, M. K. (2004). Neural network approach for diagnosis of grinding operation by acoustic emission and power signals. *Journal of Materials Processing Technology*, Vol.147, No.1, pp. 65-71

Kwak, J. S. & Song, J. B. (2001). Trouble diagnosis of the grinding process by using acoustic emission signals, *International Journal of Machine Tools & Manufacture*, Vol.41, No.6, pp. 899–913, ISSN

Lezanski, P. (2001). An intelligent system for grinding wheel condition monitoring, *Journal of Materials Processing Technology*, Vol.109, No.3, pp. 258-263, ISSN 0924-0136

Liu, Q.; Chen, X. & Gindy, N. (2005). Fuzzy pattern recognition of AE signals for grinding burn, *International Journal of Machine Tools & Manufacture*, Vol.45, No.7-8, pp. 811–818, ISSN 0890-6955

Liu, Q.; Chen, X. & Gindy, N. (2006). Investigation of acoustic emission signals under a simulative environment of grinding burn, *International Journal of Machine Tools & Manufacture*, Vol.46, No.3-4, pp. 284-292, ISSN 0890-6955

Malkin, S. (1989). *Grinding technology: theory and applications of machining abrasives*, Ellis-Horwood, ISBN 0872634809, Chichester, England

Nathan, R. D.; Vijayaraghavan, L. & Krishnamurthy, R. (1999). In-process monitoring of grinding burn in the cylindrical grinding of steel, *Journal of Materials Processing Technology*, Vol.91, No.1-3, pp. 37–42, ISSN 0924-0136

Nuttal, A (1997). Performance of power-law processors with normalization for random signals of unknown structure, Naval Undersea Warfare Center, *NPT Technical Report 10,760*

Nuttal, A. (1996). Detection performance of power-law processors for random signals of unknown location, structure, extent, and strength, *AIP Conference Proceedings*, Vol.375, No.1, pp. 302-324, ISBN 1-56396-443-0

Oppenheim, A. V.; Willsky, A. S. & Hamid, S. (1997). *Signals and systems*, 2nd. Edition, Prentice Hall Signal Processing Series, ISBN 0138147574, Upper Saddle River, New Jersey, USA

Pham, D. T. & Pham, P. T. N. (1999). Artificial intelligence in engineering. *International Journal of Machine Tools & Manufacture*, Vol.39, No.6, pp. 937-949, ISSN 0890-6955

Shaw, M. C. (1996). *Principles of Abrasive Processing*, Clarendon Press, ISBN 0198590210, Gloucestershire, England

Spadotto, M. M.; Aguiar, P. R.; Souza, C. C. P.; Bianchi, E. C. & Nunes, A. N. (2008). Classification of burn degrees in grinding by neural nets, *The IASTED International Conference on Artificial Intelligence and Applications, AIA 2008*, pp. 175-180, ISBN 978-0-88986-709-3, Innsbruck, Austria, February 11-13, 2008

Tönshoff, H. M.; Jung, M.; Männel, S. & Rietz, W. (2000). Using acoustic emission signals for monitoring of production process, *Ultrasonics*, Vol.37, No.10, pp. 681-686, ISSN 0041-624X

Wang, Z. (1999). *Surface grinding monitoring by signal processing of acoustic emission signals*, Master Thesis, University of Connecticut, 1999

Wang, Z.; Willet, P; Aguiar, P. R. & Webster, J. (2001). Neural network detection of grinding burn from acoustic emission, *International Journal of Machine Tools & Manufacture*, Vol.41, No.2, pp.283-309, ISSN 0890-6955

Webster, J.; Dong, W. P. & Lindsay, R. (1996). Raw acoustic emission signal analysis of grinding process. *CIRP Annals – Manufacturing Technology*, Vol.45, No.1, pp. 355-340, ISSN 0007-8506

Webster, J.; Marinescu, I.; Bennett, R. & Lindsay, R. (1994). Acoustic emission for process control and monitoring of surface integrity during grinding, *CIRP Annals – Manufacturing Technology*, Vol.43, No.1, pp. 299-304, ISSN 0007-8506

Zadeh, L. A. (1965). Fuzzy sets, *Information and Control*, Vol.8, No.3, pp.338-353, ISSN 0019-9958

Acoustic Emission (*AE*) for Monitoring Stress and Ageing in Materials, Including Either Manmade or Natural Structures, and Assessing Paroxysmal Phases Precursors

Giovanni P. Gregori[1], Gabriele Paparo[1,2], Maurizio Poscolieri[1], Claudio Rafanelli[1] and Giuliano Ventrice[3]
[1]*CNR-IDASC - Institute of Acoustics and Sensors "O. M. Corbino", Rome,*
[2]*Italian Embassy, Buenos Aires,*
[3]*P.M.E. Engineering, Rome,*
[1,3]*Italy*
[2]*Argentina*

1. Introduction

A preliminary assessment of the basic logical framework is essential in order to avoid misunderstandings and/or misconceptions.

1.1 "Elasticity" vs. "plasticity" - "continuity" vs. "quantum bonds" – the space scale and detail

A few definitions have to be recalled: "solid", "liquid", "continuity", and "homogeneity".

"Solid" means that the atomic or molecular bonds prevail on all other forces (thermal, gravitational, etc.). "Liquid" means that gravity forces prevail. Natural reality never fits with either one such an abstraction. Therefore, it is customary to refer to "plastic" materials, or to "viscous" fluids. By this, we afford in keeping a "simple" scheme based on an arbitrary abstraction, and, at the same time, we can approximately fit observations by means of a few *ad hoc* corrections. "Elasticity" is a concept that derives from - and applies to - either an ideal "solid" or an ideal "liquid" body. An ideal "elastic" body is such that some potential "elastic" energy is transformed into "kinetic" energy and vice versa, while this entire process strictly implies no energy loss of any kind. The concept of "elasticity" is closely related to the concept of matter "incompressibility", which is usually taken for granted, although it should deserve consideration being an additional key abstraction.

"Continuity" is related to the abstraction implied by "infinite" and "infinitesimal" quantities that *per se* are not a requirement of natural reality. Until the past XIX century, science was a daughter of Newton and Leibniz, due to their differential and integral calculus.

The infinity of the universe is debated. In contrast, since the discovery of quantum effects, it is well assessed that infinitesimal quantities do not exist. For instance, the van der Waals equation is a simple correction that introduces in some approximate way the intrinsic non-vanishing volume of a gas molecule. This simple correction, however, can only justify some metastable trend of a gas-liquid mixed system, while a correct description of its equation-of-state must unavoidably appeal to quantum phenomena.

Owing to this same reason, classical electromagnetism alone (Maxwell laws, Faraday, Coulomb, Ampère, Gauss, etc.) has never permitted the exploitation of modern electronics, which rather relies on solid state physics. Quantum bonds play a key role. Quantum phenomena are a fundamental aspect of natural reality. They can even be managed by avoiding formal reference to the Schrödinger equation, Hilbert spaces, Feynman graphs, etc. The cleavage plane of a crystal – hence the ageing of a material - shall never be justified by any model based on "continuity", rather it must rely on atomic bonds. Atoms are incompatible with "infinitesimal" quantities, hence with "continuity" (in its strict sense of mathematical analysis). "Homogeneity" is an additional abstraction aimed at fitting the requirements for "simplicity". It correctly applies to manmade structures. But – when dealing with subsystems of the natural environment - it can be applied only within suitable approximations. "Homogeneity" is closely related to the damping of "elastic" waves, which can be propagated through a medium only as far as the wavelength is larger than the mean gaps in the material (these gaps violate homogeneity and continuity).[1]

The present paper deals with records of acoustic emission (*AE*) typically inside a manmade structure – such as a machinery, a building, a bridge, a pipeline, a dam, a rail or road embankment. But *AE* monitoring can also be carried out in the field, for monitoring crustal stress, volcanic precursors, land slide hazards, etc.

As a standard every application is biased by some perturbations that affect the raw *AE* records. Therefore, a few somewhat different criteria are required while approaching data analysis and interpretation in every specific application. Upon concrete exploitation of this whole approach, the results were found to be heuristically very effective for different algorithms, when they are comparatively applied to different physical systems – either in the laboratory or in the field. The comparison between the different applications resulted very useful for understanding the physics of phenomena that – when considered in every single application alone – could be hardly understood. The present discussion refers therefore to *AE* records measured both in the laboratory and/or in the field.

In general, the intensity of the measured *AE* signal depends (*i*) on the intensity of its primary original (either known or unknown) source, and (*ii*) on the damping of the signal – hence on the acoustic impedance, which is different for different probes, and for different *AE* frequencies, temperatures, etc. Let us recall that the acoustic impedance $Z = \rho V$ of a

[1] Communication between different scientific communities is important. A specification on terms is therefore important. Consider a sample of matter, and an energy input to it. If the input and output energies are compared with each other, it is found that some energy was lost. Therefore, it is claimed that some energy has been "absorbed", or that the signal has been "damped" in the process, or that some energy has been "radiated" outside by "scattering loss", or it was released by any other speculated mechanism, etc. All these terms are equivalent in Earth's sciences. In other scientific communities specific "technical" distinctions eventually result from undeclared agreements. This evidently seems to be a matter of semantics, while it has little concern when dealing with the scientific interpretation of phenomena. Engineers distinguish damping and absorption, while this distinction has less meaning to Earth's scientists.

material is defined as the product of its density ρ and its acoustic velocity V. This is important for the determination of wave transmission and reflection at the boundary of two materials, for assessing the wave absorption, and for designing ultrasonic transducers.

In the last respect, however, it should be pointed out that laboratory measurements are easily carried out in a liquid. For instance, one can measure the acoustic impedance inside water and determine its dependence on temperature, frequency, pressure, etc. In contrast, this is impossible in the case of most solid media. The reason is that the propagation of an "elastic wave" inside a solid object - hence its apparent acoustic impedance - is critically controlled by the geometrical shape of the body, due to wave reflection (and refraction) across the outer boundaries of the object. In the case of water these measurements are carried out inside a pool, with an *AE* source located at its centre. But this is not possible for a solid medium.

The concept of density ρ implies the continuum approximation. The concept of velocity V refers to physical effects observed at different times and different sites. The acoustic impedance $Z = \rho V$ relies on the same assumptions. In contrast, fracturing events rely on quantum effects.

The propagation or damping of an elastic wave through a medium is related to how much longer the wavelength is, compared to the typical scale size of the heterogeneities of the medium. The Earth is approximately homogenous for seismic waves (*~0.5-1 Hz*), because the spatial scale of heterogeneities in the Earth's body cannot affect these wavelengths. At shorter wavelengths, such as ultrasounds, the Earth appears much more heterogeneous, and ultrasound can be transmitted only through rocky bodies, etc. That is, transmission and damping are related not only to ρ but also to frequency. Therefore, transmission critically depends, more or less, on the ideal homogeneity of the medium. In addition, the temperature of the medium makes more or less effective the crystal bonds, hence it affects the capability to transmit an elastic wave.

The density ρ influences absorption, hence also the acoustic impedance Z. Otherwise, one could show that V is exactly proportional to $1/\rho$, by which $Z=\rho V$ is constant. According to a treatment in terms of elementary approximations, it can be shown (by an abstraction) that for a homogenous infinite medium it is $V = \sqrt{const/\rho}$, where *const* depends on whether one considers P-waves, or S-waves, or surface waves at the boundaries of the medium. In reality these conditions can hardly be attained even in the laboratory, while in the field the granular structure of the system raises serious problems, such as e.g. the difficult implications of the sand pile theory, etc. It is therefore concluded that also $Z= V\rho \sim \sqrt{const \cdot \rho}$ depends on ρ.

All this appears as a mere academic exercise, while physical reality behaves according to its specific laws, which are much more complicated compared to any human schematization by means of a "simple" model. This also originates misunderstanding between different scientific communities. A correct interdisciplinary communication therefore requires a fundamental clarification.

Let us point out that the meaning of "transmission", "conduction", "transport", and "advection" are very similar, and sometimes closely related to one another. In Earth sciences the choice of any one term is determined by intuitive criteria, in order to give some feeling about the nature of the speculated underlying process. In contrast, engineers, who are concerned with manmade systems, agree on some often undeclared conventional choices, by which they use these terms only with specific meanings. For instance, it is customary to talk about "thermal energy conduction" (which is different e.g. from advection), or about "electric energy conduction" - hence about "electrical conductors" etc. Thermal advection, charge transport by motion of an ionized medium, or transport of sound inside a moving mass of matter such as the noise of a train listened when we are down-wind with respect to the train, etc., all these phenomena imply a "conduction" of energy, but they are better described as "transport" phenomena.

Engineers claim that an elastic wave is "transmitted" through a medium, Earth's scientists may sometimes consider the transmission of an elastic wave much like en electric signal that goes through a "conductor", etc. The motion of an ion is a transport phenomenon, i.e. an advection phenomenon. But it is the same as an electric current. All ions move through the atmosphere and they compose the atmospheric electrical circuit. But also a lightning strike is an electric current. However, it is different to deal with the gentle wave of the surface of a pond, and with a catastrophic tsunami. The physics are the same, but energy and implications are much different. Terms can be different. But processes are clear. Therefore, the authors apologize to a reader who is not familiar with these matters of lingo, and kindly ask her/him to consider the physics of the process, rather than the terms used for explaining it.

In terms of the physics of the process, a fundamental point is that this entire item is closely related to the space domain of phenomena, which determines the required number of points where *AE* recordings are to be carried out. For clarity purposes, let us refer to a few case histories.

Consider a bridge or a building. Everything in natural reality is always subjected to stress, hence to fatigue and ageing. But some specific structural knots of the bridge or of the building experience a comparatively greater stress, hence they are likely to suffer by a comparatively earlier loss of performance. Every structural component of a bridge or of a building, however, is composed of material (either steel or concrete) that is a good *AE* conductor. Hence, it is sufficient to monitor a limited number of points of the entire bridge, depending on its length and structural details. The same argument applies to a pipeline, where, however, the concern is more about the time scale, rather than about the space scale (see section 1.2).

A different concern deals with a rail embankment or with land-slide hazards. In general, soil is insufficiently compact for ensuring a good transmission of the *AE* signal. Hence, there is need for monitoring some "solid" structure - such as a house, or a building of any kind, or a pall, etc. - which is embedded into the "loose" soil material (e.g. for a rail embankment one can use the palls for the power supply to the railway).

In either case, the spatial detail that is effectively attained by *AE* monitoring is defined by the number of monitoring points that are located inside the given system. That is, it depends on the number of "solid" probes that can be used. If they are insufficient for the required needs, some newly built "solid probes" should be set up.

Concerning a volcanic edifice, the *AE* array can be reasonably small, as the primary agent, i.e. the time-varying pressure of the endogenous hot fluids, is likely to affect simultaneously a large fraction of the entire volcanic edifice, and a few measuring sites can therefore be sufficient to monitor the entire system.

Comparatively much more complicate is the case history of seismic monitoring. The tectonic setting, and structural details, of the interested area play a critical role in the choice of the *AE* array. In general, these features are also poorly known. Hence, a careful multidisciplinary modelling is required in order to choose the hopefully "optimal" location of the *AE* array.

Summarizing, the space scale of the required *AE* monitoring depends on the specific application of concern, and on the spatial detail that is required.

1.2 The time domain – the *AE* trigger

Figure 1 is a cartoon that shows the time dependence vs. frequency of the observed *AE*. The physical principle is that every former "solid" sample - which originally has almost no flaws - suffers by the formation of some small flaws, which are associated with some comparatively high frequency (*HF*) *AE*. On the occasion of every additional stress - which is subsequently applied to the same sample - some new flaws will be originated. This will occur preferentially close to the crystal bonds that have already yielded. At these points the material is comparably weaker.

Hence, new comparatively larger flaws will be generated by coalescence of the previous smaller flaws. The process can be illustrated in terms of a progressive implosion of small flaws to generate larger flaws, almost like in a chain reaction. An analogy is the well known game of domino tiles that drop one after the other in a time sequence.

According to this physical rationale, we know with certainty that - owing to specific physical reasons - we would expect to observe different phenomena involving first some comparatively *HF AE* time series, and subsequently *LF* (low frequency) *AE*.

A key aspect of concern - about the time scale - deals with the primary driver of the process, which originates the *AE*.

Some phenomena are characterized by some "long" time scale. Indeed, some phenomena certainly occur – and therefore ought to be detected in some way - during some "long" time lag preceding the incoming "catastrophe" of the system (this remark applies either to an earthquake, or also to simple crustal stress propagation, or to a land slide, or to the loss of performance of a bridge or of any other manmade structure due to material ageing, etc.).

In contrast, sometimes some *AE* can be originated only by an action that lasts a few ten seconds (such as e.g. when operating an electric drill on a solid sample). This requires a much higher time resolution for *AE* data acquisition, as there is need to monitor the rapid evolution of the system until the opening of the drill hole.

But even much more rapid *AE* data acquisition is sometimes required, if we want e.g. to monitor the time evolution of the performance of the system while it is disrupted by an explosion. For instance, consider a pipeline that is broken by a cold chisel struck by a hammer triggered by explosive.

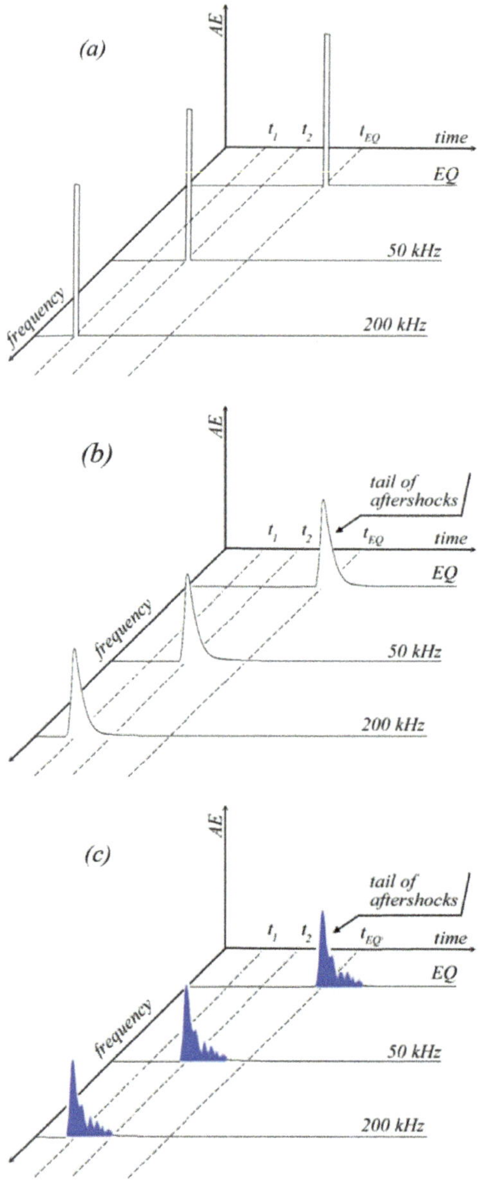

Fig. 1. Cartoon showing how the recorded *AE* signals are first released at comparatively *HF*, corresponding to the yield of smaller flaws, and subsequently to progressively lower frequencies. The signals are not Dirac δ-function distributions *(a)*, rather lognormal distributions *(b)*, which are eventually modulated such as by tide effects *(c)*, e.g. observed on the volcanic edifice of Vesuvius, displaying some apparently damped oscillation, just analogously to the aftershocks of an earthquake (here denoted by *EQ*). See text. After Paparo & Gregori (2003)

In this case it is important to focus on the need for specifying - as a needed logical prerequisite in every data analysis – the target of concern. Long time scales (compared to the human standards) are to be considered when one wants to monitor the evolution of the system towards its possible "catastrophe" in order to intervene. In contrast, when the typical duration of a phenomenon is much shorter compared to the typical human standards (e.g. in the case of an explosive event), the required information is to assess where and when the "catastrophe" already occurred in order to repair the damage.

Summarizing, the time scale of concern depends on the application required by the *AE* information.

Independent of whether the primary *AE* trigger has a long or short time scale, in general some trigger always exists, being the ultimate reason by which everything is ageing in the universe.

As far as any kind of manmade structure is concerned, thermoelasticity is an important trigger for *AE*. Figure 2 shows the case history of records collected on Gran Sasso (Paparo et al., 2002; Gregori & Paparo, 2004). It is the largest mountain massif in the central Apennines, composed of dolomia and limestone. The diurnal variation of the recorded *AE* signal (for operative details refer to section 2) shows a maximum in the very early hours of the day (local time) when the mountain is cooling. In fact, when the comparatively outer layers of rocks warm up, they expand with respect to the inner cooler layers. In contrast, when the mountain cools, the outer layers contract over their respective internal warmer – hence more expanded – layers. Therefore, some cracks occur. It is the same principle by which in an old country house, with no heating, during cold nights the furniture releases noisy cracks in the early hours of the day.

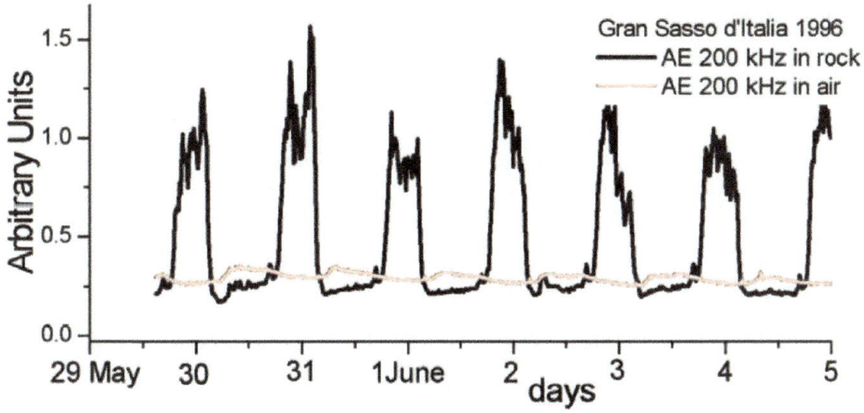

Fig. 2. *AE* recorded on the Gran Sasso mountain (Apennines, central Italy). The large diurnal variation has a maximum during night time when the mountain cools. An identical transducer suspended in air measures a negligible signal (light grey line). The difference between different days very likely depends on different meteorological and insolation conditions, i.e. the observed effect shows all features consistent with a mere thermoelastic effect. See text. Figure after Gregori & Paparo, 2004 and (simplified) after Paparo et al. (2002)

Thermoelasticity is the main cause for weathering for rocks in Sahara or other desert areas. But it is also a steady and permanent feature e.g. in a bridge, or in a concrete or steel building in open air. This fact is quite useful for monitoring the different response - hence the progressive ageing and loss of performance - of the materials that compose that structure.

Other triggers for *AE* - e.g. on a bridge - can be the transit of a train or of a heavy vehicle, or a slow land slide affecting its basement, etc. The same applies e.g. to a possible unexpected rupture of a pipeline.

In general, one can envisage either a steady, real-time, monitoring of the ageing and loss of performance of a structure, or - whenever the loss of performance is very slow - one can envisage some periodical check of the structure e.g. of a crane, of a rail car, etc.

In the case of steel structures it is even possible - at least in principle - to take advantage of the magnetic hysteresis cycle of its ferromagnetic components. For this, one can use - apart the concern about a suitable intensity of the applied external field - a device that in principle is similar to a standard metal detector that has an oscillating circuit with a resonance frequency that depends on the inductance, which is altered by the presence of a metal object. The Weiss domains of the metal structure experience a hysteresis cycle. But, while they change the inductance of the oscillating circuit, they also cause an *AE* release from the metal. Owing to the action-reaction principle, if the metal object causes an effect on the metal detector, the metal detector causes an effect on the metal object, and this can be monitored by *AE*. Upon applying suitable algorithms to this signal (see section 3), one can evaluate the stage of ageing of the material. This technique can apply e.g. to making a periodical check of the performance of specific ferromagnetic components of machinery, a rail car, etc.

Summarizing, the time scale to be considered depends either on the ultimate target of the application of *AE* monitoring, or on the typical time scale of the primary trigger of the observed *AE*.

1.3 Lognormality

Unless otherwise stated, let us refer to one given *AE* frequency at a time. A key point of concern deals with the trend of the observed *AE* record. The signal is the result of some "*fog*" of primary *AE* sources, which are identified with the flaws of some specific size that release that observed *AE*.

Let us begin by an abstraction. The definitions of entire mathematics, geometry, fractality, set theory, abstract algebras, group theory, probability theory, etc., all rely on abstractions or analogies. The same logics apply to many different material systems. All these analogies are therefore crucial for understanding different phenomena in terms of a unique rationale shared by different systems and processes.

The *AE* rationale is the same as the typical justification of a distribution of the Kapteyn class, known as "lognormal distribution". It was defined by Kapteyn (1912) and Kapteyn et al. (1916). It was also included in some textbooks (e.g. Arley & Buch, 1950). But apparently it was generally almost forgotten in the subsequent literature (Paparo & Gregori, 2003; Gregori & Paparo, 2010).

Its leading motivation is shared by several phenomena, typically by every public service, i.e. in every case history when the probability that a user takes advantage of that service is proportional to the number of users that already use it.

The same rationale is shared by every physical phenomenon, whenever the occurrence of an event is proportional to the number of events that are already occurring. Typical examples (refer to Gregori & Paparo, 2010 for more extensive discussion and references) are the hypsometric curve (i.e. number of points at a given height above sea level) of the Earth or of a planetary object, or a geomagnetic storm (Campbell, 1996), or a magnetospheric substorm (which is an "elementary" component of a geomagnetic storm; Akasofu, 1968), or with a financial crisis when the psychological impact determines the lognormality of the event, etc.

The same rationale can be applied to the *AE* release from the *"fog"* of elementary *AE* emitters inside a given "solid" structure, where crystal bonds are progressively broken with a higher probability of rupture wherever the structure is comparatively weaker, etc.

Historically, the first discipline where this important logical feature was clearly assessed was geomagnetism and polar auroras. Hence, let us borrow their terms for applying them to other disciplines. For future reference, let us therefore recall that a "geomagnetic storm" is assessed by the North-South horizontal ("H-component") geomagnetic field (Chapman & Bartels, 1940). In contrast, a "magnetospheric substorm" - which is shown to be some kind of more elementary component of a magnetospheric (or geomagnetic) storm – was clearly recognized by Akasofu (1964) by means of polar auroras, as the geomagnetic signal is excessively perturbed for unambiguous assessment. That is, one and the same phenomenon involves one and the same physical system, but it eventually requires to be monitored by different diagnostic tools.

In the case of a magnetospheric substorm – which typically elapses ~ *2-3 hours* - the process is the consequence of the progressive lack of particle supply from the particle reservoir represented by the plasmasheet inside the tail of the magnetosphere (Gregori, 1998, 1999, 2002): hence, the progressive exhaust of particle inflow from the tail is reflected in the lognormal trend of a substorm.

Similarly, concerning a geomagnetic storm - which typically elapses ~ *a few days* - it reflects the time variation of the flux of solar wind. As mentioned above, a storm is composed of a formal disordered sequence of overlapping substorms. When the solar wind flux exhausts its "anomalous" flux, the decaying trend of the storm will result lognormal due a progressive fading off of its primary trigger. In the meantime the magnetosphere attains a new equilibrium state, after taking advantage of the availability of particles in the reservoir of its plasmasheet.

The same logical sequence of "storm" and "substorm" can be applied (see below) to *every* phenomenon, even to the *AE* technological or environmental applications here of concern.

That is, lognormality is a frequent mathematical feature observed in a large variety of phenomena. This is just a matter of a mere observational fact. And also the *AE* monitored at a given frequency is expected to occur according to a lognormal trend. Its basic physical interpretation and implications, however, are to be better specified after discussing its fractality (see section 1.4). For the time being, let us point out that the *AE* sources operate according to the aforementioned rationale, hence the *AE* signal - at every given frequency - will display a lognormal distribution, as shown in figure 1(b).

In addition, the tail of the lognormal distribution will eventually be modulated by some external action. For instance, at the lowest frequency - i.e. at ~*0.5 – 1 Hz* which is the typical frequency of a seismic shock - the tail of the lognormal distribution [figure 1(c)] will be identified with the classical aftershock sequence. Seismologists have recognized only the tail of this distribution. They call it the Omori law. The lognormal distribution is comparatively sharper when the typical timing of the evolution of the system is faster. Stein and Liu (2009) found that the faster the slip-rate along a fault (in the range *0.1 - 100 mm year*$^{-1}$), the shorter the aftershock time series (in the range *a few – a few thousand years*).

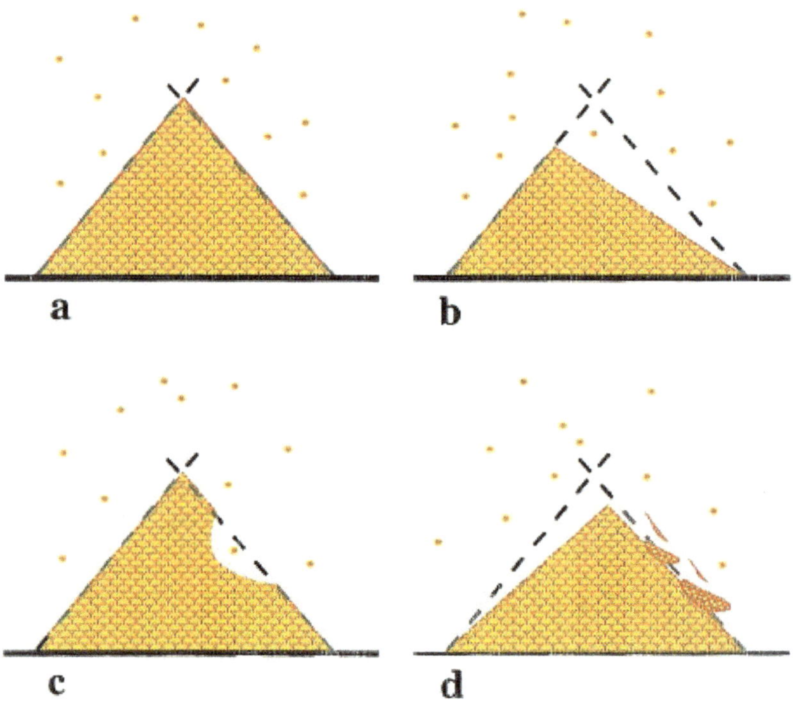

Fig. 3. "*Grains of sand (small dots) being added slowly to a sandpile.* (a) *All sides of the sandpile have reached the angle of repose whereby additions of sand result in instabilities, i.e., avalanches, of various sizes.* (b) *A large avalanche has taken place along one small range of azimuths of pile taking that zone out of a self-organized critical state and making it incapable of being the site of a large avalanche for a long time (until grains of sand are added to it to bring its slope back to the angle of repose; large avalanches can still occur at any time along other azimuths.* (c) *A small avalanche occurs along one azimuth but does not affect its entire downdip slope; small avalanches can still occur along other portions of slope either up or downdip of that small avalanche.* (d) *Moderate-size avalanches occur as a given azimuth approaches or reaches a state of instability prior to a large avalanche. Large and small avalanches correspond to large and small earthquakes …*". Figure and captions after Sykes et al. (1999)

Consider another example pertinent to the present discussion: a sand pile which is in a state of critical equilibrium (Buchanan, 1997; Coontz, 1998; Sykes et al., 1999). Once in a while, some part of the pile falls down, and some small sand slide occurs here and there (figure 3). The system is said[2] to be in a state of "self-organizing criticality" (*SOC*). This critical state will last until the sand pile has attained its final equilibrium.

In terms of an analogy with the geomagnetic storm - which as mentioned above is composed of an irregular sequence of overlapping magnetospheric substorms - it is possible to liken the entire process, of the collapse of the sand pile towards its final equilibrium, to a "storm" of the system. In addition, this "storm" is composed of the irregular sequence of overlapping "substorms", each identified with a partial sand slide of the pile. Altogether the system, which implies a "storm" and its "substorms", is a *SOC* process.

Concerning the applications, one has first to define the time scale of the process (section 1.2): even a structure, which is destroyed by a huge explosion, has its "history" and timing of every different yielding component, although the time scale of the evolution of the explosion, i.e. of "catastrophe" of the system, is not comparable with the typical human time scale.

After having defined and assessed the time scale of concern, the yielding process has some typical temporal evolution of every part of the system – e.g. consider a building made of concrete, or a Medieval tower made of bricks or stones, or a bridge made either of steel or of concrete, or any kind of manmade object. This evolution will reflect some progressive loss of performance of either one part, or "elementary", component of the system. All this appears to occur similar to the aforementioned sand pile theory.

An engineering-designed object, such as a bridge or a building - compared to the case history of a sand pile - will respond to a more deterministic and less erratic trend. But, within suitable approximations, one can liken this to a *SOC* process with the progressive loss of performance of every ageing component of a bridge or building.

Summarizing - upon borrowing the same terms and concepts used in geomagnetism - an observed *AE* event (for definition of "event" see section 3) can be likened to a *"stress storm"* (at every given *AE* frequency) provided that - as mentioned above - the process can be expected to occur according to some ideal situation to be approximately depicted by a lognormal phenomenon. In any case, the entire phenomenon, i.e. the *"stress storm"*, will appear like a *SOC*, characterized by some irregular superposition of overlapping *"stress substorms"*.

From a comment by Cliff Ollier, whom we deeply thank, the term *"stress storm"* can sometimes be misleading, as *"stress" per se* is not *"strain"*. Hence, one should call *"crustal storm"* one phenomenon that occurs inside the Earth's crust etc., or one could use the term "strain storm" instead of "stress storm". On the other hand, ideal "elasticity" is only an abstraction. Hence, in natural reality every non-ideal elastic deformation always causes the yield of some crystal bonds, hence some *AE* release. Therefore, for simplicity, let us use either term *"stress storm"* or *"strain storm"*, or briefly *"storm"* that can be applied to every case history, either in Earth sciences, engineering, or other disciplines.

---

[2] It was formerly defined by Bak et al. (1987). Or see also e.g. Sornette et al. (1990), Sornette & Davy (1991), Cowie et al. (1993); Main (1995), Cowie (1998), Cello (2000), and references therein. It is even reported by science popularization books (e.g. Barrow, 2009).

Thus, the purpose of analyzing an observed *AE* time series - or set of different frequency *AE* time series - is to recognize, within the observed *AE* records, both *"AE storms"* and their constituent *"AE substorms"*.

1.4 Fractality and SOC

Fractality is a crucial aspect of an *AE* time series. It is the result of an abstraction, similarly also is lognormality. Concerning the general and more exhaustive discussion of lognormality and fractality refer to Gregori and Paparo (2010).

Every time series of events, either of *AE* or any other kind, in general is found to be more or less randomly distributed in time (concerning algorithms etc. refer to section 3 and references therein). In the case of "perfect" or "ideal" randomness, the fractal dimension D_t of the *AE* series is equal to *1*. This is just a matter of mere definition, or just one possible rigorous way of formally defining whether a time series of events is randomly distributed or not.

In contrast, suppose that some event has a "memory" of the previously occurred events – or equivalently that it has "memory" of the events that are to come in the future; in fact for our purposes either one possibility is the same. It shall be found that the fractal dimension must be $D_t < 1$. In the case of "total memory" - i.e. in the case that all events occur altogether at the same instant of time - it must be found $D_t = 0$. These are just matters of mathematical definition, in a strict sense.

Physically, we monitor a system and we collect a time series of *AE* events. We evaluate its D_t by means of a specific *AE* data series that spans some pre-chosen and given total time interval. The D_t will reveal whether during that given time lag the *AE* sources are activated more or less randomly. The greater is the randomness, the less "aged" is the material that releases the observed *AE*. The less randomness means that the *AE* sources are closer to their final cleavage micro-plane, hence the more "aged" is the material sample, and the closer it will be to its final "catastrophe". For instance, a steel bar, monitored in the laboratory (see below), was found to be close to final rupture when $D_t \sim 0.45$.

In reality, however, when referring to data measured while the "solid" structure is actually yielding, the computed D_t refers to *AE* emitters while they are fully yielding, i.e. the actual contribution of every "elementary" *AE* emitter in general is to be expected (in the aforementioned case history of a steel bar) to be $Dt << 0.45$.

An additional fundamental concept, which is important for understanding our observations, is the distinction between two different logical models, i.e. whether it is supposed *(i)* that the observation refers to *one* single *AE* source, or *(ii)* to a *"fog"* of some *huge number* of *AE* sources operating together, although not synchronously.

Another result of the data analysis (see section 3) is the so-called *"hammer effect"*, which is expressed by an index *H*. When dealing with a single *AE* emitter, it is possible to recognize - on an instant basis - whether the system (*H*=+*1*) is releasing *AE* because it suffers by some action applied from its exterior, or rather (*H*=-*1*) it experiences a transitional evolution towards a new equilibrium state, after having suffered some external trigger.

Differently stated, we know that the *AE* signal is likened to a lognormal distribution (section 1.3), although this is an ideal situation. In contrast, in general the *AE* records will appear highly perturbed compared to this simple logical scheme. On the other hand, we can evaluate H objectively, on an instant basis. We can therefore state – and this can be shown by formal mathematics applied to its specific algorithm - that during the *rising* stage of the (instant) lognormal distribution it is $H=+1$, and during its *tail* stage it is $H=-1$.

Based on this argument, it can be claimed - when dealing with *one single AE* emitter - that whenever $H=+1$ the material is subject to some external action, and whenever $H=-1$ it is during its recovery stage.

In natural reality *one single AE* emitter is an ideal condition that can be only approximately simulated in the laboratory, while it is seldom observed in the field. A more realistic model is therefore to be considered in terms of some suitable *subdomain* of the physical system. Every subdomain is composed of a "*fog*" of some huge number of "elementary" *AE* emitters. Suppose that - during some given time interval of some finite duration - the flaws of a given size inside this subdomain collectively experience their decay, while they coalesce towards larger size flaws. During that given time interval, the measured *AE* will be the result of the sum of several "elementary" *AE* signals, all having some very small D_t, let us say $D_t \sim 0$. But, the number of these "elementary" *AE* emitters will decrease *vs.* time, according to a lognormal trend (see section 1.3). So, we call this entire mechanism a "*lognormal fog*" of "elementary" *AE* emitters.

That is, every aforementioned subdomain is responsible for an appearance - in the observed *AE* series – of a sequence that recalls some approximate lognormal distribution (in the D_t *vs.* t plot it will appear like a lognormal trend with a *reversed* ordinate axis). This feature depends on the decrease *vs.* time of the population of "elementary" *AE* emitters, rather than – as it occurred in the case of *one single AE* emitter – on the ageing of its material.

In general, at different times, different subdomains will experience a similar phenomenon. The final effect is that all their respective outputs will appear to overlap with one another, in some apparently erratic way, due to the huge number of unknown degrees of freedom of the system – hence due to some erratic number of overlapping different subdomains. Their trend can therefore be said to occur according to a *SOC* process.

Differently stated, everyone of the aforementioned "reversed" lognormal trends, associated with one given "*lognormal fog*" of "elementary" *AE* emitters, will more or less erratically overlap one with the other, almost like "*stress substorms*". They will compose the final general observed morphology of the "catastrophe" or "*storm*" that affects the system.

Therefore, in order to recognize a "*substorm*" of the system, it appears more effective to appeal to D_t. It is much better than to appeal to the raw *AE* signal, because D_t is a physically expressive parameter, which is independent of the arbitrariness of the amplitude of the original raw *AE* datum. Also the index H can be sometimes heuristically more effective.

Summarizing, by means of an analogy and by using the same terms used for the geomagnetic field and magnetosphere, or for the sand pile (see section 1.3), we can call "*stress storm*" an entire paroxysm observed in the raw *AE* record at a given and pre-chosen fixed frequency. Then, we can claim that every "*stress storm*" is the result of some apparently erratic and disordered superposition and overlapping of "*stress substorms*", all being associated with some suitable (though unknown) aforementioned physical subdomain.

In addition, let us recall that a geomagnetic storm can be recognized by means of geomagnetic data, unlike its "elementary" components i.e. the magnetospheric substorms, which are recognized by polar auroras. That is, the physical system and the observed phenomenon are always the same, although different morphological features can be more or less clear, detectable, and reliable for diagnosing its state and its evolution.

In a similar way, a *"stress storm"* is recognized by means of the raw *AE* records, while its component parts, i.e. the *"stress substorms"*, can be better recognized by D_t.

From the physical viewpoint, the distinction between *"stress storm"* and *"stress substorms"* also has a relation with the space domain of this process. In fact, a *"stress storm"* appears to be some large-scale occurrence, i.e. it involves the entire system of concern. In contrast, a *"stress substorms"* involves some much more limited and specific structural fraction of the whole system. The same holds also for the Earth's magnetosphere, where the scale size of the trigger of a "storm" is much different compared to the case of a "substorm".

A final assessment is therefore that - when dealing with any kind of phenomenon either in Earth's sciences, or about financial crises, or about a public service, or about psychology, etc. - different facets of one and the same set of logics can be considered. The logical facets are named "lognormality", "fractality", *SOC*, and *"storm"* and *"substorm"*. Every facet can be monitored and diagnosed by means of different observational information. But the ultimate logic is unique and all the same.

2. The data base

The records that were used by the authors over several years - mainly for field measurements but also for laboratory tests - rely on *LF AE* (typically *25-30 kHz*) and on *HF AE* (typically *150-200 kHz*). Different data acquisition procedures are used in specific applications, depending on the total duration of the experiment, and the required time resolution of records.

An acoustic transducer, tuned at a given pre-chosen frequency, is posed in close contact with the "solid" object to be monitored. When operating in the field, a hole (about *30-50 cm* deep) is drilled into a rocky outcrop. A glass bar is put inside the hole, which is then filled with concrete. The *AE* transducer is fixed on top of the glass bar (one bar for *HF* and one for *LF*). But, when operating on a manmade structure, in general there is no need for drilling the sample.

The outcrop is the terminal on the Earth's surface of some generally much extended "natural probe" crossing underground through some unknown volume, eventually displaying rapid changes of *AE* transmission capability (or briefly let us vaguely state that the acoustic impedance of the "natural probe" suffers by step-wise time variations). In any case, our experimental setting in the field always resulted in an effective probe, because eventual cracks in the rocky body are filled with water, which is an excellent medium for *AE* propagation.[3]

A linear preamplifier is applied to the output of the transducer, before its transfer to a data logger, where amplification occurs before data storage. When operating in the field a *GSM* connection with a remote acquisition centre permits data analysis at any desired time.

---

[3] On several occasions we observed (see Gregori et al., 2010) step-like changes of the *AE* signal that we called *MFE* (minor fracture events) and that we interpreted in this way.

The measured datum is transformed into a *rms* amplitude, and in general it is averaged over *3 msec*. When operating in the field, the data logger was set in order to store one average datum (of the *3 msec* signals) every *30 sec*. When operating in the laboratory, one datum is stored at a time rate of a pre-chosen and given Δt, the range of Δt being chosen, say, between *50 msec - 200 msec*. For more rapid events, such as when they are triggered by an explosion, the time resolution must be pushed at least to, say, $\Delta t \sim 1$ *msec*.

If the concern is about monitoring the Earth's tide spectrum, or the free oscillations of the Earth, Δt = *30 sec* is certainly exceedingly long, as the typical tidal frequencies are defined with a precision higher than *30 sec*.

An important and generally unnoticed twofold warning deals with over-saturation and under-saturation of the recorded *AE* signal. In fact, owing to the unknown and arbitrary time changes of the acoustic impedance of the "natural probe" (see above), the amplifier in the data logger is arbitrarily set, in order to get some output signal to be recorded typically ranging, say, e.g. up to ~*10 V*, with a sensitivity of the order of ~*1 mV*.

Some equivalent intuitive criterion is also used in laboratory experiments. When carrying out e.g. a drill into a bar, the recorded signal eventually gets into over-saturation. The optimum amplification has to be attained by trial and error. In any case, the algorithms for data analysis resulted to be reasonably robust with respect to this drawback.

But in the field, if the amplification setting is carried out during "quiet" conditions, some important *AE* information during perturbed periods can be lost by over-saturation, and this must be avoided. On the other hand, if one sets a lower amplification in order to avoid over-saturation during strong perturbations, the opposite drawback has to be eventually faced, i.e. under-saturation. Indeed, according to the authors experience, under-saturation was often encountered in field records, although the algorithms used for data analysis were still robust even for very weak *AE* signals. The optimum instrumental setting, perhaps, ought to foresee a change of the amplification of the signal depending on the average *rms AE* signal of the preceding day.

There are different physical implications for *HF AE* as compared to *LF AE*.

According to the aforementioned physical rationale, *HF AE* reflect a much earlier "ageing" of the sample (Gregori et al., 2005, Paparo et al., 2006, Poscolieri et al., 2006b). This fact was confirmed by the expected effects associated with the so-called *"loading tide"*. This effect was clearly observed in *HF AE*, while *LF AE* was more perturbed by local tectonic features, which are much more sensitive to a comparably later ageing stage of the Earth's crust.

The *"loading tide"* is the cause of the well known almost steady increase of the length of the day (*l.o.d.*). This effect is of tidal origin, but its main component does not derive from the Earth's tide. Rather, the lunar plus solar tide moves a large mass of *ocean* water that piles up on continental shelves. Hence, it pushes on continental masses, and this originates a torque that slows down the observed spin rate of the Earth. This mechanism implies violent stress propagation through the crust. Poscolieri et al. (2006a) and Gregori et al. (2007) report a seasonal variation of the crustal stress, detected by *AE* records in central Italy (Orchi) and at Cephalonia (Greece), which are synchronous at the two sites. The average background trend also appears to display large oscillations with a peak-to-peak time lag apparently compatible with the leading lunar tide modulation. A final confirmation of this interpretation could be checked by some *AE* station located somewhere in central Asia, by which the propagation speed of crustal stress could be monitored.

In contrast, as mentioned above, the *LF AE* appear to be comparatively more affected by the *regional* or *local* tectonic evolution, i.e. they are much better suited for diagnosing the evolution of the crust at some time closer to its final eventual "catastrophe", i.e. an earthquake. They give better and more reliable information about the specific actual state of the crust inside some limited lithospheric slab, rather than referring to some planetary scale feature.

All these aforementioned inferences resulted from repeated investigations carried out during over a decade in many different tectonic settings.

A special mention is deserved about a very remarkable confirmation of this regular modulation which is an updating of Gregori et al. (2010). Superconducting gravimetry (*SG*), and its complementary absolute gravimetry (*AG*), is a comparatively recent achievement that is beginning to give unprecedented evidence of some lesser effects related to several phenomena of different origin (de Linage et al., 2007; Crossley & Hinderer, 2008, 2009; Longuevergne, 2008; Wiese et al., 2009). They are exhibiting great modelling skills at correlating many different observational inputs, atmospheric, hydrological, tidal, etc. They also correlate their records with *InSAR* monitoring and with the records of the *GRACE* satellite (Crossley et al., 2005). They monitor all gravitational effects associated with transport of solid and fluid matter in the Earth system originated by any kind of driver. They find temporal trends (Zerbini et al., 2007; Wilmes et al., 2009; Rosat et al., 2009) that look very similar to the aforementioned *HF AE* seasonal modulation that Poscolieri et al. (2006a) envisaged as a loading tide effect.

Compared to *AE*, their observations are different. *SG* monitors a time change of gravity, while *AE* monitor a time change of the stress field. Stress is expected to be maximum when the *speed* of the gravity related deformation is maximum, rather than the *intensity* of the gravity field.

But several irregularities are observed both in the *SG* and *AE* series - e.g. unexplained background trends, changes of phase or of synchronism, *LF AE* appear heavily perturbed compared to *HF AE*, etc. These facts are clearly indicative of phenomena associated with tectonic setting and geodynamic actions.

SG and *AG* envisage sub-continental size phenomena (Crossley et al., 2005) that cannot be planetary-scale tidal phenomena, but rather they appear more likely related to crustal stress effects. By similar methods also coseismic effects were evidenced, by means of *GRACE* data, of the Sumatra-Andaman *Dec 26, 2004* earthquake (de Linage et al., 2009).

It appears therefore very important to be capable to recognize "storm" and "substorm" time phenomena in either *SG* and/or *AE* records, in order to achieve a more detailed morphological classification and assessment of events during their trigger and evolution.

A wide and important perspective of unprecedented ways of monitoring crustal phenomena is thus being envisaged. The key role ought to be stressed of the mutually complementary information provided by *SG, AG, InSAR, GRACE* and *AE,* aimed at monitoring different observational facets of one and the same system and phenomenon.

In addition, crustal stress propagation ought not to be expected to be isotropic, being rather more effectively controlled by serpentinization and serpentosphere (Judd & Hovland, 2007).

3. The data analysis

The details already appeared in several papers (Gregori & Paparo, 2004; Gregori et al., 2002, 2005, 2007, 2008; Paparo et al., 2001; 2002; 2006; Poscolieri et al., 2006a, 2006b) and, owing to brevity purposes, they are not repeated here. Only some highlights are presented. A standard software was progressively implemented and improved during past years. The present data handling can be synthesized by the acronym *OFTH*, and is explained as follows.

"F" denotes a weighted running average of the raw *AE* datum, over a given pre-chosen time lag. The weight is defined by a triangular system function, aimed at reducing the perturbation originated by the side lobes of a simple non-weighted running average. For field applications we used a time lag of *24 hours* in order to reject all effects associated with diurnal variation, including mostly the aforementioned thermoelastic effects (see section 1.2), and also some fraction of tidal effects, although not all of them.

"T" denotes loss of performance *vs.* time or "ageing", being quantitatively estimated by means of the aforementioned D_t. It is computed by the standard box counting method (refer to any book on fractal analysis).

"H" is the aforementioned parameter of the *"hammer effect"* (Gregori et al., 2007). It derives from a combined analysis of the raw database, of its *"F"* and of its residual, and their respective time derivatives. It is formally shown that, when the system is subject to some forcing originated from its exterior, an index *H* is computed being necessarily $H=+1$. In contrast, when the system is recovering towards a new equilibrium state after having been subject to some external perturbation, it must be $H=-1$.

The data series of the instant value *H* is somewhat scattered, although in reality the result appears much more stable than expected, thus denoting a robust physical significance. Some derived parameters are e.g. (in field *AE* records) the hourly means or the *24 hour*-means of the *H* instant values, or equivalently the percent number of $H=+1$ evaluated inside a given running time interval, or one can re-define a new index = ± 1 depending on whether the prevailing *H* inside a given running time lag is either *+1* or *-1*, etc. Analogous criteria can be applied to any other kind of application. In every case, the physics of the information of the instant *H* is always the same. The concern is rather about its graphical representation, and about smoothing the possible scatter of the *H* instant data series. That is, the optimum choice is arbitrary and it depends on the specific application of concern.

"O" denotes outliers. An outlier is a datum that does not fall into a Gaussian distribution defined by its nearby data. Consider a running time lag, and analyze the distribution of all *AE* records contained inside it. Finally reject the data that manifestly do not fit with a Gaussian distribution. This can be achieved by formally evaluating a suitable parameter. Refer to the aforementioned papers for details. The outliers are rejected twice: a first time by referring to the data series of the raw *AE* records; the next time by referring to the residual signal, after subtracting the *"F"* series from the original raw datum. The two *"O"* series being thus obtained generally differ only by a few percent. But it is worthwhile to repeat the

"*O*" evaluation as this is helpful for getting rid of the drawback from possible over-saturation of the signal. The subsequent data handling can be carried out separately *(i)* on the given data series, after being "cleaned" from its outliers, and *(ii)* also on the data series of the rejected outliers, as they have a different, though significant, physical meaning.

Outliers are not only simply concerned with isolated point-like perturbations. In fact, the algorithm is like a logical sieve that selects unusual objects, and it operates after having arbitrarily defined the size of the sieve holes. We expected that, when the size of the holes is sufficiently large, the number of outliers rapidly damps off. In contrast it was found that a conspicuous number of outliers always remains. The physical reason is that the measured *AE* is composed of asymmetric "elementary" events, every one reminding about the lognormal distribution (as in figure 1). In contrast, the aforementioned test for Gaussianicity is concerned with a *symmetric* distribution. Hence, the *asymmetric* tail will always give a large number of formal outliers.

At present, the outliers were mainly useful for analysing *AE* data collected on volcanoes, as every volcano unexpectedly seemed to be almost like a high precision stopwatch for monitoring Earth's tides (see section 4.1). More in general, a suitable standard software (in preparation) can transform every *AE* station into a station for monitoring both the spectrum of Earth's tides, and the free oscillations of the Earth.

Whenever we implement some given technological monitoring device, every different output of the *OFTH* set is to be evaluated and physically interpreted. The physical information is, however, entirely contained only inside the original raw *AE* data series of observations. Different logical tools are applied in order to provide evidence of different aspects of a given natural phenomenon. Then, after exploiting a suitable exhaustive physical interpretation, the optimum *OFTH* parameters - which are specifically best suited for that given application - are to be selected for routine operation, eventually in real time. In general, however, when dealing with different applications different optimal *OFTH* parameters are to be used.

For completeness sake, we later realized that, in his classical study carried out on micro-tremors records recorded on smoked paper, Imbò (1954) applied a similar rationale that can be, at least partly, generally likened to part of the aforementioned *OFTH* approach. An extensive historical account is in preparation.

4. Case histories

It is impossible to mention here all the different case histories that were investigated, including volcanic phenomena, and seismic events. They were important and effective tests for improving algorithm and software performance. Here we mention only a few studies that are more closely analogous to possible extrapolations toward technological applications.

Just for completeness sake, let us mention (see Gregori et al., 2010 for details) that - concerning field measurements carried out in Italy and on the Cephalonia Island (Greece) - the Italian peninsula is (as expected) tectonically very distinct compared to the Balkan peninsula. Refer to Table 1. During *1996-1997* a violent "*stress storm*" crossed the Italian peninsula, and two strong earthquakes occurred during this period. No information is available during *1998-2001*. But in *2002* the Molise earthquake occurred. Then from *2002* through the very end of *May 2008* the Italian peninsula experienced a period of great quietness. This period was later interrupted by a violent "storm" beginning at the end of *May 2008*, and by the end of *2009* it was still ongoing. During this period the L'Aquila earthquake occurred. A few related case histories are here recalled in the following.

Name	Date	Start time (GMT)	Lat. (N) of epicentre	Long. (E) of epicentre	Magni-tude	Depth (km)
Potenza	3 Apr 1996	13: 04: 35	40.67°	15.42°	4.9	8
Colfiorito, Italy (only shocks with M ≥ 5.0)	*26 Sep 1997*	*00: 33: 12.89*	*43.022°*	*12.891°*	*5.7*	*3.5*
	26 Sep 1997	*09: 40: 26.73*	*43.014°*	*12.853°*	*6.01*	*9.9*
	03 Oct 1997	*08: 55: 22.02*	*43.042°*	*12.824°*	*5.25*	*12.05*
	06 Oct 1997	*23: 24: 53.23*	*43.027°*	*12.846°*	*5.46*	*3.91*
	12 Oct 1997	*11: 08: 36.87*	*42.906°*	*12.920°*	*5.22*	*0.05*
	14 Oct 1997	*15: 23: 10.61*	*43.898°*	*12.898°*	*5.65*	*7.3*
Molise	*31 Oct 2002*	*10: 32: 58*	*41.695°*	*14.925°*	*5.8*	*10.0*
Lefkas	*14 Aug 2003*	*05: 14: 03*	*38.81°*	*20.56°*	*6.3*	*10.0*
L'Aquila	06 Apr 2009	01: 32: 39	42.334°	13.334°	6.3	8.8
Albania	06 Sep 2009	21: 49: 42	41.49°	20.43°	5.5	3.0

Table 1. Data from six different seismic events cited in the text (values in Italic taken from the catalogue **cpt08_1991-2006)**

4.1 Model with one *AE* source alone – the "Potenza effect"

Consider the observational case histories that can be interpreted according to the logical model (see section 1.3) of *one single source* alone.

Consider just one steel bar (Zanini, 2004; Biancolini et al., 2006). Bend it, say, *10,000 times*. Then, bend it once more while monitoring its *AE* (at one and always the same given frequency) and evaluate its D_t. Bend it anew, say, *10,000 times*, and repeat the same procedure, etc. Finally plot D_t *vs.* the number of times the bar was bent. The gentle decrease of D_t reveals the "ageing" of the steel that composes that bar. When D_t *~0 .45* the bar is found to be close to final rupture. Figures 4 and 5 refer to two groups of measurements. "*The first test was performed applying a load of* 48 N, *corresponding to a bending moment equal to* Mf=3.35 Nm, *obtaining nominal stress* σnom = 370 MPa *and maximum stress* σmax = 521 MPa. *The specimen fatigue life was estimated approximately to* 700000 cycles. *The second test was performed applying a load of* 63 N, *corresponding to a bending moment* Mf = 4.4 Nm, *estimating nominal stress to* σnom = 498 MPa, *maximum stress* σmax = 700 MPa *and the fatigue life to* 200000 cycles."

A curious effect is to be pointed out. After a while, the steel temporarily recovers (instead of steadily losing) its performance. This is the consequence of a transient re-adjustment of the micro-crystals of its alloy, in such a way that for a while they ameliorate the steel performance, soon before, however, experiencing the final evolution towards the final "catastrophe".

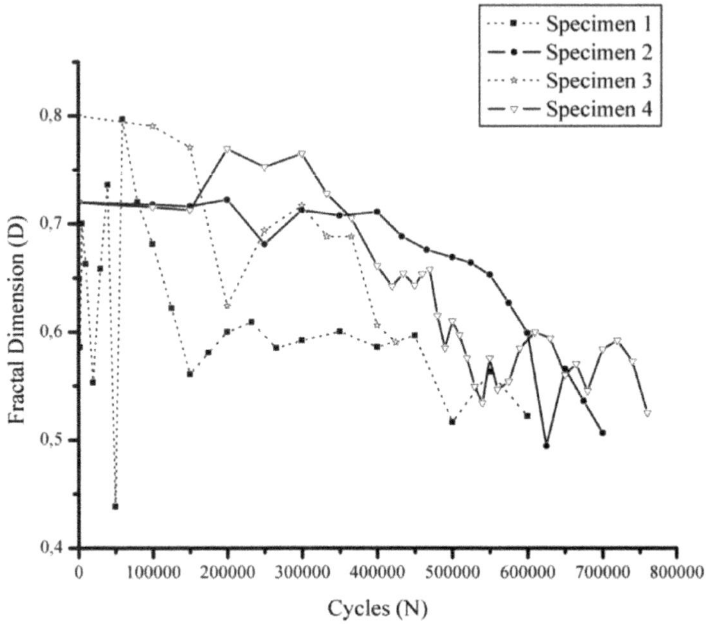

Fig. 4. Fractal dimension – Fatigue cycles (group 1). See text. After Biancolini et al. (2006)

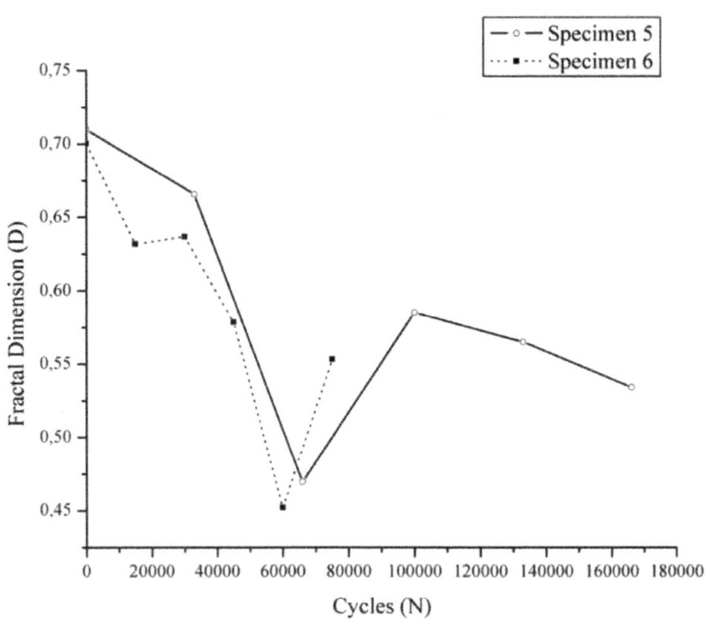

Fig. 5. Fractal dimension – Fatigue cycles (group 2). See text. After Biancolini et al. (2006)

In addition, *AE* were recorded on a steel blade of the *VIRGO* super-attenuator, while the blade was forced to its standard operative charge. The blade is casted by some special alloy etc. It was found (Braccini et al., 2002) that it is possible to assess, by means of *AE*, how many times the blade was bent after its casting.

A very intriguing case history, which seems to fit well with a *one single source* model, was measured in *1996* at the station of Giuliano (close to Potenza). See figure 6. During several weeks before the shock (*M=4.5*, with epicentre located at *~29.7 km* from the *AE* station; see Table 1) it was found that D_t decreased from $D_t \sim 1$ down to $D_t \sim 0.45$, when the shock occurred. Hence, this can be called *"Potenza effect"*.

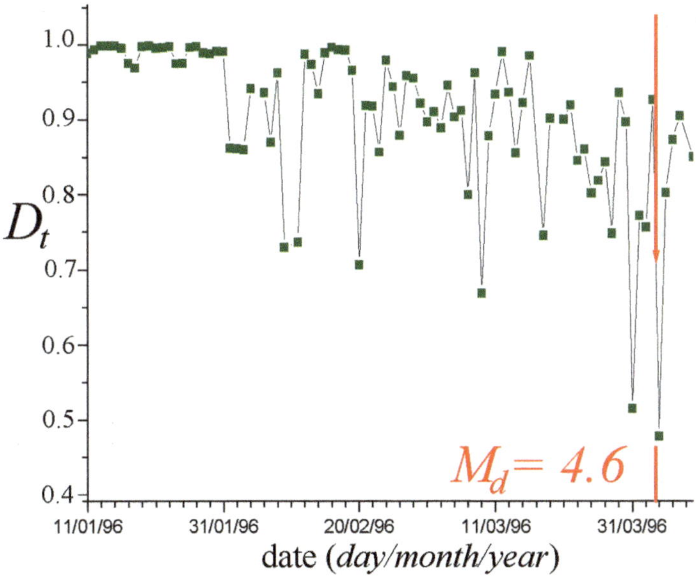

Fig. 6. Fractal dimension D_t of *LF AE* records carried out at *~29.7 km* from epicentre of the Potenza earthquake. See text. Redrawn after Paparo et al. (2006)

We observed another similar case history in *2009* in the Cephalonia (Greece) records on the occasion of the Albania earthquake (see Table 1, and figure 7).

The guessed physical interpretation is that the strain that precedes an earthquake is observed only in a limited region around the epicentre - unlike the much wider set of phenomena associated with a *"stress storm"*, which is *not* necessarily associated with a *local* final fracture of a fault. Hence, if the *AE* station is located inside this limited region of the crust, it monitors what precedes the final shock. The size of this limited region is, however, different in different areas, depending on the specific respective tectonic setting. In fact, while referring to the two aforementioned case histories, the distance between *AE* station and epicentre was *29.7 km* at Potenza, and *369 km* for the *2009* Albania event.

The model in terms of *one single source* alone is likely to be observed in every technological application, whenever the concern is about one single structural element alone. In contrast, sometimes the system is composed of several potential *AE* emitters that overlap one with

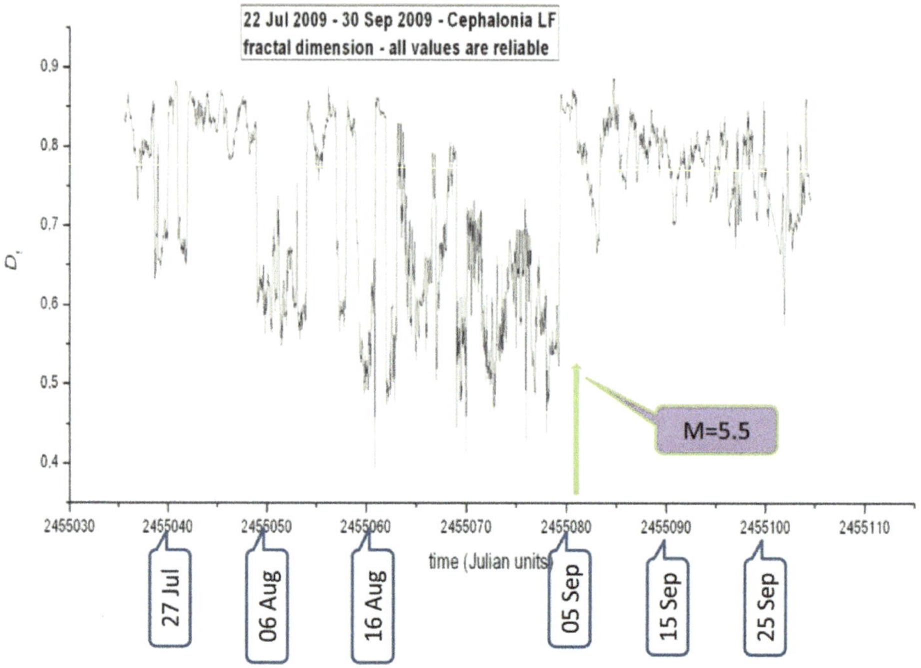

Fig. 7. "*Potenza effect*" observed at Cephalonia (Greece) preceding the Albania earthquake (see Table 1)

the other. For instance, a bridge is composed of a finite set of structural elements, which are all potential *AE* emitters. A bridge is therefore an intermediate situation. The extreme case history - opposite to the single source alone – is found when a "*fog*" of *AE* emitters has to be considered. This is the case history that is found as the standard for Earth's crust, and in some laboratory experiments. This is the concern of the next section.

4.2 A "fog" of *AE* sources

One case history of this kind appears to be a precursor of the L'Aquila earthquake. In this respect, it should be strongly emphasized that *AE* provides no earthquake "*prediction*". Rather *AE* is a *diagnostic* tool, much like in medical sciences one or several different tools can diagnose the health of a patient, although they cannot forecast the time of her/his passing away. In addition, it should be stressed that phenomena are not repetitive. Every earthquake is a different case, either because it occurs in a different tectonic setting, or because - even when it occurs in the same area - the general environmental conditions have evolved: the direction of the arrow of time is always the same, and everything is permanently ageing (recalling a famous adage by Heraclitus).

This case history is significant for envisaging some possible applications to technological or security problems, whenever we need to monitor the timing evolution of performance loss of a system.

Figure 8 shows D_t evaluated *vs.* time both for *HF AE* and *LF AE*, monitored by the *AE* station of Valsinni (province of Matera, Basilicata, southern Italy, located at *354 km* from the epicentre of the L'Aquila earthquake). The large ellipse in figure 8, denoted as *HF*, shows the reversed lognormal distribution associated with the *"lognormal fog"* of the flaws that released the *HF AE*. At some later time, the analogous *"lognormal fog"* associated with the *LF AE* is also shown by another large ellipse in figure 8.

If *AE* records had been available at some additional frequencies (e.g., say, at *20, 15, 10, 5, 2.5, 1 kHz, ...*) analogous evidence for every available frequency should be expected to be found corresponding to a time series of *"lognormal fogs"*.

In any case, the temporal evolution of the system was such that, in the case history of the L'Aquila earthquake, the typical frequency for an earthquake - i.e. ~ *0.5-1 Hz* - was attained approximately close to the completion of the *LF AE "lognormal fogs"* (at *25 kHz*).

In general, however, it must be stressed that this is not a condition that is expected to be repeated on the occasion of *every* other earthquake. Other case histories can have different speeds of evolution. Consider that this is a precursor phenomenon that lasts several weeks (or maybe even longer). If the (unknown) physical system remains in an approximately steady and "quiet" state during a sufficiently long time interval, the full temporal "evolution" can complete its transformation - which precedes the earthquake and determines the precursor morphology. In this way the final earthquake will occur after the full exhibition of the precursor. But if this approximately steady and "quiet" state is perturbed by some newly occurred changes of its boundary conditions, the temporal "evolution" of its transformation will be abruptly interrupted. In this case, some earlier fraction of the entire precursor will be observed, but no earthquake will occur. That is, one can claim that, in general, some much irregular overlapping of *"crustal substorms"* can sometimes cancel the simple regularity shown in figure 8. In a laboratory experiment, however, the operative conditions are much better - and more regular and controlled - compared to field *AE* records.

An earthquake is a *local* event: the focal volume is just a small amount of cubic kilometres. But it occurs only due to a *local* yield, following some *large scale* phenomenon, to be identified with a *"stress storm"* involving some *much larger* slab of lithosphere and crust. A *"storm"*, however, is composed of *"substorms"*, which are phenomena spanning – compared to a *"storm"* - a much shorter total time lag, and they involve some limited fraction of the aforementioned large slab. In addition, compared to the very strong although much general evidence provided by a *"storm"*, the possibility to recognize a time series of *"substorms"* detected by progressively decreasing *AE* frequencies sometimes permits to guess the expected time of a forthcoming earthquake.

It should be pointed out that this *AE diagnostic* tool is unsuited for envisaging the *location* of the epicentre, which should rather be inferred upon considering e.g. the elastic energy *locally* stored at different sites (this is a classical topic, the object of a wide range of literature and not here of concern).

Similarly, the magnitude of a forthcoming earthquake cannot be estimated by *AE* records. The intensity of the *AE* recorded signal depends both on intensity of the source, and on the unknown characteristics of the "natural probe" (see section 2). The physically significant information given by the *OFTH* algorithms is rather concerned with the *timing* of the *AE* signals, not with its intensity.

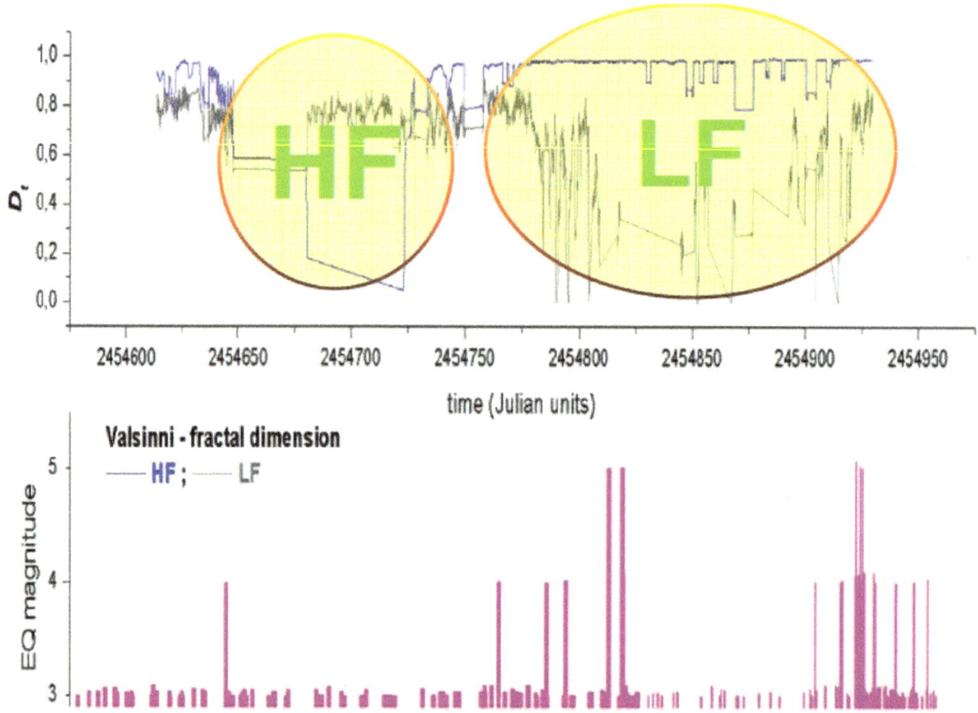

Fig. 8. Fractal dimension D_t of both *HF AE* and *LF AE* records carried out at Valsinni, at *354 km* from the epicentre of the L'Aquila earthquake. The *"stress substorm"* at either *HF AE* or *LF AE* is denoted by a large ellipse. The L'Aquila earthquake occurred at the completion of the *LF AE "substorm"*, although in general different earthquakes can have a different speed of evolution in time. See text.

Consider e.g. the aforementioned experiment carried out on a steel bar (see section 4.1; Biancolini *et al.*, 2006), where the bar was, say, *~20 cm* long. But one can imagine repeating the identical experiment either with a tiny steel needle, or with a huge steel girder of a bridge. The trend of the fractal dimension D_t is the same, independent of the size of the object. But the amount of elastic potential energy is much different, which is stored inside the object and is eventually released when the catastrophe occurs.

Differently stated, an earthquake ought to occur every time the same precursor phenomenon is observed, including its complete temporal evolution. But if the earthquake magnitude is small, no information is given by seismologist or mass media. So it appears to be an unsuccessful precursor, although the precursor was correct. But nobody knows the physical extension of the physical system that is yielding, which is the source of the observed *AE* signal.

This basic drawback has distinct implications, for earthquakes, volcanoes, landslides, or technological applications.

In the case of earthquakes, it is essential to use an array of *AE* recording stations, in order to determine the spatial extension of the *AE* sources and the general tectonic setting of the area.[4] A planetary *AE* array is necessary for understanding crustal stress, much like a meteorological array is needed for synoptic meteorology.

Global meteorological charts are provided, and the different national meteorological services use them for issuing a forecast for their respective region of interest. Flight assistance uses the global maps for security purposes. But, if one wants to issue an exact forecast for specific purposes (sports issues or others), in general this can result practically impossible. Much in the same way, a real time chart of the global pattern and evolution of crustal stress can be used by national services for issuing more or less accurate seismic forecast for the regions of respective concern.

In general, earthquake precursors can be distinguished by separating concerns about *(i)* magnitude, *(ii)* site, and *(iii)* time. *AE* are suited for the time concern.

Earthquake prediction is impossible in a strictly deterministic sense. However, it has to be correctly taken into account like a probabilistic estimate according to the same criterion used for insurance.

In addition, owing to economic, humanitarian, and catastrophe management implications, it is a deontological obligation to recognize that - much like it is certainly possible to assess whether a person is a child or an old man - it is *certainly* possible to get a reliable monitoring and diagnosis of the state of the Earth crust, by operating *AE* arrays combined with other information.

Compared to earthquakes, volcanoes are a much simpler problem in two respects. The site is known. In addition, the primary cause is the long-lasting increase of the pressure of endogenous hot fluids, and this is associated with well recognisable *AE* effects. Volcanic forecasting is therefore realistically and reliably feasible.

In the case of a landslide, in general, monitoring can be very effectively carried out. But the prediction reliability depends on how detailed is the knowledge of the geological structure of the hazard area, and on the number of available *AE* recording points. An advantage is the much slower process that precedes the occurrence of a catastrophe.

In the case of technological applications, predictability is much easier, because the composition, the size, and the structure of the system, and the timing of the primary drivers are well known.

Concerning technological applications, suppose e.g. to monitor a bridge, or some old building, or a rail embankment, or an unstable slope of a slow land slide, etc. The "catastrophe" of the system will happen during a "*stress storm*". But, for practical purposes in order to monitor the timing and evolution of the structure towards its "catastrophe" - it is useful to detect the occurrence of "*stress substorms*". For instance, when a bridge is in

[4] The maximum magnitude of an earthquake that can hit a given area can be inferred by a fractal analysis applied to the 2D fault distribution on the Earth's surface (by a method first exploited by the late Giuseppe Cello). The greater the crustal fracturing, the less the maximum magnitude is of an occurring shock. Refer to mentions and references in Gregori *et al.* (2010). This analysis, however, requires the availability of long secular historical seismic data series.

yielding hazard, a *"stress substorm"* could be triggered just due to the diurnal thermoelastic deformation (section 1.2), or when some heavy truck or a train is crossing over it. The comparative morphology of the *"stress substorm"* is likely to be much different depending on the ageing of the system. The same rationale can be applied to other systems, e.g. to the response of a building located on the slope of a hill that is sliding down, or of a rail embankment that is yielding at some site, etc. That is, the *"storm"* and *"substorm"* rationale is a logical abstraction that applies to different case histories, and it helps with interpretation of the observations. But the real phenomena depend on the specific physical nature of the system.

But an occurrence is often characterized by a *"lognormal fog"* of *AE* emitters, which is very different compared to the aforementioned aspect. In the aforementioned physical case history the system is a *"fog"* of *AE* emitters that coalesce and evolve towards flaws of larger size, which emit lower frequency *AE*. That is, the system is in a state of *passive evolution* towards a new equilibrium state.

In contrast, a much different physical phenomenon occurs in the opposite case history, when the system is subject to an *increasing* action by some external trigger.

Let us consider the case history of a volcano. A much different logic is applied. In fact, volcanism is strongly affected by the exhalation - under pressure - of endogenous hot fluid, and this pressure can either increase or decrease.

When the endogenous heat supply increases, the pressure from hot fluids also increases, and a large number of flaws will yield within the volcanic edifice. The hot fluids diffuse in 3D, implying a corresponding 3D disordered distribution of "elementary" *AE* emitters. Hence, it shall be found $D_t \sim 1$. It can be said that the volcano is "inflating". A large number of *instrumental* seismic events will then be detected.

When the opposite effect occurs, the volcanic edifice collapses under its weight, while it is no longer sustained by the pressure, which is decreased. Hence, the *AE* will be released along micro-cleavage planes of the structures that collapse under the weight of the volcanic edifice. A micro-cleavage planes is *per se* derived from a 2D micro-process, and $D_t < 1$. It can be said that the volcano is "deflating". A comparatively much smaller number of seismic events will be observed, although with comparatively larger intensity, which is due to the rupture of 2D cleavage planes.

"Inflation" and "deflation", and their seismic association, were observed by *AE* on Vesuvius (Paparo et al., 2004, 2004a). The same concept was however already mentioned in the classical paper by Sassa (1936).

Every volcano is a much different "animal" in the "zoo" of all volcanoes of the world. The volcano Peteroa (on the Andes, at the border between Argentina and Chile) - owing to its comparatively much greater weight associated with a much thicker lithosphere or crust - is such that its edifice operates much like the weight of a security valve of a pressure cooker. The result is that it operates like a high precision watch for monitoring several spectral lines of the Earth's tide (Ruzzante et al., 2005, 2008).

Stromboli appears very intriguing, when we extrapolate its observational evidence and use it for technological applications. The progressive evolution of its system was nicely monitored by *AE* records (figure 9) until its final unusual and violent paroxysm occurred at the end of *December 2002* (Paparo & Gregori, 2001; Paparo et al., 2004; Gregori & Paparo, 2006).

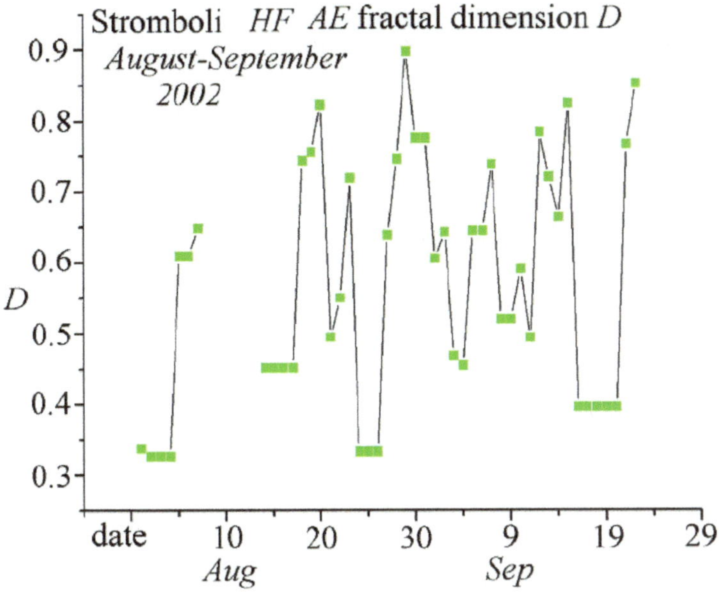

Fig. 9. Fractal dimension D_t computed on the point-like process that results after subtracting the ± *12 hours* running-average from the unsmoothed raw data set (i.e. every raw datum is an average over *15 min*). The point-like process is defined by choosing only "YES" events that result above some given threshold. Every D_t value was computed for one day. Whenever the number of "YES" events available for one given day resulted insufficient, several subsequent days were considered altogether, and they are here plotted like several days with the same D_t. The significant circumstance ought to be pointed out by which, when the volcanic edifice is "deflating" and D_t results comparatively low (see text), the *AE* events in the point-like process seldom occur, due to a temporary reduction, or lack, of any primary energy breeding to the volcano. The measurements were unfortunately interrupted when a lightning discharge stroke the recording system late in September. Figure after Paparo et al. (2004) and Gregori & Paparo (2006).

Its heat supply violently increased during several months. The number of newly generated flaws - within a 3D space distribution – continuously increased. The system had intrinsic oscillations, the primary cause of which is still unclear, being possibly an effect associated with varying meteorological pressure, or just some intrinsic variations related to its energy balance (much like for a pressure cooker, etc.). But, apart from this intrinsic oscillation, the maximum D_t progressively increased towards $D_t \sim 1$. When it reached its maximum threshold, i.e. $D_t = 1$, the "catastrophe" occurred.

In any case, whatever the origin of this intrinsic oscillation is, the system responds differently in time to the trigger created by the increasing primary cause. Therefore, every oscillation can be likened to a "*substorm*", which characterizes the evolution of the system towards its final "catastrophe". It should be pointed out that we are unaware of any alternative diagnostic tools capable of monitoring the precursor stage and evolution of Stromboli towards it paroxysm. Only some gas exhalation within wells could be correlated with *AE* records, although *AE* data are much more reliable, having a comparatively much better signal-to-noise ratio (see references given above).

A much similar result was observed by stressing concrete cubes, *15 cm* size (figure 10; Guarniere, 2003) until their final collapse. In this experiment, a *"substorm"* is defined by every addition of a constant weight on top of the concrete cube. As this externally applied trigger increases, the system responds differently depending on its ageing.

In either the case for Stromboli or for a concrete cube, the final collapse, i.e. the *"storm"*, is composed of *"substorms"*. In the concrete experiment, we know that every *"substorm"* is the obvious consequence of the newly added weight. But when referring to any kind of physical

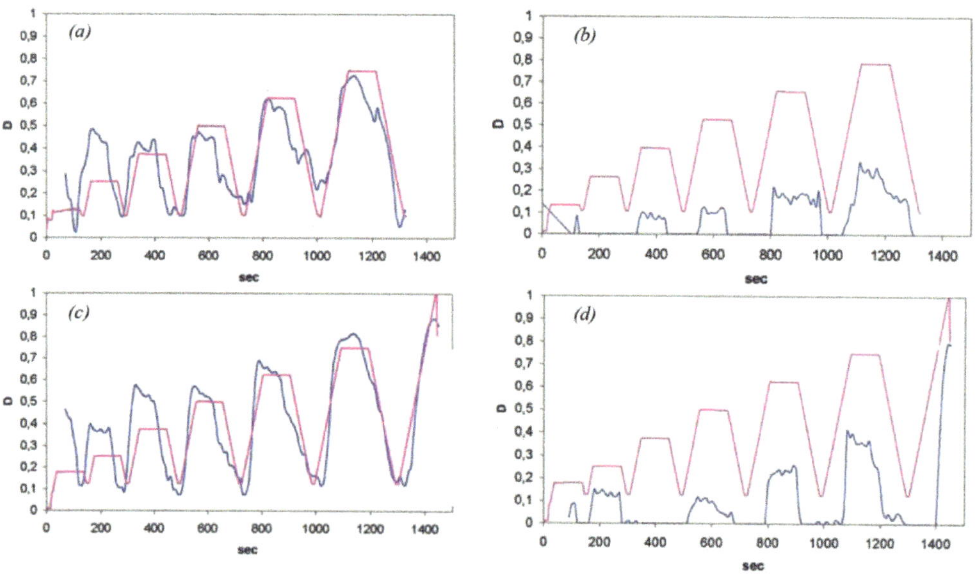

Fig. 10. a,b,c,d. Average fractal dimension D_t measured on two sets, of *10* elements each, of concrete specimens (cubes of *15 cm* side). All specimens are identical, made of identical concrete. Every specimen was stressed by adding progressive loads *50 kg* each, with unconfined uni-axial compression. The line with straight segments (in red) shows the applied load, normalized to rupture loading. The wavy line (in blue) is the average of the *10* computed D_t. The top row refers to the set of specimens that were not brought to final collapse, the bottom row refers to another set of specimens that were stressed until their final rupture. The left column refers to *HF AE* (*200 kHz*) and the right column refers to *LF AE* (*25 kHz*). The *HF* information refers to a much more preliminary stage of the fatigue process, as it deals with the rupture of much smaller crystal bonds, while the *LF* information deals with the yield of comparatively much larger flaws, i.e. when the specimen already suffered by some previous severe fatigue. The *HF* plot appears better representative of the starting process of deterioration of the solid structure of the specimen. After Guarniere (2003) and Gregori & Paparo (2006). The similarities appear impressive between figures 9 and 10. See text.

system that is being stressed by any kind of externally applied trigger, we lack the specific information about the primary cause of this applied trigger. Thus, the D_t trend is an effective diagnostic tool for assessing two bits of information: *(i)* when the system is suffering by a "*substorm*", and *(ii)* how the morphology of a "*substorm*" changes depending on the ageing of the structural elements of the system.

Summarizing, when we deal with a physical system that can be modelled by a "*fog*" of "elementary" *AE* emitters (better than in terms of *one single AE* emitter alone), we have to distinguish two different and competing case histories: *(i)* either the system responds to an internal evolution of its "*lognormal fog*"; or *(ii)* the system responds to the increase of an applied external trigger. In this case, the assessment and separation of its different "*substorms*" derives from some physical cause that determines some kind of more or less regular apparent "oscillation" of the system.

In either case, *AE* monitoring - and its *OFTH* parameters - are effective tools for monitoring the timing and evolution of the loss of performance of the system.

For the sake of completeness - and as a response to a comment by a referee, whom we thank - a mention is to be given to a closely related study in progress since over a decade (by one of the authors, Giovanni P. Gregori (GPG)). "Climate" is the environment where the biosphere can develop and survive. It is a condenser with an upper plate (ionosphere and magnetosphere) supplied by an electric current generator (the solar wind). The lower plate is much irregular, with spikes underground, and it is supplied by an electric power generator (the Earth's geodynamo; Gregori, 2002). All "climate" phenomena can be interpreted by this model. This includes long-range relations between observations of several different phenomena, such as endogenous heat generation and release, volcanism, orogenesis, geodynamics and seismicity, serpentinization, geomagnetism, astronomical motion of the Earth, transient luminous events, atmospheric electrical phenomena including lightning discharge activity, precipitation phenomena, carbon cycle, water cycle and balance, trigger of hurricane formation, and the fundamental role of the interaction within the biosphere, either supplied by solar energy or by endogenous energy, etc.[5]

A worldwide array of *AE* recording stations could be of paramount importance for monitoring stress propagation and teleconnection on the planetary scale providing one additional presently lacking key element of knowledge in the chain of cause-and-effect, within the whole aforementioned multidisciplinary scenario.

5. Conclusion

A comparison is heuristically useful between *AE* records measured in the field and *AE* records measured in the laboratory or in some technological application. Since most often

[5] This study also relies on a sum of several partial, much helpful, private inputs by outstanding scientists. GPG feels deeply indebted to them. In particular, a few of the most relevant inputs in alphabetical order are: Dong R. Choi, Ugo Coppa, Dong Wen-Jie, Gao Xiao-Qing, Raymond Hide, Martin Hovland, Alberto Incoronato, Tamara V. Kuznetsova, Louis J. Lanzerotti, Bruce A. Leybourne, Iginio Marson, Helmut Moritz, Nils-Axel Mörner, Cliff D. Ollier, Eugene N. Parker, Michel Parrot, John M. Quinn, Karsten M. Storevedt, Fumio Tsunoda, and Carlo Forese Wezel, ... and the late Baron Paul Melchior, and Wilfried Schröder. But this list should be much longer.

case histories found in the field can rarely be reproduced in the laboratory, this comparison is an effective and concrete way of looking at different physical aspects of a difficult and multi-faceted problem.

The ultimate target is recognizing "*stress substorms*" or "*strain substorms*", or briefly "*substorms*" which are the same, since their timing is a more precise tool for monitoring the temporal evolution of the system compared to a general assessment of a vague occurrence of a "catastrophe" observed at the end of some "*storm*".

In the final analysis, when dealing with different case histories, the physics is always the same – i.e. all phenomena depend on the yield of atomic bonds inside "solid" structures. But the physical systems can be much different, as well as their respective primary *AE* drivers. The *OFTH* analysis appears very effective for applying some logical tools aimed at focusing on different morphological features of the observations.

When referring to some given physical system, however, different physical aspects can be more or less relevant depending on the specific type of information that is most useful for any given application.

In addition, it is possible to recognize "*stress substorms*" manifested at every given *AE* frequencies. Different *AE* frequencies permit the recognition of a time sequence of "*substorms*", hence a much better definition of the timing and evolution of the performance loss of the system.

A remarkable similarity with the *AE* records with several other lines of evidence should be pointed out for Earth's science applications. It is documented by time series records from superconducting gravimeters (*SG*) combined with absolute gravimeters (*AG*), and also correlated with *DInSAR* and *GRACE* records that with skilful analysis and data handling they are useful for modelling and subtracting several tidal and environmental phenomena. There is evidence that some relevant residuals appear likely related to tectonic and geodynamic actions.

A synergy of all these monitoring techniques - and the assessment of "*storm*" and "*substorm*" time – can clarify several morphological features of crustal phenomena and their time-space evolution.

This relevant potential achievement applies, however, to environmental investigations. Concerning engineering topics, *SG* appears to be of no practical help. In contrast, however, other monitoring techniques can be used in engineering structures that cannot be applied in the field. It should be emphasized that a multi-parametric monitoring of the same physical system can lead to an understanding of phenomena in a way that every technique alone could never attain.

Some harder thinking, additional criteria, and continued software development will certainly allow better evidence for some additional physical aspects of the *AE* records, and allow a better understanding of the specific physical drivers which will narrow the focus on observational aspects that are best suited for monitoring the hazard of concern in every given application.

6. Acknowledgements

We feel particularly grateful and indebted with Iginio Marson and with Martin Hovland, for calling our attention, respectively, on the important and much recent SG results, and on the crucial role of serpentinization in crustal stress propagation.

The comments by the referees are gratefully acknowledged. In particular, we greatly appreciated the comments and the careful revision of the English editing carried out by Bruce Leybourne. We also feel indebted for several discussions (among others) with Ugo Coppa, Alberto Incoronato, Igino Marson, Maria Isabel Lòpez Pumarega, and José Ruzzante. The Cephalonia *AE* data, used for drawing figure 7 as an anticipation of a more systematic study, were collected by Vassilis A. Sakkas, in the frame of a cooperation with CNR-IDASC, in progress since 2003 altogether with Evangelos Lagios and Issaak Parcharidis. They are gratefully acknowledged for providing us with these data for inferring this preliminary evidence. We feel obliged with them also for several discussions dealing with *AE*. In addition, we thank Mr and Mrs Raponi for kindly hosting the *AE* monitoring apparatus in the basement of their house at Orchi (Foligno, PG, Italy).

7. References

Akasofu, S.I. (1964). The development of the auroral substorm. *Planet. Space Sci.*, Vol. 12, No. 4, pp. 273-282

Akasofu, S.I. (1968). *Polar and magnetospheric substorms*, D. Reidel Publ. Co., Dordrecht, Netherland

Arley, N. & Buch, K.R. (1950). *Introduction to the theory of probability and statistics*, Science Editions, John Wiley and Sons, Inc., New York, USA

Bak, P., Tang, C. & Wiesenfeld, K. (1987). Self-organizing criticality. An explanation of *1/f* noise. *Phys. Rev. Lett.*, Vol.59, No.4, pp. 381-384

Barrow, J.D. (2009). *Cosmic imagery: key images in the history of science*, W. W. Norton & Company, New york, USA

Biancolini, M.E., Brutti, C., Paparo, G. & Zanini, A. (2006). Fatigue cracks nucleation on steel, acoustic emission and fractal analysis. *Int. J. Fatigue*, Vol.28, No.12, pp. 1820-1825

Braccini, S., Casciano, C., Cordero, F., Frasconi, F., Gregori, G.P., Majorana, E., Paparo, G., Passaquieti, R., Puppo, P., Rapagnani, P., Ricci, F., & Valentini R. (2002). Monitoring the acoustic emission of the blades of the mirror suspension for a gravitational wave interferometer. *Phys. Lett. A*, Vol.301, pp. 389-397

Buchanan, M. (1997). One law to rule them all, *New Sci.*, Vol.8, pp. 30-35

Campbell, W.H. (1996). Geomagnetic storms, the *Dst* ring-current myth and lognormal distributions. *J. Atmos. Terr. Phys.*, Vol.58, No.10, pp.1171-1187

Cello, G. (2000). A quantitative structural approach to the study of active fault zones in the Apennines (peninsular Italy), *J. Geodyn.*, Vol.29, pp. 265-292.

Chapman, S. & Bartels, J. (1940). *Geomagnetism*, 2 vol., Oxford Univ. Press, (Clarendon), London, UK., New York, USA

Coontz, R. (1998). Like a bolt from the blue. *New Sci.*, Vol.2155, pp. 36-40.

Cowie, P.A. (1998). A healing-reloading feedback control on the growth rate of seismogenic faults. *J. Structural Geol.*, Vol.20, No.8, pp. 1075-1087

Cowie, P.A., Vanneste, C. & Sornette, D. (1993). Statistical physics model for the spatiotemporal evolution of faults. *J. Geophys. Res.*, Vol.98, No.21, pp. 809, 821

Crossley, D., Hinderer, J. & Boy, J.P. (2005). Time variation of the European gravity field from superconducting gravimeters. *Geophys. J. Int.*, Vol.161, pp. 257-264

Crossley, D., & Hinderer, J. (2008). The contribution of GGP superconducting gravimeters to GGOS, In: *Observing our changing Earth*, M. G. Sideris, (Ed.), *IAG Symposia,* Vol.133, 841-852

Crossley, D. & Hinderer, J. (2009). A review of the GGP network and scientific challenges. *J. Geodyn.*, Vol.48, No.3-5, pp. 299-304

de Linage, C., Hinderer, J. & Rogister, Y. (2007). A search for the ratio between gravity variation and vertical displacement due to a surface load. *Geophys. J. Int.*, Vol.171, No.3, pp. 986-994

de Linage, C., Rivera, L., Hinderer, J., Boy, J.P., Rogister, Y., Lambotte, S. & Biancale, R. (2009). Separation of coseismic and postseismic gravity changes for the 2004 Sumatra-Andaman earthquake from 4.6 years of GRACE observations and modelling of the coseismic change by normal-modes summation. *Geophys. J. Int.*, Vol.176, No.3, pp. 695-714

Gregori, G.P. & Paparo, G. (2004). Acoustic emission (*AE*). A diagnostic tool for environmental sciences and for non destructive tests (with a potential application to gravitational antennas), In: *Meteorological and geophysical fluid dynamics,* W. Schröder, (Ed.), 166-204, *AGGKP*, W. Schröder/Science, Bremen, Germany

Gregori, G.P., Lupieri, M., Paparo, G., Poscolieri, M., Ventrice, G. & Zanini, A. (2007). Fatigue, ageing, and catastrophe of solid structures. *Nat. Hazards Earth Syst. Sci.*, Vol. 7, pp. 723-731

Gregori, G.P., Paparo, G., Coppa, U. &Marson, I. (2002). Acoustic emission in geophysics: a reminder about the methods of analysis. *Boll. Geofis. Teor. Appl.*, Vol.43, No.1-2, pp. 157-172

Gregori, G.P., Paparo, G., Poscolieri, M. & Zanini, A. (2005). Acoustic emission and released seismic energy. *Nat. Hazards Earth System Sci.*, Vol.5, pp. 777-782

Gregori, G.P., Paparo, G., Poscolieri, M., Ventrice, G., Rafanelli, C. &de Simone, S. (2008). Crustal stress and acoustic emission (*AE*), Poster presented at the *3rd World Stress Map Conference,* Heidelberg Academy of Sciences and Humanities, GFZ Potsdam, Germany, October 15-17, 2008

Gregori, G.P. (1998). The magnetosphere of the Earth. A theory of magnetospheric substorms and of geomagnetic storms, In: *From Newton to Einstein,* W. Schröder, (Ed.), 68-106, *Mitteilungen des AGG der DDG,* Vol.17, No.3/4, Science Edition - IDCH-IAGA -*AKGGKP*, Bremen-Roennebeck and Potsdam, Germany

Gregori, G.P. (1999). The external magnetic sources over the polar caps. Feasible modelling *vs.* unrealistic expectations. *Annali di Geofisica,* Vol.42, No.2, pp. 171-189

Gregori, G.P. (2002). Galaxy – Sun – Earth relations. The origin of the magnetic field and of the endogenous energy of the Earth, with implications for volcanism, geodynamics and climate control, and related items of concern for stars, planets, satellites, and other planetary objects. A discussion in a prologue and two parts. *Beitr. Geschichte Geophys. Kosmisch. Phys.*, Vol.3, No.3, 471 pp

Gregori, G.P. & Paparo, G. (2006). The Stromboli crisis of *28÷30 December 2002. Acta Geod. Geophys. Hung.*, Vol.41, No.2, pp.273-287

Gregori, G.P., & Paparo, G. (2010). Lognormality and fractality. A unique rationale for different phenomena. Preprint

Gregori, G.P., Poscolieri, M., Paparo, G., de Simone, S., Rafanelli, C. & Ventrice, G. (2010). *"Storms of crustal stress"* and *AE* earthquake precursors, *Nat. Hazards Earth Syst. Sci.*, Vol.10, pp. 319–337

Guarniere, S. (2003). L'emissione acustica come strumento diagnostico di strutture a varia scala. *Unpublished PhD Thesis*, 144 pp., University of Messina, Messina, Italy

Imbò, G. (1954). Sismicità del parossismo vesuviano del marzo 1944. *Ann. Oss. Ves.*, VI Ser., Vol. 1, I-XXXI, pp. 1-211

Judd, A. G. & Hovland, M. (2007). *Submarine fluid flow, the impact on geology, biology, and the marine environment,*, Cambridge University Press, Cambridge, UK

Kapteyn, J.C. (1912). Definition of the correlation - coefficient. *Month. Not. Roy. Astr. Soc.*, Vol.72, pp. 518-525

Kapteyn, J.C. & van Uven, M.J. (1916). *Skew frequency-curves in biology and statistics*, Hoitsema Brothers, Groningen, Netherland.

Longuevergne, L. (2008). *Contribution à l'hydrogéodésie*, Ph. D. Thesis, University Pierre et Marie Curie, Paris VI, France. Available from http://tel.archives-ouvertes.fr/docs/00/31/92/05/PDF/ these_longuevergne.pdf

Main, I.G. (1995). Seismogenesis and seismic hazard, *Proceedings of VIII Summer Course Earth Planet. Sci. on "Plate tectonics: the first twenty-five years"*, pp. 395-419, Siena, Italy

Paparo, G. & Gregori, G.P. (2001). Volcanoes and environment and the natural laboratory of Stromboli, In: *The Bridge between the big bang and biology - Stars, planetary systems, atmospheres, volcanoes: their link to life*, F. Giovannrelli, (Ed.), 289-309, CNR, Rome, Italy

Paparo, G. & Gregori, G.P. (2003). Multifrequency acoustic emissions (*AE*) for monitoring the time evolution of microprocesses within solids. *Rev. Quantitative Nondestruct. Evaluation*, Vol.22, pp. 1423-1430, *Proceedings of AIP Conference.*, D. O. Thompson & D. E. Chimenti, (Eds.)

Paparo, G., Gregori, G.P., Angelucci, F., Taloni, A., Coppa, U. & Inguaggiato, S. (2004). Acoustic emissions in volcanoes: the case histories of Vesuvius and Stromboli, *Proceedings of SCI 2004, 8th World Multi-Conf. Systemics, Cybernetics and Informatics Meeting*, Orlando, Florida, USA, July 2004.

Paparo, G., Gregori, G.P., Coppa, U., De Ritis, R. & Taloni, A. (2002). Acoustic Emission (*AE*) as a diagnostic tool in geophysics. *Annals Geophys.*, Vol.45, No.2, pp. 401-416

Paparo, G., Gregori, G.P., Poscolieri, M., Marson, I., Angelucci, F. & Glorioso, G. (2006). Crustal stress crises and seismic activity in the Italian peninsula investigated by fractal analysis of acoustic emission (*AE*), soil exhalation and seismic data, In: *Fractal analysis for natural hazards*, G. Cello &B. D. Malamud, (Eds.), 47-61, Geol. Soc. London, Special Publ., Vol. 261, London, UK

Paparo, G., Gregori, G.P., Taloni, A. & Coppa U. (2004a). Acoustic emissions (*AE*) and the energy supply to Vesuvius – 'Inflation' and 'deflation' times. *Acta Geod. Geophys. Hung.*, Vol.40, No.4, pp.471-480

Poscolieri, M., Gregori, G.P., Paparo, G. & Zanini, A. (2006a). Crustal deformation and *AE* monitoring: annual variation and stress-soliton propagation, *Nat. Hazards Earth Syst. Sci.*, Vol.6, pp. 961-971

Poscolieri, M., Lagios, E., Gregori, G.P., Paparo, G., Sakkas, V.A., Parcharidis, I., Marson, I., Soukis, K., Vassilakis, E., Angelucci, F. & Vassilopoulou S. (2006b). Crustal stress and seismic activity in the Ionian archipelago as inferred by combined satellite and ground based observations on the Kefallinìa Island (Greece). In: *Fractal analysis for natural hazards*, G. Cello &B. D. Malamud, (Eds.), 63-78, Geol. Soc. London, Special Publ., Vol. 261, London, UK

Rosat, S., Boy, J.-P., Ferhat, G., Hinderer, J., Amalvict, M., Gegout, P. & Luck, B. (2009). Analysis of a ten-year (1997-2007) record of time-varying gravity in Strasbourg using absolute and superconducting gravimeters: new results on the calibration and comparison with GPS height changes and hydrology, *J. Geodyn*, Vol.48, No.3-5, pp. 360-365

Ruzzante, J., Lòpez Pumarega, M.I., Gregori, G.P., Paparo, G., Piotrkowski, R., Poscolieri, M. & Zanini, A. (2008). Acoustic emission (*AE*), tides and degassing on the Peteroa volcano (Argentina), In: *Acoustic emission,* Vol.1, *Microseismic, learning how to listen to the Earth…*, J. Ruzzante & M.I. Lòpez Pumarega, (Eds.), 37-68, CNEA, Buenos Aires, Argentina

Ruzzante, J., Paparo, G., Piotrkowski, R., Armeite, M., Gregori, G.P. &Lopez I. (2005). Proyecto Peteroa, primiera estaciòn de emisiòn acustica en un volcàn de los Andes, *Rev.Uniòn Iberoamericana Soc. Fìs.*, Vol.1, No.1, pp. 12-18

Sassa, K. (1936). Micro-seismometric study on eruptions of the volcano Aso. *Mem. Coll. Sci. Kyoto Imp. Univ.*, Vol.19, No.1, pp. 11-56 (+4+3)

Sornette, D. & Davy, P. (1991). Fault growth model and universal fault length distribution. *Geophys. Res. Lett.*, Vol.18, pp. 1079-1081

Sornette, D., Davy, P. &Sornette, A. (1990). Structuration of the lithosphere as a self-organized critical phenomenon. *J. Geophys. Res.*, Vol.95, No.17, pp. 353-361

Stein, S. & Liu M. (2009). Long aftershock sequences within continents and implications for earthquake hazard assessment. *Nature,* Vol.462, pp. 87-89

Sykes, L.R., Shaw, B.E. &Scholz, C.H. (1999). Rethinking earthquake prediction. *Pure Appl. Geophys.*, Vol.155, pp.207-232

Weise, A., Kroner, C., Abe, M., Ihde, J., Jentzsch, G., Naujoks, M., Wilmes, H. &Wziontek, H. (2009). Gravity field variations from superconducting gravimeters for GRACE validation. *J. Geodyn.*, Vol.48, No.3-5, pp.325-330

Wilmes, H., Wziontek, H., Falk, R. &Bonvalot, S. (2009). AGrav - the new international absolute gravity database of BGI and BKG and its benefit for the Global Geodynamics Project (GGP). *J. Geodyn.*, Vol.48, No.3-5, pp.305-309

Zanini, A. (2004). Nucleazione di cricche di fatica in acciai, emissione acustica ed analisi frattale, *Unpublished Thesis*, University of Roma-Tor Vergata, Rome, Italy

Zerbini, S., Richter, B., Rocca, F., van Dam, T. &Matonti, F. (2007). A Combination of space and terrestrial geodetic techniques to monitor land subsidence: case study, the Southeastern Po plain, Italy. *J. Geophys. Res.*, Vol. 112, No. B5, B05401

Permissions

The contributors of this book come from diverse backgrounds, making this book a truly international effort. This book will bring forth new frontiers with its revolutionizing research information and detailed analysis of the nascent developments around the world.

We would like to thank Wojciech Sikorski, for lending his expertise to make the book truly unique. He has played a crucial role in the development of this book. Without his invaluable contribution this book wouldn't have been possible. He has made vital efforts to compile up to date information on the varied aspects of this subject to make this book a valuable addition to the collection of many professionals and students.

This book was conceptualized with the vision of imparting up-to-date information and advanced data in this field. To ensure the same, a matchless editorial board was set up. Every individual on the board went through rigorous rounds of assessment to prove their worth. After which they invested a large part of their time researching and compiling the most relevant data for our readers. Conferences and sessions were held from time to time between the editorial board and the contributing authors to present the data in the most comprehensible form. The editorial team has worked tirelessly to provide valuable and valid information to help people across the globe.

Every chapter published in this book has been scrutinized by our experts. Their significance has been extensively debated. The topics covered herein carry significant findings which will fuel the growth of the discipline. They may even be implemented as practical applications or may be referred to as a beginning point for another development.

The editorial board has been involved in producing this book since its inception. They have spent rigorous hours researching and exploring the diverse topics which have resulted in the successful publishing of this book. They have passed on their knowledge of decades through this book. To expedite this challenging task, the publisher supported the team at every step. A small team of assistant editors was also appointed to further simplify the editing procedure and attain best results for the readers.

Our editorial team has been hand-picked from every corner of the world. Their multi-ethnicity adds dynamic inputs to the discussions which result in innovative outcomes. These outcomes are then further discussed with the researchers and contributors who give their valuable feedback and opinion regarding the same. The feedback is then collaborated with the researches and they are edited in a comprehensive manner to aid the understanding of the subject.

Apart from the editorial board, the designing team has also invested a significant amount of their time in understanding the subject and creating the most relevant covers. They scrutinized every image to scout for the most suitable representation of the subject and create an appropriate cover for the book.

The publishing team has been involved in this book since its early stages. They were actively engaged in every process, be it collecting the data, connecting with the contributors or procuring relevant information. The team has been an ardent support to the editorial, designing and production team. Their endless efforts to recruit the best for this project, has resulted in the accomplishment of this book. They are a veteran in the field of academics and their pool of knowledge is as vast as their experience in printing. Their expertise and guidance has proved useful at every step. Their uncompromising quality standards have made this book an exceptional effort. Their encouragement from time to time has been an inspiration for everyone.

The publisher and the editorial board hope that this book will prove to be a valuable piece of knowledge for researchers, students, practitioners and scholars across the globe.

List of Contributors

Paul Ziehl and Adrian Pollock
University of South Carolina and Mistras Group, Inc., USA

Kristián Máthis and František Chmelík
Department of Physics of Materials, Faculty of Mathematics and Physics, Charles University, Prague, Czech Republic

Narayan R. Joshi, Stephen D. Russell, Ayax D. Ramirez and David W. Brock
HCC-NW, Houston, Texas and SPAWARSYSCEN PACIFIC, San Diego, California, USA

Amir Refahi Oskouei
Islamic Azad University, East Tehran Branch, Iran

Andrea Zucchelli and Giangiacomo Minak
University of Bologna, Italy

Mehdi Ahmadi
Amirkabir University of Technology (Tehran Polytechnic), Iran

Wojciech Sikorski
Poznan University of Technology, Poland

Waldemar Ziomek
University of Manitoba, Canada

Rúnar Unnþórsson
University of Iceland, Iceland

Hiroshi Inoue
Osaka Prefecture University, Japan

Alexander Zakupin, Virginia Mubassarova, Galina Kachesova and Boris Borovsky
Research Station RAS in Bishkek, Kyrgyzstan

Leonid Bogomolov
Institute of Marine Geology and Geophysics, FEB RAS, Russia

Salem Al-Lababidi, David Mba and Abdulmajid Addali
Cranfield University, UK

A.S. Paipetis and D.G. Aggelis
Department of Materials Science and Engineering, University of Ioannina, Greece

Sheng-xing Wu, Yan Wang, Ji-kai Zhou and Yao Wang
Hohai University, People's Republic of China

Yasushi Fukuzawa and Shigeru Nagasawa
Department of Mechanical Engineering, Nagaoka University of Technology, Nagaoka, Japan

Jordi Cusidó i Roura, Miguel Delgado Prieto, Jose Luis Romeral Martínez
MCIA Group, Technical University of Catalonia, Spain

Deng Aidong and Jiang Zhang
National Engineering Research Center of Turbo-Generator Vibration, Southeast University, China

Albert Leksovskii
Ioffe Physical - Technical Institute, Russian Academy of Sciences, St. Petersburg, Russia

Paulo Roberto de Aguiar, Eduardo Carlos Bianchi and Rubens Chinali Canarim
University Estadual Paulista Julio de Mesquita Filho (UNESP) - Bauru campus, Brazil

Giovanni P. Gregori, Maurizio Poscolieri and Claudio Rafanelli
CNR-IDASC - Institute of Acoustics and Sensors "O. M. Corbino", Rome, Italy

Gabriele Paparo
CNR-IDASC - Institute of Acoustics and Sensors "O. M. Corbino", Rome, Italy
Italian Embassy, Buenos Aires, Argentina

Giuliano Ventrice
P.M.E. Engineering, Rome, Italy